Deep Ecology, Business Ethics and Personal Responsibility

Frontiers of Business Ethics

Series Editor

Laszlo Zsolnai
Business Ethics Center
Corvinus University of Budapest

Volume 13

PETER LANG
Oxford • Bern • Berlin • Bruxelles • New York • Wien

Deep Ecology, Business Ethics and Personal Responsibility

Selected Papers (1988–2020)

by

KNUT J. IMS

PETER LANG

Oxford • Bern • Berlin • Bruxelles • New York • Wien

Bibliographic information published by Die Deutsche Nationalbibliothek
Die Deutsche Nationalbibliothek lists this publication in the Deutsche Nationalbibliografie; detailed bibliographic data is available on the Internet at http://dnb.d-nb.de.

A catalogue record for this book is available from the British Library.

Library of Congress Cataloging-in-Publication Data

Names: Ims, Knut Johannessen, 1951- author.
Title: Deep ecology, business ethics and personal responsibility : selected papers (1988-2020) / Knut Ims.
Other titles: Papers. Selections
Description: Oxford : New York : Peter Lang, [2022] | Series: Frontiers of business ethics, 1661-4844 ; vol. 13 | Includes bibliographical references and index.
Identifiers: LCCN 2021028374 (print) | LCCN 2021028375 (ebook) |
ISBN 9781800792296 (hardback) | ISBN 9781800792302 (ebook) |
ISBN 9781800792319 (epub)
Subjects: LCSH: Business ethics. | Social responsibility of business. |
Sustainable development. | Deep ecology.
Classification: LCC HF5387 .I47 2022 (print) | LCC HF5387 (ebook) | DDC 174/.4—dc23
LC record available at https://lccn.loc.gov/2021028374
LC ebook record available at https://lccn.loc.gov/2021028375

Cover design: Judit Kovacs, Createch Ltd, Hungary

ISSN 1661-4844
ISBN 978-1-80079-229-6 (print)
ISBN 978-1-80079-230-2 (eBook)
ISBN 978-1-80079-231-9 (ePub)

Published by Peter Lang Ltd, International Academic Publishers,
52 St Giles, Oxford, OX1 3LU, United Kingdom
oxford@peterlang.com, www.peterlang.com

This publication has been peer reviewed.

Contents

Figures/Tables

Preface

The intention of this volume is to present selected works from almost forty years as a scholar with the Norwegian School of Economics (NHH) as a basecamp. Two undercurrents run like a red line through the selected articles: the concern for deep ecology and personal responsibility. Even if these seemingly are two different substantive areas, the reader might discover a certain unity among the different works manifested as business ethics, ecology, and the challenges related to business ethics as a teaching subject, in line with a gestalt-oriented view that the whole is more than the sum of its parts. I hope this single work will demonstrate that business ethics, deep ecology, caring for students, and personal responsibility are not independent of each other, rather the contrary. When we dig into deep ecology, we find that personal responsibility is a necessary condition. One argument for making a single volume of works that are published in different journals and many books via different publishers is that readers may discover new and surprising patterns and connect the different elements in new ways. The aspiration of the selected structure of the volume is that the reader might gain new insights. The usual scholarly way of writing is to focus on different parts. But challenged by deep ecological thinking, we should not stop our reasoning with the study of single parts, but see the different elements as part of a habitat. This requires a shift in mindset and opening up to an "eco-self." The unit of survival is not organism alone, but organism plus environment. To sum up, deep ecology is both a philosophical approach and a campaigning platform. It is a process of reflection that should lead to action. The element of personal responsibility is essential in deep ecology. Any alternative personal worldview is called an ecosophy; that is, a philosophy of ecological harmony. This kind of *Sophia* or wisdom is openly normative and contains policy wisdom, prescription, and hypothesis. Each deep ecologist has a personal responsibility to work out his/her own ecosophy. Mine is as follows.

Why is deep ecological thinking so urgent today? The environmental challenges are unprecedented in the history of this planet. Life on Earth faces the greatest mass extinction since the end of the dinosaur age sixty-five million years ago. Hundreds of species a day are becoming extinct, and air, water, and soil, the natural resources that sustain life on this planet, are being polluted or depleted at alarming rates. The world's wilderness areas – mountains, forests, wetlands and grasslands – are being "developed" – paved, drained, burned, overgrazed, and exploited. My main assumption is that this is a man-made disaster, with the man-made greenhouse effect threatening the atmosphere and the climate of the planet itself.

The very notion *deep* ecology as a philosophical approach builds upon the belief that the current environmental crisis can be traced to deep philosophical causes. The cure for the crisis must, therefore, involve a radical change in our philosophical outlook, including both personal and cultural transformations, which will affect basic economic and ideological structures. The problem has been characterized as "the denial of nature" by philosophers. We have an illusory belief that we can live independent of nature and that there are no limits to the resources that humans need now or in the future.

The fundamental wisdom in deep ecology is that humans need non-human nature to fully develop, and that nature has intrinsic value. We should fear that our technological world becomes the second nature, which would drastically reduce the richness of the human experiences. We debase ourselves by debasing outer nature. As one of the strong promoters of deep ecology, Arne Næss writes: "If we destroy nature, we destroy ourselves."

There are more challenges that should be mentioned where deep ecology strongly confronts mainstream thinking. Our western worldview was developed at a time of disengagement from the world (Ims, 1988; Vetlesen, 2015). The new enlightenment paradigm was thinking about the world and knowing about the world, thereby invalidating unreflective experiences. As Vetlesen (2015) writes, the paradigm we are ingrained in has created a monological, autistic kind of being where we consult our theories rather than turn our attention to the world. The first hand "unbefangen" has no prestige. Abstraction becomes the principal tool of enlightenment. Sensuous knowledge like *this* tree that can be seen, touched, smelled, and even heard when the wind blows through the branches, has been discredited. The purpose was to free our minds from superstition to give the method and procedures primacy over objects, form over content, and emphasize the general over the particular (Vetlesen, 2015, pp. 56–57).

Deep ecological thinking challenges this way of looking at the world with its emphasis on spontaneous experiences and gestalt thinking (Ims, 2018). Deep ecological thinking assumes a relational, total-field perspective and rejects the "man-in-environment image", and has important similarities with a feminine perspective in the tradition in Gilligan (1977). Gilligan underscores a relational self, the value of personal experiences, emotion, and cooperation, and a non-violent approach. Gilligan's view challenges traditional "masculine-oriented" ways of teaching ethics, which led me to experiment with role-play in the teaching of ethics for business students. Finally, the grand hope is that the harm and violence toward nature and society that today's mainstream global business represents is challenged and inspired by deep ecological thinking. We need to minimize the harm and violence to save the Earth from human destruction and avoid committing ecocide (Zsolnai, 2021).

The structure of the book

Part I consists of topics related to business ethics. It starts with two articles that scrutinize ethical guidelines as a frequently used tool to discipline business practice, followed by a presentation of works on the need to ground mainstream economics in an organic worldview. The article "competition or cooperation in a mechanic or organic worldview" is followed up by several works on the urgent need to clarify the metaphysical assumptions in economics, drawing upon A. N. Whitehead's process theory. Another article demonstrates that economic incentives might easily destroy social, ecological, and existential values. Henrik Ibsen's character, Peer Gynt, is presented and discussed as a modern businessman with profit maximation as his leading motto, and unscrupulous conduct. The discussion problematizes self-realization in business. What is self-realization in the ultimate sense? The last work in Part I shows paradigmatic examples of business leaders and entrepreneurs who care for society as well as nature.

Part II presents articles that provide insight into the causes, nature, and potential solution to major ethical problems in contemporary business and society. Most of the articles are directly related to deep ecology and, also, positive psychology. By combining Aristotelian virtue ethics and modern positive psychology we find surprising similarities on how to facilitate human flourishing. The contributions share some fundamental features that reflect rethinking of central elements of business ethics in theory and practice, such as the need to apply a humanistic virtue-oriented perspective to economics and to search for the long lines in the history of ideas across different disciplines. The work *Computer and Science as the Bearer of Culture*, written in 1988, points forward and indicates the main direction in my academic journey; that is, my fascination with the existential issues of man, (ir)rationality, and science, and the leap in the human mindset from the Middle Ages to the modern and postmodern age. This shift in mindset created the preconditions for the need to protect and preserve nature as having intrinsic value; that is, the urgent need for a deep ecological approach.

My context has a western bias, even if I have been open to challenging perspectives from sources based in Eastern cultures. Laszlo Zsolnai vitalized my interests in Buddhist economics and the important similarities with a deep ecological approach. Deep ecology is an illustrative example of Eastern influence since the roots and sources of deep ecology are partially drawn from an Eastern perspective where Gandhi and Buddha are important inspiring sources. In addition, one of my favorite scholars, Aristotle, grounds his works on an organic worldview.

Part III addresses the role that alternative learning theories may play in business education. In particular, experiential learning theory inspired by David Kolb plays a decisive role in several papers. One of the articles argues that reading fiction is an essential source of leadership development. Reading high-quality literature can train and increase ethical sensitivity and augment the capacity to react to an ethical challenge. Furthermore, several works on the importance of personal responsibility in organizations are presented. One core case used is the unethical actions of Adolph Eichmann. Even if he pretended to have been "only a cog in a big machine", carrying out the orders of his leader, he was finally sentenced to death and held personally responsible for contributing to the Holocaust. The last article presents the fascinating story of the most important entrepreneur in Norwegian history, Hans Nielsen Hauge (1771–1824). This story also demonstrates the dark side of a successful life project.

Acknowledgements – working in a context of friendship

In 1983 I got a tenure position at NHH. My primary obligation was to teach and do research within the field of information systems. Since then I have had the delight to interact with and learn from competent and generous colleagues and thousands of students from Norway and, over the last twenty years, also from international colleagues and students. But I should not forget my reading "and interacting with" classic philosophers. My pleasure of "dialoguing" with the texts of Socrates, Plato, and Aristotle cannot be overestimated. This may be an important reason why I was invited by rector Arne Kinserdal to start my "second career" at NHH in 1988. The new responsibility was to establish a course in business ethics, and I was obligated to deal with the grand questions of life – how to do good and minimize harm? This meant to unify more fully my life with my work. Most of the articles are from the past twenty years where I have been promoting business ethics at NHH and via CEMS – The Global Alliance in Management Education. Often the chapters are written as invitations to a special conference and/or as a collaboration with colleagues who later became my personal friends. *In retrospect, the most surprising finding when looking back at my context of writing is that most of the articles have been written in a genuine friendship-based, social and intellectual context.*

Laszlo Zsolnai, the chairman of the ethics group in CEMS, generously included me in his comprehensive international network, his portfolio of projects and even his personal life. Laszlo's impact on my career and my personal life during the past twenty years is beyond words. Laszlo is also the president of the European SPES Institute in Leuven, Belgium. In this international context I have also met new friends and colleagues: Luk Bouckaert, Rita Ghesquiere, Hendrik Opdebeeck, Carlos Hovel, and Peter Pruzan. Within CEMS I have enjoyed close collaboration with a number of new colleagues: Eleanor O'Higgins, Nel Hofstra, Zsolt Boda, Antonio Tencati, Aloi Soppe, Gabor Kovács, and Andras Ocsai.

At the same time I have the pleasure of working with my "older" Norwegian friends: Ove Jakobsen, Johannes Brinkmann, and Lars Jacob Thynes Pedersen. Lars Jacob was one of my students in the classroom. Now he is my close colleague at NHH. Those friends have stimulated my professional life and enriched my personal life immensely! I would not have been able to write the articles without their generous collaboration! I would also like to thank Maria Prestmo, Morten Huuse, Judith White, and Magne Supphellen who have made a decisive contribution to the articles in this selection. And I should not forget my elder mentors Odd Langholm and Kjell Grønhaug, who have been my teachers and colleagues for more than twenty years, and my recent colleague, Bram Timmerman. Other colleagues and friends whom I have had the privilege to interact with and learn from through formative periods in my professional life are Sten Jønson, Thomas Polesie, Richard Boland Jr, Leif Methlie, Tore Holmesland, Jonny Holbek, Johannes Brinkmann, Morten Huse, and John Fundingsrud.

I would also like to thank members of the staff at NHH who have provided necessary support and assistance in general, and for this work in particular: Elaine Pettersen, Andela Culjak, and student assistant, Synne. My student assistant Axel Bøe has done comprehensive and excellent work in formatting the book, and the head of our department, Leif Hem, has in a pleasant way supported the book project. I would also express my gratitude to Helle Snellingen and Bianca Snellingen for their highly competent translation and editing of the article *Computer and Science as Bearer of Culture* included in this volume. I am grateful for the generous financial support from my alma mater, the Norwegian School of Economics, for this project as well as to all the conferences and sabbatical research stays during the last forty years, which has been the foundation for my academic journey so far.

Acknowledgements

Acknowledgement is made to the collaborative authors and publishers cited below for permission to republish the following copyrighted works.

Part 1

A Conflict Case Approach to Business Ethics (Journal of Business Ethics 2004 53(1–2)). Reprinted by permission of the publisher and the co-author, Johannes Brinkmann.

Good Intentions Aside: Drafting a Functionalist Look at Codes of Ethics (Business Ethics. A European Review 2004 12(3)). Reprinted by permission of Blackwell Publishing Ltd and the co-author Johannes Brinkmann.

Cooperation and Competition in the Context of Organic and Mechanic Worldviews: A Theoretical and Case-Based Discussion (Journal of Business Ethics 2006 66(1): pp. 19–32). Reprinted by permission of the publisher and the co-author, Ove D. Jakobsen.

Consumerism and Frugality: Contradictory Principles in Economics? (In L. Bouckaert, H. Opdebeeck and L. Zsolnai (eds) Frugality: Rebalancing Material and Spiritual Values in Economic Life. Peter Lang Publishing, Oxford, 2008). Reprinted by permission of the publisher and the co-author, Ove D. Jakobsen.

Holistic Problem Solving (In Laszlo Zsolnai and Antonio Tencati (eds) The Future International Manager: A Vision of the Roles and Duties of Management. Palgrave, 2009). Reprinted by permission of the publisher and the co-author, Laszlo Zsolnai.

Competition or Cooperation? A Required Shift in the Metaphysics of Economics (In Antonio Tencati and Laszlo Zsolnai (eds) The Collaborative Enterprise: Creating Values for a Sustainable World. Peter Lang Academic

Publishers, Oxford 2010). Reprinted by permission of the publisher and the co-author Ove D. Jakobsen.

Self-Realization in Business: Ibsen's Peer Gynt (In Rita Guesquiere and Knut J. Ims (eds) Heroes and Anti-heroes. European Literature and the Ethics of Leadership. Garant, 2010.). Reprinted by permission of the publisher and the co-author, Laszlo Zsolnai.

How Economic Incentives Destroy Social, Ecological, and Existential Values: The Case of Executive Compensation (Journal of Business Ethics, August 2014 123(2): pp. 353–360). Reprinted by permission of the publisher and the co-authors, Lars Jacob Thynes Pedersen and Laszlo Zsolnai.

Product as Process: Commodities in Mechanic and Organic Ontology (Ecological Economics February 2015 110: pp. 11–14). Reprinted by permission of the publisher and the co-authors Ove D. Jakobsen and Laszlo Zsolnai.

The Future of Business Ethics: A Structured Dialogue Between the Participants (In Knut J. Ims, and Lars Jacob Tynes Pedersen (eds) Business and the Greater Good. Rethinking Business Ethics in an Age of Crisis. Edward Elgar, 2015). Reprinted by permission of the publisher and the co-author Lars Jacob Thynes Pedersen.

Social Innovation and Social Development in Latin America, Egypt, and India (In Georges Enderle and Patrick E. Murphy (eds) Ethical Innovation in Business and the Economy. Edward Elgar, 2015). Reprinted by permission of the publisher and the co-author, Laszlo Zsolnai.

Land Ethics in the Context of Economics and Commercialization Effects. Paper presented on the 14th GCGI International Conference and the Fourth GCGI and SES Joint Conference, "OUR SACRED EARTH: Spiritual Ecology, Values-led Economics, Education and Society Responding to Ecological Crisis", Lucca, Tuscany August 28 – September 1, 2018. The paper was sent to the Globalisation for the Common Good Initiative (GCGI) in the UK, but has not yet been published in their journal.

Caring Entrepreneurship and Ecological Conscience: The Case of Patagonia Inc. (In Ora Setter and Laszlo Zsolnai (eds) Caring Management in the New Economy – Socially Responsible Behaviour Through Spirituality. Palgrave, London, 2019). Reprinted by permission of the publisher.

Part 2

Man, Machines and Science. From Intuition to Calculation. (Translated from "Datamaskin og vitenskap som kulturbærere" (In Tore B. Holmesland and Knut J. Ims (eds) Bedriftsøkonomiens helhet. Spenning mellom analyse og humaniora. Festskrift til Odd Langholm, Alma Mater Forlag AS, 1988)). Reprinted by permission of the rights holder, Fagbokforlaget, Bergen, Norway.

Mismeetings. An Essay about Money, Urban Life, and the Ethics of Proximity (In Tore B. Holmesland, Knut J. Ims and Ansgar Pedersen (eds) Essays in Marketing and Management. A Festschrift in Honour of Kjell Grønhaug. Fagbokforlaget, 1995). Reprinted by permission of the publisher.

Shallow Success and Deep Failure (In Laszlo Zsolnai and Knut J. Ims (eds) Business within Limits: Deep Ecology and Buddhist Economics. Peter Lang, Oxford, 2006). Reprinted by permission of the publisher and the co-author Laszlo Zsolnai.

Deep Ecology (In Luk Bouckaert and Laszlo Zsolnai (eds) The Palgrave Handbook of Spirituality and Business. Palgrave Macmillan, 2011). Reprinted by permission of the publisher.

Deep Authenticity: An Essential Phenomenon in the Web of Life (In Antonio Tencati and F. Perrini (eds) Business Ethics and Corporate Sustainability. Edward Elgar, MA, 2011). Reprinted by permission of the publisher and the co-author, Ove D. Jakobsen.

Deep Ecology and Personal Responsibility (In Laszlo Zsolnai (ed.) The Spiritual Dimension of Business Ethics and Sustainability Management. Springer, 2015). Reprinted by permission of the publisher.

Peace in an Organic Worldview (Contributions to Conflict Management, Peace Economics and Developement, Vol. 24, 2015). (In Luk Bouckaert and Manas Chatterji (eds) Business, Ethics and Peace, 2015). Emerald Group. Reprinted by permission of the publisher and the co-author, Ove D. Jakobsen.

Happiness and Human Flourishing (In Peter Rona and Laszlo Zsolnai (eds) Economics as a Moral Science. Springer, 2017). Reprinted by permission of the publisher.

Quality of Life (In Ove Jakobsen and Laszlo Zsolnai (eds) Integral Ecology and Sustainable Business. Emerald, 2017). Reprinted by permission of the publisher and the co-author Ove D. Jakobsen.

Quality of Life in a Deep Ecological Perspective. The Need for a Transformation of the Western Mindset? (Society and Economy, 2018 40(4): pp. 531–552). Reprinted by permission of the publisher.

Nature, Economics and Scream (In Luk Bouckaert, Knut J. Ims, and Peter Rona (eds) Art, Spirituality and Economics. 2018). © Springer International Publishing AG 2018; all rights reserved. Reprinted by permission of the publisher.

Deep Ecology and Its Relevance to GNH and Bhutan (In Dasho Karma Ura and Dorji Penjore (eds) GNH. From Philosophy to Praxis. Proceedings of the Sixth International Conference on Gross National Happiness, November 4–6, 2015, Paro, Bhutan. Center for Bhutan Studies & GNH, 2017). Reprinted by permission of the publisher.

Part 3

Learning Ethics in a Social Context: A Practical Guide Using Experiential Learning Theory (Proceedings of the International Association for Business and Society 1993 4: pp. 725–735). Permission from the publisher and the co-author, Judith A. White.

The Consultant-Client Relationship: Personal Autonomy and Development Through Dialogue (In Heidi Von Weltzien Hoivik and Andreas Føllesdal (eds) Ethics and Consutancy: European Perspectives. Kluwer, 1995). Reprinted by permission of the publisher.

Community in the Classroom: The Philosophy and Practice Behind a Course in Ethical Action (Proceedings of the International Association for Business and Society 1998 9: pp. 1077–1088). Permission from the publisher and the co-author, Morten Huse.

Dialogue-Based Problem Solving: A Discussion Based on Teamwork Among Ethics Students (In Willy Haukedal and Bård Kuvaas (eds) Creativity and Problem-Solving in the Context of Business Management. Bergen

Fagbokforlaget, 2004). Reprinted by permission of the publisher and the co-authors, Lars Jacob Tynes Pedersen and Maria Prestmo.

Faculty Members' Attitudes Towards Ethics at Norwegian Business Schools: An Explorative Study (Journal of Business Ethics 2005 62: pp. 299–314. Reprinted by permission of the publisher and the co-authors, Ove D. Jakobsen and Kjell Grønhaug.

Take it Personally (In Laszlo Zsolnai and Knut J. Ims (eds) Business within Limits: Deep Ecology and Buddhist Economics. Peter Lang, Oxford, 2006). Reprinted by permission of the publisher.

From the Art of Reading to the Art of Leading (In Hendrik Opdebeeck and Laszlo Zsolnai (eds) Spiritual Humanism and Economic Wisdom. Garant, Antwerpen and Apeldoom, 2011). Reprinted by permission of the publisher and the co-author Lars Jacob Tynes Pedersen.

Personal Responsibility and Ethical Action (In Laszlo Zsolnai (ed.) Handbook of Business Ethics – Ethics in the New Economy. Peter Lang International Academic Publishers, Oxford, 2013). Reprinted by permission of the publisher and the co-author, Lars Jacob Tynes Pedersen.

Personal Responsibility for the Greater Good (In Knut J. Ims and Lars Jacob Tynes Pedersen (eds) Business and the Greater Good. Rethinking Business Ethics in an Age of Crisis. Edward Elgar, 2015). Reprinted by permission of the publisher and the co-author, Lars Jacob Tynes Pedersen.

The Importance of Calling in Realization of Life Projects: The Case of Maverick and Serial-Entrepreneur Hans Nielsen Hauge with Implications for Business Education (In Luk Bouckaert and S. van den Heuvel (eds) Servant Leadership, Social Entrepreneurship and the Will to Serve. Palgrave Macmillan, 2019.). Reprinted by permission of the publisher and the co-authors, Magne Supphellen and Truls Liland.

PART I

JOHANNES BRINKMANN AND KNUT J. IMS

A Conflict Case Approach to Business Ethics

Introduction

Business ethics as an academic field has two main functions. On the one hand it challenges self-satisfied business people by inviting moral criticism and self-criticism of business practices. On the other hand business ethics is potentially helpful when it comes to analyzing, handling, and preventing conflict in business contexts, with a focus on moral aspects (cf. as drafts of such a perspective French and Allbright, 1998, pp. 177–178, with further references, or Brinkmann, 2002b, pp. 161–162). This paper suggests taking a better look at such a potential conflict management function of business ethics.

The use of moral conflict cases in business ethics teaching and research

Moral conflict cases are the most popular way of teaching business ethics, consisting of more or less complex and authentic conflicts without an easy self-evident solution. Business ethics casebooks are readers of business life, of conflict histories, and issues (see for example Beauchamp, 1997; Donaldson and Gini, 1995; Harvey et al., 1994; Hoffman et al., 2001; Jennings, 2002). While full-format cases are meant as representatives of real-life conflict complexity, teaching sometimes (and research normally) uses less representative and more focused short versions of conflict cases, often called "scenarios" or "vignettes" (see for example Bain, 1994; Brinkmann, 2002a; Chonko, 1995; Peck et al., 1994; Weber, 1992). Such conflicts or dilemmas, short ones or long ones, real ones or constructed ones, are normally designed as a hopeless choice between contradictory responsibilities where at least one stakeholder will be hurt. The follow-up question is usually in the format of "what would you do if you were person X?" or "which conflict party would you side with, and how would you justify your choice?" or "identify and clarify main issues, parties and stakeholders, options and wisest solutions".

Cases and case teaching market ethics as useful tools for analyzing and handling understandable and interesting moral conflict stories trigger standpoint taking and discussion. There is a danger, however, that cases can be "too" entertaining and too superficial as a genre. Students and even teachers or researchers can easily forget that moral conflict cases are examples; that is, are not about themselves or interesting in themselves only.[1] The challenge is to reach a compromise between respect for uniqueness and desirable generalizability. On one hand one needs to be loyal to the single case (in the classroom or in the real business world); that is, to come up with a best possible (or at least good enough) solution. On the other hand the question is what one can learn from this one case for more or less similar cases, learn about the potential relevance of moral philosophy, and learn about moral conflict management in general.

More complex checklists for moral conflict case analyses, such as the seven-point list suggested by van Luijk (1994, pp. 8–9) or the twelve-point list suggested by Nash (1989, p. 246), have a common denominator. They all require a combined analysis of facts and of norms, or a situation and normative analysis. Instead of a focus on practical suggestions and rules of thumb for case teaching and conflict-case question formats we want to address, more principally, the possible strengths and weaknesses of a moral conflict case focus. Our draft of a further elaboration of a moral conflict focus borrows from social science and from philosophy. By social science we think mainly of conflict research terminology. Our philosophical elaboration exploits primarily casuistry as a kind of practical, inductive argumentation with a focus on paradigmatic examples.

Moral conflict case analysis as conflict analysis

In order to prevent case teaching and case research from being too quick and too superficial, a delayed judgment approach seems fruitful (Lustig and Koester, 1996, pp. 333–336, suggest in their textbook an acronym, a "D-I-E"-approach). Instead of jumping from a quick case description to a quick recommendation for how to handle the case, Lustig and Koester suggest asking for a sufficient description ("D") and understanding (interpretation, "I"); that is, for a deliberately delayed judgment (evaluation, "E"). Description and understanding require relevant and useful concepts, either as a language in which one can describe and understand conflict cases as stories (cf. van Luijk, 1994, pp. 4–5) or for asking individuals questions and understanding their answers about such conflict situations. Since there is a risk of over-representing big cases with media attention one can often defend a further detour and delay by asking individuals questions about their conflict perceptions before describing cases as a whole.

When mapping individual conflict experience in professional or business contexts one would probably ask individuals, in an open question, for any conflict definitions and/or conflict examples. Perhaps one would ask, in addition, questions about individual conflict handling experiences and ideas. Most respondents would probably understand such a question in one of the following formats:

- Are you experiencing any conflict in your work situation, as an observer or as a party, and if yes related to which issues, how frequently and how seriously?
- Can you recall any (recent, serious, or just any …) conflict in your work situation, as an observer or as a party, and if yes can you describe it briefly in your own words?
- If you experience or believe you experience a conflict in your work situation, how do you typically react?

Such questions about conflict concepts and conflict experiences are a good entry to professional morality studies. Coding the answers to such (more or less open) questions, however, requires a theoretical conflict concept and at least a minimum of indicators to look for and to compare by. Five conceptual distinctions could serve as a start:[2]

- conflicts as units versus conflict as a social system property,
- conflict attitudes, conflict behaviours and conflict contents,
- conflict of interest versus conflict of values,
- conflict versus conflict management perspectives and
- conflict (case) outcomes versus conflict modification.

In questionnaire-based research some respondents might refer to one specific conflict (or to several conflicts), while others might read the question as a question of whether or not the workplace is more or less "full" of conflict. For grasping this subtlety one could distinguish, in methodology language, between a unit and a property concept of conflict. In the first case one would think of *the* conflict *x*, taking place in social context *y*, during time-span *z*. One given conflict is studied as a specific time-space unit, often with a focus on properties such as issue types (such as value dissensus, or incompatible interests related to scarce resources), the number of parties involved or power-relationships. In the second case, conflict denotes a system state or an actor relationship state, such as a level of degree of conflict. There can be, for example, much conflict in a post merger organization. This means that conflict (and conflicts in the plural form) can be thought of in the indefinite form, grammatically speaking. There can be much or little destructive or productive, basic or superficial conflict. Ralf Dahrendorf's widely quoted dichotomy between conflict and consensus models of society (1958, 1959) applies a property continuum or variable from consensus to conflict when talking about societies or organizations, with conflict almost as the opposite of consensus. The unit and property concepts of conflict are related. Conflicts-as-units are symptoms or manifestations of conflict-as-system-property. In most organizations there is some latent conflict; for example, about profit/wage ratios or about proper degrees of workplace democracy. Such latent or built-in quite normal conflict manifests itself from time to time in identifiable conflicts-as-units; that is, conflict processes or episodes.

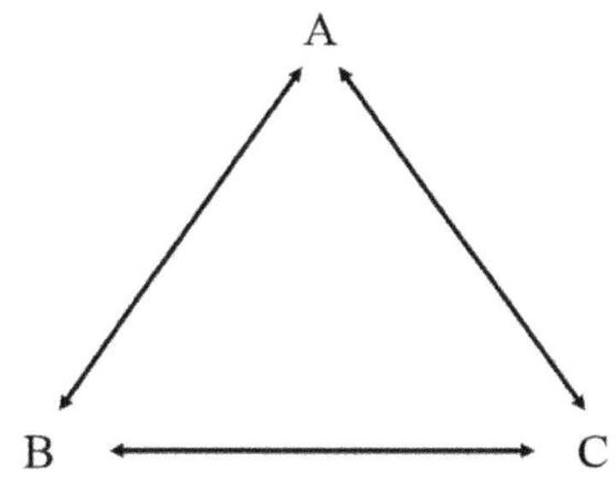

Figure 1. Galtung's "conflict triangle". (A) conflict attitudes, (B) conflict behaviour, (C) conflict as incompatibility.

Conflict attitudes, conflict behaviours and conflict contents

A second distinction could be between what the conflict is about, if and/or how the conflict is perceived by the parties, and if and/or how the parties show any overt signs of conflict-related behaviour. In other words, one could distinguish between attitudinal, behavioural and content aspects (or concepts) of conflicts, as suggested by Johan Galtung in various lectures and papers (see for example 1989, pp. 2–4; as a visualization as a triangle, cf. Figure 1).

When coding responses, the question would be if the answers refer to self-other images, to events or actions, or more specifically to conflict issues; that is, more or less abstract, underlying, incompatible goals, right-versus-right, and right-versus-wrong choices (Kidder, 1996) and interests or values.

Conflict of interest versus conflict of values

A third distinction relates closely to business ethics and to the core of this paper. It is a dichotomy by primary conflict theme, between conflicts of interest and conflicts of values. The main idea is that the most important conflict issues are competing interests versus moral disagreement, respectively, or with a lengthy quotation from Aubert's article where this dichotomy is suggested (1963, pp. 27–30, our italics):

> A *conflict of interest* between two actors stems from a situation of scarcity. (Both) ... want the same thing, but there is not enough available for each to have what he wants. In this general sense the basis for a conflict is present in all trading transactions. The seller would like to have more money than the buyer is willing to part with ... This conflict potential is eliminated through the operation of the market, usually so smoothly that no overt signs of conflict appear. If a conflict comes into the open, the solution will often be a compromise ... It is a type of social interaction in which it seems that solutions are reached by discouraging the actors from getting morally involved in a major aspect of the interaction, the condition being that the interests are not diametrically opposed ... Competing or contrasting interest does not in itself imply any disagreement between Ego and Alter concerning values. It may even be claimed that a conflict of interest presupposes a consensus, at least on the value of the good, which is sought after by both parties ... A *conflict of value* is based upon a dissensus concerning the normative status of a social object .... (The) illicit nature of compromise on the level of value and of empirical truth makes it hard to discuss matters quite candidly ... It is especially when ... questions of factual responsibility, of guilt and merit, become parts of value-conflicts ... that a solution through compromise becomes so difficult.

Different conflict management perspectives

A fourth distinction relates to differences in conflict management perspectives. A first question asks if conflict is looked at from a conflict management perspective, or if conflict-handling institutions are looked at from a conflict perspective. In the first case, conflict or conflicts are treated as a problem. Conflicts as problems need to be handled and asked for a solution (or resolution, cf. Figure 2; cf. social science terminology (Galtung, 1965, p. 355)) in order to avoid potential negative effects of unresolved conflict.[3]

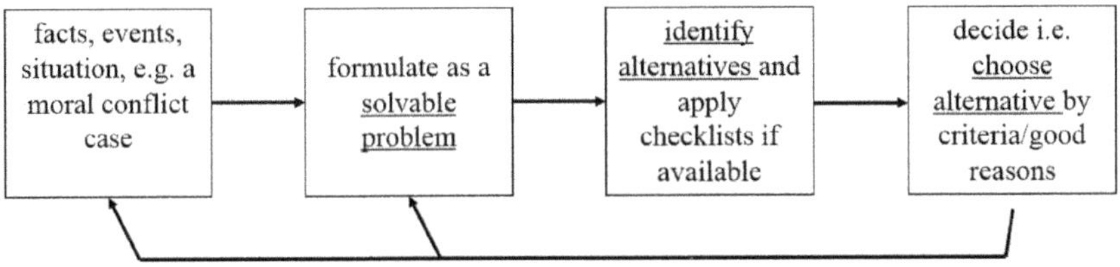

Figure 2. Solving conflicts as problems.

Or in Weiss' words (1996, p. 170):

> The goals of conflict management are, first and foremost, to prevent negative or dysfunctional conflict from occurring while, at the same time, encouraging healthy conflict that stimulates innovation and performance. If prevention does not work, eliminating the dysfunctional conflict is a second goal, and if this fails, minimizing or decreasing it becomes the fall back position …

The opposite perspective would ask critically if conflicts and if the parties' standpoints have been addressed thoroughly and constructively in their own right. Sometimes there is a suspicion that institutions primarily try to justify themselves as well functioning (if necessary, by redefining conflicts to fit with the possibilities of the institution).

The second question about *how* conflict is managed is not independent of the first one. Rewarding or persuading parties respects the parties' conflict ownership more than coercing them (cf. Kriesberg, 2003, pp. 110–124, in particular his diagram on p. 111 which is shown here as Figure 3).

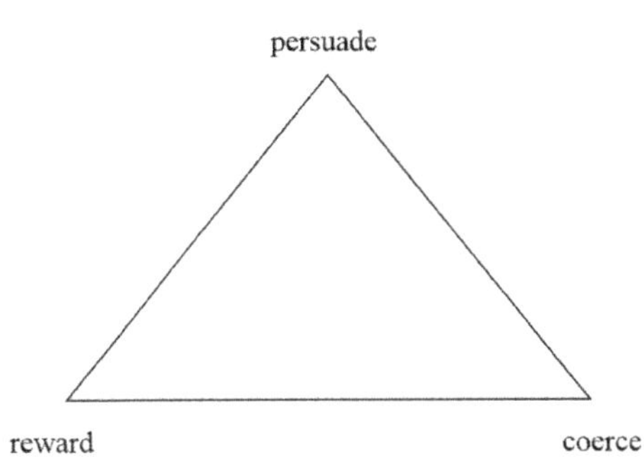

Figure 3. Kriesberg's conflict management style types (2003, p. 111).

Conflict (case) outcomes versus conflict modification

A fifth aspect can be seen as a follow-up to the first distinction and to the fourth one. Conflicts-as-units are almost defined, as processes, by their history; for example, with pre-history, emergence, development and termination; that is, they have an *outcome* (or end), and consequences (cf. for example Kriesberg, 2003, with a figure on p. 23). Conflict-as-property typically *changes or not,* that is, increases, stagnates or decreases, either by itself or by management in the above-mentioned sense. Unless one employs a purely attitudinal and/or behavioural conflict definition, the core criterion in the conflict-as-unit case and of a possible conflict-as unit solution is *incompatibility* or a contradiction of party interests and/or moral positions. The following fivefold typology (Figure 4, source: Galtung, 1965, p. 351 and Galtung, 2003, pp. 11, 26 – authors' simplification) can serve as an illustration.

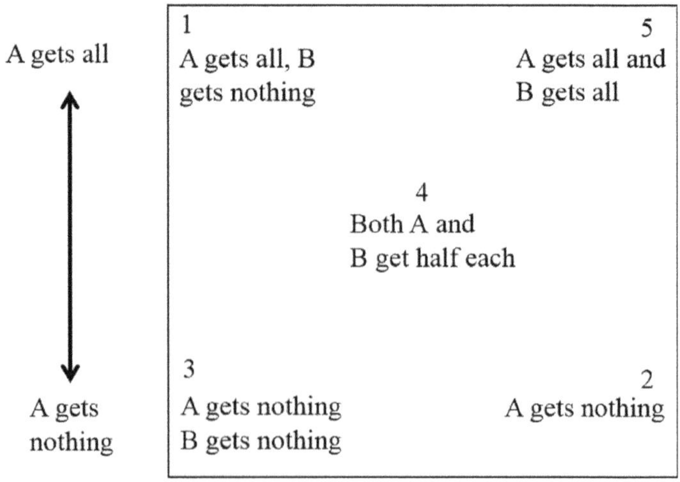

Figure 4. Galtung's conflict outcome typology.

For the conflict-as-property concept tradition one can refer to another fivefold typology, of conflict *management styles* (cf. widely quoted Thomas, 1976, quoted here after Weiss, 1996, p. 171):

Moral conflict defined

> Moral conflicts are conflicts between moral claims that may face persons or groups or communities or governments representing them, when individually or collectively they deliberate about what to do … between conflicting obligations …, purposes, ends, goals, or values …, moral codes or systems or world views …, different kinds of moral claims (for example consequentialism, deontology, partiality, authors' add.) … (Lukes, 1991, pp. 5–9)[4,5]

In terms of the five conceptual distinctions presented above one can, more simply, consider moral conflict and moral conflict management as special cases of conflict and conflict management; that is, intra- or inter-party conflict situations related to moral standpoint differences (or incompatibilities, or incommensurabilities, cf. Lukes, 1991, pp. 9–17). Instead of, or in addition to, a conventional definition of moral conflict, it can often be as fruitful to treat moral conflict as a conflict ideal type or counter type as suggested by Aubert (1963, cf. once more the quotation above and our Table 1 which repeats Aubert's dichotomy).

With or without such a typology in mind, one should leave as an open empirical question if (and how) moral conflict bases or moral significance

Table 1. A comparison of moral versus non-moral conflict

	Non-moral conflict	Moral conflict
Focus	Incompatible interests	Incompatible moral positions
Action	Ends-rational	Value-rational
Perception	Rational perception	Tendency towards judgmental perception
Outcome	Negotiable	Often non-negotiable, hence imposed sentence or segregation

of conflicts affect conflict history – how conflict cases, conflict levels, conflict attitudes, behaviours, and contents, or conflict management develop if self-administered moral conflict management is more difficult (as claimed by Aubert), or not.

Five assumptions about moral conflict in (business) organizations

Related to the distinctions presented above, we see five fruitful assumptions when it comes to understanding conflict in business organization contexts with a weaker or stronger moral component:

1. Conflict as system property and single conflict cases are normal rather than exceptional in organizations. Moral conflict can often be a sign of cultural and moral diversity.
2. Conflict or conflict cases, moral ones or not, should be dealt with constructively, as long as such conflict management is not biased and respects the given conflict(s) on their own premises.
3. Moral conflict can represent a (productive) test of principles and identity, that is, it can provoke and engages often more than non-moral conflict (for a similar reason, moral blaming and moralizing can create conflict escalation and often function destructively).
4. Ethics represents a chance to handle intra organizational conflict of different kinds (that is, moral or non-moral conflict) in a civilized and constructive manner.
5. The question of conflict outcomes repeats the ambiguity between conflict cases (units) and conflict as system property. Conflict cases can end, that is, have outcomes, while conflict as property cannot "end", but continues, increases or decreases, modified or not.

The last assumption suggests that conflict management should be sustainable. The typologies with different conflict management styles and

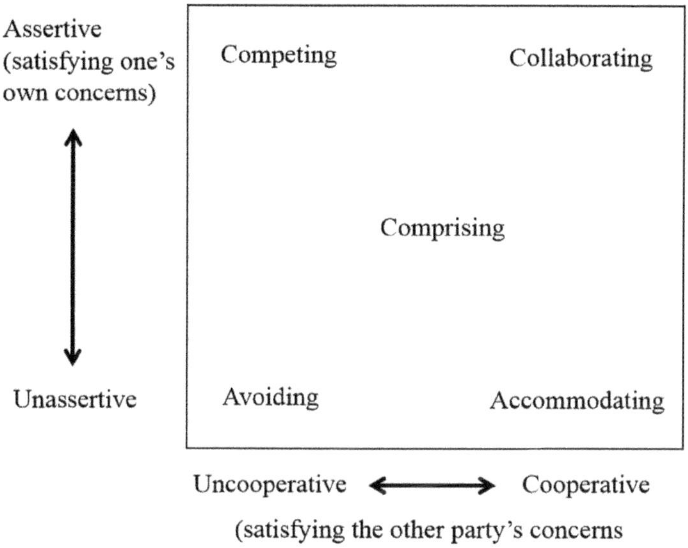

Figure 5. Thomas' conflict management style typology.

outcome types (cf. Figures 3–5) are less a question of truth than of usefulness for further conflict analyses (or not).

Sooner or later conflict analysis will turn into an analysis of incompatibilities and of incompatibility removal (cf. still Bergstrom, 1970, especially his figure on p. 213 and, once more, Galtung's recent book, 2003). To Galtung, the proper handling of a conflict is ideally a question of conflict transcendence, In other words, any conflict outcome is sub-optimal as long as better, more fruitful conflict transcendence is possible (following Aubert, 1963, one could assume that value conflict transcendence is more difficult than a transcendence of measurable interests).

A conflict case solution means often the same as a transcendence of a *logical* contradiction towards a contradiction-free final statement. As a contrast conflict-as-property; for example, in the conflict-and-consensus-antinomy, represents a *dialectical* contradiction.[6] In an older paper Pruzan and Thyssen (1990) offer an interesting elaboration of the conflict versus consensus antinomy, which is of special interest here since it builds a bridge to another important antinomy of morality versus ethics:

> In a pluralistic society there is no agreement as to what is morally right and wrong. Each subculture maintains its own values and therefore its own discriminatory norms as part of its identity. This creates a variety of morals. No formal arguments can substantiate one subculture's moral principles and deny the validity of another's ... The question is whether it is possible to develop a set of values which are shared among the subcultures and which can contribute to replacing such confrontation within a political culture, which respects conflicts and differences and still is able to create consensus ... When subcultures cannot justify their own rules for right and wrong via intuition, or an appeal to, universally valid rules, what is required if groups with different moral rules are to coexist can be considered as a *second order morality*. We will call this second order morality ethics ... Ethics is distinguished from jurisprudence by its search for the *legitimate* rather than just the *legal*. An action or decision is legitimate if it can be rationally accepted by all stakeholders. Ethics is also distinguished from morality. Moral rules are rules for dissolving substantive conflicts *within* a subculture. In a pluralistic society ethics leads to value-oriented communication aimed at dissolving conflicts in the social relationships *between* subcultures. A gap arises between the moral substance, created by the tradition of a sub culture, and the ethical form, created by the need for non-violent coexistence of many traditions and sub cultures .... (1990, pp. 136–137)

One could try to simplify the two authors' way of reasoning by still another simple typology (see Table 2).

Table 2. A comparison of morality and ethics

Morality	Ethics
Subcultural	Intercultural
Internal	External
Particularist	Universalist
Potentially furthering conflict	Potentially furthering conflict solution (consensus)
Moral substance	Ethics as legitimate form (in contrast to *legal* forms)

For the remainder of the paper the focus will be on conflicts as units, that is, conflicts as cases, shortened or not, possibly with additional information available from the individual parties in the case. As already indicated above, the suggestion is that moral conflict cases should, as a start, be described and understood as conflicts:

- with a focus on attitudinal, behavioural and incompatibility aspects whenever appropriate,
- as conflicts or interest, of values or as a combination of both,
- which balance between loyalty to the conflict and conflict management considerations,
- and not least which make learning from the handling of the single conflict relevant to future conflict handling.

Casuistry – Case-focused moral philosophy

If one's ambition is to outline how moral philosophy can make sure that a given moral conflict case is not only described and understood on its own premises, but also evaluated on its own premises, there is no way around casuistry as a philosophical tradition. The challenge is first to take a critical look at the potential strengths and weaknesses of such an approach. In a next step, one could ask how ethical casuistry could function without the power standing behind legal casuistry (following from law positivity, "legality").[7] In our opinion, such legitimacy could and should follow from combining casuistry with a discourse ethics approach, claiming that the parties own their conflict case.

A few references

A quick literature search indicates that there are only a few business ethics sources, which provide a more comprehensive presentation of casuistry as a useful and important method for our field. The first source is Ciulla's (1994) short but thorough history of philosophy presentation and literature review of casuistic thinking, printed in a business ethics anthology. Calkins (2001) addresses the compatibility of casuistry and the business case method by describing the key features of casuistry and the case method, not least as inductive and practical methods of reasoning with a focus on particular settings and real life situations. A similar, more general professional ethics focus is found in Toulmin's paper (1973), who claims that professional or applied ethics, in this case medical ethics, "saved the life of ethics" by forcing it back to real life moral conflict diagnosis and prescription. Boeyink (1992), who is concerned with journalism ethics, discusses casuistry as a method and a "middle ground" between practice and principle. As a third applied ethics example one could refer to another piece of work of Calkins, showing how casuistry could handle the triangle conflict between GM-food proponents, GM-food opponents and the farmers in between (2002). Calkins is optimistic when it comes to potential synergies between casuistry and virtue ethics.[8] Our contention is that there is a need for "much more" casuistic business ethics; that is, we claim that casuistry represents an important and underutilized[9] potential for both academic and practitioner business ethicists. As a first step towards a substantiation of such a claim we will now draft a critical and constructive overview of the potential benefits and weaknesses of the casuistic method.

Casuistry – a negative or a neutral term?

One standard definition of casuistry can be found in the Oxford English Dictionary: "Casuistry is that part of ethics which resolves cases of conscience, applying the general rules of religion and morality to particular *instances* in which *circumstances* alter cases or in which there appears to be a conflict of duties" (our italics).[10] This definition is standard in its reference to conflict situations, but narrow in its unnecessary reference to specified systems of norms, such as legal rule systems or Christian moral theology. Another and even more important definition criterion of the casuistic method should be its inductive, analogical, and dialectical form of *argumentation,* independently of such specified systems. We should therefore rather like to follow the definition of casuistry suggested by Ruyter (1995, p. 9) as a:

> [C]ase-oriented and example-based way of argumentation (…) Morally relevant similarity between good examples and the situation of moral doubt can be used as an argument for similar treatment of the new case, while morally relevant differences can be used as an argument against such similar treatment. Arguments for and against need to be balanced which in turn requires good judgment. Usually such an approach can lead to a preliminary conclusion with more or less hypothetical and acceptable solutions. Such a way of argumentation furthers dialogue and aims at researching a consensus …. (Authors' free translation from Ruyter, 1995, p. 9)

The label matters, too. Instead of the dyslogism[11,12] of "casuistry" we should prefer the German/Norwegian/Scandinavian term, *Kasuistik.* In this way one could avoid the pejorative connotation related to the abuse of casuistry in its mature period from 1650 to 1750, with moral probabilism and moral minimalism as the main blame. During this period the French philosopher Blaise Pascal started his powerful attack of the Paris Jesuits in his *Provincial Letters* (1656). Not least due to Pascal the word casuistry carries a negative connotation. But casuistry criticism is older. The founding father of Protestantism, Martin Luther (1483–1546) included the most famous casuistry textbook of his time (Angelus Claretus' *Summa Angelica)* in his bonfire of books in 1520, since he judged it to be destructive to Christian faith.[13]

"Business ethics" as casuistry

The abuse of casuistry is not unknown to contemporary business writers. The following quotation by Peter Drucker is worth sharing and further reflection:

> Business ethics … is to ethics what soft porn is to the Platonic Eros; soft porn too talks of something it calls love. And insofar as business ethics comes even close to

> ethics, it comes close to casuistry and will, predictably, end up as a fig leaf for the shameless and as special pleading for the powerful and the wealthy. (1981, p. 34)

Drucker's main concern is business ethics instrumentalization, not least by cynical profit seekers.[14] Drucker's expectation seems to be that ethical theory without virtuous habits and attitudes makes one clever rather than moral.[15] In their interesting paper, "A reply to Peter Drucker", Hoffman and Moore (1982) depart from the premise that casuistry to them "remains an important aspect of ethical reasoning". In a next step Drucker is criticized for his pejorative interpretation and use of the term casuistry, and then for blaming business ethics for being casuistry.[16] To them, casuistry has to do with "the application of general principles in specific circumstances", thus representing "the mechanical aspect of ethics" (p. 297). Klein (2000), with a focus on Drucker's more comprehensive management works, concludes that Drucker is a "business moralist", that he really "takes business ethics seriously", and that he holds an "essentially Platonic" view of business manager responsibility.

Strengths of casuistry

With reference to Jonsen and Toulmin's use of casuistry in a National Commission (for the Protection of Human Subjects of Biomedical and Behavioural Research, in 1974), Keenan (1998) considers modern casuistry as "a moral taxonomy for distinguishing acceptable from unacceptable ways of involving humans as subjects in medical or behavioural research"; that is, as different from sixteenth century casuists' use of it due to contemporary complexity being based on an incommensurable diversity of ethical systems. In a next step Keenan (1998) demonstrates the sustainability of casuistry as a consequence of its *turn to the subject* and emphasis of context dependency. Deontology and consequentialism are in effect inhibiting the beliefs of the commission members; that is, functioning as ideologies. Casuistry "on the other hand (is) a formal conveyor, a translucent mediator bringing beliefs more directly into the concrete world … Casuistry is free of such ideological biases … because … casuistry is 'pre theoretical' …" (Keenan, 1998, p. 165). The one bias of casuistry is its suspiciousness towards "ideology" and "generalities". Therefore, casuistry represents and expresses the beliefs that form us, the practitioners' way of thinking,[17] their presuppositions and presumptions. Furthermore, casuistry is "… unintelligible as an activity separated from its communal context …" (Keenan, 1998, p. 166).

Weaknesses of casuistry

Even more important, perhaps, than an awareness and exploitation of the potential strengths of casuistry is an awareness and avoidance of its weaknesses. This concerns less the traditionally bad reputation of casuistry already addressed rather than a number of more specific traps or limitations. After quoting more well-known objections such as "lack of critical distance, … methodological stringency and determinacy" and in addition mentioning a possible conservative bias, Ruyter (2003) lists altogether five limitations "worthwhile considering" (presentation shortened, in part extended, and reformulated by authors):

1. the authentic case is not as self-evident as often presented (but normally "constructed"; that is, exposed to selective perception, the authors add);
2. case solutions are typically "probable"; that is, often neither acceptable to most stakeholders, nor resolutions in a stricter sense;
3. casuistry can be narrow-minded; that is, it can overlook larger perspectives and should be supplemented by an interdisciplinary perspective;
4. there is a tendency and temptation to overstate similarities and analogies across cases; and
5. there is no easy way of identifying paradigm cases – conscious and cautious use of analogies "in order to assure relevance and significance in the comparison of context embedded cases" (p. 11).

The remedy against actual or potential weaknesses of casuistry (formulated by Ruyter or others) is perhaps simply a "virtuous user" requirement in the Aristotelian sense, so to speak, as a quality insurance (cf. also MacIntyre, 1984, pp. 152–155; Jonsen and Toulmin's view, 1988; and once more Calkins, 2002).[18]

Casuistry as a procedure for moral conflict management

Casuistry has more or less implicit theoretical assumptions. Most important, perhaps, is the assumption of inductive reasoning. Moral decision making should be organized bottom up rather than top down; or, according to Buchholz and Rosenthal (2001, p. 28), who claimed that "a sense of moral rightness comes not from indoctrination of abstract principles but from attunement to the way in which moral beliefs and practices must be rooted naturally in the very conditions of human existence". Beyond their main thesis that classical American pragmatism could link principle and case approaches, they show that moral reasoning demands a return to concrete situations as the very foundation for context-sensitive moral decision making. We have now returned where the paper started: casuistry is, essentially, a practical procedure for moral conflict management.

As such, "moral conflict management" starts with a *problematic case,* a case that contains some conflict. Keenan (1998) suggests using the threat of AIDS as an example, since it (like the threat of nuclear war) forces us to rethink our moral principles and presuppositions (p. 167). Then follows a *search for paradigm cases*; that is, resolved accounts of real-life situations or "touchstone cases that have intrinsic and extrinsic certitude" (Calkins, 2001). The interesting question is, of course, what makes a case paradigmatic. According to Aristotle, there are five paradigm types (distinguished by acknowledgement types, *endoxa*), where a case either is recognized by all, by a majority, by the wise, by most wise, by the most well-known and respectable among the wise, respectively.[19] Another alternative of paradigm legitimacy could be tradition; that is, case testing over a long time.[20] As Keenan (1998) demonstrates with AIDS as a case, it is indeed a challenge to find a paradigmatic case with fruitful and interesting parallels (relevant similarities) to the present case. On the other hand, modern information technology makes it much easier to find, classify, store, and retrieve paradigmatic cases. The next step is inductive *analogical reasoning*. Deliberation *(pro aut contra dicere)* draws analogies between the paradigm case and the present situation, attempting to identify relevant similarities and differences. Such dialectical argumentation consists in a dialogue-based attempt to promote critical ethical reflection, departing from common acknowledged meanings, through questions and answers, arguments, and counter-arguments (cf. Ruyter, 1995, p. 13, note 20). According to Ruyter (1995), this very step makes the casuistic method more action-oriented because one asks for arguments which can lead to recommendations or solutions. Brown's argumentative model has also a lot in common with casuistry (see 1990, Ch. 3).[21] Brown asserts that such a way of reasoning stimulates ethical reflection because it shows the necessity of value judgments and assumptions in making policy decisions. In other words, casuists are not satisfied with an abstract discussion of theoretical necessity and eternal conclusions. Figure 6 summarizes the casuistic way of resolving moral conflicts as problems.

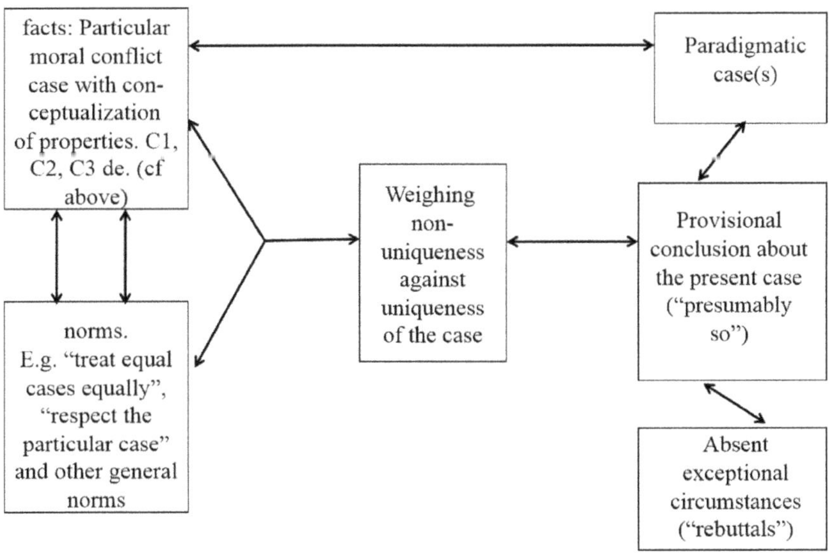

Figure 6. Revised and extended from Jonsen and Toulmin, 1988, p. 35.

Figure 6 can be read as a repetition of the main thoughts in the different sections of this paper. We claim that any shortcut from moral conflict case descriptions to checklists is problematic. Conflict theory and casuistry can, hopefully, delay and improve case description and understanding as an alternative to jumping to a quick solution, by:

- recognizing such cases as *conflict* cases (or even better as a range of different conflict types, discovered and focused on by different conceptualizations), and
- recognizing popular case analysis as a kind of *casuistry light* raising the same critical main question as casuistry, namely to weigh uniqueness and exceptions against general and generalization considerations, balancing between justice to the specific case in its context and justice more generally across cases.

In addition, one should not forget the typical ambiguity of models or approaches.[22] On the one hand a model or an approach (for example a conflict concept) may be better than none; on the other hand any focus includes a risk of overlooking non-focused aspects and narrowing rather than opening up our minds. The best remedy against such narrowness or bias is a similar kind of approach combination, which is well known from research methodology – triangulation.

Open end: Casuistic moral conflict management

What is a vice in a research report financed by the business community is a virtue at a conference where researchers meet. We are offering a number of open questions and suggestions for a discussion about desirable future work and possible research cooperation, rather than a conclusion:

- Having presented both conflict diagnosis and casuistry on their own premises prepares a desirable synthesis of these two related ways of thinking. There are also other streams of research which should be taken into consideration when trying out such a synthesis, not least discourse ethics for ensuring legitimacy of the procedure.[23]

- The possible synergy effects between conflict theory and casuistry could also be tried out in critical-empirical analyses of cases, ranging from standard length single business ethics cases to short scenarios.
- Finally, the topic of this paper could contribute to a better theoretical grounding of case teaching in business ethics.[24]

If there is any short and simple conclusion (as a repetition of a reference above), in spite of any potential weaknesses, casuistry seems indispensable as a method and a "middle ground" between practice and principle. The alternative seems to be blind practice and empty principles.[25]

Acknowledgement

We should like to thank, in addition to the reviewers, Tore Nordenstam, Knut W. Ruyter and Roberta Wiig Berg for valuable help with comments and native language proof-reading. Any remaining imperfections are, of course, the authors' responsibility.

Notes

1. Cf. for example, van Luijk (1994, pp. 6–7), who distinguishes between two dangers of case-teaching. A "minor, didactical" danger overemphasizes case-teaching (hoping that many cases *per se* generate understanding and of looking at cases as puzzles to be solved by almost guessing one solution only instead of focusing on balancing arguments and comparing cases). A "major, normative" danger overemphasizes and excuses the exceptional aspects of a case in a similar way as what is meant by the term casuistry, used in a derogative way. Donaldson talks of the case method as "not foolproof", since business cases are necessarily looked at ex post as static simplifications, often presented in a hurry with too little discussion time (Donaldson and Gini, 1993, p. 21).

2. There are further heuristic schemes and distinctions one could have referred to – for example, checklists of starting questions such as: *who* is in conflict with whom, about which *issues, why,* by acting *how,* with which *outcomes* and *consequences, regulated* by third parties or relevant institutions (or not – cf. also the checklist provided by Weiss, 1996, p. 176). In addition, one could have referred to the traditional social science distinction between micro and macro, or even micro-meso-macro *system levels.* And there are inter- and intra-group-conflict, inter- and intra organizational conflict. Inter-party-relationships often represent another important conflict dimension, such as power equality versus inequality, or party autonomy versus interdependence. If one distinguishes three relationship types (symmetry, superiority, membership) one can combine this dimension with the micro-macro dimension to a 15-cell-typology (cf. Dahrendorf, 1972, p.27, with several examples for each type).
3. Criteria or good reasons (in terms of the last diagram box) could be, for example, legitimate procedures, rules and principles, and desirable/undesirable consequences.
4. The first author's PhD thesis (*Konfliktpraxis und Rechtspraxis,* Münster and Oslo, 1975) suggests that this distinction was the core of the Habermas-Luhmann controversy in German social science in the early 1970s, cf. Luhmann, *Legitimation. durch Verfahren.,* Neuwied, 1969; and Habermas and Luhmann, *Theorie der Gesellschaft oder Sozialtechnologie …,* Frankfurt, 1971.
5. This quotation could have been continued over half a page or so, as an elaboration of the distinction that "…conflict may signify diversity, incompatibility, or incommensurability …", (ibid., p. 9). These quotations are taken from a text which is a convincing bridge builder between social science and moral philosophy (the latter looked at from outside), eventually ending up with a liberal defense of taking moral conflict seriously and making sense of it, as opposed to its "enemies", that is moral philosophers who try not to make sense of it (cf. esp. pp. 3–5 and p. 20). For a less theoretically and more practically sophisticated conceptualization of moral conflict see for example the book by Pearce and Littlejohn, 1997, which includes the following suggestion: "Although abstract definitions of moral conflicts are useful for certain purposes, they should not be pushed too far. Matters of definition involve a necessary trade-off between abstract terms that are useful for delineating categories and more specific terms that help describe actual events …" (p. 50).
6. The difference between these two contradiction types is not always clear, since finding – or guessing – a logical contradiction transcendence is one thing, while marketing it can be quite another. As indicated above, conflict and consensus almost form a dialectic opposition, where conflict and consensus concepts define one another, test and transcend one another, and offer critical evaluation criteria for one another. It has also been mentioned that the conflict versus consensus theme is one of the classical antinomies in social science. See in addition to the two classical works of Dahrendorf (1958, 1959) textbook chapters and sections such as Ritzer (1996, Ch. 7); Wallace and Wolf (1995, Ch. 3); Collins (1994).

7. There is another important difference between morality and law, which needs to be mentioned. Law has typically a strong preference for predictability, precedences, and prejudicats (cf. the *stare decisis* rule – what is decided should not be changed). Morality on the other hand is not necessarily bound by precedences, but rather a question of what is fair in the given situation, all things considered (cf. Sundby, 1978, p. 291).
8. Calkins (2002) also suggests searching for more exemplary "hero" stories – such as the one of the 1970 Noble Peace Prize winner, Nils Borlaug.
9. According to Jansen and Toulmin (1988) such "underutilization" of the casuistic method is partly due to its (pre-modern) abuse by the Jesuits and to the (modern) drive towards deductive and abstract science.
10. Cf. Ciulla's (1994) definition of casuistry, as "the art of reasoning from cases. The Latin word 'casus' means the falling away or declension of a noun. By analogy, the term 'casuistry' implies a kind of deflection or falling away from a law or principle. Casuistry serves the dual purpose of applying principles to cases and using cases to help us understand and sometimes alter principles …" (p. 172). Cf. also the short casuistry definition suggested by Calkins (200: 2), as "… a method of moral deliberation that relies on settled cases to resolve present moral dilemmas. It has users reach a decision about an ambiguous present situation by comparing that situation to previous incidents in which judgments have already been rendered." (p. 312)
11. About casuistry as an attempt to practical handling of a particular, morally difficult situation cf. Jansen and Toulmin (1988, p. 13).
12. Cf. *Collins Cobuild English Language Dictionary* (Collins, et al., 1987): "reasoning that is extremely subtle and designed to mislead people …". See Jansen and Toulmin (1988) about the word ending "-ry" in the word "casuistry", which communicates family resemblance to other dyslogistic terms such as wizardry, "sophistry", "harlotry", and "poperty".
13. According to Ruyter (1995, p. 122), Luther suggested *Summa diabolica as* a more suitable title because he considered casuistry to be more than devilish (*plus quam diabolica*).
14. Jansen and Toulmin (1988) remind us of the double roots of casuistry, both in moral philosophy and in rhetoric's traditions, the latter having the major purpose of inventing arguments likely to convince an audience.
15. Cf. similarly Williams (1982, p. 19): "Morality is primarily a way of life, an ethos, and what one does in business ethics courses is to reflect on the ethos and its implications for business practice."
16. In the same paper Hoffman and Moore also offer an insightful analysis of a case with conflicting moral obligations (Henry VIII's dilemma to divorce Catherine of Aragon versus "to govern his country wisely").
17. Cf. Toulmin's formulation, quoted by Keenan (1998, note 14): "… Catholic principles say more about Catholics than about the issue they are addressing …"

18. A virtuous person would simply not instrumentalize ethics in order to obtain disputable purposes. In particular, *phronesis* (or prudence, practical wisdom) is the most relevant virtue for casuistry. In a strict sense, *phronesis* is not an ethical but an intellectual virtue (such as *techne, nous, episteme* and *sophia),* though intimately connected to ethical virtues (cf. Aristotle, 1985; for example, Book 6, 1139 b11–b18). Cf. also Toulmin's references to Aristotelian arguments in favor of casuistry (1973, pp. 93–5), MacIntyre's illustration case for the importance of Aristotelian *phronesis* (the Wampanoag Indians' tribal land claim case, 1984, pp. 133–4), or as more general supplementary readings of Wheelwright (1935). Different terms are used in different translations. While T. Irwin uses the term "intelligence" and Wheelwright uses "sagacity", we prefer the term *phronesis,* or practical wisdom as used in Aristotles' *Nicomachean Ethics* (translated by Ross), Oxford University Press 1925, or in *Den Nikomakiske etikk* (translated by Rabbas and Stigen into Norwegian), Bokklubben Dagens bøker, Oslo, 1999. The emphasis on *phronesis* in Aristotelian ethics means that practical issues should be discussed and deliberated in "rhetorical" terms. Rhetorics and ethics are both practical fields where formal proofs or intellectual precision are replaced by real life-experience.
19. About moral probabilism, see Jansen and Toulmin (1988, pp. 164–75).
20. Such an aspect of "acknowledged meaning" builds, by the way, a bridge between the casuistic tradition and common, empirical morality – that is, a paradigmatic case solution must be acceptable in the context where it applies.
21. Brown elaborates on Stephen Toulmin's five-concept argumentative model (conclusion, data, warrant, backing, and qualification; cf. also Figure 2 in Jansen and Toulmin, (1988, p. 35).
22. Cf., almost as a paradigmatic case, Allison (1971) about the 1962 Cuba Missile Crisis, referring to three organizational "conceptual lenses or frames of reference, or perspectives" for explaining the course of events, with an emphasis on exploring [the authors add] " …the influence of unrecognized assumptions upon our thinking about events like the missile crisis …" by [the authors add] "… comparing and contrasting the three frameworks, we see what each magnifies, highlights, and reveals as well as what each blurs or neglects …" (Allison, 1971, p. v).
23. We are aware of the important differences between discourse ethics universalism and casuistic particularism. There is some complementarity, too. The acknowledgement of different views within casuistry, for example, could profit from the discourse ethics approach to reaching a consensus, cf. once more, Pruzan and Thyssen, 1990.
24. Cf., perhaps as a paradigm, Ruyter's paper about how to teach bioethics (2003).
25. Cf. C. Wright Mills' classical sociology-quotation of empty theory (without data) and blind data (without theory) in *Sociological Imagination* (Penguin, Harmondsworth, 1970) or Jonsen's metaphor of the line connecting the bicycle (practice) with the balloon (ethical theory), referred to and recycled by Ruyter (2003).

Bibliography

Allison, G. T.: 1971, *Essence of Decision. Explaining the Cuban. Missile Crisis* (Little, Brown and Company, Boston).

Aristotle: 1985, *Nicomachean. Ethics,* translated by T. Irwin (Hackett Publishing, Indianapolis).

Aubert, V.: 1963, "Competition and Dissensus: Two Types of Conflict and of Conflict Resolution", *Journal of Conflict Resolution* 7, 26–42.

Bain, W. A.: 1994, "Creating and Using Vignettes to Teach Business Ethics", *Business Ethics. A European Review* 3, 148–152.

Beauchamp, T. L.: 1997, *Case Studies in. Business, Society, and Ethics* (Prentice-Hall, Englewood Cliffs, NJ).

Bergstrom, L.: 1970, "What Is a Conflict of Interest?", *Journal of Peace Research* 7, 197–217.

Blalock, H. M.: 1989, *Power and Conflict* (Sage, Newbury Park).

Boeyink, D. E.: 1992, "Casuistry: A Case-Based Method for Journalists", *Journal of Mass Media Ethics* 7(2), 107–120.

Borisoff, D. and D. A. Victor: 1989, *Conflict Management* (Prentice-Hall, Englewood Cliffs, NJ).

Brigley, S.: 1995, "Business Ethics in Context: Researching with Case Studies", *Journal of Business Ethics* 14, 219–226.

Brinkmann, J.: 2002a, "Moral Reflection Differences among Norwegian Business Students", *Teaching Business Ethics* 6, 83–99.

Brinkmann, J.: 2002b, "Marketing Ethics as Professional Ethics. Concepts, Approaches and Typologies", *Journal of Business Ethics* 41, 159–177.

Brown, M. T.: 1990, *Working Ethics. Strategies for Decision Making and Organizational Responsibility* (Jossey-Bass Publishers, San Francisco).

Buchholz, R. and S. B. Rosenthal: 2001, "A Philosophical Framework for Case Studies", *Journal of Business Ethics* 29, 25–31.

Burton, J. and F. Dukes: 1990, *Conflict: Readings in Management and Resolution* (Macmillan, London).

Calkins, M.: 2001, "Casuistry and the Business Case Method", *Business Ethics Quarterly* 11, 237–259.

Calkins, M.: 2002, "How Casuistry and Virtue Ethics Might Break the Ideological Stalemate Troubling Agricultural Biotechnology", *Business Ethics Quarterly* 12, 305–330.

Chonko, L. B.: 1995, *Ethical Decision Making in Marketing* (Sage, Thousand Oaks, CA).

Ciulla, J. B.: 1994, "Casuistry and the Case for Business Ethics", in T. Donaldson and R. E. Freeman (eds), *Business as a Humanity* (University Press, New York), pp. 167–181.

Collins, R.: 1994, *Four Sociological Traditions* (Oxford University Press, New York).

Dahrendorf, R.: 1958, "Out of Utopia", *American Journal of Sociology* 64(2), 115–127.

Dahrendorf, R.: 1959, *Class and Class Conflict in Industrial Society* (University Press, Stanford, CA).

Dahrendorf, R.: 1972, *Konflikt und Freiheit* (Piper, Munchen).

Donaldson, T. and A. Gini: 1993, *Case Studies in. Business Ethics* (Prentice-Hall, Upper Saddle River, NJ).

Drucker, P. F.: 1981, "What Is 'Business Ethics?'", *The Public Interest* 63, 18–36.

French, W. and D. Allbright: 1998, "Resolving a Moral Conflict Through Discourse", *Journal of Business Ethics* 17, 177–195.

Fritzsche, D. J.: 1991, "A Model of Decision-Making Incorporating Ethical Values", *Journal of Business Ethics* 10, 841–852.

Galtung, J.: 1965, "Institutionalized Conflict Resolution", *Journal of Peace Research* 2, 348–397.

Galtung, J.: 1989, *Solving Conflicts* (University of Hawaii Press, Honululu).

Galtung, J.: 2003, *Både-og. En innføring i konfliktarbeid* (Kagge, Oslo).

Harvey, B. et al.: 1994, *European Casebook on Business Ethics* (Prentice-Hall, New York).

Hoffman, W. M. and J. M. Moore: 1982, "What Is Business Ethics? A Reply to Peter Drucker", *Journal of Business Ethics* 1, 293–300.

Hoffman, W. M. et al.: 2001, eds, *Business Ethics: Readings and Cases in. Corporate Morality,* 4th edn (McGraw-Hill, Boston).

Jansen, A. R. and S. Toulmin: 1988, *The Abuse of Casuistry. A History of Moral Reasoning* (University of California Press, Berkeley).

Jennings, M. M.: 2002, *Case Studies in Business Ethics* (South-Western College Pub., Minneapolis/ St Paul).

Keenan, J. F.: 1998, "Making a Case for Casuistry: AIDS and Its Ethical Challenges", in J. Wetlesen (ed.), *Hva er kasuistikk? Om Moralsk læring og Refleksjon* i *Tilknytning til Forbilder og Eksempler* (Skriftserie for HFs etikkseminar, bind 3, University of Oslo).

Kidder, R. M.: 1996, *How Good People Make Tough Choices* (Fireside, New York).

Klein, S.: 2000, "Drucker as Business Moralist", *Journal of Business Ethics* 28, 121–129.

Kriesberg, L.: 2003, *Constructive Conflicts* (Rowman & Littlefield, Lanham).

Lukes, S.: 1991, "Making Sense of Moral Conflict", Chapter 1 in *Moral Conflict and Politics* (Oxford University Press, Oxford), pp. 3–20.

Lustig, M. W. and J. Koester: 1996, *Intercultural Competence* (HarperCollins College Publishers, New York).

MacIntyre, A.: 1984, *After Virtue. A Study in Moral Theory,* 2nd edn (University of Notre Dame Press, Notre Dame, IN).
Nash, L. L.: 1989, "Ethics Without the Sermon", in K. R. Andrews (ed.) *Ethics in Practice* (HBS Press, Boston).
Pearce, W. B. and S. W. Littlejohn: 1997, *Moral Conflict* (Sage, Thousand Oaks, CA).
Peck, L. E. et al.: 1994, "Enhancing Arthur Andersen Business Ethics Vignettes: Group Discussions Using Cooperative/Collaborative Learning Techniques", *Journal of Business Ethics* 13, 189–196.
Pruzan, P. and O. Thyssen: 1990, "Conflict and Consensus – Ethics as a Shared Value Horizon for Strategic Planning", *Human Systems Management* 9, 135–151.
Ritzer, G.: 1996, *Sociological Theory,* 4th edn (McGraw-Hill Companies, New York).
Ruyter, K. W.: 1995, *Kasuistikk som Saksbasert Problemløsning* (Doctoral Dissertation, School of Theology, University of Oslo).
Ruyter, K. W.: 2003, Of Balloons and Bicycles and the Implications for Teaching Bioethics, unpublished manuscript in review.
Sundby, N. K.: 1978, *Om Normer* (Universitetsforlaget, Oslo).
Toulmin, S.: 1973, "How Medicine Saved the Life of Ethics", *Perspectives in Biology and Medicine* 25(4), 736–750.
van Luijk, H.: 1994, "Analyzing Moral Cases in European Business", in B. Harvey et al. (eds), *European Casebook on Business Ethics* (Prentice-Hall, New York), pp. 3–12.
Wallace, R. A. and A. Wolf: 1995, *Contemporary Sociological Theory* (Prentice-Hall, Englewood Cliffs, NJ).
Weber, J.: 1992, "Scenarios in Business Ethics Research: Review, Critical Assessment, and Recommendation", *Business Ethics Quarterly* 2, 137–159.
Weiss, J. W.: 1996, *Organizational Behaviour and Change* (South-Western College Pub., Minneapolis/St Paul).
Wheelright, P.: 1935(1951), *Aristotle* (The Odyssey Press, New York).
Williams, O. F.: 1982, "Business Ethics: A Trojan Horse?", *California Management Review* 24(4), 14–25.

JOHANNES BRINKMANN AND KNUT J. IMS[1]

Good Intentions Aside: Drafting a Functionalist Look at Codes of Ethics

Code concepts

"A code of ethics by most definitions is a written, distinct, formal document which consists of *moral standards which help guide* employee or corporate behaviour" (Schwartz, 2001).

Codes are *maps of expected moral conflicts,* expected or suggested solutions and, perhaps, predictable sanctions (Brinkmann, 2002a).

"A profession's code of ethics is perhaps its most visible and explicit enunciation of its professional norms. A code embodies the *collective conscience* of a profession and is testimony to the group's *recognition of its moral dimension*" (Frankel, 1989: 110).

Our italics refer to the core definition criteria.[2] The first definition[3] focuses on intention of influencing a target population, the second one emphasizes moral conflict and resolution possibilities and the last one the expression and recognition of an organization's collective conscience. Table 1 gives examples of three different types.

Before looking more closely at intended code functions as code definition criteria and instead of looking more closely at code elements[4] we want to focus on an examination of the *ethicalness* of codes in general, or of a given code in particular, since as business ethicists we find this to be the most interesting aspect.

Table 1. Three different ethical code types with examples

Advertising: (Fifteen) Issues to be careful with (guideline type)	Public relations: Four basic principles (collective conscience expression type)	Real-estate agency: (Five) Areas where conflict can occur (conflict recognition type)
• Don't exploit fear or superstition	• Openness	• Receiving and carrying out agency assignments
• Do not further violence or discrimination	• Loyalty	• Advice within one's field of competency
• Don't plagiarize	• Integrity	• Advertising and marketing of agency services
• Don't misuse quotations, statistics, research	• Credibility	• Confidentiality about client's affairs
• Be careful with children and minors… etc.		

Codes' ethicalness

There are two main ways of approaching the ethicalness of codes. The first is to take a closer look at our definition of ethics and to treat ethicalness as a question of degree, more or less related to the definitional ambiguity of moral phenomena, somewhere between moralism and ethics, law and conscience (Brinkmann, 2001, 2002b). The ethicalness question is then whether a given code is superficial or radical (moralist or ethical), and whether it appeals to the public (as legalism) and/or to private conscience. A second way would be to judge codes and code ethicalness from a consistent moral philosophy position, by the universal validity of their raw material; that is, whether they express more or less universal moral standards such as "trustworthiness, respect, responsibility, fairness, caring and citizenship" (Schwartz, 2001). We agree with Schwartz' objective of emphasizing the ethicalness of codes, but suggest a broader answering strategy, looking for a wider mix of legitimacy sources which would include at least a look at *code functions* as a source of legitimacy. Schwartz offers deontology (see also L'Etang, 1992), that is, a focus on good principles which can be generalized across cultures and settings, but since we are mainly concerned with code "dysfunctionality" our approach to the ethicalness of codes requires utilitarianist (Starr, 1983) as well as dialogue *and* virtue ethics code justifications rather than deontological justifications and criticisms alone.[5]

For *discourse ethicists* open communication climates and communication quality are primary criteria of codes of ethicalness. For Jürgen Habermas, ethicalness is almost synonymous with democracy, not as any particular institutional form but as a "self-controlled learning process".

To create such learning processes it is essential to identify suppressed interests, forced consensus and illegitimate domination. Habermas's basic criterion for evaluation is the relative closeness to imagined conditions of ideal communication. The issue is to find arrangements where actors can participate freely and equally in a discursive will formation (Habermas, 1979: 186). The primary issue and evaluation criterion of ethicalness is whether all parties affected can articulate their needs in an undistorted communication context. Through an ideal discourse the participants form and discover their interests and the norms that they will rationally accept as binding. In this *ideal discourse* there will be no force except that of the better argument. All motives except that of the cooperative search for truth are excluded. "If under these conditions a consensus ... to accept a norm arises argumentatively ..., then this consensus expresses a 'rational will' ..." (Habermas, 1975: 107).[6] Another important element in Habermas's conceptual system is his antinomy of strategic interaction and communicative action. *Strategic interaction* is (manifestly or latently) oriented towards success, treating the other actor as an opponent and rival. This kind of interaction turns typically into a game of win or lose. The actors calculate egocentrically and assume that the other actor has bad intentions. *Communicative action* is oriented to reaching an understanding, not through egocentric calculation and individual success but in the belief that the actors may pursue individual goals "under the condition that they can harmonize their plans of action on the basis of a commonly defined situation. In this respect the negotiation of definitions of the situation is an essential element of communicative action" (Habermas, 1981: 286).[7] In summary: for discourse ethics, code ethicalness requires that the code is a product of, and revisable by, a fair and open dialogue among all parties affected, and that the code furthers a communicative rather than a strategic interaction climate.

In a *virtue ethics perspective,* external rules or institutional settings count less than good personality and character.[8] Rules address people as if they were equal, but different people respond differently to rules – some with opportunism, others with integrity;[9] that is, rules are followed only if they correspond to one's self-interest.[10] Most ethical rules can be exploited or practiced "unethically", and even following the spirit of the rules can sometimes produce bad solutions when applied to new, unique and unforeseen situations. According to Aristotle's virtue ethics, a good man is good beyond an ordinary level of expectation. He has arete, virtue, and excellence.[11] Negative rule conformism, that is, not breaking the code rules, is simply not enough. The issue is to do one's best and to be sensitive to the situation as a whole.[12]

Another Aristotelian emphasis is that an application of rules requires good judgment (Aristotle's "phronesis"), and good judgment in turn requires a perception of concrete details of a situation rather than abstract theoretical formulations. In a virtue ethics perspective, individuals should be "more than" simply loyal to a code and should not stick to the code whenever this would be too narrow. Such morality transcends "conventional" expectations and sometimes even requires breaking the law. We suspect that codes assume pre-conventional or conventional morality and reproduce rather than transcend such morality (towards post-conventionalism).[13] Aristotelianism offers a much richer vocabulary for ethics than terms like right and wrong, or good and bad (Solomon, 1993). Narrow good-and-bad vocabulary hampers the type of subtle ethical judgments needed in everyday practice. To sum up, virtue ethics typically go beyond what is expressed in rules and codes. Phronesis is instrumental to good interpretation. It is based upon character and practical experience. The prejudice towards codes arises from a sound skepticism towards rules and regulations; good or virtuous people do not (necessarily) need a code, while bad people tend to instrumentalize and twist code rules rather than follow them loyally.

Code functions

Frankel lists and elaborates explicitly eight "functions" of codes (1989: 111–112)[14] as: enabling documents; as sources of public evaluation; as media of professional socialization; for the enhancement of a profession's reputation and public trust; for the preservation of entrenched professional biases; as a deterrent to unethical behavior; as a support system; and as adjudication. Similarly, Bowie (1982: 91–97) writes about "advantages" of and "objections" to codes, such as providing stable guides, structuring situations, power control, specification of business responsibilities and self-regulation instead of state interventionism. Objections to codes are that they undermine competition, they tend to be vague and amorphous, and they are difficult to enforce.

Such listings of pros and cons, of good purposes and more or less positive effects, are redolent of consequentialism and utilitarianism, at least implicitly. Although we subscribe to consequentialism here, we propose that ethical codes can be seen in terms of their *social functions* by applying the concepts of latent functions, of dysfunctions and of functional equivalents[15] (Merton, 1968; Gans, 1972). In the context of this paper it is sufficient to refer to a few key definitions and to outline how they could serve as a fruitful approach for understanding ethical codes, alone or together, with complementary approaches (for example, deontology) or with similar approaches (for example, utilitarianism or virtue ethics). We will draw on Merton's classical work (1968: 104–109), on manifest and latent functions and use as examples code functions referred to in the business ethics literature.

> *Functions* are those observed consequences which make for the adaptation or adjustment of a given system; and *dysfunctions,* those observed consequences which lessen the adaptation or adjustment of the system. There is also the empirical possibility of *nonfunctional* consequences, which are simply irrelevant to the system under consideration … *Manifest functions* are those objective consequences contributing to the adjustment or adaptation of the system which are intended and recognized by the participants of the system; *Latent functions,* correlatively, being those which are neither intended nor recognized … (Merton, 1968: 105)

> ...Once we abandon the gratuitous assumption of the functional indispensability of particular social structures, we immediately require some concept of functional alternatives, equivalents, or substitutes. This focuses attention on the *range of possible variation* in the items which can, in the case under observation, subserve a functional requirement. It unfreezes the identity of the existent and the inevitable .... (Merton, 1968: 106)

And a few pages later, Merton writes:

> It is precisely the latent functions of a practice or belief which are *not* common knowledge, for these are unintended and general unrecognized social and psychological consequences ... The introduction of the concept of latent function in social research leads to conclusions which show that social life is not so simple as it first seems .... (Merton, 1968: 122)

Such quotations suggest that intended and positive code functions are less interesting than unintended, often non-obvious negative code functions. The quotations also raise interesting follow-up questions, such as: which functions are most important, what decides whether a code has positive or negative functions and what could serve as an alternative to codes; that is, whether one should look at mixes of alternative tools rather than codes alone? As a first general introduction we propose translating Frankel's distinction between educational, regulatory, and aspirational codes (1989) into a distinction between more or less dominating code functions, and then to construct a six-cell-typology as a point of departure for further discussion (Table 2).

Table 2. Six main functions of ethical codes

	intended & positive	**latent & negative**
educational	increase individual moral awareness & behavior	assume & reinforce pre-/conventionalism, cynicism, outer-directedness
regulatory	recognize moral conflicts and help with resolving them	hinder necessary/possible learning of conflict handling by dialogue
aspirational	communicate ideals for individuals and collective conscience (moral climate)	window dressing, covering up/ concealing a disputable practice

Code functions and dysfunctions as a question of mediating conditions

Rather than judging a single code or codes in general, one should look first for conditions which determine *if* or to *what degree* a code has positive or negative functions or (most likely) a mix of both (Table 3). Some of these conditions are probably cultural. While codes seem to be universally prevalent in the US they are not at all in other cultures, for example, in France.[16] Hofstede's criteria by which cultures differ may be relevant here – for example, the higher a culture's power-distance or the higher a culture's uncertainty avoidance, the lower its masculinity, the better for the general acceptance of a code (Hofstede, 1994). Other conditioning or mediating factors could be the degree of match between postulated and factual working conditions of a code (for example, how well known the code is), as well as situational factors.

Of the conditions mentioned, the most inclusive and fruitful reference point for code functions in general and as a mediating condition in particular could be the *moral climate*[17] of the industry, the profession or the company which is addressed by a code of ethics. By referring to moral climate one can build on classic social science perspectives and on an inclusion of various business and professional ethics approaches under a common, holistic umbrella (Brinkmann, 2002a).[18]

Table 3. Conditions which determine code (dys)functions

Conditions which further positive code effects, i.e. code functions	**Types of mediating conditions (understandable as a sequence of stages)**	**Conditions which further negative code effects, i.e. code dysfunctions**
openness and honesty	organization climate	pessimism, cynicism
idealistic, inner-directed	code intentions & objectives	defensive, reactive, outer-directed
participatory, bottom-up	code creation & procedure	administrative, top-down
simple and abstract elements	code content	sophisticated, detailed elements
stimulation of dialogue	code implementation, use & administration	blocking of dialogue and problem sharing

The point of departure for a (moral) climate approach is more important than the holistic hermeneutic credo of "without understanding the parts there is no understanding of the whole and vice versa". Instead of starting from, for example, moral conflict or from codes or roles, one can start from the climate or culture of which they (and other relevant aspects) are elements, and which is made up of their interdependence and interaction. "Moral climate" is suggested as a vague and wide "umbrella term" for a company's, an industry's, or a profession's normative socialization environment. Moral climate as a socialization medium consists essentially of role norms, which are learned by future members ("anticipatory" socialization) and by new members from normative and comparative reference groups. Climate shapes people, but people also shape climate. Moral climates are produced and reproduced by their members and their practices.[19] As an umbrella term, moral climate repeats classic social science references such as "collective conscience" (Durkheim), "ideology" in a neutral and negative sense (Marx) or "value rationality" (Weber). In the business ethics literature, moral or ethical climate has been suggested as a theoretical, empirically measurable concept (Wimbush and Shepard, 1994; Wimbush et al., 1997; Vidaver-Cohen, 1998; Fritzsche, 2000). Derived from a work climate definition, moral climate has been defined as "stable, psychologically meaningful, shared perceptions employees hold concerning ethical procedures and policies existing in their organizations" (Wimbush and Shepard, 1994: 636), and further elaborated in a similar vein (Victor and Cullen, 1987: 52–57, 1988: 101–104). Ideally, the moral climate in a given company, industry or profession would be reconstructed from a combination of several different data sources and types, such as history, media, observation of board meetings or annual gatherings, professional codes of ethics or group interviews. Such qualitative data could then be combined with representative survey data collected with standard instruments, such as the ECQ referred to elsewhere (Brinkmann, 2002a).[20] A moral climate approach also has weaknesses that deserve consideration. Claims of good or good enough moral climates assume, often tacitly (and optimistically), a sufficient individual internalization and a sufficient value consensus.

Codes as functional equivalents within the ethics toolbox

The focus on codes as one moral climate improving tool among many can vary. If we are concerned with positive functions for individuals, dialogues and codes must be judged by their relative contribution. As long as optimum results count, codes or other means matter relatively less (whereas a deontological position might indicate that a code is important as a promise of non-flexibility in possible conflict situations). Secondly, from a Mertonian perspective of functional alternatives, codes always should be looked at as something replaceable or mixable, not as something indispensable (although US jurists might say the opposite, perhaps referring to the Federal sentencing guidelines). Third, one could consider codes in the context of the other ingredients of the ethics toolbox and look for possible contradictions or reinforcement possibilities among them (cf. the twelve interdependent instruments mentioned by Buscher et al. (2002), presented in Figure 1).[21]

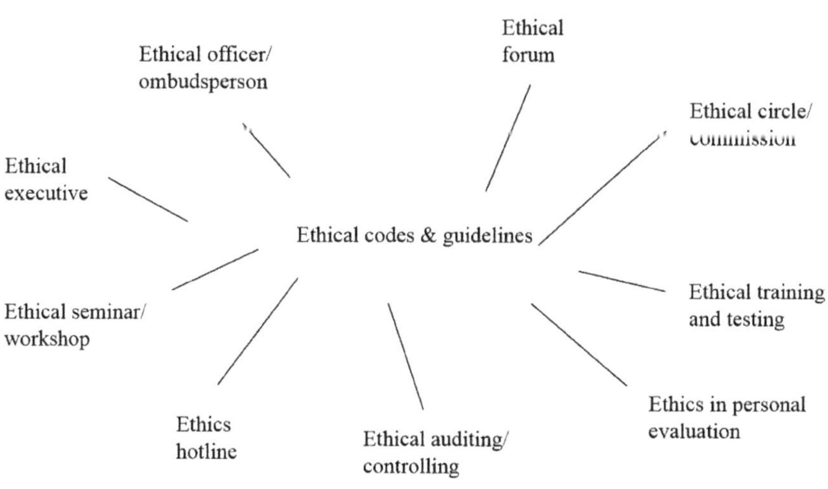

Figure 1. Codes as functional equivalents, that is one tool among many.

Empirical illustration of code effects in general and of code dysfunctions in particular

Even if this paper is mainly conceptual it is tempting to show how some code target groups voted on the acceptability of various arguments for or against codes. Table 4 summarizes one of the authors' questionnaire data gathered from 150 Norwegian advertising, 160 PR and 140 real estate professionals in the mid-1990s.

Most of the respondents seem to have a positive attitude towards codes. Codes are welcomed because of their expected contribution to role clarity and fair competition. At the same time, many respondents seem realistic or skeptical towards code efficiency and code enforcement. Instead of over-exploiting single-item answers, one should rather look into a reasonable grouping of items (for example, into a fairplay-expectation and a legitimate moral defensiveness factor (cf. Brinkmann and Tvedt, 1995).

Table 4. Target population votes on arguments in favor of codes

	Real estate mean	Advertising mean	PR mean	US mean
Professionals consider codes as a useful aid when they want to refuse an unethical request impersonally	1.7	1.7	1.5	1.8
Codes raise the ethical level of the industry	1.2	1.9	1.5	2.1
A code helps managers in defining clearly the limits of acceptable conduct	1.6	1.8	1.7	1.9
In cases of severe competition, a code reduces the use of sharp practices	1.8	2.0	2.0	2.7
People violate codes whenever they think they could avoid detection	2.8	2.4	2.9	2.5
Codes are easy to enforce	1.9	2.7	2.8	3.3
Codes protect inefficient firms and retard the dynamic growth of the industry	3.5	3.3	3.6	3.3

Question wording: "Independently of your answers to the questions above we would like to know more about your principal attitude towards codes of ethics. We have listed a number of statements about possible consequences of such a code for (information, real-estate, advertising) work. We ask for your vote on a scale of 1 (strongly agree) to 4 (strongly disagree).

US means from Becker and Fritzsche 1987: 292, Table V (multiplied by 0.8 to correct for 5-point scale) are included for comparison.

Table 5. Moral climate and code acceptance

	Code skepticism dominates	**Code acceptance dominates**
optimistic evaluation of a profession's moral climate	there is no need for a code	a code won't hurt
pessimistic evaluation of a profession's moral climate	a code won't help	a code is needed

Alternatively or additionally one could use a similar data set as in Table 4 to develop a simple typology (see Table 5).[22]

Suggested next steps

The aim of our paper has been to raise questions and to draft rather than elaborate answers. Arising from this analysis we see four topics that deserve further and deeper elaboration:

- Codes of ethics functions in a discourse ethics perspective – as institutionalized rather than communicative conflict management;
- Codes of ethics functions in a virtue ethics perspective – as catalysts for or barriers to a culture of virtue and civil courage;
- Codes of ethics as a way of structuring professional or work role duties; and
- Development and validation of a standardized instrument where code knowledge, positive and negative attitudes towards the code and improvement needs are measured.

Notes

1. Our article is based on a presentation at the European Ethics Summit/Annual EBEN Conference in Brussels, Belgium, August 29–30, 2002. The article title is

borrowed from L. Nash's book (Boston: HBS Press, 1990). We wish to thank Kjell Grønhaug and Gunnar Pettersen for useful comments.

2. Clarke (1998) claims that codes probably represent the most widely employed ethics tool. There is no shortage of academic papers about ethical codes. A search in ABI/Inform for "ethical codes" produces forty-eight hits for peer-reviewed journal papers over ten years, mostly appearing in the *Journal of Business Ethics.* For good reviews of literature and/or studies about ethical codes see, for example Schwartz (2001) and Stevens (1994).
3. Two additional examples of this (most frequent) type could be: "(Ethical codes) [...] are written documents, ranging in length from one paragraph to more than fifty pages which are *intended to impact employee behaviour.* Firms frequently attempt to manage and articulate ethics through these messages, which are designed for both internal and external audiences. Codes, then, are managerial tools for shaping change. They often demand from employees higher standards of behaviour than required by law ..." Stevens (1994: 64) and L'Etang (1992: 737), who distinguishes codes of ethics as a "fairly *short* set of ethical principles expressed in the imperative mode" from codes of conduct which are "much more *specific*" and which detail "exceptions and particular circumstances", and from codes of practice as a *quality standard* definition (our italics).
4. Cf., as an example, Doig and Wilson's tabulation of more or less frequently mentioned issues (1998: 146).
5. In addition, one could also refer to postmodern and to social science functionalist skepticism.
6. For a critique of Habermas's reasoning, see for example Lukes (1982).
7. There is more than Habermas. For a perspective on good dialogues as learning processes cf. also Isaacs (1993, 1999) who notes that consensus approaches do not have the ambition of exploring or altering underlying patterns of meaning, while dialogues may inquire openly into deeply taboo subjects (such as feelings of self-protection and anomie among professional workers and how these feelings may be counterproductive to the system which they are members of). One of the barriers to dialogue is a widespread culture of "civility, an inclination to follow the norm, to maintain 'face' by being polite and never differ publicly. In this sense people follow a rule even when it is counterproductive. They do not dig into deep layers of organizational life and come to grips with the real problems that penetrate the organization" (Isaacs, 1999: 259–61). Isaacs calls this stage "*shared monologues*", and he gives examples of how people through crises may reach a new field of breakdown where they start to "say what they think". In this stage, rule-revealing behaviour becomes the norm. The next fields (or spaces) are called "*spirit of curiosity*" and "*generative dialogue*", respectively. In the last-mentioned stage people come into an awareness of the primacy of the whole, and genuinely new possibilities emerge (cf. Isaacs, 1999: 257–87). In a consultancy role one of the authors recently observed the importance of reaching Isaacs's field three, "the spirit of curiosity". The organization's ethical code

specified sanctions whenever members of the organization committed specified unethical acts. Communication with the organization members revealed that the code promoted a stifling rather than a flourishing of dialogue and reflections. In order to change from a pattern of hostile inwardness to a pattern of reaching out the following rules of thumb were suggested: (1) doubts are acceptable; (2) not having the right answers is permitted; and (3) it is acceptable to being puzzled with taken for granted assumptions.

8. Cf. Machiavelli's "virtue ethicist" claim that it is better to live in a state with bad laws but good morality than in a state with good laws but bad morality – that is, the letter of the law does not guarantee a realization of the spirit of the law, as bad people typically exploit the letter of the law for their own bad purposes. In certain respects, Immanuel Kant is a virtue ethicist too. For Kant, a virtue is a central attitude prohibiting an unethical application of a rule or a guideline. To Kant, morality means much more than following certain rules and respecting the external positive law. In this sense virtue functions as a remedy against ethical rigor. Morality starts beyond legal duties, for example by inner rather than external coercion.
9. J. Rest talks instead of *moral character* with indicators such as "ego strength, perseverance, backbone, toughness, strength of conviction, and courage ..." (Rest and Narvaez, 1994: 24). Moral character or courage has to do with the individual self-confirmation effect of proper moral conflict handling and the consistency of feeling, thinking, and acting on the individual level. Nielsen has something similar in mind and proposes to extend the widely accepted political citizen role model to business life (1984: 160).
10. Cf. Asheim (1997: 263) who claims that interpretation and fulfilment of rules is a question of attitude.
11. According to virtue ethicist Solomon (1993) doing no harm or not doing anything wrong is not good enough. Being virtuous means excelling and not merely "keeping one's nose clean". It means to realize or to optimize one's best capabilities.
12. Cf. Warren (1993: 189), referring to an unpublished conference paper of Petrie: "codes further rule compliance at the expense of autonomy and ethical reflection". Cf. also the suggestion to treat virtues as attitudes that are conducive to moral sensitivity (Nortvedt, 1995: 176–90); that is, the claim that a virtuous person tends to see elements in a complex situation that others not striving for the good will miss.
13. Cf. Kohlberg's (1973) well-known distinction between three major levels of moral development: a *pre-conventional level* where individuals (infants) react to punishment, pleasure, and pain. A *conventional level,* with a morality of conventional role conformity. In this stage a person reacts and conforms to the expectations of peers. This is the "law and order" stage. Here the individual becomes acculturated and does what a good citizen is expected to do. According to Kohlberg most people live at this stage. Few persons reach the third level of moral development, called the *post conventional, autonomous, or principled* level. This is the level of self-accepted moral principles. Here one thinks of morality based on rights of individuals and

on agreements between consenting adults. On this level one is capable of giving reasons and a rational defense of moral principles that guide one's actions.

14. The same author mentions additional, and perhaps at least as important, code functions, without calling them functions; for example, three purpose ideal types (aspiration, education, and regulation, 110f), which could be read as aspirational, educational, and regulational "code functions" as well as dysfunctions, such as manipulating public impressions (111f) and reinforcement and redefinition of the profession's collective responsibility (113), which furthered alienation towards non-professional, broader social values, which furthered alienation between an organization's top-dogs and ordinary members. Code is used as an instrument for parts of the organization in conflict situations.
15. Functionalism is a common denominator of a variety of sociological approaches, with representatives such as (French) E. Durkheim, (Americans) T. Parsons and R.K. Merton, or (German) N. Luhmann. The key terms of the approach, "structure" and "function" are somewhat ambiguous and are combined differently. Depending on authors and contexts functions can denote purposes, meanings, effects, consequences, or indispensable activity. Structures can be translatable as totality, part, institution, pattern, essence, relation, or even other terms. A few sociological examples of such an approach could be studies of religion (Durkheim), of conspicuous consumption (Veblen), of social inequality and stratification (Davis), of penal law and punishment (Aubert), of deviance and conflict (Coser), of poverty (Gans) or even of the business ethics business (Luhmann). For additional, mostly older, references cf. also Merton (1968: 109–14) with further examples and ibid. notes 70–82 with further references.
16. Cf. F. Seidel's unpublished paper, presented at the 1994 St Gallen EBEN conference, "Translating codes of ethics from one culture to another: A case study."
17. This concept is similar to the 'moral community' concept in virtue ethics as stressed by MacIntyre.
18. Cf. also the rest of this paper which deals mainly with empirical or pre-empirical presentations of moral climates – within three marketing sub-professions. The paper also presents a visual model where a moral climate approach transcends the three other approaches.
19. For a functionalist view which warns against an executive or actor bias and claims that social system-level and moral climate factors can often be more important than micro-level and personality factors (Etzioni 1991: 356f).
20. See Brinkmann and Bakken (1996), Brinkmann and Tvedt (1995). Cf. also Ingjerd Grenne's Diploma thesis NMH 1998, reporting ten in-depth interviews with Norwegian PR managers.
21. Cf. also Weaver et al. (1999) claiming to give a "snapshot of the 'state of the art' regarding the formalized corporate ethics function" and distinguishes between seven "corporate ethics practices".

22. Cf. as a more sophisticated version of Table 5 (Brinkmann 2002a, exhibit 7), with plotted-in averages on a simple add-up index (independently of the reasons given) as a preliminary code acceptance or code skepticism measure, combined tentatively with another preliminary dimension of moral climate evaluation (for example, optimism vs. pessimism).

Bibliography

Asheim, I. 1997. *Hva betyr holdninger? Studier i dyds etikk,* Oslo: Tano Aschehoug.

Becker, H. and Fritzsche, D. J. 1987. Business Ethics: Cross-Cultural Comparison of Managers' Attitudes, *Journal of Business Ethics* 6, 289–295.

Bowie, N. 1982. *Business Ethics,* Englewood Cliffs, NJ: Prentice Hall.

Brinkmann, J. 2001. On Business Ethics and Moralism, *Business Ethics: A European Review* 10, 311–319.

Brinkmann, J. 2002a. Marketing Ethics as Professional Ethics. Concepts, Approaches and Typologies, *Journal of Business Ethics* 41, 159–177.

Brinkmann, J. 2002b. Business Ethics and Inter Cultural Communication. Exploring the Overlap between Two Academic Fields, *Intercultural Communication* 5.

Brinkmann, J. and Bakken, T. 1996. *Bransjeetikk. Møbel- og innredningsprodusentene som case (Furniture Industry Ethics, project report in Norwegian)* Unpublished Project report, Norwegian School of Management BI, School of Marketing.

Brinkmann, J. and Tvedt, H. G. 1995. A Survey of Moral Conflicts among Norwegian Public Relations Professionals, in: Ulrich, P. and Sarasin, Ch. (eds), *Facing Public Interest,* Dordrecht: Kluwer, pp. 167–181.

Buscher, M., DeGeer, H. and Marcia, D. 2002. *Ethics, Values, Value Conflicts. Professional Education for Business Professionals.* EU-Leonardo-Project, Publication in Progress, Iserlohn.

Clarke, M. A. 1998. Can Corporate Codes of Ethics Influence Behaviour? *Journal of Business Ethics* 17, 619–631.

Doig, A. and Wilson, J. 1998. The Effectiveness of Codes of Conduct, *Business Ethics: A European Review* 7, 140–149.

Etzioni, A. 1991. Reflections on Teaching Business Ethics, *Business Ethics Quarterly* 1, 355–365.

Frankel, M. S. 1989. Professional Codes: Why, How, and with What Impact? *Journal of Business Ethics* 8, 109–115.

Fritzsche, D. J. 2000. Ethical Climates and the Ethical Dimension of Decision-Making, *Journal of Business Ethics* 24, 125–141.

Gans, H. 1972. The Positive Functions of Poverty, *American Journal of Sociology* 78, 275–289.

Habermas, J. 1975. *Legitimation Crisis.* London: Beacon Press.

Habermas, J. 1979. *Communication and the Evolution of Society.* London: Beacon Press.

Habermas, J. 1981. *The Theory of Communicative Action* Vol. I. *Reason and the Rationalization of Society.* Boston: Beacon Press.

Hofstede, G. 1994. *Cultures and Organizations.* London: HarperCollins.

Isaacs, W. 1993. Dialogue, Collective Thinking, and Organizational Learning, *Organizational Dynamics,* 22(2), 24–39.

Isaacs, W. 1999. *Dialogue and the Art of Thinking Together.* London: Currency.

Kohlberg, L. 1973. The Claim to Moral Adequacy of a Highest Stage of Moral Judgment, *The Journal of Philosophy* 70, 630–646.

L'Etang, J. 1992. A Kantian Approach to Codes of Ethics, *Journal of Business Ethics* 11, 737–744.

Lukes, S. 1982. Of Gods and Demons: Habermas and Practical Reason, in: Thompson, J. B. and Held, D. (eds), *Habermas: Critical Debates,* London: The Macmillan Press Ltd.

Merton, R. K. 1968. *Social Theory and Social Structure,* Glencoe, IL: Free Press.

Nielsen, R. P. 1984. "Toward an Action Philosophy for Managers Based on Arendt and Tillich". *Journal of Business Ethics* 3, 153–161.

Nortvedt, P. 1995. *Sensitive Judgment. Nursing, Moral Philosophy, and an Ethics of Care.* Doctoral dissertation, Universitetet i Oslo.

Rest, J. and Narvaez, D. (eds). 1994. *Moral Development in the Professions,* Hillsdale, NJ: Erlbaum.

Schwartz, M. S. 2001. A Code of Ethics for Corporate Codes of Ethics, *Journal of Business Ethics* 41, 27–43.

Solomon, R. C. 1992. Corporate Roles, Personal Virtues: An Aristotelian Approach to Business Ethics. *Business Ethics Quarterly,* 2(3), 317–339.

Starr, W. 1983. Codes of Ethics – Towards a Rule Utilitarian Justification, *Journal of Business Ethics* 2, 100f.

Stevens, B. 1994. An Analysis of Corporate Ethical Code Studies: Where Do We Go from Here? *Journal of Business Ethics* 13, 63–69.

Victor, B. and Cullen, J. B. 1987. A Theory and Measure of Ethical Climate in Organizations, *Research in Corporate Social Performance and Policy* 9, 51–71.

Victor, B. and Cullen, J. B. 1988. The Organizational Bases of Ethical Work Climates, *Administrative Science Quarterly* 33, 101–125.

Vidaver-Cohen, D. 1998. Moral Imagination in Organizational Problem-Solving: An Institutional Perspective, *Business Ethics Quarterly,* Special issue 1, 123–148.
Warren, R. C. 1993. Codes of Ethics: Bricks without Straw, *Business Ethics: A European Review* 2, 185–191.
Weaver, G. R. et al. 1989. Corporate Ethics Practices in the Mid-1990s: An Empirical Study of the Fortune 1000, *Journal of Business Ethics* 18, 283–294.
Wimbush, J. C. and Shepard, J. M. 1994. Toward an Understanding of Ethical Climate: Its Relationship to Ethical Behaviour and Supervisory Influence, *Journal of Business Ethics* 13, 637–647.
Wimbush, J. C. et al. 1997. An Empirical Examination of the Multi-dimensionality of Ethical Climate in Organizations, *Journal of Business Ethics* 16, 67–77.

KNUT J. IMS AND OVE D. JAKOBSEN

Cooperation and Competition in the Context of Organic and Mechanic Worldviews: A Theoretical and Case-Based Discussion

Introduction

In this article we look more closely into the inconsistency between competition and cooperation as coordinating principles in a sustainable economy. Competition has a central standing as a regulating principle in the market economy. Distinguished scholars in economics, from Adam Smith to Milton Friedman, have argued that competition is necessary in order to promote efficiency and wealth. They look at competition as a natural state of affairs in a hostile market environment, assuming aggressive individuals motivated by self-interest. In such an environment, it is a strategic necessity for the players to obtain market power to fend off competitive threats. Effective unfolding of power requires independence and, as a consequence, companies should not rely on specific suppliers or buyers. The players see collaboration as the strategy of the weak. Due to the competitive nature of business relationships, companies will attempt "to serve their own interests to the detriment of the others, by maneuvering, manipulating or cheating" (De Witt and Meyer, 1998, p. 510). The Norwegian Competition Act (entered into force 2004 and partially harmonized with the EU's competition rules[1]) is based on this paradigm, stating that by promoting competition (within defined rules), the result will be more efficient utilization of natural and societal resources (This purpose of the Act is stated in section 1).

Within the embedded organization perspective, the preconditions are different. Business is not a war and the relations between companies are described by predominantly cooperative behaviour. Firms "intentionally embed themselves in a web of durable collaborative relationships" (De Witt and Myer, 1998, p. 511). There is a greater balance between cooperative and competitive positions in the firm's relationships toward others. Most firms have a multitude of bilateral collaborative relationships. Value-adding partnerships, strategic alliances, and joint ventures are illustrative examples. In the White Paper, NOU 2002, p. 19, we find indications that the Norwegian authorities respond positively to solutions based on cooperation. In the Norwegian Competition Act, section 10, the authorities encourage cooperation if the goal is to contribute "to improving the production or distribution of goods or to promoting technical or economic progress" (The Norwegian Competition Authority, 2004, p. 50).

Given the goals of sustainability, fair trade, and satisfying vital consumer needs it is necessary to establish suitable principles for coordinating the interplay between production, distribution, consumption, and the reprocessing of waste. In the following paragraphs we will explore which principles are best suited to coordinate the market – competition or cooperation. We claim that in order to find the answer it is necessary to examine the ontological and ethical foundations of economics.

Principles of coordination in the context of different worldviews

Textbooks in business strategy describe a continuum between hostile competitive behaviour and collaborative relationships. At one end of the continuum, the competitive game consists of atomistic players involved in a zero-sum game, and cooperation, if used, is reduced to an instrumental and tactical way of operation. When collaboration takes place it is typically a conspiracy of the strong to restrain competition and eliminate competitors. At the opposite end of the continuum, the embedded perspective, partnerships leading to a win-win situation are developed. Based on this, strategists have two alternatives, they must choose between competition and cooperation.

In our opinion, the two different principles for organizing the market are based upon different paradigms. In the following paragraph, we will shed light on the interconnection between:

- the mechanistic worldview and competition, and
- the organic worldview and cooperation.

We argue that if competition and cooperation are introduced within the wrong context – for instance, if cooperative behaviour is introduced within a context in which the mechanistic worldview prevails – the consequences may be negative.

The mechanistic worldview

The mechanistic perspective is characterized by the idea that pieces of matter are isolated individuals (atomism) related to each other only externally. In this perspective, society is nothing more than a mere mechanism based on the interplay between egocentric individuals seeking their own ends. One of the most important consequences of the mechanistic worldview is that the whole universe is completely causal and deterministic. Everything that has occurred has had a defined cause and gave rise to a defined effect, and that the future of any part of the system may – in principle – be predicted with absolute certainty if its state at any time is known in detail. There is no capacity for creativity, spontaneity, self-movement or novelty in the mechanistic worldview.

According to Georgescu-Roegen (1971) and Daly and Cobb Jr (1994), modern mainstream economics builds upon mechanistic presuppositions. Inspired by the mechanistic metaphor, players in the market are supposed to act independently of each other in order to optimize their self-interests. Market theory presupposes that economic players act autonomously in most market transactions. The assumption that economic rationality largely excludes other-regarding behaviour has deep roots in western theoretical understanding of human nature. The tendency to model economic concepts and theories in accordance with the mechanistic paradigm has become a severe handicap for the development of sustainable economics. According to Whitehead, an important precondition in the mechanistic worldview is that the coordination in nature is regulated through external rules of connections. This is called the doctrine of laws as imposed (Whitehead, 1967b, p. 113).

In much the same way, the norms regulating the interplay between individuals in an atomistic, materialistic society are based upon mechanical solidarity. Durkheim uses the term "mechanical" to illustrate that social molecules "lack any movement of their own, as do the molecules in inorganic bodies" (Durkheim, 1991, p. 84). This does not mean that the term "mechanic" indicates that the solidarity is produced by mechanical or artificial means. Instead mechanical solidarity represents an "analogy with the cohesion that links together elements of raw materials, in contrast to that which encompasses the unity of living organisms" (Durkheim, 1991, p. 84).

The mechanistic worldview does not leave much space for ethics or, for that matter, aesthetics. If nature is valueless, there is, on the one hand, no reason to feel deep respect and esteem for natural or artistic beauty. On the other hand, there is no reason to orient our practices around such values. In the field of economics, it is obvious that ethics is often reduced to purely instrumental values. In much the same way as for aesthetics, ethics is regarded as a competitive tool to increase the market value of the firm or the product (Porter and van der Linde, 1995).

Competition

Competition between autonomous actors in the market has a fundamental standing in mainstream liberalistic economics. Competition is based on the idea of conflicting interests between the actors in the market. According to Adam Smith, the logic of the competitive market implies that private virtues will be turned to public virtues by means of the market's "invisible hand".

> Every individual necessarily labours to render the annual revenue of society as great as he can. He generally, indeed, neither intends to promote the public interest, nor knows how much he is promoting it … he intends only his own gain, and he is in this, as in many other cases, led by an invisible hand to promote an end which was no part of his intention … By pursuing his own interest, he frequently promotes that of society more effectually than when he really intends to promote it. (Smith, 1976, pp. 477–478)

Today, Smith's "invisible hand" has global influence. According to the OECD, all member countries fundamentally rely on competition to organize their economic activities. The reason is that they believe that competition stimulates innovation and efficiency in the use of resources, and thereby leads to greater product diversification and lower prices. "Therefore, competitive product-markets are in the interest of all consumers." (<http://www.oecd.org/about/>) The raison d'etre of competition is linked to economic efficiency, while there is no claims of rationality connected to the ends of economic activity.

The EU is based upon a similar set of ideas, with business being a competitive game in which independent companies "selling similar products or services compete with each other on; for example, price, quality, and service to attract costumers" (<http://www.europa.eu.int/comm/competition/publications/consumer>). The reduction of the human being to a consumer exemplifies the argument that our western concept of rationality is linked to economic ends (Allinson, 2004).

The one-dimensional focus on economic ends is underlined in the following quotation from the OECD, which presupposes that "competition plays an important role in economic growth and affects economic efficiency, the allocation of resources, and can also lead to improved labour market performance" (<http://www.oecd.org/topic/>).

The organic worldview

Whitehead (1967a) argues that it is necessary to introduce an organic categorical scheme to establish a foundation for a holistic worldview based on ethical values. As a start, we have to rethink the status of life in nature.

The mechanistic paradigm presupposes a dualistic separation of "nature" and "life", whereas in the organic concept of nature "life" and "mind" are interwoven with matter and motion. It is the essence of life that it exists for its own sake, as an intrinsic value. The point is that we can neither understand the physical nature or life unless we fuse them together as essential factors in the composition of the whole universe.

Capra's (1982) interpretation of the organic worldview as living systems having a high degree of "nonlinear" inter-connectedness has many similarities with Whitehead's philosophy of organism. Both agree that the interconnectedness is nonlinear in the sense that freedom is considered as the claim for self-assertion. Spontaneity and originality of decision are the supreme expressions of individuality. In a civilized society the general end is that the variously coordinated groups should contribute to the complex pattern of community life, presupposing that individual freedom within each group should be possible without the destruction of the ends of the whole society. In this perspective, the individual and the community both make each other and require each other at the same time.

One important consequence of the organic worldview is that the market could not be reduced to parts in a mechanical system governed by law and scientific rationality. Instead, the market consists of interconnected partners integrated in a living natural and cultural system. A more complex and dynamic framework takes into consideration that economic behaviour is both multi-faceted and context dependent. For example, it is obvious that contextual factors including collective beliefs in ethical norms and pro-social orientation of economic behaviour will contribute to evade some of the fallacies of misplaced concreteness in mainstream economics.[2]

Within the organic perspective it is reasonable to replace mechanistic solidarity with organic solidarity based on coordination and cooperation. According to Durkheim, societies based on organic solidarity are based upon "a system of different organs each of which has a special role, and which are themselves formed of different parts" (Durkheim, 1973, p. 69). Organic thinking is based on the concept of culture as a collective phenomenon, not as the sum of individuals. Economy is like a living organism, which means that its order, structure, and function are not imposed by the environment, but are established by the system itself. Following this reasoning we can conclude that the doctrine of "laws as immanent" (Whitehead, 1967b, p. 112) represents an important condition in the holistic and teleological perspective of the organic worldview.

Cooperation

Cooperation means that different participants or partners try to find solutions in which all relevant information and values are considered. In an economic context, the participants are typically defined as stakeholders. Freeman defines the organization's stakeholders as "any group or individual who can affect or is affected by the achievement of the organization's objectives" (Freeman, 1984, p. 46). In practice, different stakeholders have their own values and aims, and the organization has to interact with the stakeholders for mutual benefit. Values are not reduced to preferences (weak evaluations), since the stakeholders' fundamental values (strong evaluations) are also taken into consideration (Taylor, 1985). In cooperative decisions the ideal is to establish a platform of consensus; that is, to find solutions all stakeholders can agree upon. This way of thinking is different from voting, in which a simple majority wins all power, whereas by cooperating all stakeholders start with a common attitude; that is, "You may be right and I may be wrong" (Habermas, 1990). Table 1 summarizes some of the consequences of the two different coordinating principles in the market.

Table 1. Mechanic versus organic worldview

Mechanic worldview	*Organic worldview*
The market consists of isolated, egocentric actors (atomism)	Individuals and society are inter-woven (holism)
The market is regulated through instrumental, externally defined norms	Internalised values and norms are essential
Competition	Cooperation
Competition anchored in productivity and efficiency From private vices to public virtues	Communicative action to reach common goal

It can be reasonably argued that the principle of competition is insufficient to establish solutions based upon a long-term social and environmental perspective. Welford emphasizes that "productive cooperation … always (will) be superior to blind competition and recognizing cooperative opportunities are part of recognizing interconnectedness" (Welford, 2000, p. 141). Hence, Welford's argumentation is based on the presupposition stating that the market cannot be defined as an aggregate of autonomous actors; instead, the market must be considered an integrated whole. Korhonen lends support to this reasoning by stressing that "Competition is … a barrier of the efforts of increasing stakeholder cooperation and cooperation between the firms and its suppliers or the local community actors" (Korhonen, 2002, p. 70).

This indicates that the focus shifts from merely means to include ends. Through cooperative processes the members might agree on the priority of different ends as well as of the use of available means. Cooperation based on dynamic dialogue allows more integrated solutions than the mechanisms of an atomistic and competitive economy. Equality and mutuality among the involved actors are necessary conditions for constructive cooperation. When competition is replaced by cooperation as the main principle for interaction in the market, the development of solutions based upon the common good will gradually take place. Cooperation presupposes that the partners disclose relevant and valid information without strategic action (Habermas, 1982, pp. 263–271).

Theoretical discussion

We assume that a mechanistic worldview implies competition between actors in the market, as exhibited in cell 1 in Table 2. Mainstream economic literature presupposes that this constellation will lead to the most efficient use of natural and human resources. Both Adam Smith and Milton Friedman, representatives from two different epochs in western history, presupposed a mechanistic worldview, claiming that competition between actors in the market will lead to resource efficiency. In this perspective, competitive power is based upon the actors' preferences, or weak evaluations (Taylor, 1985). The most dangerous threat to this efficiency is that many companies are tempted by the opportunity to: (i) earn high short-term profits, (ii) avoid performance competition with each other, and (iii) attempt to set their own rules for the game, thereby exploiting their power base in order to pressure the other actors, be it a firm or a customer, into accepting the companies premises.

Looking at cell 3 in Table 2, we can see a constellation between the mechanistic worldview and cooperation, which may be exemplified with different types of collusion. Collusion is illegal activity because it may lead to price-fixing or the carving up of markets between different companies which, on the surface, seem to compete. This kind of secret "cooperation" has an overriding objective: to expand the competitive power of the involved actors to the detriment of the other actors in the market in a zero sum game.

In this perspective, globalization of economic activity requires common legal action to ensure that competition is fair. Global players must not be able to do as they please just because they escape any single government's control. The EU wants the World Trade Organization (WTO) to take the lead in obtaining agreement on certain basic principles.

In cells 2 and 4 in Table 2 we assume an organic worldview. In these constellations the partners are perceived as integrated through dialogical processes, and they share common long-term values and interests. The organic paradigm presupposes cooperation (cell 4) which leads to "fair trade", just as competition is the natural principle within the mechanistic paradigm (cells 1 and 3). The organic worldview involves cooperative

Table 2. Principles of coordination, in the context of different worldviews

	Mechanic Worldview	*Organic Worldview*
Competition	(1) Free trade	(2) Disintegration
Cooperation	(3) Collusion	(3) Fair trade

communication processes between the partners being essential inherent parts of the market (network economy).

With reference to nature, in which different organisms cooperate in the nutrition cycle, we advocate that the economy must be based on similar principles. In nature, producers, that is organisms which transform sunlight to chlorophyll through photosynthesis, represent the existence basis for all kinds of primary consumers (herbivors). The secondary consumers (carnivors) eat the meat produced by the herbivors. The decomposers eat the "waste" from all other life forms. Through decomposition processes, new soil is made to create the means of life for the producers.

In addition to synergic co-existence, different life forms living in the same niche in the ecosystem compete for the same food. The natural principle of "the survival of the fittest" decides which individuals and species win the competition for food. Interpreted in an economic context this means that cooperation is the fundamental principle for coordination of activities, while competition has a subordinate function. When competition takes over as the one fundamental principle for co-ordination, as in cells 1 and 2, the different players on the market enter into a fierce competition, leading to serious problems.

We claim that the organic worldview provides a better and often more accurate description of interplay in the market. Hence, market behaviour based on competition will often lead to disintegration and egocentric behaviour. Within an organic worldview the use of one-sided power – or suggestive competition – is destructive (cell 2 in Table 2).

Based on the above argumentation, we reach the conclusion that the discussion concerning cooperation or competition as guiding principles in the market economy should not be isolated from a discussion concerning the contextual worldview.

Cases

In the following paragraphs we present two cases that illustrate our main thesis. First, "fair trade" is illustrated and discussed through a presentation and discussion of Max Havelaar, a paradigmatic case of how business might function in an economy based upon solidarity and sustainability – not on hostility. Second, we illustrate and discuss some problems connected to "collusion" by presenting the case of TINE, a Norwegian grocery corporation engaged in collusion in order to eliminate a small competitor from the market. We will see TINE as an illustrative example of how a company may use fraudulent behaviour in order to eliminate its competitors and win the game.

The focus on particular settings and real-life situations anchors the use of "fair trade" as a case of casuistry as a method (Brinkmann and Ims, 2004, pp. 129–133). Casuistry illustrates an inductive and practical way of reasoning based upon exemplary cases in a middle ground between practice and principle. They act as "touchstone cases that have intrinsic and extrinsic certitude" (Calkins, 2001). Such paradigmatic cases might be used to establish a moral taxonomy for distinguishing acceptable from unacceptable ways of doing business. Use and interpretations of such cases need a virtuous user in the Aristotelian sense (Ruyter, 1995). An interesting question is what makes a case paradigmatic, that is an example that others should follow?

A paradigmatic case in a business setting is a model stimulating rethinking of business principles and presuppositions. It should be easy to see fruitful and interesting parallels to the exemplary case – not through an abstract discussion of theoretical necessity and eternal conclusions, but by inductive reasoning with a focus on particular settings and real-life situations. The tradition of use of exemplary cases has similarities to narrative ethics, which recites the time, manner or consequences of an action. In the next section our two cases are organized as narratives. A narrative is a story and, in the sense of being paradigmatic, the question is, according to Aristotle, whether the case is recognized by all persons in the relevant community, by a majority of those persons, by the wise, by most of the wise, or by the most well-known and respectable among the wise. This tradition is closely connected to Aristotelian virtue ethics, and an important assumption concerns whether the potentials of the persons can be realized in a paradigmatic case.

On the other hand, there are also "bad cases", by which we mean cases stirring moral feelings, and thereby stimulating moral reflection and moral talk (Gustafsson, 1988). These types of cases may be called "issue intensive cases" (Jones, 1991; Frey, 2000). By issue intensive cases we mean cases that violate the norms and values of common morality[3] in a distinct manner. The intensity of issues will vary according to different factors, such as the significance of consequences, the proximity to the consequences of the action, temporal immediacy, and the concentration of effects. Such aspects signal the intensity of the issue as perceived by the actors, which means that the intensity of the issue is not an objective fact, but has to be framed and recognized by the actors as a moral issue. The phenomenon of not recognizing the issue as morally relevant is a phenomenon characterized as moral muteness (Bird and Waters, 1989). It is a state in which the actors do not describe the case in moral terms in the context of their work setting – even if their actions have obvious moral consequences. Instead the actors reframe their motives and actions through different use of legitimization strategies, such as arguing that they do something for the sake of efficiency in their organizations, or due to organizational interests in general. These actors fear that moral talk would provoke confrontation and make choices more time consuming and difficult within their organizations.

Case 1: Max Havelaar – An example of how global solidarity leads to "fair trade"[4]

Max Havelaar is a "trading partnership, based upon dialogue, transparency, commitment, and respect, which seeks greater equity in international trade" (Moore, 2004, p. 73). It is a cooperation between producers living in the south and consumers living in the north.

The paradigmatic example of Max Havelaar is not based upon either an invisible or a visible hand, but more accurately upon a "continuous handshake" that co-ordinates the relations between consumers and producers, and ensures that producers are not exploited, but earn a decent living. We consider solidarity to be an organic principle because solidarity can be defined as unity or communion of interests and responsibilities among nations or mankind in general. Solidarity may be characterized as compassion put into system. Max Havelaar is a paradigmatic case of building a cooperative business institution within a market system.[5] It demonstrates that both producers and consumers include ethical considerations in their evaluations of what to produce and what to consume.

"We want trade, not aid"[6]

The market of coffee is characterized by a value chain that, upstream, includes poor smallholders working as farmers and, subsequently, traders, exporters, and roaster companies that brand their own coffee. Downstream, it includes the supermarkets with their customers. However, during recent years it has become generally accepted that there is a crisis in the coffee market – for several reasons. The supply of coffee has increased due to new coffee-producing countries such as Vietnam, which has now become the world's second largest coffee producer, as well as mechanization of processes in the leading coffee producing country, Brazil.

In addition, the demand for coffee in several markets has decreased due to an increased demand for soft drinks such as coke. Thus, the coffee market is in a state of huge imbalance and the vulnerable poor coffee farmers cannot even cover their production costs when selling their coffee beans. These coffee farmers typically possess no bargaining power in a market characterized by oversupply, and they have to accept the low, often unjust, prices they are offered. On the other hand, multinational corporations (four huge coffee companies or roaster companies dominate the global market: Kraft, Nestle, Procter & Gamble, and Sara Lee are the main buyers. Under such circumstances, the small farmers are not able to negotiate better terms of trade.

The movement of Fairtrade gives the farmers hope in this situation.[7] Max Havelaar is one paradigmatic example of this movement. It forms a partnership between consumers and producers, which implies an ethical awareness on the part of the consumer that presumes that the southern producers obtain a decent living by the terms of trade. The producers are granted respect through receiving a living wage, and they are not treated merely as means but as ends in themselves.

Practical implications – Cooperation based on fair trade

The Max Havelaar movement can be dated back to 1986 when a group of Mexican farmers contacted a Dutch aid organization (Solidaridad). The farmers asked for direct access to the international market arguing that "we want trade, not aid". "We do not need charity; we are not beggars. If we receive a fair price for our products, we are able to be self-sufficient without charity." With direct access to the markets in the rich countries, they would become independent of any aid. Based on these principles, the organization "Max Havelaar" was established, and within a few years it had spread to twenty-one other countries.

On a global base, more than one million farmers and workers are included in the Max Havelaar trade partnership, in which the producers received an additional price of 100 million USD for their products in 2004, and the sales value of all Fairtrade labelled products was 1 billion USD.[8] On a worldwide basis the market share for the different products (coffee, tea, cocoa, bananas, oranges, apple juice, rice, honey, sugar, rice, wine etc.) has grown at an average rate of 30 percent during the last years. In Norway, Max Havelaar was formed in 1997.

Norway is one of the highest coffee consumers in the world, and the growth rate of Fairtrade coffee has been substantial: 305 tons in 2003; 362 tons in 2004; and 424 tons in 2005. Nevertheless, Max Havelaar labelled coffee accounts for only 1 percent of the Norwegian coffee market. Norwegians drink about 4 billion cups of coffee yearly, even if the number of inhabitants is only 4.5 million. The potential for fairtrade coffee is immense.

The FLO (Fairtrade labelling organization) plays a vital part in this partnership. FLO certifies and gives licenses to the partners in the value chain, requiring the partners to satisfy certain criteria, such as granting the farmers a minimum price for their products. It also encourages the farmers to work in a sustainable ecological way using a minimum of pesticides. FLO is a kind of umbrella organization consisting of twenty-one national initiatives and twenty-two countries, which use the Fairtrade standards for farmers and traders. In Figure 1, a principled overview of some important elements in the organization of the Max Havelaar partnership is shown.

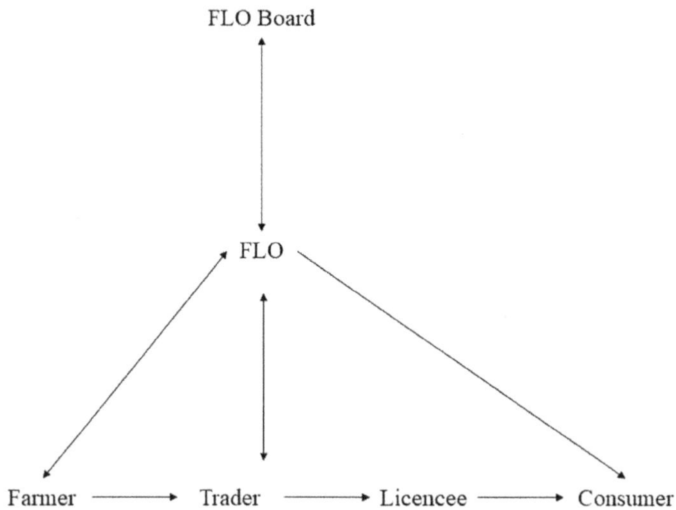

Figure 1. The organization of Max Havelaar's national initiatives.

FLO registers all the producers who satisfy the standards set by them, as well as monitoring the farmers, traders, and licensees on a periodic base. In order to set standards as well as monitor them, FLO is separated into two legally different entities in order to make the Fairtrade certification operations more transparent, and to fully comply with ISO 65 (a worldwide quality standard for certifications organizations based upon the philosophy of "say what you do, do what you say." Fairtrade standards are proposed to the FLO Board, a body with a wide stakeholder representation, for approval by the Standards and Policy Committee (see the Fairtrade annual report 2003/2004). Only the licensees can use the Fairtrade brand (in casu Max Havelaar) to sell products to consumers.

The operations of Max Havelaar in Norway are mostly financed by NORAD, the Norwegian Agency for International Development.[9] In addition, the licensees pay a fee for the right to use the brand and the twelve Norwegian organizations that founded Max Havelaar in Norway also contribute to Max Havelaar yearly with a symbolic payment. The international Fairtrade criteria for coffee must satisfy the following rules:

1. A decent price is at present defined as USD 126 cents per pound (a minimum price of USD 121 cents per pound plus a premium of 5 cents).
2. A democratic organization. The farmers must belong to cooperatives or organizations that are transparent and democratically controlled by its members.
3. Direct trade and long-term relationships. The traders (importers) must buy coffee directly from Fairtrade-certified producers and accept the establishment of stable long-term relationships.
4. Access to credit. If the farmers ask for it, the importers should give the farmers as much as 60 percent of payment in advance.
5. Protecting the environment. The farmers must implement Integrated Crop Management (ICM) plans for taking care of the environment. Price incentives encourage the farmers to work towards ecological production, and some environmental standards must be fulfilled by the farmers.[10]

Response in the Norwegian market

In 1999, only 6 percent of the Norwegian population were aware of the existence of Fairtrade. In 2002, 30 percent, and in 2004, 36 percent, had knowledge of Max Havelaar, even though 75 percent of all retail chains in Norway sold Max Havelaar products. Several movements in Norway (Korsvei and "Green congregations") are engaged in testing whether the local shops in Norway sell and promote Max Havelaar products.[11] The results of these "shop tests" are published in the media and local newspapers. Because of this, the promotion of Fairtrade products has increased because most Norwegian franchise takers do not want to be viewed as lagging behind as regards ethical considerations. Now and then "Action Weeks" have been arranged by a number of NGOs in order to increase awareness of Max Havelaar products.

Case 2: TINE – An example of how local solidarity turned to collusion

TINE BA is "the sales and marketing organization for Norway's dairy cooperative and is responsible for product development, quality assurance, production and distribution planning, marketing and the export of TINE products" (<http//www.tine.no>). In the past, milk products in Norway were prepared on small farms. At the beginning of the twentieth century a great expansion in dairy farming led to an increased need for cooperation within the industry. The Norwegian Dairy Association, founded in 1881, was the first nationwide organization. The forerunner of TINE, the Norwegian Dairies' Export Company, was established in 1928. Its aim was to organize the export of Norwegian dairy products. TINE (Norwegian Dairies) was introduced in 1992 and, after extensive reorganization in 2002, TINE BA became the official name of the company.

"We will be Norway's most important contributor to value creation"[12]

TINE BA is a cooperative owned by 18,500 dairy farmers. Each of the farmers own a share in the cooperation's five dairy companies, to which they deliver 1.5 billion litres of cow's milk and about 20 million litres of goat's milk every year. TINE BA is by far Norway's largest dairy processor with a total of 5,500 employees and an annual turnover of about 2 billion euros for 2004.

TINE is responsible for products from the fifty-three plants that are run throughout Norway. The Norwegian dairy industry is based on modern principles, using advanced technology. Some of the plants are fully automated. The milk is processed into a product spectrum of more than 200 products that can be found on shop shelves bearing the TINE trademark. The trademark "TINE" is strongly positioned in the consciousness of Norwegian consumers and has always enjoyed a high reputation.

TINE claims that environmental responsibility represents a fundamental base for all decisions concerning product development and assessing new production technology. The focus on society and environment is articulated in the corporation's objectives; "The dairy industry must strive to be a profitable and environmentally friendly industry characterized by pure products of high quality, a clean environment, and use of environmentally friendly technology" (<http//www.tine.no>).

TINE's formulated vision in the "Annual report and accounts 2004" emphasizes the importance of the triple bottom line, that is, integrating social, environmental and financial values. TINE announced that; "We shall prove ourselves worthy of trust in all parts of the value chain" (<http//www.tine.no>). In other words, TINE dairies draw their values from a fusion of "nature, (agri-)culture and economy".

Practical implications – cooperation based on collusion

TINE is the market leader in all of the segments it operates, but in December 2004 it was accused of collusion. A small private grocery company, Synnøve Finden (SF), which was established in 1928, had recently increased their market share in the cheese market. In December 2004, SF was abruptly excluded from one of the largest retailers in Norway, ICA.[13] For SF, ICA was a major outlet for their products, and to be excluded meant a substantial drop in their sales. Then SF claimed publicly that TINE had paid "slotting fees", or "supplier-bonuses", to the retailer ICA in order to obtain exclusiveness for TINE products and thus eliminate SF, the small but awkward competitor.

After some weeks with almost daily news about SF's claim that TINE had colluded, followed by a strong refutation by TINE in a story unfolding like a thriller, it finally emerged that SF's claim held water. In February 2005, the top manager in TINE had to admit that they had offered ICA, 2 million euros in order to obtain "exclusiveness in the retailer chain" through "extra facing" for sixteen different TINE products.[14] Concerning the ICA case, TINE acknowledged that they were the party to blame – not ICA, and offered ICA a public apology, unilaterally accepting responsibility for paying a bonus to the retailer chain. In ethics in general the briber and the bribed have to share the responsibility, but this case shows clearly that the retailer chain has the most power in the value chain.

Response in the Norwegian market

In a short space of time the media announced that the market share for SF had increased dramatically, while the market share of TINE had fallen substantially.[15] In one week TINE's sales of cheese slumped by 25 percent, while SF sales shot up by 73 percent.[16] The cases of TINE, ICA, and Synnøve Finden provide several lessons.

First, consumers in Norway are concerned about ethical issues. Goliath, in the shape of "TINE", tried to destroy David "SF" by the use of collusion and, in so doing, provoked most of Norway's consumers. When eliminating the competitor and focusing on short-term profits, the leaders of TINE actually undermined their long-term goals of making values for the whole of society. They had undermined one of their most important visions and preconditions to reach these goals, that is, the goals of "proving themselves worthy of trust in all parts of the value chain" (<http//www.tine.no>).

The use of collusion and economic power to get rid of market competitors was not accepted by the consumers who showed no loyalty to the brand TINE. They were willing and capable of turning their backs on TINE and switch to SF. Instead of increased profits, TINE's economic results worsened.

Second, in the long run the owners of the cooperative, the Norwegian farmers, as well as the environment, would be the losers if TINE were to lose their dominant position in the Norwegian market for dairy products. A reduction of the cooperative's power could lead to a situation in which the solidarity between Norwegian farmers was undermined. TINE's attempt to capture new markets by using unfair methods actualized the fact that the success of small-scale farming in Norway depends on solidarity and fair trade.

Representatives from different NGOs argued that the consumer boycott of TINE's products, if successful, would have a negative impact on the environment. The leader of the NGO "Nature and Youth" claimed that "the production of food could not be separated from questions concerning the use of recourses and environmental protection".[17] She argued that the dairy cooperatives were an important contributor to eco-friendly, small-scale farming all over the country. In the perspective of environmental responsibility, small-scale farming is favorable due to reduced transportation, and hence less pollution, and better food quality.

An immediate result of TINE's collusion admittance was that two top leaders in the corporate management had to resign from their positions. The new leader of TINE immediately announced that the most important task to be achieved was to re-establish the trust between TINE and the stakeholders in the market. In addition to finding solutions to the problems related to collusion and economic power, she was well aware that TINE had to implement the corporation's environmental goals in practice in a trustworthy way.

Discussion

What are the similarities and differences between the two cases? We will compare experiences from the two cases with the theoretical expectations regarding the different principles of coordination and their particular context of worldviews, respectively. We will present our understanding of why the case of Max Havelaar is characterized by fair processes and good economic, societal, and environmental results, whereas the case of TINE is a "bad, issue intensive case" with negative consequences along different dimensions, as shown in Table 3.

We can find several similarities between the two cases. Both Max Havelaar and TINE were established to take care of the farmer's interests. Being the weakest part in the value chain for food and dairy products, the farmers were forced to accept deals that were socially unjust and environmentally harmful. Max Havelaar and TINE are both based upon the idea that cooperative integration leads to several advantages, particularly if we integrate different values. In addition to economic profit, both cooperatives integrate societal and environmental responsibility in their fundamental values. They also agree that small-scale farming represents an important means of reaching these goals. As indicated in the cases, powerful and wealthy actors on the retailer side represent a threat to these multi-value

Table 3. Dimensions of analyses

	Max Havelaar	TINE
Goal	Fair trade based upon partnership between consumers in the First World and farmers in the Third World. Eco-friendly production	Take care of the farmers' interests and contribute to scattered population in Norway. Eco-friendly production
Values	International solidarity	National solidarity
Ontology	Organic/holism	Mechanic/atomism
Market coordination	Cooperation through open discourse	Competitive power through collusion
(Dys-)functional results	Better life quality for 1 million farmers in the Third World	Decreased trust in the market leads to reduced sale of TINE's products and insecurity for Norwegian farmers

goals because of a relatively one-dimensional focus on prices and profit in the short run.

Decisions and actions on the market arena indicate the main differences between the two cases. On the one hand, Max Havelaar integrates all stakeholders through communicative action based upon ethical values like openness and fairness. On the other hand, TINE attempts to attain competitive advantage through strategic actions based upon secret agreements between selected actors in the market (Habermas, 1982).

We think it might be clarifying to analyse the cases in the context of different worldviews. Max Havelaar represents an example of successful cooperation in the context of an organic worldview (cell 4 in Table 2) whereas TINE is an example of how bad things can turn out when one actor in the competitive market engages in secret agreements for fraudulent purposes (cell 3 in Table 2). Within the context of a mechanistic worldview cooperation is just a euphemism for collusion, and is not based upon the interest of "fair trade" for the common good. Fraudulent behavior destroys the trust between the actors, which is supposed to obtain efficiency in a competitive market. Using a mechanistic worldview, competition between autonomous actors regulated by imposed norms and laws (mechanical solidarity) is the guarantee for efficient behavior.[18] In a competitive market based upon a mechanical perspective, fraudulent behavior breaks the external rules governing the interplay between the actors.

TINE might be described as an internal cooperative concern with a defined task to take care of the farmers' interests (in a wide sense), acting within an external context characterized by competition between autonomous actors motivated by their self-interests. When TINE uses unfair means to reach short-term profit goals the result, in a long-term perspective, will have a negative impact on the farmer's interests, society, and the environment. We believe that the two cases strongly indicate that the coordination of different values connected to the production of food in general and dairy products in particular cannot occur by making actors in the market independent like atoms. However, these assumptions are seemingly presupposed in mainstream economics and by the Norwegian Competition Act (2004) in order to increase competition in the market.

Based on this, we claim that there is a need for changes in the worldview. In accordance with Whitehead (1967a, 1967b, 1985) we believe that the mechanistic worldview, dominating western culture during the past three centuries, is far too narrow to serve as a satisfactory categorical scheme for understanding the world of today. We maintain that "an alternative organic doctrine" is needed that is adequate for interpreting the phenomenon of life as a whole. Welford (1995, 2000) supports our beliefs, maintaining that it is impossible to implement sustainable solutions integrating economic, environmental, and societal values within an atomistic, competitive perspective.

Max Havelaar is a paradigmatic case showing how sustainable solutions based on a holistic, organic worldview has positive consequences for the different partners at the individual as well as the societal level. The case of Max Havelaar also indicates that the different stakeholders in an expanded perspective (both in time and space) have common interests because they are integrated parts in the global community. We claim that it is morally unacceptable and economically hazardous to establish the world economy based upon principles defined by the powerful minority on behalf of the poor majority. We find the cooperative solution of Max Havelaar to be a showcase of how "the common good" is related to a combination of cooperation as an organizing principle between interrelated stakeholders and an organic worldview.

If cooperation is based upon an atomistic, mechanistic worldview, the behaviour will typically lead to group egotistical consequences that will undermine the long-term common good, not only in terms of environmental and societal effects, but also in an economical sense (at least the long-term interests). The reason is that cooperation in a mechanistic worldview often has an exclusive character, which leads to collusion and unfair trade. In order to encourage market behaviour that integrates economic, societal, and environmental values we find that the case of Max Havelaar indicates that cooperation within a context of an organic worldview takes care of important intrinsic as well as instrumental values.

Notes

1. The Norwegian Competition Act has a long history in Norway and may at least be traced back to the Trust-commission's recommendation in 1921, where Wilhelm Thagaard, characteristically called Lex Thagaard, played an essential role. The Norwegian Price Law of June 26, 1953 also intended to stop businesses that would damage the common good or use unfair practice against another party.
2. Whitehead argues that our use of abstractions based on limited worldviews is dangerous because we tend to forget that they are abstractions. Even worse, we also tend to mistake the abstractions for the concrete actuality. Many of our abstractions try to claim a completeness that they do not possess. Whitehead calls this error "the fallacy of misplaced concreteness" (Whitehead 1967a, b, p. 51).
3. There are several Norwegian authors who presuppose a common morality. A recent work is Tranøy (1999), and the Norwegian pioneer in business ethics, Lunde (1967, 1982), describes in several of his works the ethical limit of common morality as the "norm horizon".
4. Boda (2001) demonstrates that the concept of fair trade is multifaceted and discusses four different ways upon which the concept can be interpreted.
5. Allinson (2004) gives a thorough description of some ways to build ethics into the market system.
6. In this paper we chose to analyse the market for coffee. In fact, Max Havelaar in Norway is based upon producers in several markets; for example, tea, bananas, cacao, etc. There is an important difference between the coffee and the tea market. Much of the tea market consists of large plantations (Ceylon is one example of being a huge tea producer which produces tea on plantations) in which the

embedded context may have traits of slavery (see The Norwegian (NGO) Journal *Folkevett* no. 1 2005, pp. 10–18).

7. It may be pertinent to differentiate between Max Havelaar and ethical sourcing. By ethical sourcing we mean a supply chain member which emphasises the social and environmental criteria in its purchase decisions in order to support a group of people or a certain practice. This has led to a regime of certification of social and environmental criteria (for example the quality standard ISO 14001). Ethical sourcing assumes that there is a corporation (or group of corporations) that is the customer – not an individual customer. Thus, the buyer has massive power in the negotiations of terms for trade. Those corporations are, however, only bound by their own ethical guidelines, which may be too flexible in several situations.
8. Fairtrade labelled products is a more inclusive category than Max Havelaar. For example, the Norwegian Cooperative does not use the label or brand Max Havelaar on their products, even if some of their products have ideological similarities to Max Havelaar. In the same manner there are several chains in Norway that market and brand their own Fairtrade products.
9. The payment by NORAD to the FLO in Norway amounted in 2004 to about 280,000 EUR.
10. The sources used are from the website: Max havelaar.no
11. The majority of shops are organized as franchise systems in which the individual franchise taker has a certain freedom in the supply of products.
12. TINE's Annual report and accounts 2004 (<http//www.tine.no>).
13. CA is a chain of retailers organised as a franchise system.
14. VG 05.02.05 (VG is the largest national newspaper in Norway).
15. In April 2005 it is announced that Synnøve Finden's sales during the first three months of the year has not increased, but had fallen by 3 percent. Some people now claim that the market is being deceived.
16. Aftenposten 22.02.05. (Aftenposten is a national newspaper in Norway).
17. Vårt Land 25.02.2005. (Vårt Land is a national newspaper in Norway).
18. This type of behaviour is typically termed "solution" in the vocabulary used by economists. It is a solution only in a narrow technical sense presupposing that effective demand is equal to supply. It is not a solution from the perspective of poor people that have basic needs and are lacking money.

Bibliography

Allinson, R. B.: 2004, "Circles within a Circle: The Condition for the Possibility of Ethical Business Institutions within a Market System", *Journal of Business Ethics* 53, 17–28.

Bird, F. and J. Waters: 1989, "The Moral Muteness of Managers", *California Management Review* 32, Fall, pp. 73–88.
Boda, Z.: 2001, "Conflicting Principles of Fair Trade", *Business Ethics Papers* No. 3 (Business Ethics Center, Budapest University of Economic Sciences).
Brinkmann, J. and K. Ims: 2004, "A Conflict Case Approach to Business Ethics", *Journal of Business Ethics* 53, 123–136.
Calkins, M.: 2001, "Casuistry and the Business Case Method", *Business Ethics Quarterly* 11, 237–259.
Daly, H. E. and J. Cobb, Jr, 1994, *For the Common Good – Redirecting the Economy Toward Community, the Environment, and a Sustainable Future* (Beacon Press, Boston).
De Witt, B. and R. Meyer: 1998, *Strategy, Process, Content, Context,* 2nd edn (An International Perspective. International Thomson Business Press, London).
Durkheim, E.: 1973, *On Morality and Society* (The University of Chicago Press, Chicago).
Durkheim, E.: 1991, *The Division of Labour in Society* (MacMillan Press Ltd, London).
EU: <http://www.europa.eu.int/comm/competition/publications/consumer>.
Frey, B.: 2000, "The Impact of Moral Intensity on Decision Making in a Business Context", *Journal of Business Ethics* 26, 181–195.
Georgescu-Roegen, N.: 1971, *The Entropy Law and the Economic Process* (Harvard University Press, Cambridge, MA).
Gustafsson, C.: 1988, *Om Føretag, Moral, och Handling* (Studentlitteratur, Lund).
Habermas, J.: 1982, "A Reply to My Critics", in J. Thompson and D. Held (eds), *Habermas, Critical Debates* (Macmillan Press, Cambridge, MA).
Habermas, J.: 1990, *Moral Consciousness and Communicative Action* (The MIT Press, Cambridge, MA).
Jones, T.: 1991, "Ethical Decision Making by Individuals in Organizations: An Issue – Contingent Model", *Academy of Management Review* 16, 366–395.
Korhonen, J.: 2002, "The Dominant Economics Paradigm and Corporate Social Responsibility", *Corporate Social Responsibility and Environmental Management* 9, 67–80.
Lunde, J.: 1967, *Etiske normer i næringslivet. Forbrukerorientert markedsetikk* (Universitetsforlaget, Oslo).
Moore, G.: 2004, "The Fair-Trade Movement: Parameters, Issues and Future Research", *Journal of Business Ethics* 53, 73–86.
Norges offentlige utredninger: 2002, *Avfallsforebygging – En visjon om livskvalitet, forbrukerbevissthet og kretsløpstenkning* (NOU 2002: 19, Oslo).
Norwegian Competition Authority: 2004, *Konkurransemessig vurdering av ordninger for produktgjenvinning* (Oslo). OECD: <http://www.oecd.org/about/>.

Porter, M. and C. van der Linde: 1995, "Green and Competitive – Ending the Stalemate", *Harvard Business Review* 73 (5), Sept. –Oct., pp. 120–134.

Ruyter, K. W.: 1995, *Kasuistikk som saksbasert problemløsning* (Doctoral Dissertation, School of Theology, University of Oslo).

Smith, A.: 1976, "An Inquiry into the Nature and Causes of The Wealth of Nations", in E. Cannan (eds), *Two Volumes in One (first edition 1776, the citations are from the 5th edn)* (The University of Chicago Press, Chicago).

Taylor, C.: 1985, *Human Agency and Language, Philosophical Papers 1* (Cambridge University Press, Cambridge)

TINE: <http//www.tine.no. >

Tranøy, K. E.: 1999, *Det åpne sinn. Moral og etikk mot et nytt årtusen (The Open Mind. Morality and Ethics Towards a New Millennium* (Universitetsforlaget, Oslo).

Welford, R.: 1995, *Environmental Strategy and Sustainable Development* (Routledge, London).

Welford, R.: 2000, *Corporate Environmental Management – Towards Sustainable Management* (Earthscan, London).

Whitehead, A. N.: 1967a, *Science and the Modern World* (The Free Press, New York).

Whitehead, A. N.: 1967b, *Adventures of Ideas* (Macmillan, New York).

Whitehead, A. N.: 1985, *Process and Reality* (The Free Press, New York).

KNUT J. IMS AND OVE D. JAKOBSEN

Consumerism and Frugality: Contradictory Principles in Economics?

Introduction[1]

Western economies grew as much in percentage terms in the twenty years between 1950 and 1970 as they had done in the thousand years between 500 and 1500 AD. Today the growth of world GDP[2] exceeds 3 percent annually, and many researchers within different fields of science agree with Pearce, who asserts that, "the man-made bacterium that threatens the life of the planet is the growth imperative at the heart of modern economics" (Pearce, 2001, p. 26). Consumerism is an important mantra which suits "mainstream" economics well.

The huge growth in production and consumption over the last decades has actualized the following challenges. First, many local, regional, and global environmental and societal problems are connected to an oversized economy. Second, the global economy is characterized by unfair distribution of resources and goods, both between different countries and within countries. Third, the production of *waste*, both from industry and from households, is growing faster than ever. Fourth, the technical capabilities at our disposition in the West have an inclination to manifestation. Surplus of capabilities of man leads to an ever-increasing production of means, even when material needs are satisfied (Zapffe, 1996).

To manage these vital questions it is not enough to adopt changes at a system level, including more restrictive laws and rules. In addition to external political regulation, we argue that it is necessary to change attitudes and values from inside. In order to attain a sustainable scale of production and consumption, fair distribution of resources and wealth, and efficient resource allocation, it is necessary to rethink the economy within a worldview characterized by integration, dynamics and holism.

We argue that frugality represents an interesting gateway to meet these challenges, presupposing a context in which sustainability and life quality are the overriding goals. Within the context of mainstream economics, which gives priority to economic growth, frugality represents a problem; in fact, it can even be seen as a threat. To illuminate and illustrate our arguments we will concentrate, in the next paragraphs, on a comparison between mechanical and organic worldviews on the one hand, and consumerism and frugality on the other.

A question of different worldviews

To support our ideas and understanding of the two different worldviews, we have chosen Whitehead's "philosophy of organism". We agree with Whitehead who argues that all human thought has categorial schemes as an underlying presupposition. Although "categorial schemes" are often unconscious, they represent a fundamental condition for understanding the world we live in. Just like Kuhn's paradigms, the categorial schemes are of great importance in shaping actual concepts, hypotheses, and theories in different scientific disciplines. A consequence of this reasoning is that the underlying worldview influences our theoretical understanding and our practical action.

Whitehead's position corresponds with the arguments of Heidegger and Taylor when they explained the success of the mechanical worldview by referring to the separation between "the physical world" and "the life-world". We find a parallel to this bifurcation in Greek antique society with Aristotle opposing Democritus' atomistic theory. The atomists thought that everything in existence was atoms and space, and all the different objects consist simply of different collections of atoms in space. Aristotle saw the true essence of any object as consisting not in the matter of which it is made, but in the organic function it performs.

The fallacy of bifurcation is deeply rooted in European philosophy from the beginning of the seventeenth century. Combined with the ideas characterising the mechanical paradigm, human beings have no experience of the intrinsic value of life and they perceive their actions as more or less meaningless. It follows that modern society finds itself in a state of decline. In our context the decline is exemplified by the expanding ecological and social crises connected to the growth ideology, characterising mainstream economics.

The mechanical worldview

The mechanical worldview presupposes that physical matter in its movements is reality, and that everything can be explained in terms of physical laws. In other words, mechanism is a worldview claiming that physical matter is reality, complete and total. This theory is in accordance with Democritus' statement 2,500 years ago, stating that everything in the universe can be explained in terms of imposed physical laws:

> It became natural, to conceive of the world as made up of discrete components, which fit together like the parts of a machine. The behaviour of atoms was conceived as tiny bouncing balls whose behaviour could be predicted, as could the behaviour of more complex objects assembled from them. (Derfer, in Xie, Wang and Derfer, 2005, p. 87)

One of the most important consequences of the mechanical worldview is that the whole universe is completely causal and deterministic. "All that happened had a definite cause and gave rise to a definite effect, and that the future of any part of the system could – in principle – be predicted with absolute certainty if its state at any time was known in detail" (Capra, 1997, p. 120). This doctrine of nature as a self-sufficient, meaningless complex of facts is central in the mechanical paradigm. A consequence of this perspective is that "dead" nature can provide no reasons, and it aims at nothing.

Bergson drew a similar demarcation line between the "dynamic" and the "static" perspective. "True change can only be explained by true duration; it involves an interpenetration of past and present, not a mathematical succession of static states. This is what is called a 'dynamic' instead of a 'static' view of the world" (Russel, 1979, p. 763).

Science based on the mechanical perspective is characterized by the idea that bits of matter are isolated individuals (atomism) related to one another purely externally. Through natural laws society represents no real unity in itself. Society is nothing more than a mere mechanism based on the interplay between egocentric individuals seeking their own ends.

Today, the mechanical worldview still forms the basis of many scientific disciplines, including economics. In lifeless nature, problems can be solved within the framework of physical laws. It is, at least in principle, possible to arrive at unambiguous solutions for this kind of convergent problem. Schumacher argued that convergent problems have nothing to do with self-consciousness or life functions. According to Georgescu-Roegen (1971) and Daly and Cobb Jr (1994), this assertion still holds validity for modern mainstream economics.

Mainstream economic theory builds on the mechanical presuppositions in which all questions converge into certain solutions. Inspired by this metaphor, agents in the market are supposed to act independently of one another in order to optimize their own interests. Market theory presupposes that economic agents act autonomously in most market transactions. The assumption that economic rationality largely excludes other-regarding behaviour has deep roots in western theoretical understanding of human nature. Another presupposition is that the dominating value in economics is "profit". Today there is expanding pressure for higher short-term sales rates and profit maximisation. This is justified by economists such as Friedman who argue that "few trends could so thoroughly undermine the very foundations of our free society as the acceptance by corporate officials of social responsibility other than to make as much money for their stockholders as possible" (Friedman, 1963, p. 133).

Consumerism

In our context, we refer to consumerism as the idea that growth in production and consumption of goods and services are important driving forces in sound economic development. As an illustration, growth in GNP is one of the most important goals of the World Bank. A consequence of these attitudes is that little attention is paid to the true needs of the consumers, durability of the products, product origin, or the environmental consequences of manufacture and disposal. Huge amounts of money and much entrepreneurial talent are (mis-)used on advertising designed to increase the sales of all kinds of products and to create a desire to follow trends. An intended consequence of this is to discard what might give an image of being old, recycled, or second-hand products because products with such characteristics are out of fashion.

Consumerism interferes with the workings of society by replacing the normal common-sense desire for an adequate supply of life's necessities, community life, a stable family, and healthy relationships with an artificial, ongoing, and insatiable quest for things, and the money to buy them, with little regard for the true utility of what is bought. At aggregate level economic growth is determined by the rate of utilization of the factors of production – capital and labor – and the efficiency of their use. In mainstream economics, growth is described as an important goal connected to the economy's ability to provide increasing levels of standard of living to its population. A negative consequence of this development is that consumerism can lead to a more materialistic-oriented society.

The organic worldview

The philosophy of organism constitutes an alternative worldview, based on a holistic perspective and inherent values. The motivation for introducing values and purposes into nature is the realisation that science based on materialism cannot give an adequate account of the emergence of life. In accordance with philosophy of organism we argue that we have to rethink the status of life in nature and to accept that integrated organisms are units emerging for their own sake.

As mentioned earlier, the mechanical paradigm presupposes a dualistic separation of "nature" and "life". The idea of including human nature as an element in nature results in notions expressing that value and freedom can no longer be excluded from the descriptions of nature. According to this reasoning different starting points could exist which are relevant for framing an organic categorial scheme. "In principle it does not matter which field of study one uses as a starting point since each discipline is a window through which one may view the general structures and categories constituting reality as such" (Fowler, 1976, p. 57). The term *life* refers to the enjoyment of things such as self-enjoyment, freedom, creativity, purpose, and subjectivity, derived from the past and aimed at the future. Thus, the characteristics of life are absolute self-enjoyment, creative activity, and purpose.

Nothing in nature could be what it is, except as an integrated ingredient in nature as a dynamic whole. According to the philosophy of organism, an isolated event is not an event because every part is a factor in a larger whole and is significant to the process of the whole:

> The isolation of an entity in thought, when we think of it as a bare "it", has no counterpart in any corresponding isolation in nature. Such isolation is merely part of the procedure of intellectual knowledge. (Whitehead, 2000, p. 142)

Hence, it seems reasonable to assert that Whitehead's ontology can be compared with Spoinoza's "psycho-physical parallelism", or Leibniz' "theory of monads" (Woolhouse, 1993). Within the organic concept of nature "life" and "mind" are interwoven with matter and motion. It is the essence of life that it exists for its own sake, as an intrinsic reaping of value. The point is that we can neither understand the physical nature nor life unless we fuse them together as essential factors in the composition of the whole universe.

Referring to Capra's (1982) interpretation of the organic worldview as living systems having a high degree of "nonlinear" interconnectedness, we conclude that the individual and the community both make each other and require each other at the same time. Thurow points in the same interpretative direction of the organic paradigm when he writes the following:

> Societies are not merely statistical aggregations of individuals engaged in voluntary exchange but something much more subtle and complicated. A group or community cannot be understood if the unit of analysis is the individual taken by himself. A society is clearly something greater than the sum of its parts. (Daly and Cobb Jr, 1994, p. 7)

This indicates that a conversion to the organic worldview has far-reaching consequences for both economic theory and behaviour. A more complex and dynamic framework takes into consideration that economic behaviour is both multi-faceted and context-dependent. For example, it is obvious that contextual factors including collective beliefs in ethical norms and pro-social orientation of the economic behaviour contribute to avoiding some of the fallacies of misplaced concreteness in mainstream economics. Whitehead argues that our use of abstractions based on limited worldviews are dangerous, because we tend to forget that they are abstractions. Even worse, we also tend to mistake the abstractions for the concrete actuality. Many of our abstractions try to claim a completeness that they do not possess. Thus, we do a blunder, "the fallacy of misplaced concreteness" (Whitehead, 1967, p. 51). These fallacies flourish because the economists are socialized to think in abstractions without being aware that they are abstractions. Organic thinking is based on the concept of culture as a collective phenomenon, not as the sum of individuals.

In accordance with this organic perspective, Schumacher argues that problems within economics must be handled as divergent because the precondition of lifeless nature is transcended. "Without the characteristic human element of self-consciousness, one easily slips back into the one-dimensionality of order without freedom, thus reducing all problems to converging problems" (Opdebeeck, 2006, p. 6).

Schumacher argues that divergent problems "are resolved in the most optimal manner by reaching economic practice on the macro-level from the point of departure of economic practice on the micro-level" (Opdebeeck, 2006, p. 7).

Frugality in the Aristotelian Christian tradition

As a theoretical concept and in practical conduct, frugality refers to an ideal and a way of attaining high life quality *involving* a low level of consumption (Opdebeeck, 2006). Frugality means that individuals practice restraint in both the acquisition and use of economic goods and services in order to achieve lasting and more fulfilling goals. At aggregated level, frugality emphasizes a low level of consumption meeting long-term personal, familial and communal needs. Another intended effect of frugality is that it leads to reduction of waste through changing consumption habits. Different cultural traditions, also in the West, consider frugality a virtue or a spiritual discipline. Today frugality is a central value in Buddhist economics.

We argue that even in the Christian tradition we can find important inspiring sources of an economy of frugality. The thoughts of Aristotle were, during Medieval times, interpreted and synthesized by Thomas Aquinas (1225–1274). Thomas was a Christian and a theologian, and as Tranøy (1957) writes, a rare crossbreed of a saint and a scientist.

In the Christian tradition one important notion is that the use of external goods has a *natural limit*. Material wealth is needed to a certain extent, but should be used only as an instrument. The highest good *beatitudo*, is of a spiritual character. The better you come to know the highest good, the more you will love it. On the contrary, external things are like wind or ordinary water. Thomas uses the example of the Lord who told the woman of Samaria: "Whosover drinketh of this water shall thirst again" (John 4.13). When the appetite for a material good is quenched, the oppressed materialist will continue to search for other things. Such appetite is vicious because one ends up as a slave, not a master of wealth. Thus, it is not wealth in itself that is evil, but the individual's attitudes to wealth (Langholm, 1992, p. 209).

This view corresponds to Aristotle's recommendation of the golden mean, which in several ways can stand as a model of an economy based on frugality, since he condemns all kind of excesses. "Evil belongs to the class of unlimited and good to that of the limited" (NE, II.6 1106 b27).

Virtue is also a kind of mean as it aims at what is moderate. And good must be something final and self-sufficient. Wealth is evidently not the good we are seeking; for it is merely useful for the sake of something else (NE, I.6 1096 a5).

Aristotle is skeptical to the viewpoint that an economy should be the purpose of a society. A fear was that the instruments would enshadow the purpose of the good life *eudamonia*, which consists of an active life developing one's best abilities, thereby becoming a virtuous human being, blossoming within a community. Moral virtue is a state of character. A realization of one's abilities is not possible without the security, economic resources, and education opportunities present in a politically well-organized society. A virtuous life presupposes friendship and family life as well as intellectual activities. In order to realize his/her virtues as a generous and just person, the virtuous person needs a moderate, but necessary, amount of money.

A closer look at the concept of wealth in relationship to the final purpose may prove worthwhile here. Thomas distinguishes between two types of external wealth, natural, and artificial (Aquinas, 2.1). Natural riches consist of the things that humans collect and use because they lack them. This is food and drink, clothes, houses, and means of transportation. The highest good cannot be connected to such things because they are imperfect and serve merely as *instrumenta* in man's search for virtue and *beatitude* (Langholm, 1992, p. 208). Accordingly, natural riches are not meant to be a purpose, only a means. In particular, this reasoning applies to artificial riches such as money, which only satisfies human needs in an indirect manner. As a consequence, the desire for natural richness will not be infinite. But the avidity for artificial richness is infinite because, according to Thomas, it serves the sexual drives, which are not tamed. Thus, the inclination that the means gradually become the final purpose is obvious and dangerous (Aquinas, 2.1).

No abundance of riches is necessary for the exercise of virtue. In fact, too much wealth may hinder the development and cultivation of virtue. But the real danger of having too many riches is that this impedes the love and fear of God. When entrapped by avarice, a man makes riches his God, because God is that in which one places one's end. By placing one's end in an infinite quest for worldly riches, one redirects one's attention from God to wealth.

One important environmental consequence of mainstream economics is that economists generally fail to recognize that the economy is merely one aspect of "living" ecosystems. The ecosystems are composed of continual interaction between all life forms, including human beings. According to Costanza et al., continuous economic growth is impossible within the Earth's finite and non-growing ecosystems.

> [T]he economy cannot grow forever (at least in a material sense)... ultimately some sort of sustainable steady state is desired. This steady-state is not necessarily absolutely stable and unchanging. Like in ecosystems, things in a steady-state economy are changing constantly in both periodic and aperiodic ways. (Costanza et al., 1997, p. 62)

Frugality is a way to counteract an economy based upon the goal of growth in production and consumption that is neither consistent with sustainability nor life quality. The basic idea of frugality is that life quality depends on more factors than high consumption rates. Life quality depends on the possibilities to develop and realize the potential of human virtues. In other words life quality concerns values not visible within the mechanical worldview.

Frugality versus consumerism

Economics is defined as the discipline dealing with production, distribution, consumption, and redistribution of wealth (Ingebrigtsen and Jakobsen, 2006). Hence, it is reasonable to assert that economic models and theories will always be based upon certain values and norms. As we have commented on earlier, mainstream economics is based upon monetary values related to profit and GDP. In contrast, Buddhist economics is based on the notions of "sustainable development" and "life quality" (Schumacher, 1973; Welford, 2006). The Christian tradition, as represented by Thomas Aquinas, is based upon the natural limitation of external goods, and moderation in acquiring and using external commodities.

We argue that the market cannot be reduced to parts in a mechanical system governed by law and scientific rationality. Instead, the market consists of interconnected participants integrated in a living natural and cultural system. Economy is like a living organism, "which means that its order and structure and function is not imposed by the environment but is established by the system itself..." (Capra, 1982, p. 269). Hence, a well-functioning market must be based upon values like confidence and trust between the participants.

Within mainstream economics frugality is easily thought upon as utopian and in conflict with the nature of the "economic man". In Table 1 we illustrate the connection between worldview and consumption attitudes. The concepts consumerism and frugality lead to different conclusions depending on the context.

We find that growth in production and consumption is translated to the positive concept "welfare" within a mechanical perspective. Growth in welfare is an important goal in most countries today. If we change the focus to frugality, we find that this concept has connotations for life quality and well-being.

To comprehend the connection between frugality and life quality, we use the field of positive psychology as a gateway. In positive psychology the attention is focused on attributes and traits that constitute psychological strengths and individual qualities that give life meaning and purpose, not on pathologies and deficits (Seligman and Csikszentmihalyi, 2000). Positive psychology, in accordance with philosophy of organism, is based upon a

Table 1. Worldview and attitudes to consumption

	Consumerism	**Frugality**
Mechanical Worldview	Welfare	Stagnation
Organic Worldview	Excess	Life quality

worldview that living systems are self-organizing and oriented towards increasing differentiation and complexity. Life quality cannot be attained as a state of being. In accordance with the philosophy of organism, positive psychologists argue that life-quality is a process of becoming:

> [A] good life cannot be attained as a permanent state, … (it) is a continuous development of the individual's strengths and values. A good life is, therefore, a life in continuous development or growth … At their best, individuals, furthermore, act both in concert with their own premises and capacities and in concert with the surroundings. (Jørgensen and Nafstad, 2005, pp. 890–891)

Positive psychology argues, on the one hand, in the same way as Aristotle and in opposition to the pragmatics, that there are some common, preprogrammed dispositions in the human nature. On the other hand, positive psychology accepts the idea that human nature is also a product of history and cultural environment. In this perspective life quality is closely connected to the person's search for:

> [I]ncreasingly complex challenges in associated activities and will improve his or her skill accordingly. This process has been defined as cultivation; it fosters the growth of complexity not only in the performance of the flow activities but in the individual behaviour as a whole. (Massimini and Delle Fave, 2000, p. 27)

We argue that the strong interest in positive psychology is connected to the societal and ecological challenges of today. It is important that the shift in perspective leads to an increased interest in the ontology of human beings, nature, and culture.

Realizing the necessity of giving priority to sustainability and life quality, it is obvious that, from an organic point of view, frugality constitutes an essential value. One important question to be discussed is how to handle frugality within the context of integrated dynamic wholes. Taking into consideration that individual humans perform actions, it is more convenient to discuss frugality from an individual point of view. At the same time the cultural element is essential in constituting the subjective values and truths.

Modern marketing is based upon the idea that the goal is to stimulate consumption by putting forward new varieties of scarcities that are to be "tackled with as much utility maximization as possible" (Opdebeeck, 2006, p. 1). Consumption is not a bad thing; we have to consume to survive, and the world's poorest people have to consume more to improve both "welfare" and "well-being". But whenever increasing consumption becomes an end in itself, it threatens not only the natural and social environment and the life quality of future generations, it presents a serious problem for individual consumers here and now, as well.

If the levels of consumption in rich countries were adopted on a global scale, "the impact on our water supply, air quality, forests, climate, biological diversity, and human health would be severe" (Gardner et al., 2004, p. 3). Attempts to try to solve environmental, individual and social problems within the existing mechanical worldview seem very problematic.

Notes

1. We would like to thank Luk Buckart, Hendrik Opdebeeck, and Laszlo Zsolnai for the inspiration to write this chapter as well as for their constructive comments on an earlier draft.
2. Gross Domestic Product (GDP) measures the total value added of enterprises operating in the country concerned, regardless of whether they are owned by residents or non-residents.

Bibliography

Aquinas, Thomas (ca 1271) *Summa theologiae I-II*, 2, 1, ad (and) 3.

Capra, Fritjof (1982) *The Turning Point – Science, Society and the Rising Culture.* London: Flamingo.

Capra, Fritjof (1997) *The Web of Life – A New Synthesis of Mind and Matter.* London: Flamingo.

Costanza, Robert, Cumberland, John, Daly, Herman, Goodland, Robert, and Norgaard, Richard (1997) *An Introduction to Ecological Economics.* Boca Raton: St Lucie Press.

Daly, Herman E. and Cobb, John B., Jr (1994) *For the Common Good – Redirecting the Economy toward Community, the Environment, and a Sustainable Future.* Boston: Beacon Press.

Derfer, (2005) in Xie, Wang, and Derfer (eds) *Whitehead and China – Relevance and Relationships.* Frankfurt: Ontos Verlag.

Fowler, Dean R. (1976) Alfred North Whitehead. *Zygon*, vol. 11, no. 1 (March), pp. 50–68.

Friedman, Milton (1963) *Capitalism and Freedom.* Chicago: University of Chicago Press.

Gardner, Gary, Assadourian, Erik, and Sarin, Radhika (2004) The State of Consumption Today. In: *State of the World 2004*. The World Watch Institute.

Georgescu-Roegen, Nicholas (1971) *The Entropy Law and the Economic Process.* Cambridge, MA: Harvard University Press.

Ingebrigtsen, Stig and Jakobsen, Ove (2006) Circulation Economics – A Turn towards Sustainability. *International Journal of Social Economics,* vol. 33, no. 8, pp. 580–593.

Jørgensen, Ingvild S. and Nafstad, Hilde Eileen (2005) Positive Psychology: Historical, Philosophical, and Epistemological Perspectives. *Tidsskrift for Norsk Psykologforening*, vol. 42, pp. 885–896.

Langholm, Odd (1992) *Economics in the Medieval Schools: Wealth, Exchange, Value, Money and Usury According to the Paris Theological Tradition, 1200–1350.* Køln: E. J. Brill.

Massimini, F. and Delle Fave, A. (2000) Individual Development in Biocultural Perspective. *American Psychologist*, vol. 55, pp. 24–33.

Opdebeeck, Hendrik (2006) *The Urgency of a Frugality Based Economic Paradigm – The Legacy and Spirituality of E.F. Schumacher (1911–1977).* Paper presented at the International Workshop "Spirituality and the Economics of Frugality" the Centre for Ethics, University of Antwerp, April 20–21, 2006.

Pearce, Joseph (2001) *Small Is still Beautiful.* London: HarperCollins Publishers.

Russel, Bertrand (1979) *A History of Western Philosophy.* London: Unwin Paperbacks.

Schumacher, Ernst Friedrich (1973) *Small Is Beautiful: A Study of Economics as if People Mattered.* London: Blond & Briggs Ltd.

Seligman, Martin E. P. and Csikszentmihalyi, Mihaly (2000) Positive Psychology – An Introduction. *American Psychologist,* vol. 55, no. 1, pp. 5–14.

Tranøy, Knut Erik (1957) *Thomas av Aquino som moralfilosof.* Oslo: Universitetsforlaget.

Welford, Richard (2006) Tackling Greed and Achieving Sustainable Development, in Zsolnai, L. and K. Johannessen Ims (eds) *Business Within Limits: Deep Ecology and Buddhist Economics.* Oxford: Peter Lang.

Whitehead, Alfred North (1967) *Science and the Modern World.* New York: The Free Press.

Whitehead, Alfred North (2000) *Concept of Nature*. New York: Cambridge University Press.
Woolhouse, Roger S. (1993) *The Concept of Substance in Seventeenth Century Metaphysics*. London: Routledge.
Xie, Wenyu, Wang, Zhihe, and Derfer, Geaorge E. (2005) *Whitehead and China – Relevance and Relationships*. Frankfurt: Ontos Verlag.
Zapffe, Peter Wessel (1996) *Om det tragiske*. Oslo: Pax Forlag (First Published in 1941).

KNUT J. IMS AND LASZLO ZSOLNAI

Holistic Problem Solving

Existing challenges and deficiencies

Today's management practice tends to reduce every problem to the *technical dimension*. This often results in the "Error of the Third Kind" (E3), which means *solving the wrong problem precisely* (Mitroff, 1998). The controversial case of the World Bank encouraging polluting industries to move to Third World countries is an insightful illustration to this point (Box 1).

Box 1 The World Bank Environmental Policy Proposal

In the early 1990s, some economic advisors of the World Bank were proposing that the organization should encourage increased migration of dirty industries to less-developed countries. The argument was as follows:

The measurement of the costs of health-impairing pollution depends on the foregone earnings from increased morbidity and mortality. From this point of view a given amount of health-impairing pollution should be done in the country with the lowest cost, which will be the country with the lowest wages. (…) The costs of pollution are likely to be non-linear as the initial increments of pollution probably have very low cost. (…) The demand for a clean environment for aesthetic and health reasons is likely to have very high income-elasticity. The concern over an agent that causes a one-in-a-million in the odds of the prostate cancer is obviously going to be much higher in a country where people survive to get prostate cancer than in a country where under-5 mortality is 200 per thousand. Also, much of the concern over industrial atmospheric discharge is about visibility-impairing particulates. These discharges may have very little health impact. Clearly, trade in goods that embody aesthetic pollution concerns could be welfare enhancing. (*The Economist*, February 8, 1992, p. 66)

One should take a broad perspective on problem solving in order to avoid the strong tendency to solve economic and organizational problems *as if* they were technical. It is mostly due to the bias of disengagement in people's real needs, which also lead to abstract knowledge, theories and principles. Theories, models and beliefs are deeply held, as long as nobody challenges them. The bias toward technical solutions in management is partly explained by professional training and experience.

Crucial in problem solving is *who* defines the problem and what kind of professional training he or she has. Professional training colours the values of the decision maker and influences the formulation of problems in a non-conscious way. When dealing with important problems one should take different perspectives on the same problem in order to evaluate the fruitfulness and relevance of different problem formulations.

Referring to a vast body of literature we argue that human beings and organizations have serious limitations in their capacity to operate in a complex way to match a type of environment that is dominated by dynamic changes. As a frame of reference we draw upon the Carnegie school, whose prominent scholars were Herbert A. Simon, James March, and Johan P. Olsen.

Bounded rationality

Our purpose is to make suggestions on how organizations in a dynamic environment might be viewed in order to perform their functions efficiently. More precisely, our purpose is to indicate under what conditions the organization might attain its goals in an efficient way, given that the organization has to work in a manner deviating from objective rationality. We take an empirical point of departure focusing on the limits which characterize humans and organizations as decision making units – contrasted with the ideal type of objective rationality, which clings to an idealistic picture and an omniscient view of the rational man. The idealized picture of rational economic man requires a stable, closed system of variables that we seldom find in actual business situations where significant indirect effects are always present (Simon, 1976).

The view represented by Simon and others is called *bounded rationality*. We maintain that human limitations are controllable in the sense that awareness of the limits may lead to transcending and altering the limits. Whenever one knows that he or she has a technical bias in search for information, he or she may alter his or her behaviour by broadening the search, for example in the direction of existential aspects of the situation.

A vast body of documented evidences indicates that there are a number of limits on each member of the organization. These include (i) limits of skills, habits, and reflexes which are not in the realm of the consciousness, (ii) limits of values and conceptions of purpose which influence making decisions, and (iii) limits of knowledge relevant to the given situation (Simon, 1976, p. 40). For example, decision makers have only an imperfect and incomplete knowledge about the consequences of their decisions, which may cause serious systemic effects that unfold in the future. The values of different alternatives can only be anticipated in an imperfect manner. Bounded rationality tells us that we should be focusing on those factors that are most "closely connected with the decision in cause and time" (Simon, 1976, p. 82).

Traditional economic theory postulates an "economic man" who in the course of being "economic" is also "rational". This man is assumed to have knowledge of the relevant aspects of his environment which, if not absolutely complete, is at least impressively clear and voluminous. He is assumed also to have well-organized and stable system of preferences and a skill in computation that enables him to calculate the alternative courses of action that are available to him, and which of these will permit him to reach the highest attainable point on his preference scale (Simon, 1955).

A firm is not an omniscient system. It is a system of units in which organizational members come from different professions with different goals. Different professions have a learned incapacity to see, define, understand, and solve problems. Professionals see things differently in the same situation because they see through their conceptual lenses, which are coloured by the special language they have appropriated through their professional training. They have to make decisions, and thereby they have to collect, interpret, analyse, and use information which concerns the problems to be solved. It may be appropriate to distinguish between the *primary decisions* – the decisions that have to be done, and the *secondary decision* – the decision concerning which information sources are appropriate. In order to make reasonable decisions we have to acquire adequate information. This is evident in important investment decisions, or in public policy decisions which require impact analysis in order to take into consideration aspects that otherwise might have been ignored. Rational decision making requires adequate information. Then the rational decision maker has to be conscious of the way he or she searches for information.

The administrative man

The rational model of economic man is strongly attacked by scholars who believe in another kind of model, the so-called "administrative man". The administrative man shows another kind of rationality, *bounded rationality*, which involves how members of different units of the organization perceive problems in specialized ways. While the sales departments of a company may define the problem as an underuse of PR, the production department may look upon the "same" problem in technical terms – asking for a new machine to make the parts and parcel stronger.

Organizational problem solving displays many similarities to individual problem solving. For one thing, organizations use individual members of the organizations as instruments. In general, organizations give priority to simple search rules in seeking to reduce the stress of handling complexity. This means extensive use of *programmed decisions*; that is, do what you have done before if it means a limited search in a well-structured way to solve the problem. Another characteristic of organizations is that they use "*rule of thumb*" decision rules, even for complex decisions. This deviates in a strong sense from what is prescribed in rational decision models. One example is the making of budgets. When forecasting next year's cost, one adds 10 percent to this year's budget. Another phenomenon is called *local rationality*. Compared with objectively rational behavior, we find that the human mind has a limited capacity to formulate and solve complex problems. Local rationality implies that decision units have limited horizon in problematic situations. In particular, the environment of the decision bodies is viewed as set, even when the environment has undergone important changes.

Another phenomenon is that organizations are "*uncertainty avoiders*", selecting strategies that tend to minimize uncertainties. One well-known strategy is to try to stabilize the environment and the practice of the unit through standardization. This means selecting one practice of pricing as well as of employment, even if it may have been better to have a contingency-approach in the same situation. Another implication of avoiding uncertainty is that organizations usually adapt a short-term perspective. This involves short run solutions to reduce the immediate pressure of problematic situations. For example, in an equipment replacement situation, a one-piece-at-a-time replacement rather than a more holistic renovation may be made, even when the latter would be more rewarding in terms of long-run profits.

In organizational as well as individual behaviour, problem solving mostly consists of "*problemistic search*", which means that the search is motivated and stimulated by a problem. It means, too, that the *search is not proactive, but reactive*. Search is a pragmatic activity; that is, members of the organization search until they believe they have solved the problem, but not because they want a deeper understanding of the phenomena: they want to control the problem, not understand it. Motivated search also means that solutions are searching for problems. Pet projects can look for crises and sellers of technical equipment search for opportunities so that their equipment can find commercial opportunities (March and Olsen, 1976). Motivated search and bounded rationality encompass the idea that the search stops when it has reached a level of satisfaction. "Good enough" solutions are quite different from maximizing ones. This way of thinking dictates that if you are searching for a needle in a haystack and find one, you stop searching. It may not be cost-effective to continue the search for a sharper needle. The notion of "satisficing" replaces the notion of "maximizing".

Problemistic search assumes that the search starts with a simple model of causality and evolves into a more complex model, but only if the search does not solve the problem. Two simple rules employed are: (i) search in the neighbourhood of the problem symptom, and (ii) search in the neighbourhood of the current alternative. The assumption is that the cause lies close to the effect and the new solution will be similar to the old solution. Thus, problemistic search starts near the symptoms of the problem and typically employs a narrow process that leads to myopic solutions.

A serious weakness of problemistic search is the tendency for waiting until bad things happen. This reactive approach is not always fruitful because it often means only treating the symptoms – resources could have been better used preventing the "fire" and proactively planning for the future. The chief problem with such a "fire station strategy" is the tendency to assume that one solves the problem by extinguishing the fire. It is often more fruitful to think of the fire as the symptom of the real problem, with the important challenge being to define the core of the problem. Much of the theory of problem solving is about solving predefined problems. However, *problem finding* – the art and skill of apprehending and conceptualizing problems – is more important, more difficult, and more challenging.

New approaches and good practices

What is a problem? A problem is a gap between an "ought" and an "is"; the difference between an ideal, something which is desired and the actual situation as it is perceived. The problem is solved when one perceives that the gap has disappeared. The gap may be closed either through adaption, which leads to a reduction of the level of aspiration toward the existing situation, or through action that brings the situation closer to the ideal.

Important problems typically appear as "messes", which present themselves with no clear indication on how to solve them. It is neither easy to know what the ideal is, nor easy to state what the actual situation is. Human feeling and intuition are then called upon to begin the process of apprehending and conceptualization the problem. We believe that this intuition is, to a large extent, a product of professional training. In the following, we present a framework that will help in taking a more holistic view on problems. We think that the formulation of problems, and ultimately the decisions we make, depend in part upon the questions asked.

Perspectives of the problem

As a frame of reference, we present a model developed by *Ian Mitroff* (1998), which shows four dimensions of any problem: scientific/technical, interpersonal/social, systemic, and existential.

The *technical/scientific* perspective is the dominant perspective in today's western culture and favors technical solutions to most of the problems, even if other solutions are more appropriate. This *technocentric bias* is partly explained by the status and prestige that engineers and economists enjoy in our culture.

The *interpersonal/social* perspective concerns the way the problem might be looked upon from social, group, and family points of view.

The *systemic* perspective takes into consideration the long-term consequences of how the problem is solved. We will define this kind of thinking as involving the perspective of future generations as well as other stakeholders – even if they are not heard or seen. This perspective goes beyond geographic borders and narrowly defined time limits. It also takes into account dysfunctional or boomerang effects caused by decisions.

The *existential* perspective emphasizes the lives and fates of individual human beings and their lifeworlds. Human feelings, dignity, and meaning are important aspects in this perspective. The important existential questions asked are as follows: Who am I? Who do I want to be? How do my actions influence my life project, which gives meaning and purpose to my life? Ultimately, all important decisions and acts influence the self of the decision maker, and reasonable beings as such should care about themselves.

It is important to emphasize the existential perspective because it is usually ignored or even denied in management literature and thinking. However, when we take this perspective seriously we get a new and different view of many business problems. For example, it may explain why conventional reward systems based upon the doctrine that "more money is better than less money" do not always work.

Familiar examples showing the importance of the existential dimension of problems are blood donation and garbage handling in Denmark. Research found that after offering blood donors money for their blood contributions, a new pattern developed. Relatively poor people, like drug-users, increased their donor frequency, and even more people were recruited from this population. In contrast, among people that had a higher socio-economic status, donor frequencies decreased, and some of these people stopped their donations altogether.

When a garbage-handling project run by a Danish community proved successful, the municipality realized that they could make a profit by recycling the garbage. Therefore they decided to pay the locals for their garbage handling. But then the municipality got a surprise. The collection of garbage was reduced, not increased due to the money involved. An activity based upon the conviction of doing the right thing had an intrinsic value was transformed into a market transaction such that many of the locals lost their willingness to contribute to the common good.

Both examples show that "material incentives destroy rather than supplement moral incentives" (Goodin, 1980). There are some little known "interaction effects" between economic and normative input. We argue that this phenomenon is strongly connected with the existential aspect of life. People often oppose financial incentives when their primary motive is to be a person possessed of dignity and self-worth. They do not want to look upon themselves as greedy sharks. Having self-respect one does not enter into actions that may cripple self-esteem.

We illustrate the multidimensionality of business decisions with the complex case of downsizing and focus on "surveyor's sickness" (Box 2).

Box 2 Survivor's Sickness in Downsizing

Downsizing is a problem which is usually understood and solved within the technical/scientific perspective. The main reason for downsizing a company is that it does not generate enough profit. Managers assume that profit can be increased by moving parts of the company from a high-cost country to a low-cost country, so typically some production plants are moved to developing countries. However, after a certain period of time (9–12 months), data shows that the "scorecard of downsizing is negative" (Karake-Shalhoub, 1999). In some situations downsizing might be seen as a case of solving the wrong problem precisely.

By "downsizing" we mean the reduction of the workforce through either voluntary or involuntary means, or a combination of the two. The fundamental assumption is to reduce the costs, and the most obvious factor to be reduced is the number of employees. This implies that people who do not want to leave the company are forced to leave. Without work they become depressed because they miss the opportunity to work and earn money, as well as missing the daily company of their colleagues. Such a change for the now unemployed is expected. However, there are also unexpected changes. The motivation of the remaining employees is reduced, so the impact of downsizing within the company goes beyond immediate economic considerations. It not only affects organizations in economic terms, it also affects the level of achievement of the surviving workforce. In short, rather than increasing, morale among the survivors decreases.

Existential dimension: The retained employees suffer from a loss of meaning and dignity and reduced self-esteem. In a sense they have taken the plight of their colleagues upon themselves and blamed the managers for the treatment of those who have lost their jobs. Those who still have jobs become alienated and feel emotionally uneasy.

Social dimension: A downsizing usually breaks up social relations, informal friendships and networks. It impacts the life of families by contributing to economic uncertainty and forced relocations, which means that children may end up changing schools and environments.

Systemic consequences: One typical consequence is less innovation because the problems are moved to a foreign country, typically to a developing country where the law is less restrictive in terms of taxation, environmental protection, and human rights. Thus, one ultimate result is environmental dumping, which in turn results in a boomerang effect. Weaker environmental laws mean some industries use more pesticides and produce more dangerous emissions. The systemic consequences, in turn, have effects on social factors, such as the creation of mistrust and anguish among the workers who were spared by the first round of downsizing. This has a negative impact on productivity, and the company reacts with still more downsizing. Finally, the company which began life in a developed country finds itself operating fully in a developing country, creating new problems for vulnerable stakeholders without a strong voice.

There are also systemic influences on the local community. When workers have lost their jobs as a result of downsizing, they move to new areas and house prices fall. This creates a difficult financial situation for families when they have to re-establish themselves in a new place. That fewer families remain in the local community also means a reduction of welfare institutions, like schools and kindergartens. When worker morale is weakened due to feelings of guilt, stress, and overwork, the costs within the downsized company are not really cut because many of the same persons are taken on as consultants at a higher salary.

The main problem with downsizing, as typically exercised, is that it does not focus on the critical issue: the structures that have to be changed. A one-dimensional outlook is required to believe that a reduction in one factor, the size of the company, is in itself the problem and that changing one variable does not have effects on other dimensions such as social, existential, and systemic aspects.

Critical thinking is needed

Managers need critical thinking to solve the right problem. Even an approximate solution to the right problem is better than an elegant solution to the wrong one. Assumptions managers hold about the stakeholders should be scrutinized. They should pick the right stakeholders, and this generally means expanding the set of stakeholders. To manage the important problems in a fruitful way, holistic thinking – that is, the use of systems methodology – is indispensable to avoiding the "Error of the Third Kind" (E3) – solving the wrong problem precisely.

The systems methodology developed by *Russell Ackoff* and his colleagues at *The Wharton School* states that as long as the assumptions underlying the management action are explicit, there exists the chance to learn from experience and improve the quality of our practice. This is the way to produce, and successively improve, informed choices. No problem or solution is context free. Context itself changes continuously, rendering a solution obsolete in time. Therefore, managers should be empowered with the systems methodology and the "why questions" so they can redefine the problem and redesign the solution if need be (Ackoff Center Guiding Principles, 2004).

The performance of any system is determined by the adequacy of the basic assumptions upon which it is functioning and the way the system is capable of pursuing its objectives. Problems indicate that the basic assumptions are inadequate and/or the system is incapable of performing adequately.

Managers should *dissolve* rather than solve problems by redesigning the relevant system to eliminate the problems and preclude their reappearance. They should conceptualize problems and opportunities in the largest context possible over which they have some control. They should take into account the effects of any proposed actions on all of the relevant stakeholders. Finally, managers are responsible to provide their organization with an increased ability to do the *right things* as well as do *things right* (See Box 3).

Box 3 The Interface Corporation

The Interface Corporation in Atlanta is a leading innovator in carpet industry. The traditional old-fashioned broadloom carpet is environmentally damaging and produces a lot of waste. In the late 1990s Interface launched a transition from selling carpet to leasing floor-cover services and providing the necessary maintenance services. Contrary to the traditional solution, Interface, not customers, owns the carpeting and maintains responsibility for keeping it clean and fresh, for a monthly fee. Whenever needed, Interface replaces 10–20 percent of the carpet tiles that show 80–90 percent of the wear. This reduces the amount of carpet material required by about 80 percent, because the unworn part of the carpet is left in place. This solution also provides better service at a reduced life-cycle cost. Because the carpet is laid in the form of tiles, glue fumes are also significantly reduced, or possibly eliminated. So far so good: saving in materials, energy, and money.

But Interface's technological innovation goes even further by turning waste into savings. Other manufacturers "down-cycle" nylon and PVC-based carpet into lower-level use thus losing the embodied energy. Interface has instead made a novel polymeric material, called Solenium, that can be completely remanufactured back into itself. The production process thus becomes simpler and less wasteful. The new product also provides superior benefits; it is stain-resistant, easily cleaned with water, 35 percent less material intensive, and yet four times as durable. It uses sevenfold less mass flow per unit of service than standard carpeting. It is suited to renewable feedstock and is acoustically and aesthetically improved. Compared to standard nylon broadloom carpet, the combination of Solenium's improved physical attributes and the service lease encompasses a reduction in the net flow of materials and embodied energy by 97 percent. Its net climate impact is zero.

> The Interface Corporation applies a comprehensive, holistic approach to running its business. Its chief executive, Ray Anderson, summarizes their philosophy thus:
>
> At Interface, we are on a quest to become the first sustainable corporation in the world, and then we want to keep going and become the first restorative company … It means creating technologies if the future – kinder, gentler technologies that emulate nature … We look forward to the day when our factories have no smokestacks and no effluents. If successful, we will spend the rest of our days harvesting yesterday's carpet, recycling old petro-chemicals into new materials and converting sunlight into energy. There will be zero scrap going into landfills and zero emissions into the ecosystem. Literally, it is a company that will grow by cleaning up the world, not by polluting or degrading it. (Hawken et al., 1999, pp. 139–141, 168–169)

The required roles and duties of managers

Managers need scientific and technical knowledge, but they also need a better understanding of the existential conditions of human beings to avoid the fallacy of defining most problems narrowly and solving them in purely technical ways.

The notion of holistic problem solving does not equate with the kind of omnipotent behaviour prescribed in rational choice theories. Our point is that human rationality is bounded. This gives us a less heroic behaviour than prescribed in idealized normative theories. We have given some description of the kind of behaviour that characterizes human beings in general, and organizations in particular.

Given several biases in sensing, seeing, analyzing, and understanding organizational problematic situations, a number of tools can improve our constrained decision making processes and thus be helpful in most decision making situations. We have formulated biases as *practical limits* to *human rationality*, and emphasized that these limits are not static, but can be transcended. The manager's task is to increase his or her consciousness about human abilities and develop adequate tools that reduce some of the most significant limits.

We have assumed a description of the world which is consistent with the concepts of administrative man and bounded rationality. We believe that "*satisficing*" behavior is much more characteristic of human beings and organizations than maximizing behavior. This means that we, as administrative men and women, should be aware of and respect our limited capacities for reasoning and calculation by making some decisions with relatively simple rules of thumb.

These rules should not, however, reduce important problems to consideration of only one stakeholder group or only one dimension of a problem. It is essential that organizations broaden the set of relevant stakeholders, giving particular attention to the impact on future generations and the environment. In the same vein, the organization has to be conscious of the wise strategy of looking upon important problems as having at least four dimensions. We believe that actual, mundane behaviour constitutes a rational platform for using helpful tools and hinting at improvements.

Bibliography

Ackoff Center Guiding Principles. (2004) The Wharton School, University of Pennsylvania.

Goodin, R. E. (1980) "Making Moral Incentives Pay." *Policy Sciences*, vol. 12 (August), pp. 131–145.

Hawken, P., Lowins, A. B. and Lowins, L. H. (1999) *Natural Capitalism. The Next Industrial Revolution*. Earthscan, London.

Karake-Shalhoub, Z. A. (1999) *Organizational Downsizing, Discrimination and Corporate Social Responsibility*. Quorum, Westport.

March, J. G. and Olsen, J. P. (1976) *Ambiguity and Choice in Organizations*. Universitetsforlaget, Bergen.

Mitroff, I. (1998) *Smart Thinking for Crazy Times. The Art of Solving the Right Problems*. Berrett-Koehler Publishers, Inc. San Fransisco.

Simon, H. A. (1955) A Behavioural Model of Rational Choice. *The Quartely Journal of Economics*, vol. LXIX (February), pp. 99–118.

Simon, H. A. (1976) *Administrative Behaviour. A Study of Decision-Making Processes in Administrative Organization.* 3rd edn, Expanded with New Introduction, The Free Press, A Division of Macmillan Publishing Co., Inc. New York.

KNUT J. IMS AND OVE D. JAKOBSEN

Competition or Cooperation? A Required Shift in the Metaphysics of Economics

Economists often suffer from a kind of metaphysical blindness, assuming that economics is a science of absolute and invariable truths without any presuppositions. Some go as far as to claim that economic laws are as free from metaphysics or values as the laws of physics. Ormerod argues that conventional economics offers a misleading view of how the world actually operates (Ormerod, 1994). The absence of metaphysical questions in mainstream economics has dodged fundamental conflicts among economy, nature, and culture. These conflicts have led to a whole range of negative symptoms, such as climate change and financial crises. Instead of questioning the metaphysical assumptions, mainstream economics is mostly concerned about short-term solutions aimed at reducing the most pressing symptoms. By using increasing doses of the prescribed old medicine, negative symptoms are muted so that the pathological causes remain indistinct.

What mainstream economics needs is a profound metaphysical critique. In this chapter we analyze and discuss some of the metaphysical presuppositions in modern economics, and we suggest deep changes in the frame of reference for economic theory and practice. On the one hand, we argue that competition between autonomous actors makes sense only if the mechanical worldview is in accordance with the real world conditions. However, if the organic worldview is more relevant, cooperation between integrated actors is appropriate. On the other hand, we state reasons for claiming that, within the mechanic worldview, cooperation between actors can lead to collusion while competition between actors in an organic worldview results in disintegration. In other words, it is impossible to argue for or against competition or cooperation without making the metaphysical presuppositions explicit.

In order to look into this question, we describe the differences between mechanic and organic worldviews and argue, in accordance with Norton (1991), that real world conditions more closely resemble organisms than machines. It is therefore plausible to rebuild economic theory and practice based upon an organic frame of reference.

The chapter is divided into two sections. Firstly, we illuminate the most relevant differences between mechanic and organic worldviews. Secondly, we explore some of the implications of these two worldviews, arguing that the mechanical worldview fits well into individualistic competitive market economics, while the organic worldview leads to cooperation as the main coordinating principle of the economy.

Metaphysical presuppositions

The mechanical worldview presupposes that physical matter in its movements makes up reality, and that everything can be explained in terms of physical laws. In other words, the claim is that physical matter is reality – complete and total. This theory concords with Democritus' assertion 2,500 years ago that everything in the universe can be explained in terms of imposed physical laws:

> It became natural, to conceive of the world as made up of discrete components, which fit together like the parts of a machine. The behaviour of atoms was conceived as tiny bouncing balls whose behaviour could be predicted, as could the behaviour of more complex objects assembled from them. (Derfer in Xie, Wang, and Derfer, 2005, p. 87)

Democritus believed that everything in the universe was composed of atoms, which were physically indivisible and indestructible. Between the atoms there was empty space and the atoms always kept moving. All observable changes were reduced to changes in the con-figurations of these particles. The atomists tried to explain the world without introducing the notion of purpose or final cause.

In the seventeenth century, Newton (1642–1727) included a precise description of the force acting between these material bodies. The force was very simple, depending only on the masses and the mutual distances of the bodies. According to Russel, "the atomists asked the mechanistic question, and gave a mechanistic answer" (Russel, 1979, p. 84). One of the most important consequences of the mechanical worldview was that the whole universe was interpreted as completely causal and deterministic. According to Newton, the giant cosmic machine was seen as being completely determinate and governed by immutable laws. "All that happened had a definite cause and gave rise to a definite effect, and (…) the future of any part of the system could – in principle – be predicted with absolute certainty if its state at any time was known in detail" (Capra, 1997, p. 120).

Before Newton, Descartes (1596–1650) had argued that the physical world is a material plenum where all change can be described as local motion of various parts. "Since the motion conforms to simple, mathematically stable laws, it can be predicted with complete certainty" (Jones, 1969, p. 177). Whitehead asserts that in the mechanical worldview everything is pre-decided and determined by the configurations of masses. This doctrine of nature as a self-sufficient, meaningless complex of physical facts is central in the mechanical paradigm. A consequence of this perspective is that "dead" nature can provide no reasons, and it aims at nothing.

Although most people experience continuity, endurance, and value the mechanic worldview presupposes that the world is a static, lifeless machine. Therefore, the static view of reality that considers the world as composed of unchanging substances with changing attributes is counter-intuitive, according to Whitehead (1967a).

Bergson draws a similar demarcation line between the "dynamic" and the "static" perspectives. "True change can only be explained by true duration; it involves an interpenetration of past and present, not a mathematical succession of static states. This is what is called a "dynamic" instead of a "static" view of the world" (Russel, 1979, p. 763). It is interesting to note that Bergson claims that mechanism and teleology suffer from the same defect, since both suppose that there is no essential novelty in the world. "Mechanism regards the future as implicit in the past, and teleology, since it believes that the end to be achieved can be known in advance, denies that any essential novelty is contained in the result" (Russel, 1979, p. 757).

Converted to social science, the mechanical perspective leads to the idea that individuals are isolated bits of matter, related to one another purely externally. Through natural laws, society represents no real unity in itself. Society is nothing more than a mere mechanism based on the interplay between egocentric individuals seeking their own ends.

Today, the mechanical worldview still forms the basis of many scientific disciplines. Schumacher argues that convergent problems "in the fields of physics, chemistry, astronomy, and also in abstract spheres like geometry and mathematics, or games like chess" (Schumacher, 1977, p. 125) can be solved within the framework of physical laws. He argues that convergent problems have nothing to do with self-consciousness or life functions. Economists like Samuelson and Frisch argue that physics was an ideal in neoclassical economics, and they considered physics to be their ideal science (Ingebrigtsen and Jakobsen, 2009). According to Georgescu-Roegen (1971) and Daly and Cobb Jr (1994) this assertion still holds validity for modern mainstream economics.

Mainstream economic theory builds on mechanical presuppositions in which all questions converge into certain solutions. Agents in the market are supposed to act independently of one another in order to optimize their own interests. Economic theory presupposes that agents in the market act autonomously in most transactions. The assumption that economic rationality largely excludes alternative ways of regarding behavior has deep roots in the western theoretical understanding of human nature. Another presupposition is that the dominating value in economics is "profit". Today, there is expanding pressure for higher priced, short-term sales and profit maximization. This is justified by economists such as Friedman (1970), arguing that few trends could so thoroughly undermine the very foundations of free society as the acceptance by corporate officials of a form of social responsibility that goes beyond making as much money for their stockholders as possible.

Individualism in the social sciences is often presented as a theory of sociological explanation. It advocates, in accordance with the deterministic worldview, that social processes and events should be explained by being deduced from principles governing the behaviour of the participating individuals and descriptions of their situations.

According to Whitehead (1967a), the ontology and methodology of modern science (including economics) teaches that the world is "bleak, purposeless and barren", and that human beings live their lives isolated from other people and the surrounding environment. In this perspective, society resembles an aggregate of autonomous individuals. In much the same way, mainstream economics is based on the assumption that agents in competitive markets seek their own goals as societal atoms.

This means that explanations belonging to the mechanic world view claim that every biological or social event is a pattern of non-biological occurrences. This formulation cannot be interpreted to mean that all organisms resemble man-made machines. To avoid some of these problems we use the term mechanism more broadly than machine.

Whitehead (1967b) argues that ontological individualism represents a gap between modern science and human experience. Instead, he introduces a scheme of interpretation in which integrated society exhibits truth, beauty, adventure, art, and peace. It is important to stress that Whitehead does not eliminate subjectivity from his "collective" perspective. We regard it reasonable to interpret this phenomenon in accordance with Aristotle's formulation of "man-in-community". In other words, the individual and the community make each other and require each other at the same time.

As mentioned earlier, the mechanic worldview presupposes a dualistic separation of "nature" and "life". The idea of including human nature as an element in nature results in the notion that value and freedom can no longer be excluded from descriptions of nature. Following this line of reasoning, a different starting point is relevant in order to provide the framework for an organic worldview. "In principle it does not matter which field of study one uses as a starting point since each discipline is a window through which one may view the general structures and categories constituting reality as such" (Fowler, 1976, p. 57).

Just like mechanism, organic philosophy has its roots in Greek antiquity. According to Aristotle, matter contains the essential nature of all things. He argues that matter and form are two sides of the same process, from potentiality to actuality. The term "life" refers to states like "self-enjoyment", "freedom", "creativity", "purpose", and "subjectivity", derived from the past and aimed at the future. Thus, the characteristics of life are absolute self-enjoyment, creative activity and purpose (Georgescu-Roegen, 1971).

Whitehead's (1985) "philosophy of organism" constitutes a worldview based on a dynamic and processual philosophical perspective. Reality is understood in terms of integrated networks whose properties cannot be reduced to the aggregate of atoms. In a perspective where relations are more real than atoms, the world is seen as a network of integrated wholes rather than a collection of independent parts.

Whitehead follows Aristotle's arguments and claims that there is a creation of cosmic continuity in the world. Since each atom makes up an integrated part of every system, atomism does not exclude complexity and universal relativity. Hence, the philosophy of organism represents an integrated whole without eliminating the antique idea of "atoms" as the smallest parts. In the context of "philosophy of organism" the atoms cannot be studied or understood without referring to the integrated context in which they are parts. In other words, no entity can be considered in abstraction from the universe, and no entity can be divested of its own individuality. The traditional logic overstressed the notion of individual character. The notion of "any" frees us from individual character, but no entity is merely *any* (Whitehead, 1991, p. 678).

The final building blocks in the philosophy of organism are termed "actual entities". There are several different types of societies of entities integrated in space and time. Both the single entity and the societies of entities always find themselves in a self-created process of "becoming", making it reasonable to talk here about life processes. If we examine the "events" of natural or human history in detail, it soon becomes clear that a long and complicated process is involved. Each case is characterized by interconnected processes which "dissolve into a manifold of processes which themselves dissolve into further processes" (Rescher, 1996, p. 29). Whitehead argues that the world is composed of integrated living processes of becoming. In other words, "becoming" is more real than "being".

Whitehead argues that reality is made up of entities that are brief pulses of creative activity. "Reality is in each moment something dynamically and creatively alive, something constantly developing in reaction to what has been and to what might be" (Hosinski, 1993, p. 23). In other words, the world is dynamically alive and consists of integrated living entities with intrinsic value.

By their very nature processes are interrelated and interactive. We are in the world and the world is in us. "If we stress the role of the environment, this process is causation"; if we, on the other hand, "stress the role of *my* immediate pattern for active enjoyment, this process is self-creation" (Whitehead, 1977, p. 42).

Nothing in nature could be what it is, except as an integrated ingredient in nature's dynamic evolving web of life. According to "philosophy of organism", an isolated event is not an event because every part is a factor in a larger whole and has significance for the process of the whole. "The isolation of an entity in thought, when we think of it as a bare 'it' has no counterpart in any corresponding isolation in nature. Such isolation is merely part of the procedure of intellectual knowledge" (Whitehead, 2000, p. 142).

The motivation for introducing values and purposes into nature recognizes that materialism cannot give an adequate account of the interplay between the different actors in the market and the social and natural environmental context. Based on this line of reasoning, we argue for the need to rethink the status of life in nature and for an acceptance of integrated organisms possessing inherent value. Nothing in nature could be what it is, except as an integrated ingredient in nature as a dynamic whole. Hence, it seems reasonable to assert that Whitehead's ontology can be compared with Spinoza's "psychophysical parallelism", or Leibniz's "theory of monads" (Woolhouse, 1993). Within the organic concept of nature, "life" and "mind" are interwoven with matter and motion. The essence of life exists for its own sake as an intrinsic reaping of value. The point is that we can neither understand physical nature nor life itself unless we fuse them together as essential factors in the composition of the whole universe.

The static worldview explained change as a mathematical succession of different states. In the organic perspective, however, dynamics is explained as duration through interpretation of the past and present. In the philosophy of organism, dynamics is described and testified with reference to a mode of perception where present meets experience from the past and visions of the future.

Referring to Capra's (1982) interpretation of the organic worldview as living systems with a high degree of "nonlinear" interconnectedness, we conclude that the individual and the community both make each other and require each other at the same time. Thurow points in the same interpretative direction of the organic paradigm:

> Societies are not merely statistical aggregations of individuals engaged in voluntary exchange but something much more subtle and complicated. A group or community cannot be understood if the unit of analysis is the individual taken by himself. A society is clearly something greater than the sum of its parts. (Daly and Cobb Jr, 1994, p. 7)

This indicates that a conversion to the organic worldview has far-reaching consequences for both economic theory and behavior. A more complex and dynamic framework considers economic behavior as both multifaceted and context dependent. For example, it is obvious that contextual factors, including collective beliefs in ethical norms and prosocial orientation of the economic behavior, contribute to avoiding some of "the fallacies of misplaced concreteness" in mainstream economics.

Organic thinking is based on the concept of culture as a collective phenomenon, not as the sum of individuals. Within this complex and dynamic framework individual behavior is both multifaceted and context dependent. Hence, accepting the organic worldview has far-reaching consequences for the interpretation of the individual as an economic agent.

In accordance with the organic perspective, Schumacher argues that problems within economics must be handled as divergent because the precondition of lifeless nature is transcended. Without the characteristic human element of self-consciousness, one easily slips back into the one-dimensionality of order without freedom, thus reducing all problems to converging problems (Opdebeeck, 2008, p. 190). According to Opdebeeck (2008) divergent problems are resolved in the optimal manner by reaching economic practice on the macro-level from the point of departure of economic practice at micro-level.

Competition or cooperation

It is impossible to argue for or against competitive economics without making the metaphysical presuppositions explicit. As a conceptual context for the discussion we will give a brief presentation of Maslow's interpretation of Ruth Benedict's description of "low and high synergy societies". In low synergy societies institutions "encourage the development of jealousy, envy, resentment, distance, and finally a real likelihood of enmity" (Maslow, 1971, p. 196). According to Benedict, the reason for this is that "the structure provides for acts which are mutually opposed and counteractive (…) and the advantage of one individual becomes victory over another, and the majority who are not victorious must shift as they can" (Maslow, 1971, p. 194). High synergy societies are characterized by "generosity, mutual reciprocity relationships, co-operative techniques (…) and so on" (Maslow, 1971, p. 196). The reason for this nonaggressive behaviour is "not because people are unselfish and put social obligations above personal desires, but (because) social arrangements make these two identical" (Maslow, 1971, p. 194).

In our opinion competitive markets have much in common with institutions leading to low synergy solutions because they are based on the idea of conflicting interests among the actors. Economists use the paradigm of perfect competition as the exemplary model of the market. The language used is mechanic – dealing with small autonomous actors who cannot influence the market. Concepts such as "price mechanisms' and "market forces" amplify the close connection to mechanism. The agents do not relate to each other; they are only connected explicitly through price mechanisms. Since the image given is very abstract and depersonalized it is easy to lose contact with real humans in the real world.

According to Adam Smith, the logic of the competitive market implies that private vices will be turned to public virtues by means of the market's "invisible hand":

> Every individual necessarily labours to render the annual revenue of society as great as he can. He generally, indeed, neither intends to promote the public interest, nor knows how much he is promoting it.(…) [H]e intends only his own gain, and he is

> in this, as in many other cases, led by an invisible hand to promote an end which was no part of his intention (…). By pursuing his own interest he frequently promotes that of society more effectually than when he really intends to promote it. (Smith, 1976, pp. 477–478)

The message is clear. What counts is the output of the process – to make most out of what is – that is, efficiency – not "wasting productive resources". This means that the primary function of the competitive system is the efficient allocation of goods or the maximization of consumption. This all sounds very rational and well-thought-out. It also means that the role of the means used to reach the output is not of particular interest. It represents a brutal way of thinking, and does in fact contain many flaws.

Knight (1997) makes a fundamental critique of the individualistic western competitive economic system. He argues that a worldview where competition plays such an important role has obvious associations with mechanistic ideas. Knight goes further and argues that efficiency is not the most important aspect of an economic system; what matters are the kinds of people the system fosters. He criticizes mainstream economic theory for taking a reductionist (atomistic) approach and is able to foresee some of the societal effects of such an abstracted economic theory. A competitive system glorifying free trade and a strong affinity for simplistic solutions to complex problems is not suitable for solutions based on long-run responsibility.

Today Smith's invisible hand has global influence. According to OECD, all member countries rely fundamentally on the free market to organize economic activities. The reasoning here goes that competition stimulates innovation and efficiency in the use of resources, thereby leading to greater product diversification and lower prices. "Therefore, competitive product-markets are in the interest of all consumers" (OECD, 2010). The rationalism of competition is linked to economic efficiency, without any rationality claims linked to the ends of economic activity. The EU is based upon similar ideas, maintaining that business is a competitive game where independent companies "selling similar products or services compete with each other on, for example, price, quality, and service to attract costumers" (EU, 2010). The reduction of human beings to abstracted consumers exemplifies the argument that our western concept of rationality is not linked to economic ends (Allinson, 2004).

Competition between companies in national and international markets is one thing; competition within corporations might have even more severe social consequences. Internal competition was a part of life at Enron, the seventh-largest corporation in the US before it collapsed in 2002. To stimulate the competitive spirit among the employees they used PRC (Performance Review Committee) processes twice a year to assess bonuses and promotions (Fox, 2003). According to Fox, one ex-employee explained that the people you were competing against for bonuses were sitting next to you. It could get hairy, especially at the end of the year as bonuses were on everyone's minds. Enron practiced competition based upon a bell curve approach, determining who would stay and who would be fired. The employees ranked in the bottom 10–20 percent were fired. This clearly created tensions. The leader, Skilling, also encouraged competition between groups because he felt it would bring out the best in people and lead to the best solutions. In practice, "this competition often solidified a compartmentalization of the company into contending fiefdoms" (Fox, 2003, p. 83). This provides an illustrative example of low synergy society institutions "which set us against each other, making us into rivals necessarily, which put us into a situation where we must scrap for a limited amount of goods" (Maslow, 1971, p. 199).

According to Callahan (2004) the Enron scandal initiated a debate on "the cheating culture" in business. The thesis was that a society characterized by inequality and a winner-take-all philosophy produces the cheating that has been noted in many different spheres of life: academia, medicine, journalism, as well as business. The cultural climate favors power and money over and above personal integrity. Multiple examples are given of incentive-driven structures, such as stock options and production-based pay that do not promote productivity and "fair play", but reward deception. Cheating has created a dynamic between a class of winners, so influential that they in reality are beyond most rules, and an anxious class, members of which often cheat when they are confronted with serious problems like downsizing.

Large scale cheating is most prevalent among the the Winner class, in spite of high salaries, due to constant comparison with those who have amassed more. The cheating culture is explained as a combination of the individualistic "Me" generation and the free trade principles espoused by the neoliberals. Callahan argues that, after the deregulations in 1980, the new "Market Era" created new pressures in a very competitive economy "[W]here success and job security can't be taken for granted, it's increasingly tempting to leave your ethics at home every morning" (Callahan, 2004, p. 20). The temptations to cheat grow – given the morality of the free market. "If competition is good – if even greed is good – then maybe questionable cut-throat behaviour is also good" (Callahan, 2004, p. 22).

In high synergy societies the actors cooperate in order to find solutions where *all* relevant information and values are considered. Social institutions are arranged in such a fashion that "the people within the organization are coordinated with each other and (...) made into colleagues and teammates rather than into rivals" (Maslow, 1971, p. 199). In an economic context, the participants are typically defined as stakeholders. Freeman defines the organization's stakeholders as "any group or individual who can affect or is affected by the achievement of the organization's objectives" (Freeman, 1984, p. 46). In practice, different stakeholders have their own values and aims, and the organization has to interact with the stakeholders for mutual benefit. Values are not reduced to preferences (weak evaluations), since the stakeholders' fundamental values (strong evaluations) are also taken into consideration (Taylor, 1985). In cooperative decisions the ideal is to establish a platform of consensus; that is, to find solutions all stakeholders can agree upon. This way of thinking is different from voting, in which a simple majority wins all power. In cooperation, all the stakeholders share a common starting attitude saying, "You may be right, and I may be wrong" (Popper, 1983; Habermas, 1990).

It is reasonable to argue that the principle of competition is insufficient to establish solutions based upon a long-term social and environmental perspective. Welford emphasizes that "productive cooperation (...) always (will) be superior to blind competition and recognizing cooperative opportunities are part of recognizing inter-connectedness" (Welford, 2000, p. 141). Hence, Welford's argument is based on the presupposition that the market cannot be defined as an aggregate of autonomous actors. Instead, the market must be considered an integrated whole. Korhonen offers support to this line of reasoning, stressing that "[c]ompetition is (...) a barrier to the efforts of increasing stakeholder cooperation and cooperation between the firms and its suppliers or the local community actors" (Korhonen, 2002, p. 70).

This indicates a shift of focus from merely means to ends. Through cooperative processes the members might agree on the priority of different ends as well as on the use of available means. Cooperation based on dynamic dialogue allows more integrated solutions than the mechanisms of an atomistic and competitive economy allow. Equality and mutuality among the involved actors are necessary conditions for constructive cooperation. When competition is replaced by cooperation as the main principle for interaction in the market, the development of solutions based upon the common good will gradually takes place. Cooperation presupposes that the partners disclose relevant and valid information without strategic action (Habermas, 1982, pp. 263–271).

Max Havelaar – Fairtrade is one example of how to implement organic high synergy principles in an economic setting. In the following paragraph we will use the international coffee trade as an illustrative example in which we find cooperative market behaviour. Our main focus in the following are the processes of producing, storing, distributing, roasting, drinking, and recirculating the coffee, coordinated through principles based on Max Havelaar – Fairtrade. (See Moore (2004) and, in particular, Nicholls and Opal (2005) for an overview of the fair trade movement.)

The customer-value hierarchy applied for Fairtrade coffee is given the following content. At the first level, as a customer we want a core benefit, and in drinking Fairtrade coffee we see caffeine as an example of a core benefit. For many coffee drinkers, this is the main function of coffee, a drink that stimulates the nervous system. At the next level, the basic "product" might be coffee produced by farmers living in a specified community, who have received a fair price for their work and enjoyed decent working conditions in order to create first-class coffee beans. So the basic "product" is much more than a product or beverage with a certain color.

The third level, the expected "product" might consist of the process of tasting the coffee. The large coffeehouses such as Kraft, Nestle, Procter & Gamble, and Sara Lee all use huge and creative advertising in order to promote their own brands as the preferred coffee. The brand Max Havelaar Fairtrade tries to mobilize customers on the basis of solidarity with poor and vulnerable workers, their families, and communities. The fourth product level is the augmented "product", which could emphasize the ecological production, focusing upon how farmers care for their soil and harvest using natural fertilizers in order to produce crops that are healthy for all living organisms in the soil, and for human beings who will end up drinking the coffee. The mode of transportation and the distance from the place the beans are harvested to the final consumer should be specified. How is this transportation organized? Which kinds of externalities are involved?

The fifth level is the potential "product". In the Fairtrade case we might formulate this as high synergy society. We might think of this level as an image of organic cooperation between producers and consumers, seeing one another as connected and interwoven in a large web of life.

Implications

Table 1 "Metaphysical assumptions" sums up some characteristic elements of the underlying assumptions in the two different metaphysical presuppositions. They have important implications for the actors, their perceptions, and their relationships. The enlightenment was based upon the rise of "the new mechanical philosophy" but the rise of this philosophy was a movement into rationalism and empiricism. The result is that nature and experiences were seen in a very limited way. The abstractions that physicists or economists found adequate for their purposes resulted in the fallacy that the materialistic view of the ultimate units of nature is fully described.

Ims and Jacobsen (2006) analyzed some consequences of the mechanic and the organic worldview in the market (see pp. 49–74), assuming that a mechanical worldview is connected to competition, while an organic worldview is connected to cooperation. While mainstream economics presupposes that competition among actors will lead to the most efficient distribution of natural and human resources, this argument is

Table 1. Metaphysical assumptions

Mechanical Worldview	Organic Worldview
Atoms	Relations
Deterministic	Creativity
Substance	Process
Individualism	Collectivism
Objectivism	Subjectivism

not valid. Within a mechanical worldview many actors might act within narrowly defined self-interests, and will be inclined to exploit the other actors in the market (De Witt and Meyer 1998). In Ims & Jakobsen, 2006, Table 2, p. 57 (in this book) we define different types of collusion as unethical activities because it gives unfairly advantages to certain actors. Collusion is cheating and in business terms it will not lead to resource efficiency and is illegal. This kind of deceitful "cooperation" has the purpose of increasing the competitive power of the involved actors in order to obtain private gains to the detriment of other actors' interests and is a zero-sum game.

This view is supported by assuming that the actors in the market act like "atoms" and do not care about the relations to the other actors. To the contrary, when we adopt a relational ontology, a completely different rationality occurs. The self-interests doctrine often supported by Adam Smith's invisible hand argument, is not rational within an organic worldview. We agree with Jim March who states that the most important task in the future is to undermine the myth of the importance of self-interests in our economy and society. Self-interested actors have incentives to overexploit the given resources and are destructive for the earth in the long-term. The belief that individual pursuits of vices will lead to collective virtues is a destructive myth and unfit within an organic worldview.

The difference between the mechanic- and organic worldview has also implications for how we define success in business and in our lives. It should not be Gross national product but flourishing and wellbeing in the Aristotelian sense of Eudamonia. Happiness is a lifelong project.

The organic worldview opens for the existential-spiritual dimension as a strong force in business and society. Ian Mitroff's (1998) four-dimensional frame of reference is fruitful to gain new insight into business and economics. Building market institutions based upon cooperation will lead to a high synergy society. According to Maslow the wealth in high synergy societies tends to "spread around, it gets siphoned off from the high places down to the low places. It tends one way or another, to go from rich to poor, rather than from poor to rich" (Maslow, 1971, p. 195). In other words, the high synergy society has much in common with the principles of Max Havelaar's Fairtrade.

Based on this line of reasoning we argue that the organic worldview gives a better and more accurate description of the metaphysical conditions of economy. Hence, market behaviour based on competition will often lead to disintegration and egocentric behaviour. The use of one-sided power is destructive. From these arguments we draw the conclusion that cooperation is a better guiding principle for the economy, if we want the result to be fair trade in a high synergy society.

Conclusion

In this chapter we have emphasized that economic theory and practice cannot be isolated from its metaphysical preconditions. In accordance with Maslow's low synergy society, Knight was right when he argued that the competitive order was partly "responsible for making emulation and rivalry the outstanding quality of the character of the western people – who have adopted and developed it" (Knight, 1997, p. 39). Within the context of organic high synergy societies, it is important to look for the value of participation in economic activities as a sphere of self-expression and creative achievement. In this perspective, jealousy, envy, cheating, and disintegration are symptoms of a mismatch between worldview and principles for organizing the economy.

We are convinced it is high time to challenge the mainstream way of thinking in economics. As a number of distinguished philosophers and economists have claimed, the economic system and its institutions are not value neutral. Thus, we should be much more aware of what kind of people the economic systems foster; that is, the systems' effects on the character of the people in it. Callahan (2004) hits the nail on the head when he argues that the Chicago School economists in the 1970s, and policymakers in the USA, have over-valued competition, consumption, and deregulation. Economic man is a fiction, and treating the individual as a datum is a fallacy, because individuals learn and evolve. The economic system should not be reduced to a purely mechanical system regulated through market forces and price mechanisms. Instead we should focus upon the importance of building institutions based upon cooperation using an organic worldview.

We strongly believe that the social and ecological dimensions should have a higher value than material consumption in the western world – a world that creates a massive overuse of resources with very grave consequences for the future of our planet. The Living Planet report for 2008 provides evidence that western economies have a huge ecological overshoot. USA, which to the largest extent has been associated with the strongest competitive economic institutions, uses 600 percent more than its fair ecological share. Thus, there is an urgent need to scrutinize the fundamental assumptions that are often taken for granted in our present-day economic system.

Bibliography

Allinson, R. B. (2004): "Circles Within a Circle: The Condition for the Possibility of Ethical Business Institutions Within a Market System", *Journal of Business Ethics*, vol. 53, pp. 17–28.

Callahan, D. (2004): *The Cheating Culture: Why More Americans Are Doing Wrong to Get Ahead.* Boston, MA; New York, NY: Harcourt Trade Publishers.

Capra, E. (1982): *The Turning Point – Science, Society and the Rising Culture.* New York, NY: Flamingo.

Capra, E. (1997): *The Web of Life – A New Synthesis of Mind and Matter.* New York, NY: Flamingo.

Daly, H. E. and J. Cobb, Jr (1994): *For the Common Good – Redirecting the Economy toward Community, the Environment, and a Sustainable Future.* Boston, MA: Beacon Press.

De Witt, B. and R. Meyer (1998): *Strategy, Process, Content, Context. An International Perspective.* London, UK: International Thomson Business Press.

EU (2010): <http://ec.europa.eu.int/comm/competition/publications/consumer>.

Fowler, D. R. (1976): "Alfred North Whitehead," *Zygon*, vol. 11, no. 1, pp. 50–68.

Fox, L. (2003): *Enron: The Rise and Fall.* New York, NY: John Wiley and Sons, Inc.

Freeman, E. (1984): *Strategic Management – Stakeholder Approach.* Boston, MA: Pittman.

Friedman, M. (1970): "The Social Responsibility of Business Is to Increase Its Profits," *New York Times Magazine,* September 13.

Georgescu-Roegen, N. (1971): *The Entropy Law and the Economic Process.* Cambridge, MA: Harvard University Press.

Habermas, J. (1982): "A Reply to My Critics", in: Thompson, J. and D. Held (eds): *J. Habermas, Critical Debates.* Cambridge, MA: Macmillan Press.

Habermas, J. (1990): *Moral Consciousness and Communicative Action.* Boston, MA: The MIT Press.

Hosinsky, T. E. (1993): *Stubborn Fact and Creative Advance – An Introduction to the Metaphysics of Alfred North Whitehead.* Boston, MA: Rowman and Littlefield Publishers Inc.

Ims, K. J. and O. D. Jakobsen (2006): "Cooperation and Competition in the Context of Organic and Mechanic Worldviews – A Theoretical and Case Based Discussion", *Journal of Business Ethics*, vol. 66, pp. 19–32.

Ingebrigtsen, S. and O. Jakobsen (2009): "Moral Development of the Economic Actor", *Journal of Ecological Economics*, vol. 68, pp. 2777–2784.

Jones, W. T. (1969): *A History of Western Philosophy – Hobbes to Hume.* New York, NY: Harcourt Brace Jovanovich publ.

Knight, F. H. (1997): *The Ethics of Competition.* With a new introduction by Richard Boyd. New Brunswick, NJ: London Transaction Publishers.

Korhonen, J. (2002): "The Dominant Economics Paradigm and Corporate Social Responsibility", *Corporate Social Responsibility and Environmental Management*, vol. 9, pp. 67–80.

Maslow, A. H. (1971): *The Farther Reaches of Human Nature.* New York, NY: Penguin Compass.

Moore, G. (2004): "The Fair Trade Movement: Parameters, Issues and Future Research", *Journal of Business Ethics,* 53, pp. 73–86.

Nicholls, A. and C. Opal (2005): *Fair Trade. Market-Driven Ethical Consumption.* London: Sage Publications.

Norton, B. (1991): "Ecological Health and Sustainable Resource Management", in: Costanza, R. (ed.): *Ecological Economics: The Science and Management of Sustainability.* New York, NY: Colombia University Press.

OECD (2010): <http://www.oecd.org/about/>.

Opdebeeck, H. (2008): "The Urgency of a Frugality-based Economics", in: Bouckaert, L., H. Opdebeeck, and L. Zsolnai (eds): *Frugality – Rebalancing Material and Spiritual Values in Economic Life.* Oxford, UK: Peter Lang AG.

Ormerod, P. (1994/2001): "The Death of Economics, Faber and Faber, London", in Pearce, J. (ed.): *Small Is Still Beautiful.* London, UK: HarperCollins Publishers.

Popper, K. (1983): *The Logic of Scientific Discovery.* London, UK: Hutchinson.

Resche, N. (1996): *Process Metaphysics – An Introduction to Process Philosophy.* New York, NY: State University of New York Press.

Russel, B. (1979): *A History of Western Philosophy.* London, UK: Unwin Paperbacks.

Schumacher, E. F. (1977): *A Guide for the Perplexed.* New York, NY: Perennial Library.

Smith, A. (1976): "An Inquiry into the Nature and Causes of the Wealth of Nations", in: Cannan, E. (ed.): *Two Volumes in One* (first edition 1776, the citations are from the 5th edn). Chicago, IL: The University of Chicago Press.

Taylor, C. (1985): *Human Agency and Language, Philosophical Papers 1.* Cambridge, UK: Cambridge University Press.

Welford, R. (2000): *Corporate Environmental Management-Towards Sustainable Management.* London, UK: Earthscan.

Whitehead, A. N. (1967a): *Science and the Modern World.* New York, NY: The Free Press.

—— (1967b): *Adventures of Ideas.* New York, NY: Macmillan.

—— (1977): *Nature and Life.* New York, NY: Greenwood Press.

—— (1985): *Process and Reality.* New York, NY: The Free Press.

—— (1991): "Mathematics and the Good", in: Schilpp, P. A. (ed.): *The Philosophy of Alfred North Whitehead.* Chicago, IL: Open Court Publishing Company, pp. 666–681.

Woolhouse, R. (1993): *The Concept of Substance in Seventeenth Century Metaphysics.* London, UK: Routledge.

Xie, W., Z. Wang, and G. E. Derfer (2005): *Whitehead and China – Relevance and Relationships.* Frankfurt, Germany: Ontos Verlag.

KNUT J. IMS AND LASZLO ZSOLNAI

Self-Realization in Business: Ibsen's *Peer Gynt*

Introduction

This paper focuses on Henrik Ibsen's dramatic poem *Peer Gynt* in order to discuss what self-realization means in business. Does it mean to use one's faculties to gain power, prestige and money or to realize one's faculties in a virtue ethics sense, performing excellent actions? We will take excerpts from the poem to illustrate some of its implications for modern-day business leadership.[1] We believe that Ibsen's poem gives interesting clues that illuminate important aspects of human condition in general and business in particular.[2]

The poem ends when Peer is a ruined businessman, reflecting on his life, and discovering that his life-journey was without purpose and meaning. Finally he begins to pose existential questions like: What does it mean to be oneself? Who have I been in all my life?

We will explore Ibsen's concept of the self, and try to interpret some of his most enigmatic sentences like, "To be oneself, the self must die." The sentence has often been interpreted as a Christian life-norm. But comparing this concept of the self with Buddhism, we also find support for the denying of one's self. The problem for Peer is that he has never been himself in the true sense. Throughout his life he has lived out his "lower" self. He has been a "troll"; that is, lived in a greedy, selfish way. We will indicate the implications of Peer's search to find his true and whole self. One of the aims of our text is to stimulate business leaders to raise questions about their own identity and purpose.

Plot summary

Peer Gynt as a literary figure appeared in 1867, clearly inspired by, but not copied from, Norwegian folk tales. The narrative is rich and complicated. As a character Peer is ambiguous, neither completely good nor bad, which implies that readers in general recognize some familiar features in him. In order to understand the character more deeply, we might take a glance at *Brand* (1866), the dramatic poem that Ibsen wrote immediately before *Peer Gynt*. The main character in *Brand* is the priest Brand – a visionary thinker and an uncompromising leader with a powerful will. Brand is strictly centered upon his life-project, the salvation of the people living in a little isolated village surrounded by a very harsh nature. The pastor Brand is called by God – a strong Jehova – not the merciful Christ from the New Testament. Thus, in the name of Jehova, Brand requires iron-hard demands and sacrifices from the people, but far the strongest demands to himself! Brand demands wholeness, a genuine will to choose and sacrifice, and demonstrates his will by even sacrificing his only child's life – because his call requires it. Brand's motto is "*All or nothing*". No compromise! Everything is given up for his principles.

Ibsen wrote that Brand represented his (= Ibsen's) own character in his best moments, while Peer Gynt depicts Ibsen in his worst moments. One common interpretation is that Brand and Peer Gynt are characters that are either complementary or polar opposites. While Brand is the robust ascetic and ethicist with an uncompromising will, Peer Gynt is characterized by fantasy, opportunism, and vitality. As Bloom (1994) writes, the ambiguousness of Peer is his charm and his vitality also makes him into a "warmth bringer". In contrast to Brand, Peer is living a pure aesthetic life without making any commitment and sacrifice. Both figures are generally interpreted within an idealistic frame of reference where the purpose of their life is to realize their essence. In this sense *the demand is to be or become oneself*. However, Peer's tragedy is that he does not become himself, but only "enough" of himself, which means he is living out his "lower" self. Clearly Ibsen has some sympathy for Peer – the dreamer, the liar, the charmer, the storyteller! At the end of Act five, Peer is given a last chance to become himself – "his true and whole self".

The dramatic poem *Brand*, written in Italy, became Ibsen's artistic and commercial breakthrough. For the first time, 38 years old, Ibsen was acknowledged as one of the most important poets in Scandinavia, and received a life-long artist pension from the Norwegian Government. Ibsen rapidly changed his dress, and even his handwriting. The bohemian Ibsen became a man of the world. Ibsen wrote *Peer Gynt* in Ischia and Sorrento in Italy. Far away from Norway, Ibsen was probably more liberated than ever before, and his talents and imagination flourished.

The poem *Peer Gynt* also characterized as an odd comedy, and expressed to a large extent the inner life of the literary figure. The background changes from a little mountain valley where Peer grows up to Act four which opens on the coast of Morocco where the middle-aged Peer acts as a businessman without any scruples, proceeds into the Sahara Desert, and ends in a Cairo madhouse. As an old man Peer is shipwrecked on his way home to Norway, he survives, but is ruined. In the final Act (five) the supernatural pervades and Peer is confronted with his earlier life in an ugly way through different mystical characters, like the Dovre Master and the Button-Molder.

Our analysis will focus mainly on Act four – Peer Gynt as a corrupt business man. This act has often been ignored in the stage productions of the poem, partly because it breaks with the national romantic tradition, and consists of a number of new figures from different nationalities. Even the composer Edvard Grieg, who was asked by Ibsen to compose music to the play, was not able to do it for Act four due to the multicultural diversity of this act. We consider Peer Gynt as a prototype of a modern businessman who uses his time, energy, and talents in a global world with one overriding purpose – to make his life into an adventure, trying to exploit as many opportunities as possible in order to become rich and powerful. Finally, the quest for meaning becomes so strong that Peer wonders whether his drive for money and power was a worthy life project.[3]

The drama has some autobiographical ingredients (see for example, Meyer, 1987, the introduction p. 23). Peer Gynt was like Ibsen himself, the son of a businessman who went bankrupt. Peer's mother Åse, was not able to reconcile with such a downward socio-economic shift, and she dreamed that Peer would regain the family's wealth. When life was harsh for her

Peer was told fairy tales, and most fairy tales have a happy ending. In the Norwegian folk tales a poor young man, the Ask Ladd always wins the princess and the kingdom in the end. Peer developed his fantasy and created new fairy tales, and typically he became the hero in his own stories.

Early in the poem Peer happened to be at a wedding. Peer steals the bride and flees with her up in the mountains. Then Peer is declared an outlaw. He has to stay in the mountains where he meets a number of extraordinary characters: the Dovre Master, the troll mountain king, and her daughter. The Dovre Master gives Peer a choice: "If you marry my daughter, you will become a troll" and you will receive part of my kingdom. And the Dovre Master asks Peer: What is the difference between troll and man? The Dovre Masters own answer is: "*Man, be thyself*" and "*Troll, to thyself be – enough*." Even in Norwegian this "enough" is a very strange expression, which is often interpreted as a self-centered, egoistic attitude. Peer also meets a mystical creature, the Bøyg – the "Great in Between". In Norwegian folklore the Bøyg is an invisible, gigantic creature in the shape of a large snake, and most interpreters regard the Bøyg to be the embodiment of the spirit of compromise. When Peer asks "who are you?", the Bøyg replies: "Myself". The Bøyg's motto "*Go around*" becomes Peers motto for the rest of his life. We may interpret "going around" as not having the courage to be true to oneself – favoring self-deception (for surprising interpretations of the Bøyg see Jahr Larsen, 2006).

In the wedding, Peer also meets Solveig, who has grown up in a Christian family.[4] She falls in love with Peer, even if he is transgressing the norms of the small community via the bride prey and being declared an outlaw. Solveig commits herself to Peer, but Peer seemingly cannot handle her love, and leaves her with the excuse: "I've something heavy to carry." "I must bear it alone … Wait for me", whereas Solveig answers "Yes, I'll wait." Peer leaves Solveig and sets out on his life journey which lasts for many decades.

In Act four Peer happens to enter into a local madhouse in Egypt. Peer understands that all the patients live in their own worlds, and that they in one sense are themselves to the extent that no one cares for anybody. As a young man Peer had a vision of being an emperor. In the madhouse he finally becomes – *the emperor of the "self", a perverted and closed little self.*

As an old man Peer returns to the valley where he was born, and Peer is again confronted by the Dovre Master. Peer also encounters a new figure, the Button-Molder, who demands that Peer's soul must be melted down unless he can explain when and where in life he has been "himself". This is the strongest existential confrontation of Peer in the poem. But Peer claims that he has been himself all through his life. The Dovre Master however, states that Peer has been a troll, not a man, most of his life.

Peer finally reunites with Solveig – now an old woman. She has been waiting for Peer all her life, and expresses a deep love for Peer. She excuses his "sins": "You have not sinned at all. You have made my life into a song." And Peer asks her in a humble way:

> Where was my self, my whole self, my true self?

Solveig answers:

> In my faith, in my hope, in my love.

The play ends by Peer hiding himself in Solveig's lap, naming her his mother, while Solveig sings for Peer. As in many of Ibsen's dramas, the reader is left with an open ending. We do not know how the story finally turns out.

Is Peer Gynt a Liar?

In the beginning of the play we meet Peer Gynt – as a 20-year-old man and his mother Aase.

> AASE: Peer, you're a liar!
> PEER: No, I'm not!
> AASE: Then swear it's true - look in my eyes!
> PEER: Why should I swear?
> AASE: Just what I thought!

Nothing but lies - hot air and lies!
PEER: I tell you - every blessed word is true!

This is the opening of Ibsen's drama, and in the very beginning Ibsen strikes the tone. Peer Gynt lies – but he denies it. Is Peer Gynt a victim of self-deception because his distinctions between facts and fantasies are blurred? Or is Peer a conscious, cynical, and a notorious liar about what he has done, and about himself?

Later in the drama Ibsen digs deeper into Peer Gynt's self-concept. As an old man, Peer Gynt starts to reflect on his life. Has he been true to himself and realized his highest potential? Has he been a man who has accomplished something for the common good, or has he been an ego tripper who has always followed his lower instincts in a relentless drive for money? Central in our analysis will be the Button-Molder, who is melting those souls that did not achieve what the Master intended with their life.

The Gyntian self

The self of Peer Gynt is an all desiring and never unsatiated ego.

PEER: The Gyntish Self-it is the host
of wishes, appetites, desires,-
the Gyntish Self, it is the sea
of fancies, exigencies, claims,
all that, in short, makes my breast heave,
and whereby I, as I, exist.
But as our Lord requires the clay
to constitute him God o' the world,
so I, too, stand in need of gold,
if I as Emperor would figure. (Act 4, scene 1)

Peer Gynt's dream and life-project is to be Emperor as he confesses it to his travelling companions (or fellow travellers).

MR. COTTON: Beyond all doubt, you have a goal;
and that is-?
PEER: To be Emperor.
ALL FOUR: What?
PEER [NODDING]: Emperor!
THE FOUR: Where?
PEER: O'er all the world.
MONSIEUR BALLON: But how, friend-?
That plan is not at all a new one;
PEER: By the might of gold!
it's been the soul of my career.
There is a text, or else a saying,
somewhere, I don't remember where,
that if you gained the whole wide world,
but lost yourself, your gain were but
a garland on a cloven skull.
That is the text-or something like it;
and that remark is sober truth.
VON EBERKOPF: But what then is the Gyntish Self?
PEER: The world behind my forehead's arch,
in force of which I'm no one else
than I, no more than God's the Devil.

The Gyntish self refers to the pursuit of money, material possessions, and power. Aiming at realizing his self, Peer Gynt becomes a global entrepreneur engaged in dirty businesses. He explains his trade activities to his travel companions in a rather cynical way.

MR. COTTON: What did you trade in?
PEER: I did most
in Negro slaves for Carolina,
and idol-images for China.
MONSIEUR BALLON: Fi done!
TRUMPETERSTRALE: The devil, Uncle Gynt!
PEER: You think, no doubt, the business hovered
on the outer verge of the allowable?
Myself I felt the same thing keenly.
It struck me even as odious.
But, trust me, when you've once begun,
it's hard to break away again.
At any rate it's no light thing,
in such a vast trade-enterprise,
that keeps whole thousands in employ,
to break off wholly, once for all.
That "once for all" I can't abide,
(…)
What could I do? To stop the trade
with China was impossible.
A plan I hit on-opened straightway
a new trade with the self-same land.
I shipped off idols every spring,
each autumn sent forth missionaries,
supplying them with all they needed,
as stockings, Bibles, rum and rice.
MR. COTTON: Yes, at a profit?
PEER: Why, of course.
It prospered. Dauntlessly they toiled.

For every idol that was sold
they got a coolie well baptized,
so that the effect was neutralized.
The mission-field lay never fallow,
for still the idol-propaganda
the missionaries held in check.

MR. COTTON: Well, but the African commodities?

PEER: There, too, my ethics won the day.
I saw the traffic was a wrong one
for people of a certain age.
One may drop off before one dreams of it.
And then there were the thousand pitfalls
laid by the philanthropic camp;
besides, of course, the hostile cruisers
and the wind-and-weather risks.
All this together won the day.
I thought; Now, Peter, reef your sails;
see to it you amend your faults!

Finally, Peer Gynt sells his plantations, and on the parting day he serves the workers (slaves?) gratis grogg (alcohol) so "men and women all got drunk". Then he boasts that he believes that my "misdeeds" are "balanced by my virtues". It is clear that Peer Gynt was a relentless egoist in doing business. He employed various *moral disengagement strategies.* These strategies are the psychosocial maneuvers by which moral self-sanctions become disengaged, giving free way to a variety of misbehaviours without carrying any moral concern. Self-sanctions can be disengaged by euphemistic labelling, reconstructing the conduct, obscuring personal causal agency, misrepresenting or disregarding the injurious consequences of one's actions, and vilifying the recipients of maltreatment by blaming and devaluating them (Bandura et al., 2000).

Peer Gynt's case clearly illustrates the finding of contemporary moral psychology that the more a person is concerned with self-enhancement goals, the more he is inclined to resort to mechanisms that permit him to justify transgressions when his self-interest is at stake (Caprara and Capanna, 2006).

What is self-realization?

What does it mean to realize one's self? A popular notion is that the business person who has become wealthy and got much power and prestige has been realizing himself or herself. He or she is the modern hero in our society, admired for personal achievements. A different notion is to be found in the artist who, through all his or her life, has been faithful to the calling of artist. The artist is working not just for the money, but to create something he or she believes is of vital importance. To realize oneself, the artist might sacrifice affluence and follow personal intuition that tells him or her that one should do something noble.

There exists a classical notion of self-realization. It has a long-term perspective, as it points on the person who has obtained the good life (Gr. = eudaimonia). In ancient Greece good life presupposes realizing one's virtues, which means developing one's best abilities – thereby flourishing as a human being within a community, a polis.

In psychological textbooks self-realization is conceptualized as a hierarchical model in which it is important to satisfy several deep physical needs to climb up the steps of the hierarchy and finally be free, autonomous, and realize one's full potential. At the top of the hierarchy the self-realized person enjoys "peak experiences" – extraordinary self-transcendent movements that feel qualitatively different from ordinary life experiences (Maslow, 1964).

A refinement of these steps often shows that at the lowest level of the pyramid it is important to satisfy needs as safety, health, food, and shelter, while the next level concerns personal attachment to other people, and the top level the possibilities to achieve. Shakespeare writes that "Joy's soul lies in the doing", and the logic is that we get more pleasure from making progress toward our goals than we do from achieving them (Haidt, 2006, p. 221).

Confronted with the Button-Molder Peer learns that he has never been himself.

THE BUTTON-MOLDER: Yourself you never have been at all;
then what does it matter, your dying right out?
PEER: Have I not been-? I could almost laugh!
Peer Gynt, then, has been something else, I suppose!
No, Button-Molder, you judge in the dark.
If you could but look into my very reins,
you'd find only Peer there, and Peer all through,-
nothing else in the world, no, nor anything more.

Peer Gynt asks the Button-Molder what is it "being oneself".

THE BUTTON-MOLDER: To be oneself, Peer Gynt, the self must die.
But on you that answer is doubtless lost;
and therefore we'll say: to stand forth everywhere
with Master's intention displayed like a signboard.
PEER: But suppose a man never has come to know
what Master meant with him?
THE BUTTON-MOLDER: He must divine it. (Act 5, scene 9)

The usual interpretation of this is that the Button-Molder is representing the Christian life-norm and teaches Peer Gynt what it is to be oneself in a Christian sense. We find a direct link here to the Bible: "If anyone wants to come with me, he must forget himself. (…) For whoever wants to save his own life will lose it, but whoever loses his life for my sake will find it. For what will it profit them if they gain the whole world but forfeit their life?" (Matthew 16, 24–25)

In the Christian tradition we find a key to interpret this kind of self-sacrifice. There is a lot of evidence that Ibsen read the Bible. He admitted that *the Bible was the only book he read while writing Peer Gynt*[5] and Peer Gynt's forerunner, *Brand.* In the play of *Peer Gynt*, the church bells are used as a signal of Peer's awakening, and at the end of the play Peer Gynt appears at the time of Pentecost.

We find several persons in the play of *Peer Gynt* who were real and true selves by making a sacrifice. Solveig offered most of her life by waiting for Peer. A young farmer cut his own finger to avoid joining the army in order to take care of his family. This is beautifully expressed in the priest's funeral:

> "But there's one thing that shines above the law,
> As truly as the bright tent of Glitretind
> He was a poor patriot. To state
> And Church, an unproductive tree. But there
> On the brow of the hill, within the narrow
> Circle of family, where his work was done.
> There he was great, because he was himself." (Act 5, Scene 3, OWC, p. 137)

In the Buddhist perspective the sentence "to be oneself, the self must die" gives/yields another meaning. Buddhists have a different view of self, the *no-self (annatta),* which means that there is no individuating core we can point to and say "Here I am!" *Annatta* specifies the absence of a supposedly permanent and unchanging self in any one of the psychophysical constituents of empirical existence. What is normally thought of as the "self" is an agglomeration of constantly changing physical and mental constituents, which gives rise to unhappiness if clung to as though this temporary assemblage represented permanence. The *annatta* doctrine attempts to encourage people to detach themselves from the misplaced clinging to what is mistakenly regarded as self. One implication is that a person is not a fixed and separate entity.

What does it mean? We are informed by Peter Pruzan (2004), writing about spiritually based leadership, that in the eastern tradition selflessness and non-attachment are important concepts. There is a notion of detached involvement. The idea is that work can be transformed to selfless service freeing us from personal desires and ambition, and implies an indifference to the outcomes. In this way work may give a sense of interconnectedness and imply a transcendence of the lower, ego dominated self. Pruzan writes that one's service is a sacrament of devotion to one's duty. This is clearly linked to the notion of the servant–leadership, which is the antithesis of the power-seeking manager who attempts to control others without sensitivity to their needs.

Pruzan stresses that self-realization is "the direct experience of the *Self* or the *atma*. It involves developing knowledge of one's true self – realizing the essence of one's being. The spark of the divine purpose to life." The paradox is clear. Self-realization can only be achieved if "the person whose ego has been tamed/ignored and who is truly selfless, does not seek rewards for his deeds. The selfless leader who is not attached to the fruits of his actions does not only achieve spiritual growth, peace of mind. and freedom from fear. He also becomes an exemplar for his employees and his surroundings ..." (Pruzan, 2004, p. 25).

Man versus Troll

DOVRE-MASTER: The humans say, "Man, be yourself",
But the Trolls say,
"Troll, to yourself be – enough". (Bloom 1994, p. 335)

For the trolls the Norwegian expression used by Ibsen is "Troll, vær deg selv - nok". One possible interpretation of "enough" is that you should be generous and kind to your individual self in order to serve your "lower" self. Literally the human versus the troll's mottos are "Be yourself" versus

"Be yourself – enough." Those expressions are cryptic and mysterious even in Norwegian. Some translators fear that the literal translation would be too murky for an English speaking audience. The penetrating word "enough" alarms you, but it must be in your coat of arms. The OWC (Fry and Fillinger, 1989, p. 42) gives this translation;

> "To thine own self be true."
> But here, in the world of trolls, we say
> "To thine own self be-all-sufficient!"

The king of the trolls claims that Peer was a troll in most of his life. But Peer vehemently denies this.

PEER:	I am not a troll -
	I had been in strong opposition to trolls all my life.
DOVRE-MASTER:	No that is not true.
	Our slogan was emblazed on your heart.
	And since then your whole life has not been
	Anything but a tribute to our doctrine.
	You've kept it secret, but you've been a troll –
	the phrase I taught you was the very pole
	You claimed yourself to make yourself a millionaire.
	But the troll credo was what you professed
PEER:	A mountain troll! Me?
	Self-obsessed?
DOVRE-MASTER:	You're the illustration of being
	Troll (act 5, scene 8).

In the last act of *Peer Gynt* Peer shows genuineness in responding to Solveig's waiting for him. Here Peer is capable of a kind of directness and honesty, and finally accepts his destiny.

But Peer was never able to find genuineness in business. After all, Peer could probably not realize himself in business. He took the easiest path, the path that made him an international magnate, a *self-made man* who looked upon business as an *amoral* activity. Peer enjoyed a kind of shallow success in business through his unscrupulous pursuit of self-interests. Because he never was truly and wholly himself in business, he ended up as a lonely man who finally acknowledged the sad truth about himself.

Peer Gynt's tragedy can be illuminated by the main teaching of Buddhism, which points out that even if one attains what he or she desires, greater desires always emerge. The ego mindset cannot be fulfilled and its greed for more satisfaction and recognition becomes the source of its own destruction (Ims and Zsolnai, 2006).

In business terms we may ask what it means to be a troll? After reading *Peer Gynt,* we suggest that trollish behaviour can be identified with the *uncritical pursuit* of *profit,* the "enrich yourself" mentality, which is so well-known in today's business world. The business person who is a servant of his or her lower self can destroy nature, people, and future generations. but at the same time will certainly *destroy* one's *own self.* Being truly and wholly human and realizing your own self requires *genuine selflessness,* that is, denying the instincts of your lower self and serving purposes higher than you.

Notes

1. In the text we use several translations of Ibsen's *Peer Gynt.* This is due to the fact that the authors, Ims and Zsolnai, coming from Norway and Hungary, respectively, have different tastes for different translations. Ims, who reads Norwegian, is not completely satisfied with any major translation. English translations we have used are the following; Ibsen, H. (1989) *Henrik Ibsen. Peer Gynt. A Dramatic Poem.* Translated by C. Fry and J. Fillinger with an Introduction by James Mcfarlane, Oxford World's Classics, Oxford, Oxford University Press. This translation is abbreviated OWC in the text. Another used translation is Ibsen. H. (1963) *Henrik Ibsen. Plays: Six. Peer Gynt, The Pretenders,* translated and introduced by Michael Meyer, Methuen World Classics, London, Methuen Drama. The third translation

we have used is Ibsen, H. (1992) *Henrik Ibsen Peer Gynt*, translated by G. Bamman and I. B. Berman, New York. Finally, we also used the following translation: Ibsen, H. (1956): *Peer Gynt,* translated by R. Farquharson Sharp. 1956. London, Dent.

2. As a playwright Ibsen is ranked as the most important after Shakespeare. "His plays are world classics, staged on every continent, and studied in classrooms everywhere." (Moi, 2006, p. 17). The literary critic Harold Bloom includes Ibsen in his *The Western Canon* (1994), and writes that *Peer Gynt* "is the center of Ibsen containing everything he had,[…]. The canonicity of *Peer Gynt* is to me one with its trollishness […] for the true quintessence of Ibsenism is the troll." "Trolls mattered most to Ibsen because it can be so difficult to tell them from people." (Bloom, 1994, p. 341).
3. Several interpretations of the character of Peer Gynt assert that all his life is a search for identity – a search to find his true and whole self during his egoistic journey. See for example, Jahr Larsen (2006).
4. Solveig's family is part of the "Haugianer movement" – a strong popular Christian movement that had Hans Nielsen Hauge as its inspirational leader.
5. *Peer Gynt* was written in the years 1866–1867 (published November 14, 1867). In *Brand,* published in 1866, we find more than 200 citations from the Bible, and in *Peer Gynt* more than 20. However, in *Peer Gynt,* all the citations are reformulated in a way that suits Peer's extravagant and careless behaviour.

Bibliography

Bandura, A., Caprara, G.-V. and Zsolnai, L.: 2000, "Corporate Transgression Through Moral Disengagement" in *Journal of Human Values*, no. 6, pp. 57–64.

Bloom, H: 1994, *The Western Canon. The Books and School of the Ages.* Riverhead Books, New York.

Caprara, G.-V. and Capanna, C.: 2006, "Moral Disengagement in the Exercise of Civicness" in L. Zsolnai (ed.), *Interdisciplinary Yearbook of Business Ethics.* vol. 1. Peter Lang Academic Publishers, Oxford, pp. 85–96.

Haidt, J.: 2006, *The Happiness Hypothesis. Putting Ancient Wisdom and Philosophy to the Test of Modern Science.* Arrow Books, London.

Ibsen, H.: 1966, *Peer Gynt.* Oslo, Gyldendal Norsk Forlag (Merknader ved Johan Hertzberg, Gyldendals skoleutgave (Comments by Johan Hertzberg, Gyldendal's edition for schools).

——1963, *Henrik Ibsen. Plays: Six. Peer Gynt, The Pretenders.* Translated and introduced by Michael Meyer, Methuen World Classics, London.
——1989, *Henrik Ibsen. Peer Gynt. A Dramatic Poem.* Translated by C. Fry and J. Fillinger with an Introduction by James McFarlane, Oxford World's Classics, Oxford University Press, Oxford.
——1992, *Henrik Ibsen Peer Gynt.* Translated by G. Bamman and I. B. Berman, TCG Translations, New York.
Ims, K. J. and Zsolnai, L.: 2006, "Shallow Success and Deep Failure" in L. Zsolnai and K. J. Ims (eds), *Business Within Limits. Deep Ecology and Buddhist Economics.* Peter Lang Academic Publishers, Oxford, pp. 3–24.
Jahr Larsen, T.: 2006, *Peer Gynt-Koden.* Abstract Forlag ("The Peer Gynt Code"), Oslo.
Maslow, A.: 1964, *Religions, Values and Peak-Experiences.* Ohio State University Press, Colombus.
Moi, T.: 2006, *Henrik Ibsen and the Birth of Modernism. Art, Theater, Philosophy.* Oxford University Press, New York.
Pruzan, P.: 2004, "Spirituality as the Context for Leadership" in L. Zsolnai (ed.), *Spirituality and Ethics in Management.* Kluwer Academic Publishers, Boston, Dordrecht, London. pp. 14–32.

KNUT J. IMS, LARS JACOB TYNES PEDERSEN, AND LASZLO ZSOLNAI

How Economic Incentives Destroy Social, Ecological, and Existential Values: The Case of Executive Compensation[1]

Introduction

This paper explores how relying on ever-increasing levels of executive compensation like wages, bonuses, and stock options may have an adverse effect on managerial performance. Specifically, we argue that one-dimensional economic incentives (i) may crowd out existential, social, and ecological values that influence the manager's commitment to ensure responsible business conduct (cf. Frey, 1997; Mitroff, 1998; Bouckaert, 2006) and (ii) have negative spillover effects that reduces the manager's performance (cf. Ims and Zsolnai, 2009).

Executive compensation has long been a prominent topic in the management literature. A main question that is also given substantial attention in the business ethics literature – even more so in the wake of the recent financial crisis – is whether the soaring levels of executive compensation can be justified from an ethical point of view (see for example Carr and Valınezhad, 1994; Walters et al., 1995; Nichols and Subramaniam, 2001; Rodgers and Gago, 2003; Chan, 2008). Also, the relationship of executive compensation to instances of unethical behaviour or outcomes has received considerable attention (see for example Dalton and Daily, 2001; McCall, 2004; Harris and Bromiley, 2007).

More generally, there is a substantial literature on the relationship between compensation or remuneration and motivation and, ultimately, performance. In the psychology literature Deci and Ryan (2000) have illuminated the relationship between intrinsic and extrinsic motivation; that is, motivation that is inherent in the task itself and motivation that relies on external rewards or sanctions. In the economic literature the crowding-out effect, that is instances where external rewards in fact *reduce* the motivation of the actor, have been explored both conceptually and through a number of experimental studies (see for example Frey, 1997).

Our contribution is positioned in the extension of these findings. While we build on the distinction between intrinsic and extrinsic motivation and base our discussion largely on the crowding-out effect, our conception of performance differs from the traditional view. While the existing literature on compensation and performance assumes a one-dimensional conception of performance, that is performance implies efficient economic goal attainment, we broaden the conception of performance by viewing it in multiple dimensions. Thus, we will explore the relationship between compensation and performance, and the emergence of the crowding-out effect, with respect to performance where different types of values are at stake. We aim to illuminate how compensation may crowd-out intrinsic motivation, but also crowd out existential, social, and ecological values that influence the manager's commitment to responsible business conduct. As such, we will argue that excessive compensation schemes may in effect be detrimental to the inclinations of managers and their organizations to act responsibly. The crowding-out effect is often referred to as "the hidden cost of reward" (see Deci, 1976). Our paper further explores what we might call the hidden existential, social, and ecological costs of one-dimensionality.

The economic conception of incentives and performance

Firstly, we outline the conception of incentives and performance that follow from agency theory, and the criticism directed at the model. Secondly, we explore conceptions of motivation inherent in economic theories about performance, and present theories that challenge this paradigm. Thirdly, we discuss the crowding-out and crowding-in effects.

Economic assumptions about incentives and performance

What is the relationship between the willingness to make an effort and productivity? In principal-agent models, we assume that the agent has to be compensated in order to act as desired by the principal (see for example Eisenhardt, 1989). It is well documented that external rewards influence performance (see for example Luthans, 1985; Lawler, 1990). Furthermore, linking the compensation scheme to the share price makes the compensation scheme understandable and "logical" in the sense that payment follows performance. It is easy to administer, and objective criteria makes the compensation scheme transparent and to a lesser degree marked by misunderstandings and conflict. As such, the fundamental insights from agency theory underpin the use of variable monetary compensation as incentives for performance.

Moreover, in western culture *norms* of *rationality* are strong, which means that behaviour is expected to be utility-oriented and instrumental (see for example Elster, 1986). This implies an expectation that the individual collects information about different alternatives and finally decides on the alternative that maximizes his or her goal function. The norm of rationality also implies that people experience a need for being perceived as being rational. The expectation of human rationality, and specifically economic rationality, is an important assumption that justifies the use of external rewards as motivational sources.

However, the principal-agent model has been met with fundamental criticism in recent years. For one thing, human beings have *bounded rationality* (March and Simon, 1958). This makes the incentive mechanism unpredictable, which may undermine its effectiveness. Another problem is that human behaviour is social and has to be understood in interaction with other people. As a consequence, it is not easy to predict or control behaviour (Weick and Bougon, 1986; Axelrod and Cohen, 2000). In relation to this, it is argued that since performance is a collective phenomenon, performance should be compensated collectively, for example through collective bonuses that do not distinguish between the different members of a team or organization (see for example Bogsnes, 2009).

Moreover, it has been argued that the assumption of opportunism inherent in principal-agent models, as well as the assumption that opportunism can be managed by monitoring and incentives, in fact *creates* more opportunism. In an attempt to tamper opportunism, organizations design elaborate monitoring systems, thereby implicitly communicating to employees that they are not trusted. The net effect is a self-fulfilling prophecy – the creation of the types of behaviour the system was intended to counteract (Ferraro et al., 2005; see also Ghoshal, 2005; Ghoshal and Moran, 1996).

The various criticisms directed at traditional agency theory justify an exploration of performance that goes beyond the purely rational and instrumental understanding of human behaviour. In the following section, we explore theories of motivation that attempt to account for this.

Behaviour and instrumentality

The principal's problem is to get the agent to perform at his or her maximum level. However, we cannot assume that an agent does this, in particular over time. What makes human beings perform over time is a key issue in motivation theory. There are a number of theories on worker motivation, and they differ in their assumptions about what triggers human motivation. One stream of theory conceives of motivated behaviour as the consequence of unsatisfied needs (for example Maslow, 1943; McClelland, 1987). Another is based upon the notion that human beings work towards a goal, and that this motivates some agents more than others (Locke and Latham, 1990).

Haukedal (2008: 258) maintains that a problem with most motivation theories in the context of working life is that they largely, and often completely, presume instrumentality. An instrumental understanding of behaviour implies that performance is regarded as a means in order to achieve something else. For example, working overtime can be explained as a wish to earn more money or obtain other advantages. However, this is a too narrow conceptualization of *work performance,* which can be explained in other ways, for example as a desire to continue with an activity because of enjoyment or because one relishes the challenge. This is the case for instance when we are engaged in actions that have no purpose beyond themselves; that is so-called *autotelic* action (Csikszentmihalyi, 1975).

Haukedal (2008: 258) also emphasizes different problems with traditional motivational theories, which mainly build upon a cognitivist foundation, meaning that work is seen as the result of a decision making process. In such a view, habits and feelings are not relevant, even if such psychological variables play a central role in all other types of behaviour (Langer, 1978). This critique is akin to the approach to motivation proposed by Deci and Ryan (2000), whose self-determination theory (SDT) maintains that an understanding of human motivation requires a consideration of innate psychological needs for *competence, autonomy, and relatedness.* Thwarted satisfaction of these needs results "invariantly in negative functional consequences for mental health and often for ongoing persistence and performance" (Deci and Ryan, 2000: 262). Deci and Ryan further distinguish between instrumental (external) and non-instrumental (intrinsic) motivation. Intrinsically motivated activities are those activities that individuals find interesting and would do in the absence of operationally separable consequences (Deci and Ryan, 2000: 233), while extrinsically motivated activities require such consequences in order to motivate and reward the actor. Intrinsic motivation is a powerful determinant of performance, and its presence in employees is naturally valuable to organizations, since it produces performance without a need for external reward. Regardless, organizations tend to use external rewards and incentives also, in instances where it is likely that intrinsic motivation exists. As we will see in the following, this has destructive implications.

Motivation and crowding effects

Empirical evidence shows that there exists in fact an effect that countervails the standard relative price effect on behaviour (Frey, 1997). This effect can be explained on the basis of the distinction between intrinsic and extrinsic motivation. The standard assumption is that an increase in pay (that is external incentives) will increase the agent's effort. This may well be illustrated with the supply curve of work effort. However, empirical studies have demonstrated that when external incentives are introduced when intrinsic motivation is present, a so-called *crowding-out effect* comes into play. The result is that an increase in pay in fact reduces the agent's effort because the reward crowds out the actor's intrinsic motivation. The reason for this is that the actor's self-determination, self-esteem, and possibility of expressing altruism are all impaired (Frey, 1997: 16–17).

The crowding-out effect may be stronger than the first effect (the pure relative price effect), so that the net outcome of the different effects is that the supply curve shifts to a lower level than the initial effort. However, the crowding-out effect is not always strong, since it is conditioned by several identifiable factors, such as whether the external reward is seen as controlling or supportive (ibid.: 18).

Conversely, the *crowding-in effect* is a positive effect on intrinsic motivation of an institutional factor (Deci and Ryan, 2000). This effect is less investigated than the crowding-out effect, but some findings suggest that, for example, procedural fairness, avoiding the self-serving bias, instructions, framing of socially appropriate behaviour, personal contacts, and incomplete contracts can have an important impact on performance (see Frey and Osterloh, 2005). In the case of complex interdependencies between actors, it is important to take crowding-in effects into account.

In cases where intrinsic motivation exists, variable pay for performance may undermine the willingness to contribute voluntarily to the common good – given that it is perceived by the agents as an effort to control. According to Tyler and Blader (2000), monetary incentives explain only 10 percent of the variance in performance, while obligations based on intrinsic norms explain 30 percent. Bucklin and Dickinson (2001) find that the amount of incentive pay as a proportion of fixed pay can be relatively small, but nevertheless be effective. Hence, in compensation schemes, it is important to recognize the relevant characteristics of the situation that determine whether added external incentives will crowd in or crowd out employees' intrinsic motivation.

Adverse effects of excessive executive compensation

High compensation of managers entails a distribution of wealth that provokes society. The following question is unavoidably asked: What is the justification for high executive wages? Has the performance of firms improved proportionally with managerial compensation? And is it necessarily the case that executives should be credited for the company's performance?

Based upon his authoritative survey, Murphy (1999: 2555) writes: "Although there is ample evidence that CEOs (and other employees) respond predictably to dysfunctional compensation arrangements, it is more difficult to document that the increase in stock-based incentives has led CEOs to work harder, smarter, and more in the interest of shareholders". Murphy also indicates that the way of payments also matters. It is not only a question of size – it is how the structure of the compensation scheme is designed. There are important findings that variable pay – and payment linked to performance – may have major dysfunctional impacts. Frey and Osterlof (2005) argue that payment linked to performance might in some cases explain some of the scandals in business – like the Enron and WorldCom collapses. The main argument is that pay for performance reinforces extrinsic motivation, which entails that individuals shift their "locus of causality" from inside to outside, from the activity itself to the reward. Thus, the content of the activity loses its importance and intrinsic motivation can be undermined.

Such a view stands in sharp contrast to the position of Jensen and Murphy (1990: 139), who assert that "so many CEOs act like bureaucrats rather than the value-maximizing entrepreneurs companies need to enhance their standing in world markets". Jensen and Murphy moreover assert that "It's Not How Much You Pay, But How." They argue that, if possible, CEOs should own substantial amounts of company stock. However, in the case of giant companies like IBM, General Motors, or General Electric, CEOs should have substantial cash compensation that should be structured to *"provide big rewards for outstanding performance and meaningful penalties for poor performance"* (Jensen and Murphy, 1990: 141). Jensen and Murphy assert that monetary compensation and stock ownership remain the most effective tools for aligning executive and shareholder interests. Moreover, they strongly support the link between pay and performance. In fact, they warn that non-monetary rewards "typically motivate top managers to take actions that reduce productivity and harm shareholders ... As prominent members of their community, CEOs face pressures to keep open uneconomic factories, to keep peace with labor unions despite the impact on competitiveness and to satisfy intense special-interest pressures" (Jensen and Murphy, 1990: 149). Thus, CEOs should not act like good citizens, but regard corporate social responsibility as a strategic act, in the spirit of Milton Friedman and Michael Porter (see for example Porter and Kramer, 2006).

A second problem with individual economic incentives for executives is that managers will be too concerned about achieving what they are measured on. This may lead to suboptimal behaviour as a result of goal displacement, since it becomes tempting to aim for high performance in the areas – and within the time frame – that determines bonuses. Seen in a holistic perspective, this may imply serious dysfunctions. For one thing, it may lead the organizational focus away from important issues. Moreover, it may lead the organization into emphasizing technical efficiency at the expense of responsible behaviour and sustainable business practices.

The challenge is to find a way to measure and remunerate managerial performance that facilitates a holistic long run view of the company. In this regard, performance on the stock market gives a biased focus. The strategy for compensating the manager could, alternatively, be developed to be contingent on the satisfaction of employees, the environmental performance of the firm, etc. Empirical studies demonstrate that procedural justice is a key component in order to foster group engagement and cooperation (Tyler and Blader, 2003). This implies that executive compensation that is seen as procedurally unfair – since compensation is determined by factors outside the locus of control of managers – may undermine employees' engagement and, ultimately, performance. The group engagement model suggests that the identity is more important that resources in predicting engagement and cooperation (Tyler and Blader, 2003: 355). Identification reflects the extent to which people "cognitively merge their sense of self and their evaluations of self-worth with their judgments of the characteristics and status of their groups" (Tyler and Blader, 2003: 354). Hence, compensation schemes that are perceived as unfair effectively destruct the social and cooperative fabric of the organization.

Thus far we have explored some pitfalls associated with relying on economic incentives that, respectively, indicate adverse economic (efficiency) outcomes (for example reduced work effort), systemic outcomes (for example suboptimal behaviour due to goal displacement), existential outcomes (for example reduced self-worth), and social outcomes (for example reduced employee engagement and cooperation).

Holistic framework

Mitroff (1998) proposed a holistic framework, wherein it is distinguished among four dimensions of problem solving: the scientific-technical, the interpersonal-social, the systemic-ecological, and the existential-spiritual. Without any doubt, the scientific-technical way of thinking is the dominant perspective in our western culture. It entails a simple logic: if someone expresses frustration in his or her job, compensate him or her with more money. The assumption is that money is an efficient means to increase the effort and well-being of employees in an organization. In the following we problematize the dominance of this perspective, which easily leads decision makers into a fallacy of misplaced techno-centrism (Ims and Zsolnai, 2006). We will, therefore, focus on the hidden costs of using a scientific technical mindset in situations in which alternative perspectives might be more fruitful.

The interpersonal-social dimension of problem solving is in line with the focus on the employee's identification with his or her group or organization (See for example Tyler and Blader, 2003). Trying to solve the problem of identification with more money would be solving the wrong problem precisely.

The equality policy of American organic food retailer, Whole Food Market, plays on such social dimensions (Mackey and Sisodia, 2013). Its main characteristic is an "in-it-together" mindset where performance, and therefore compensation, is seen as a collective phenomenon. Also, such an approach plays down the differences between organizational members and emphasizes the similarities between employees in the same organization. If the problem is identification, engagement, and cooperation – which is distinct from performance, yet arguably influences performance – the objective can be reached by means of other approaches altogether. For instance, value-based leadership that aims to inspire employees to take part in a vision with the company positively influences the identification of employees (cf. Pruzan, 2009).

Another important dimension of problem solving in Mitroffs framework is the systemic-ecological dimension, which highlights the fact that problems relate to a larger whole. The assumption is that everything is interconnected, which means that we must take into account how the problem is connected to the broader system. In this perspective, increasing executive compensation may generate systemic effects that exert a heavy toll on the organization in the long run. For one thing, it takes up resources that could have been used for other purposes, and perhaps serve the organization better. Secondly, it creates a precedent towards which executives – and employees more broadly – may use as a reference point when considering their own future compensation plans. Also, new conflicts may be created in the organization due to perceived unfairness, which has implications far beyond the compensation case itself. Conflicts are easy to create, but harder to manage, especially after they have mutated and spread in the organizational organism. Finally, the response from society at large may be negative, in effect causing serious damage for the reputation and the branding efforts of the company (see for example van de Ven, 2008). A contemporary example of the detrimental reputational effects of perceived unfair compensation is the case of Goldman Sachs, who paid their executives record bonuses following the turmoil of the financial crisis and the receipt of bailouts from the American government.

The most ignored problem dimension today is the existential-spiritual dimension. Within the existential-spiritual dimension we open up for understanding the human being's life-projects (Williams, 1981), dignity, search for meaning (Frankl, 1969), pride, and self-respect. When we look at man in such a framing we see man not as a utility maximizer, but more in line with the conceptualizations in modern motivational theory, where the inherent pleasure of the task (Deci and Ryan, 2000), or the experience of flow (Csikszentmihalyi, 1975) are motivators of activities that are intrinsically valuable for the actor.

As discussed by Frey (1997), intrinsic motivation is undermined when the individual's self-determination is impaired. When the individual senses an external intervention that aims to control his or her behaviour, the individual's deep-rooted responsibility for the task is also undermined. This implies that the individual externalizes the responsibility for the task or performance, effectively rendering him or her disengaged from the task and its consequences (see for example Bandura et al., 2000). The net effect of this process is that the preexisting commitment of the individual to the behaviour is displaced, and the moral or social norms that incited him or her to act are substituted for transactional or purely rational norms. Thereby, the act itself is transformed from having existential relevance for the individual to an act that is carried out in a narrow transactional perspective.

Pruzan (2009) gives insightful observations and analyses of how spiritual perspectives emerge at the individual as well as corporate level throughout the world. He writes that "once leaders have developed their own spiritual self-awareness, they naturally exercise it in some form of service beyond self-interest. The dualistic distinction between one's self and others becomes replaced by a deeply felt connectivity, and the ordinary distinction between responsibility to one's self and to others attenuates" (Pruzan, 2009: 252).

Focus on the existential-spiritual dimension indicates that we are oriented towards the person's self-concept. This means we see managers "that they are beings of the requisite depth and complexity to have an identity ... (or to be struggling to find one)" (see Taylor, 1989: 14). The organization has to address and to deal with such questions, otherwise words like visions, values, goals, and strategies will only have a shallow meaning. The answers to those questions cannot be announced via managerial decree but need to be part of the leaders and employees' identity and integrity. An "outside in" perspective is not satisfactory. What we need is an "inside out" perspective if the leaders and the organizations claim to be responsible in a genuine sense (Auster and Freeman, 2013).

But what if the existential life is perverted and the satisfaction of the innate needs are thwarted? What happens when the scientific-technical view dominates to such an extent that managers think and act like money making robots? Economic rationality may take such a serious grip on the managers' thought and behaviour that it colonizes their private lifeworld. On such a background it is possible to be very rich and poor at the same time. Economic wealth and greed do not lead to joy, deep satisfaction, and happiness. Numerous psychological studies demonstrate that the more people prioritize materialistic goals, the lower their personal well-being and the more likely they are to engage in manipulative, competitive, and ecologically degrading behaviours (Kasser, 2011).

Alternative solutions

Considering all the relevant aspects of the executive compensation problem (that is scientific-technical, interpersonal-social, systemic-ecological, and existential-spiritual) alternative solutions can be developed.

Swiss economists Bruno Frey and Margit Osterloh suggest that the firm should be looked at in terms of a bundle of common pool resources. Common pool resources are collective goods that generate a joint surplus not attributable to single actors. It is essential that the production of such collective goods depends on prosocial intrinsic incentives (Frey and Osterloh, 2005).

Alternative solutions, consistent with the view of firms are common pool resources, emerge in the practice of "conscious capitalist" enterprises that adopted a policy that caps executive pays. For example, at the Whole Food Market – a leading organic food retailer company in the USA – the total cash compensation, including bonuses, for top managers cannot be higher that nineteen times that of the average pay of the employees. In other publically traded companies in the USA, this ratio is as high as four to five hundred times (Mackey and Sisodia, 2013: p. 93).

Caps on executive compensation create a sense of fairness because the compensation system is perceived by employees to be fair. With constrained compensation executives get less money than their counterparts in mainstream business organizations, but they are usually compensated by getting higher moral satisfaction. Offering existential spiritual, social, and ecological pay-offs for business leaders, responsible organizations can get the right persons in the right position. Pre-selection occurs and the persons who are intrinsically motivated to work for high-purpose, high-mission companies happen to occupy the companies executive positions.

Summary

In this paper we explored the adverse effects on managerial performance of relying on ever increasing levels of executive compensation. Rather than conceiving of performance in a one dimensional, technical-economic sense, we illuminated how such compensation schemes may undermine holistic individual and organizational performance. Specifically, we have argued that relying on external rewards is associated with the danger of crowding out existential, social, and ecological values in executives' behaviour and performance. Our main argument is that we have to go beyond traditional economic rationality in which employees and organizations are mainly perceived as instruments to produce increased profits and financial wealth. In such an environment the managers will be motivated by external compensation alone. We need a model of compensation that encompasses other dimensions as well; a model that takes into account man as a whole being whose commitment to important existential, social, and ecological values is not only reflected in his or her performance, but is the very point from where that performance springs.

There is no doubt that the technical-scientific paradigm has been enormously successful in solving economic problems. The danger is that it has swallowed other important dimensions, and when using a scientific-technical perspective on social, ecological, and existential problem dimensions we may entail counterproductive behaviour of managers.

Mitroffs four dimensions of problem solving are not antithetical, but are mutually supportive. The major problem arises when one dimension, the scientific-technical, becomes dominant and other dimensions are ignored. The sensitivity to the needs of other beings and stakeholders are vital for corporate success and survival in a broad and long-term sense. Taking the existential, social, and ecological perspectives seriously might lead to an improved capability for reflection on the usefulness of high compensations of managers and the negative effects generated by such practices.

Note

1. The original version of the paper was developed by the first two authors. Laszlo Zsolnai joined them later. His contribution was written while he was visiting international scholar at the Jepson School of Leadership Studies, University of Richmond, Virginia, USA.

Bibliography

Auster, E. R. and Freeman, R. E.: 2013. "Values and Poetic Organizations: Beyond Value Fit Toward Values through Conversation". *Journal of Business Ethics,* 113, 39–49.

Axelrod, R. and Cohen, M. D.: 2000. *Harnessing Complexity: Organizational Implications of a Scientific Frontier.* New York, NY: The Free Press.

Bandura, A., Caprara, G.-V. and Zsolnai, L.: 2000. "Corporate Transgressions through Moral Disengagement". *Journal of Human Values,* 6, 1, 57–64.

Bogsnes, B.: 2009. *Beyond Budgeting: Unlocking the Performance Potential.* Hoboken, NJ: John Wiley & Sons.

Bouckaert, L.: 2006. "The Ethics Management Paradox". In L. Zsolnai (ed.), *Interdisciplinary Yearbook of Business Ethics, Volume 1*. Bern: Peter Lang.

Bucklin, B. R. and Dickinson, A. M.: 2001. "Individual Monetary Incentives: A Review of Different Types of Arrangements Between Performance and Pay". *Journal of Organizational Behaviour Management,* 21, 3, 45–137.

Carr, L. L. and Valinezhad, M.: 1994. "The Role of Ethics in Executive Compensation: Toward a Contractarian Interpretation of the Neoclassical Theory of Managerial Remuneration". *Journal of Business Ethics,* 13, 2, 81–93.

Chan, M.: 2008. "Executive Compensation". *Business and Society Review,* 113, 1, 129–161.

Csikszentmihalyi, M.: 1975. *Beyond Boredom and Anxiety.* San Francisco, CA: Jossey-Bass.

Dalton, D. and Daily, C.: 2001. "Director Stock Compensation: An Invitation to a Conspicuous Conflict of Interest?". *Business Ethics Quarterly,* 11, 89–108.

Deci, E. L.: 1971. "Effects of Externally Mediated Rewards on Intrinsic Motivation". *Journal of Personality and Social Psychology,* 18, 105–115.

Deci, E. L.: 1976. "The Hidden Cost of Rewards". *Organizational Dynamics,* 4, 3, 61–72.

Deci, E. L. and Ryan, R. M.: 2000. "The 'What' and 'Why' of Goal Pursuits: Human Needs and the Self-Determination of Behaviour". *Psychological Inquiry,* 11, 4, 227–268.

Eisenhardt, K.: 1989. "Agency Theory: An Assessment and Review". *Academy of Management Review,* 14, 1, 57–74.

Elster, J.: 1986. *Rational Choice.* New York, NY: NYU Press.

Ferraro, F., Pfeffer, J. and Sutton, R. I.: 2005. "Economics Language and Assumptions: How Theories Can Become Self-Fulfilling". *Academy of Management Review,* 30, 1, 8–24.

Frankl, V. E.: 1969. *The Will to Meaning: Foundations and Applications of Logotherapy.* New York: The World Publishing Company.

Frey, B.: 1997. *Not Just for the Money: An Economic Theory of Personal Motivation.* Cheltenham: Edward Elgar.

Frey, B. and Osterloh, M.: 2005. "Yes, Managers Should Be Paid Like Bureaucrats". *Journal of Management Inquiry,* 14, 1, 96–111.

Ghoshal, S.: 2005. "Bad Management Theories Are Destroying Good Management Practices". *Academy of Management Learning & Education,* 4, 1, 75–91.

Ghoshal, S. and Moran, P.: 1996. "Bad for Practice: A Critique of the Transaction Cost Theory". *Academy of Management Review,* 21, 1, 13–47.

Harris, J. D. and Bromiley, P.: 2007. "Incentives to Cheat: The Influence of Executive Compensation and Firm Performance on Financial Misrepresentation". *Organization Science,* 18, 3, 350–367.

Ims, K. J. and Jakobsen O. D.: 2010. "Competition or Cooperation? A Required Shift in the Metaphysics of Economics". In A. Tencati and L. Zsolnai (eds), *The Collaborative Enterprise: Creating Values for a Sustainable World.* Oxford: Peter Lang.

Ims, K. J. and Zsolnai, L.: 2006. "Shallow Success and Deep Failure". In L. Zsolnai and K. J. Ims (eds), *Business within Limits: Deep Ecology and Buddhist Economics.* Bern: Peter Lang.

Ims, K. J. and Zsolnai, L.: 2009. "Holistic Problem Solving". In L. Zsolnai and A. Tencati (eds), *The Future International Manager: A Vision of the Roles and Duties of Management.* London: Palgrave MacMillan.

Ims, K. J. and Zsolnai, L.: 2010. "Self-Realization in Business: Ibsen's *Peer Gynt.*" In R. Ghesquiere and K. J. Ims (eds), *Heroes and Anti-heroes. European Literature and the Ethics of Leadership.* Antwerpen – Apeldorn: Garant.

Jensen, M. C. and Murphy, K. J.: 1990. "CEO Incentives – It's Not How Much You Pay, But How". *Harvard Business Review,* 68(3) May-June, 138–153.

Kasser, T. 2011: "Materialistic Value-orientation". In Luk Bouckaert and Laszlo Zsolnai (eds), *The Palgrave Handbook of Spirituality and Business.* Basingstoke, UK: Palgrave MacMillan.

Langer, E. J. 1979. "Rethinking the Role of Thought in Social Interaction". In J. H. Harvey, W. J. Ickes and R. F. Kidd (eds), *New Directions in Attribution Theory, Vol. 2.* Hillsdale, NJ: Lawrence Erlbaum.

Lawler, III, E. E.: 1990. *Strategic Pay: Aligning Organizational Strategies and Pay Systems.* San Francisco, CA: Jossey Bass.

Locke, E. A. and Latham, G. P.: 1990. *A Theory of Goal Setting and Task Performance.* Englewood Cliffs: Prentice Hall.

Luthans, F.: 1985. *Organizational Behaviour.* New York, NY: McGraw-Hill.

Mackey, J. and Sisodia, R.: 2013. *Conscious Capitalism. Liberating the Heroic Spirit of Business.* Boston, MA: Harvard Business Review Press.

March, J. G. and Simon, H. A.: 1958. *Organizations.* New York, NY: John Wiley & Sons.

Maslow, A. H.: 1943. "A Theory of Human Motivation". *Psychological Review,* 50, 370–396.

McCall, J. J.: 2004. "Assessing American Executive Compensation: A Cautionary Tale for Europeans". *Business Ethics: A European Review,* 13, 243–254.

McClelland, D. C.: 1987. *Human Motivation.* Cambridge, MA: Cambridge University Press.

Mitroff, I. I.: 1998. *Smart Thinking for Crazy Times: The Art of Solving the Right Problems.* San Francisco, CA: Berrett-Koehler Publishers.

Nichols, D. and Subramaniam, C.: 2001. "Executive Compensation: Excessive or Equitable?". *Journal of Business Ethics,* 29, 4, 339–351.

Porter, M. E. and Kramer, M. R.: 2006. "Strategy and Society: The Link Between Competitive Advantage and Corporate Social Responsibility". *Harvard Business Review,* 85, 6, 78–92.

Pruzan, P.: 2009. *Rational, Ethical and Spiritual Perspectives on Leadership: Selected Writings.* Bern: Peter Lang.

Rodgers, W. and Gago, S.: 2003. "A Model Capturing Ethics and Executive Compensation". *Journal of Business Ethics,* 48, 2, 189–202.

Sandemose, A.: 1936. *A Fugitive Crosses His Tracks.* New York, NY: A. A. Knopf.

Taylor, C.: 1989. *Sources of the Self: The Making of the Modern Identity.* Cambridge: Cambridge University Press.

Tyler, T. R. and Blader, S.: 2000. *Cooperation in Groups: Procedural Justice, Social Identity, and Behavioural Engagement.* Philadelphia, PE: Psychology Press.

Tyler, T. R. and Blader, S.: 2003. "The Group Engagement Model: Procedural Justice, Social Identity, and Cooperative Behaviour", *Personality and Social Psychology Review,* 7, 4, 349–361.

van de Ven, B.: 2008. "An Ethical Framework for the Marketing of Corporate Social Responsibility". *Journal of Business Ethics,* 82, 339–352.

Walters, B., Hardin, T. and Schick, J.: 1995. "Top Executive Compensation: Equity or Excess? Implications for Regaining American Competitiveness". *Journal of Business Ethics,* 14, 3, 227–234.

Weick, K. E. and Bougon, M. G.: 1986. "Organizations as Cognitive Maps: Charting Ways to Success and Failure". In H. P. Sims and D. A. Gioia (eds), *The Thinking Organization: Dynamics of Organizational Social Cognition.* San Francisco, CA: Jossey Bass.

Williams, B.: 1981. *Moral Luck.* Cambridge: Cambridge University Press.

Zsolnai, L.: 2010. "Ethics Needs Spirituality". In S. S. Nandram and M. E. Borden (eds), *Spirituality and Business: Exploring Possibilities for a New Management Paradigm.* London: Springer.

KNUT J. IMS, OVE D. JAKOBSEN, AND LASZLO ZSOLNAI

Product as Process: Commodities in Mechanic and Organic Ontology

Introduction

Referring to Ormerod, who claims that mainstream economics offers a misleading view of how the world actually operates (Ormerod, 1994), we argue that the ontological presuppositions in economics need to be critically reconsidered and discussed. The problem, according to Boulding, is that "economics has rested to long in an essentially Newtonian paradigm of mechanical equilibrium and mechanical dynamics" (Boulding, 1981, p. 17). Rees argues even more forcefully, claiming that: "Modern economic society operates from an outdated mechanistic perception of the natural dynamics of the Earth" (Fabel and St John, 2007, p. 104). From these distinguished contributors in ecological economics we can conclude that our current environmental and societal problems are, in part at least, due to distorted ontological assumptions about reality. The absence of metaphysical questions has "dodged fundamental conflicts among economy, nature, and culture" (Ims and Jakobsen, 2017, p. 17).

This paper explores and interprets the product concept in two different ontologies, mechanistic and organic. In our opinion a shift in ontology for understanding commodities has crucial implications for economic theory and practice. In mainstream economics the product is understood in accordance with mechanistic ontology as a more or less fixed, atomized commodity, which is exhibited on the shelves in the shops, or is represented in marketing communications. In the organic ontology of ecological economics, the product cannot be separated from the dynamic network of relations among economy, ecology, and society involved in the whole process of the product, from the extraction of natural resources to the processing of waste.

One of the main differences between mainstream economics and ecological economics is that in mainstream economics the product is nothing more than the bundle of different properties at a certain moment in space and time. Products are characterized objectively as the static set of characteristics offered by the seller to the buyer. Whereas mainstream economics is criticized for being abstract, that is disconnected from real world conditions, the mechanistic interpretation of the product is disconnected from the ecological and social world. Values associated with the product are imposed from the outside. In the organic interpretation, the product is a dynamic phenomenon, integrated in social and ecological networks and its values thus emerge inherently.

In order to clarify the ontological distinction between mainstream economics and ecological economics we refer to Hume's philosophy of empiricism, which dominates mainstream economics, and Whitehead's philosophy of organism, characterizing ecological economics. In the following paragraphs we discuss to what extent social and environmental values are imposed from the outside into mainstream economics and are inherent in ecological economics.

In the next paragraph we will elaborate on how the differences between the ontology behind mainstream economics and the ontology behind ecological economics influence the interpretation of commodities. To elucidate the two ontologies, we firstly focus on Hume's empiricism and, secondly, on Whitehead's philosophy of organism.

The ontology of mainstream economics

"Mainstream economics is sustained by nineteenth century neoclassical theories and principles based on a mechanic worldview" (Max-Neef, 2014, p. 152). To describe the mechanic ontology of mainstream economy we choose Hume's philosophy of empiricism as he decisively influenced the writings of Adam Smith regarding both moral philosophy and economic theory. Hume claimed that all valid contents of the mind (impressions and ideas) come from sense perceptions (Jones, 1970). His main point was that ideas are ultimately copied from sense impressions. Perceptions can be distinguished as simple impressions; complex images are composed of simple impressions. Hume also suggested drawing a strict demarcation line between the impression of sensation (for example color, taste, and smell) and the impression of reflection (for example passions, moral sentiments, and other feelings).

Inspired by Hume's empiricist philosophy, mainstream economics concludes that the product is nothing but a complex idea (the product brand) composed of a variety of simple perceptions (the tangible and intangible components of the product). The central focus is on the mental construct of complex sensory perceptions based on impressions and ideas arising from the commodity exhibited in shop shelves.

Hume made a distinction between ideas produced by memory and ideas produced by imagination. Memory gives us ideas based on direct experience. For example, we can memorize what a certain car looks like or recall the taste of a certain piece of pizza. According to Hume, these ideas are more or less accurate copies of our previous sense impressions. The imagination, by contrast, combines separate impressions into new and complex ideas.

In accordance with Hume's empiricist philosophy, Bagozzi (1986) differentiates between the tangible and the intangible aspects of the product. The tangible aspect of the product consists of quantitative properties (for example the shape and weight of the product) and qualitative properties (for example, the color and taste of the product). The intangible aspect of the product refers to symbolic properties such as style, brand name, logo, and aesthetics. The brand image contains all these characteristics. The brand is the image of the product as the consumers and other stakeholders perceive it.

One of the main tasks of marketing communication is to couple the memory of many different sense impressions into complex ideas in the mind of the consumer. In marketing communication Hume's principles of association, resemblance, contiguity, and cause and effect are used to affect the consumers to associate positive attitudes with the product brand. Hume's empiricist interpretation of the world, made up of disjointed perceptions of objects with no unity, corresponds with the definition of commodities in mainstream economics.

The ontology of ecological economics

We choose Whitehead's philosophy of organism to illustrate the ontology of ecological economics because he is often referred to as a source of inspiration in ecological economics. Distinguished scholars such as Georgescu-Roegen (1971) and Daly and Cobb (1994) refer to Whitehead's philosophy in their analysis and descriptions of the principles of ecological economics. They argue that many of the environmental and societal problems connected to mainstream economics are due to the underlying mechanistic ontology. These problems can only be solved by economics based on an organic ontology. To study the distinction between these two ontologies in more depth we apply Whitehead's philosophical analysis.

According to the "philosophy of organism" (Whitehead, 1966, 1978), everything consists of combinations of actual entities. Nature is a theatre for the interrelations of activities. "All things change, the activities and their interrelations" (Whitehead, 1977, p. 15). Actual entities are small momentary drops of experience or feelings. Actual entities are contrasted with eternal objects, which are pure potentials for actualization. Whitehead named the process of connecting actual entities and eternal objects "prehension". Prehensions describe how complex organisms come into being as a combination of past experiences (actual entities) and potentials (eternal objects). According to Emmet, prehensions are "a general word for the grasping, or taking hold of one thing by another, and so connoting an active coming together" (Emmet, 2012, p. 41). Prehensions can be either positive or negative. Positive prehensions concern the process of integration: the actual entities connect into complex and interdependent nexus or societies. Negative prehensions mean blocking out those elements of the past and possibilities of the future that will not be incorporated into the present.

The new aggregated entities confront the whole past actual world and bring the world into new entities. Because of this interconnectedness in time (and space) everything that has happened in the past has some impact on the present. What happens in the present has impact on the events of the future. This is the principle of relativity: everything is related to everything else. According to Whitehead, creativity comes into existence through the interplay between the physical world (actual entities) and the ideas (eternal entities). "Reality is constituted by entities which are brief pulses of creative activity, (…) constantly developing in reaction to what has been and to what might be" (Hosinski, 1993, p. 23).

From this ontological position, it is reasonable to claim that commodities derive their character from the social and ecological networks they are integrated parts of. The process can be understood from both an individual and a systemic perspective. We can consider the formation of the system from the characterization of the individuals; "or we can characterize the individuals and conceive them as formative of the relevant process" (Whitehead, 1966, p. 98). In this perspective, everything is an integrated part of dynamic networks. Nothing can exist as an isolated atom. The interconnectedness between the product and its natural and societal environment becomes obvious. Firstly, relations become more important than objects. This means that the product cannot be understood as an isolated entity. The product depends on numerous relations to many different processes, which are interwoven with one another in the entire lifespan of the product. Secondly, since the product is integrated in social and ecological networks, values such as responsibility and fairness cannot be separated from the product. An important feature of the philosophy of organism is that everything is interdependent both temporally and spatially (Ims and Jakobsen, 2006). According to this line of argumentation, it is essential to capture the environmental, social, and ethical dimensions of the product, to broaden the time horizon both in the past and in the future, and to expand the spatial focus from local to global.

Accepting organic ontology has important implications for the economic actors. Being aware of "product as process" leads to acceptance of the actors "co-responsibility" for the whole life-cycle of the product they use or produce.

Discussion

"Since the world is not mechanic, as supposed by neoclassical economists, but organic, as understood by ecological economists, it is no surprise that (…) mainstream economics creates economists who don't understand the real world" (Max-Neef, 2014, p. 152). Using the same line of argumentation, we argue that we need an ontological shift in economics to solve some of the most serious problems in our time, the converging crises of climate, water, soil erosion, and poverty. More specifically, we argue that in order to understand and possibly solve these problems the interpretation of the product concept is of great relevance. One important consequence is that the definition of the product changes from being a "bundle of characteristics offered by one party, the seller, to another party, the buyer" (Bagozzi, 1986, p. 137) to a complex nexus of entities integrated in the flux of social and ecological networks. In ecological economics the product cannot be reduced to a static object, it is a dynamic phenomenon integrated in social and ecological processes.

To illustrate this phenomenon, we describe some important stages in the life-span of coffee. "Coffee" may be the coffee beans hanging on the trees in Brazil, or what is produced at a coffee roasting factory. Usually "coffee" as a commodity is understood as the different coffee brands exhibited on grocery shop shelves, or "coffee" may be what we are served for breakfast. In addition, we can focus on "coffee" as waste from all the functions in the value-chain. This means that "coffee" is an integrated part of nature in addition to being a part of the farmer's life, the wholesaler's life, the manufacturer's life, the people working in the groceries' life, the consumer's life, and the life of people in reprocessing and waste handling. "Coffee" cannot be separated from its economic, ecological, and social contexts.

The methods of farming, manufacturing, transportation, and distribution are important, both concerning ecological sustainability and individual and social health. In all the different stages of the value chain, the price of the coffee should be adjusted to the living conditions of the people involved, securing their possibility to live a decent life. Externalities should also be considered. The producer should earn a living wage to support his or her family, and the price should not prevent the consumers from buying the products. The cultivating, manufacturing, distribution, and consumption of coffee also impacts on the social dimension. To understand how "coffee" influences the local and global conditions of the natural, social, and individual environment we have to be aware of the temporal and spatial dynamics characterizing the product as a process. The distance between the original farming place and the shelf in the supermarket is very long, both in time and in space. In the supermarket the buyer can choose among different brands of coffee. The choice between the different brands depends on the con- text of interpretation characterizing the marketer and the coffee shopper respectively. From the perspective of mainstream economics, the branding strategy is to add specific dimensions to the product in order to differentiate the product from other competing products. Assuming that the commodity has a strong brand image, the buyer's preference is influenced and has an important impact on buying behaviour. Being satisfied, he or she can return to the shop, buy the product again and, strengthened by looking at the brand advertisements, he or she becomes convinced that his or her choice was the right one.

One interesting question is: which attributes have the strongest influence when the consumer chooses between different brands of coffee? A study conducted by Wood (2007) examined the functional and symbolic attributes of product selection by 18 to 24-year-old consumers for a range of food and toiletry products. For coffee, Wood found that flavor, aroma, and brand are ranked as the most important attributes of the product, whereas environment, ethics, and health/organic production attributes were ranked at the bottom of the scale. This study indicates that young people who buy coffee give low ranking to environmental, ethical, and health aspects related to the product. The study concludes that consumers need to develop a wider horizon and move beyond a narrow-minded, myopic perspective. Decision making in traditional mainstream economics implies a short time horizon, and customers, therefore, neglect important systemic effects of their repeated buying decisions.

In most commercials, the intangible aspects of the product dominate and the environmental and social context of the product – such as international justice and environmental sustainability – is usually not emphasized. Instead of being aware of the producers "working conditions" and their income and the way the product is cultivated, there is a tendency to present the product as an object abstracted from its dynamic context. When the managerial perspective of building a strong brand is based on short-sightedness with respect to both time and space, consumers are often unaware of the (often unintended) negative consequences of their buying behaviour. This is an example of the "tyranny of small decisions" where every single act is relatively innocent, but the accumulated impact of all the acts together can pose a serious threat to the common good.

Mainstream economics and marketing focus on market shares and profits. "Even though advertising first impacts affect and cognition, the ultimate goal is to influence consumers' purchase behaviour" (Peter and Olson, 1996, p. 538). People are treated as prospective consumers and thus exposed to a number of marketing messages in order to turn them into actual consumers who prefer to buy the firm's product. The reduction of individuals to homo is not without serious effects. When instrumental practice is used extensively, it may result in the loss of freedom for the individuals thus flattening their lifeworld and culture (Taylor, 1991).

On the one hand, mainstream economics and marketing suggest offering what the buyers want. The responsibility is thereby put upon the buyer. The study of Wood (2007) indicates that customers do not always want what they "should have wanted" from an ethical perspective. There is a difference between whether customers act from a moral horizon with important values (Taylor, 1985) which may be centered on sustainability and pro-social values, or merely act instinctively – following their first order preferences. Following Taylor further, we believe that a second-order evaluation of our preferences is of the utmost importance in a world dominated by mainstream economics and "commoditization" – there is a tendency to convert all kinds of entities into atomized products.

On the other hand, it is also possible to find examples in which economic practice lies close to ecological economics. LOHAS (the acronym for "Lifestyles of Health and Sustainability") is a group of customers who care about the world, wishing to buy products produced in a sustainable way, and caring about their inner development (Cortese, 2000). According to

Spash, biophysical and social realities are interconnected and "facts about social reality are inseparable from values" (Spash, 2012, p. 45).

Developing this argument further, let us consider the selling concept of mainstream economics and marketing, which says that "consumers and businesses, if left alone, will normally not buy enough of the organization's products. The organization must, therefore, make an aggressive selling and promotion effort. The purpose is to sell more stuff to more people more often for more money in order to make more profit" (Kotler and Keller, 2006, 15). However, from the perspective of ecological economics, the mainstream selling concept is unacceptable. The economy should serve the well-being of both individuals and society within sustainable ecosystems.

Ecological economics can be exemplified by social labels creating new brands that explicitly associate ecologically and socially responsible practices (Vogel 2005). The use of social labels to trace certain commodities back to their origin has become a well-established practice in some industrial networks. Fair trade coffee is probably the best-known example. However, the fair trade product range is actually quite broad: in addition to coffee, it includes among other things, bananas, cocoa, tea, juice, sugar, and even wine. Fair trade labeling started in the Netherlands in the 1980s when the organization Max Havelaar voiced poor Mexican farmers' interests, claiming, "we want trade, not aid!" (Moore, 2004). Today fair trade practiced by Max Havelaar has grown into an international movement of major importance.

By fair trade we mean a "trading partnership, based on dialog, transparency and respect which seeks greater equity in international trade. It contributes to sustainable development by offering better trading conditions, and securing the rights of marginalized producers and workers, especially in the South" (European Fair Trade Associations, 2001). The carpet industry provides another interesting example of a successful social label called Rugmark. In the carpet industry, where child labor is usually widespread, German activists started to organize public campaigns against the use of child labor in the 1990s. The campaign was highly effective and was then followed up by launching Rugmark, a "human rights" label, which gave children educational opportunities. Rugmark's inspection system is regarded as reasonably effective.

A special kind of brand strategy within mainstream economics is building brand identity by creating a set of associations in order to build a strong brand. In contrast to brand image, which is the brand's current associations, "the brand identity is aspirational" (Aaker and Joachimsthaler, 2000, p. 40). The idea is to change and augment the image of the product by making the product into more than a just product. The attempt is to attribute meaning and even character to the product by using the concept of identity – a concept that synthesizes and involves aspects from a number of terms like self, soul, character, and personality. Today meaning- and identity-creating processes are heavily influenced by the efforts of big corporations. Massive marketing campaigns support this practice giving the products a strong position in peoples' lives, which makes the prevalent culture extremely materialistic (Jeurissen and Van de Ven, 2008).

We find today's branding practices to be an ambiguous activity. On the one hand, some branding practices try to hide the origin of the products, while others highlight it by flagging the region or country of origin (Dimara and Skuras, 2001; Fandos and Flavian, 2006).

However, this strategy is strongly connected to mainstream economics where selling more is always better than selling less. On the other hand, there are opportunities for activities within the context of ecological economics where the focus is on quality of life and sustainable ecosystems.

Frugal marketing and conscious consumerism are growing in importance, especially in affluent western economies. This implies the voluntary self-limitation of marketing itself. Jeurissen and van de Ven (2008) argue that, sometimes, it does make good business sense to sell less.

Several consumer trends point to it – not only frugality, but also ecology and the aging of the population. The "Sense and Simplicity" campaign launched by Philips seems to understand this notion. There is a market for products that are good, but straightforward, easy to operate, and stripped of all the unnecessary gadgets and features that 99 percent of consumers never use. Simpler products can be sold at lower prices without giving up profit margins. Besides, simple and sturdy products will last longer, require less maintenance, and have fewer defects. Frugal consumers may be willing to inform themselves more about these kinds of "sensible" products. This may invite sellers to communicate more about less visible product characteristics that actually really matter, such as life expectancy, total life-cycle costs, and total costs of ownership. What people see and get in a store may change completely when greater numbers of consumers take on frugality (Jeurissen and van de Ven: 240).

Conclusion

One of the most important differences between mainstream and ecological economics is that ethical considerations in the former are imposed from the outside, while in the latter case the ethical values emerge inherently. The consequence is that all agents involved in these broad and long-lasting processes have an undeniable ethical responsibility for the environmental and social impacts of the product – both "up-stream" and "down-stream".

Referring to ecological economics we agree with Taylor (1991), who describes the authentic identity formation of the individual as a process in which the individual adopts a wider horizon of important values moving beyond his or her individual self. Such important values can include nature, history, citizenship, or fulfilling the needs of others. Such an idea of an extended self can transcend hedonism and narcissism, which are shallow, narrow, and trivial modes of being.

We have used the social labels "Fair trade" and "Rugmark" to illustrate the product as process. In these cases, the past and the prospective future are added to the product and it is integrated in a wider social and ecological context. The product obtains identity through being part of the natural and social environment.

In addition, in ecological economics the product is understood holistically, that is, the processes forming parts of the creation, distribution, use, and deposition/recirculation of the product are emphasized. Thus, we suggest an extended view encompassing the space and time dimensions of the product: When, and where was the product created? Who produces it? Which materials and energy were used? Which means of transportation were employed? What was the profit margin at different stages of the value chain? What happens with the product after use by the customer? Is it reused and transformed into new resources, or left as waste? How does the used product influence the ecology of the planet and the well-being of society at large?

Buyers and sellers anchored in mainstream economics perceive values and norms as external claims, and social and environmental responsibility do not directly make an appeal to them. A shift in the ontological assumptions may increase the actor's responsibility in the market in a profound way. From the perspective of ecological economics, both sellers and buyers look upon themselves as guardians who should serve the well-being of individuals, society, and natural ecosystems.

Here we can refer to Cox (1962) and Priddle (1994) and their classical contributions claiming that economy must be based on a network perspective. Priddle argues that, "all of us live in a community – a system of local and mutual interdependence, where our welfare de- pends directly on the welfare, cooperation, and activities of our neighbours, locally and globally" (p. 57). Instead of focusing on the product as an atomized commodity, both the total set of consequences connected to nature and society should be taken into account. Interpreting the product as process assumes "understanding the basic principles of ecology" (Capra and Luici, 2014, p. 353).

We conclude that to change the economy in a socially and environmentally responsible direction it is necessary to move beyond main-stream economics' traditional product concept. To realize that, economic actors must be held responsible for the societal and environmental impact of the products, and for what they use and produce. The mechanic ontology of mainstream economics has to be replaced by the organic ontology of ecological economics in which products are seen as processes.

Bibliography

Aaker, D. A., Joachimsthaler, E., 2000. *Brand Leadership.* The Free Press, New York.

Bagozzi, R., 1986. *Principles of Marketing Management.* Science Research Associates Inc., Chicago.

Boulding, K., 1981. *Evolutionary Economics*. Sage Publications, London.

Capra, F., Luici, P. L., 2014. *A Systems View of Life – A Unifying Vision*. Cambridge University Press, Cambridge.

Cortese, A., 2000. They care about the world (and they shop too). *New York Times*, July 20, 3–4.

Cox, R., 1962. Changing social objectives in marketing. In: Decker, W. (ed.), *Proceedings of the AMA Winter Conference*. American Marketing Associations, Chicago.

Daly, H., Cobb Jr, J. B., 1994. *For the Common Good – Redirecting the Economy toward Community, the Environment, and a Sustainable Future*. Beacon Press, Boston.

Dimara, E., Skuras, D., 2001. Association and traceability in Greece. *Eur. J. Mark.* 37 (5/6), 690–705.

Emmet, D. M., 2012. *Whitehead's Philosophy of Organism*. Forgotten Books, London.

European Fair Trade Associations, 2001. *Fair Trade Handbook: Challenges of the New Millennium 2001–2003*. See also: <european-fair-trade-association.org/efta/index.php>.

Fabel, A., St John, D., 2007. *Teilhard in the 21st Century – The Emerging Spirit of Earth*. Orbis Books, New York.

Fandos, C., Flavian, C., 2006. Intrinsic and extrinsic quality attributes, loyalty and buying intention: An analysis for a PDO product. *Br. Food J.* 108 (8), 646–662.

Georgescu-Roegen, N., 1971. *The Entropy Law and the Economic Process*. Harvard University Press, Cambridge, MA.

Hosinski, T. E., 1993. *Stubborn Fact and Creative Advance – An Introduction to the Meta-physics of Alfred North Whitehead*. Rowan & Littlefield Publishers, Boston.

Ims, K. J., Jakobsen, O., 2006. Cooperation and competition in the context of organic and mechanic worldviews – a theoretical and case based discussion. *J. Bus. Ethics* 66 (1), 19–32.

Ims, K. J., Jakobsen, O., 2010. Cooperation or competition? A required shift in the meta-physics of economics. In: Tencati and Zsolnai (eds), *The Collaborative Enterprise – Creating Values for a Sustainable World*. Peter Lang, Oxford.

Jeurissen, R., Van de Ven, B., 2008. Frugal marketing: Can selling less make business sense? In: Bouckaert, L., Opdebeeck, H., Zsolnai, L. (eds), *Frugality. Rebalancing Material and Spiritual Values in Economic Life*. Peter Lang, Oxford.

Jones, W. T., 1970. *A History of Western Philosophy*. Hartcourt Brace Jovanovich Publishers, New York.

Kotler, P., Keller, K. L., 2006. *Marketing Management*. Prentice Hall Inc., New Jersey.

Max-Neef, M., 2014. The good is the bad that we don't do – economic crimes against humanity: A proposal. *Ecol. Econ.* 104, 152–154.
Moore, G., 2004. The fair trade movement: Parameters, issues and future research. *J. Bus. Ethics* 53 (1–2), 73–86.
Ormerod, P., 1994. The death of economics, Faber and Faber, London. In: Pearce, J. (ed.), 2001. *Small Is Still Beautiful.* HarperCollins Publishers, London, UK.
Peter, J. P., Olson, J. C., 1996. *Consumer Behaviour and Marketing Strategy*. Irwin, Chicago.
Priddle, J., 1994. Marketing ethics, macromarketing, and the managerial perspective reconsidered. *J. Macromark.* 14 (2), 47–62.
Spash, Clive L., 2012. New foundations for ecological economics. *Ecol. Econ.* 77, 36–47.
Taylor, C., 1985. *Human Agency and Language – Philosophical Papers 1.* Cambridge University Press, Cambridge.
Taylor, C., 1991. *The Ethics of Authenticity.* Harvard University Press, Cambridge, MA.
Vogel, D., 2005. *The Market for Virtue – The Potential and Limits of Corporate Social Re-sponsibility.* The Brookings Institution, Washington, DC.
Whitehead, A. N., 1966. *Modes of Thought.* The Free Press, New York.
Whitehead, A. N., 1977. *Nature and Life.* Greenwood Press, New York.
Whitehead, A. N., 1978. *Corrected Edition.* The Free Press, New York.
Wood, L., 2007. Functional and symbolic attributes of product selection. *Br. Food J.* 109 (2), 108–118.

KNUT J. IMS AND LARS JACOB TYNES PEDERSEN

The Future of Business Ethics: A Structured Dialogue Between the Participants

Introduction

This book has addressed questions of relevance to contemporary business ethics. In various ways, the chapters herein are concerned with ethical issues that pertain to the relationship between business and the greater good – how the two may conflict, as well as how business can be transformed in order to align the goals of business and society. In this respect, this book is also oriented towards *the future of business* in general, and *the future of business ethics* in particular.

In this concluding chapter, we turn to the latter point. Business ethics as a field has matured and developed considerably. This is in part due to developments in the problem matter that business ethics is designed to address. As business practice becomes more globalized, as issues related to the climate become more pressing, as financial markets have a bigger – and to a significant degree detrimental – impact on business and society, business ethics develop in order to account for these developments and to be suited to address the challenges that arise from these developments. In addition, however, business ethics change as a consequence of developing needs for competence in business. The professionalization of tools and approaches related to business ethics and corporate social responsibility – ISO-standards, reporting initiatives, frameworks, and so on – have introduced a set of competences that businesses expect business students to be familiar with. Such demand side developments also influence how business ethics develop. Finally, business ethics develop as new perspectives flourish, as new connections are made to other fields, and as the influence of new theoretical, intellectual, and cultural traditions integrate into the knowledge base of the field.

The big question is, however, is the state of business ethics appropriate for the challenges that individuals, businesses, and societies face? Can business ethics achieve the aim of bridging business and the greater good – the goal of creating and protecting economic, societal, and environmental value(s) simultaneously? This leads us to ask the question of what the future of business ethics is and what the future of business ethics should be in order for it to serve its purpose of promoting ethical practices and behaviours in institutions and business.

This question was the point of departure of the structured dialogue that concluded the 7th Trans-Atlantic Business Ethics Conference (TABEC). All of the conference participants were asked to offer their vision of what the future of business ethics ought to be, and in particular what kinds of changes can create business ethics suited to face the challenges of the future. The dialogue took place as a roundtable discussion, and gave the conference a thought-provoking and forward-looking ending that pointed forward towards future themes of interest for the scholarly community associated with the TABEC conferences, as well as for business ethics more broadly.

In this chapter, we aim to recreate the structured dialogue that took place at the conference. Each of the participants was asked to transcribe the ideas they presented at the conference; however, none of the texts reproduced here are exact transcripts of what participants said in the roundtable dialogue. Rather, this chapter represents the best efforts of the authors to give an account of the ideas they propagated in the dialogue.[1]

George G. Brenkert on the future of business ethics as a practical effort

George Brenkert addresses business ethics as a practical effort. He emphasizes the need for business ethics to develop in ways that make it possible to properly embed ethics in organizations, and discusses implications for how the field should develop as well as how teaching ought to be conducted.

Business ethics is an ambiguous term encompassing a number of different activities. It can refer to the study of the ethics of business (to simplify) in either a normative and/or conceptual manner, or to the social scientific study of the actual ethics of individuals, organizations, and systems related to business. It can refer to the ethics by which businesses operate or should operate. And it may include the teaching of business ethics within businesses, or at academic institutions such as colleges and universities.

Quite frequently, business ethics have been distinguished from corporate social responsibility. On this view, business ethics refers to ethical issues between management and employees, while corporate social responsibility refers to the (ethical) relations of business with society and the environment. I think that this distinction is not helpful because both areas concern various values and norms by which individuals and businesses either do or should relate to one another. I look forward to further criticism of this distinction and its increasing irrelevance in the future.

Over the past several decades, business ethics, in its broad sense, has expanded its reach in two different ways. First, a broader range of participants has entered the discussion of business ethics topics. Initially, business ethics was greatly influenced by philosophers who took up the banner. However, soon others from social science and practitioners from business entered the discussion, both changing and enriching the discussion. Second, the topics of business ethics have broadened. From issues of normative ethics, they have expanded to include empirical and legal topics. One of the interesting and valuable developments in the last couple of decades has been the study of ethics in economic decision making, both by economists and other social scientists. These studies will continue to add to our understanding of business ethics.

I am confident that we will see other changes in the future regarding the topics of business ethics. A great amount of time has been devoted over the past decades to employee/employer issues as well as stockholder/stakeholder issues. The different positions are fairly well known, and the lines dividing different camps are well marked. However, if business ethics is a practical effort, more time needs to be spent on how these ethical views might be embedded in actual organizations. For this, too little time has been spent on how the ethics of organizations change, what the role of power is in effecting changes, and what the relation of business to other institutions in society is, for example to politics. In recent years there has been some tentative movement to address these questions. It is an area in need of further development because it is linked to a more general issue of the role of power in providing a context and a basis for ethical behaviour. It is especially here, I think, that business ethics needs to move in a much more robust manner, both theoretically (that is philosophically) as well as empirically. Another area deserving of greater attention than it has received is the impact of digital technologies on the ethics of business and social relations promoted by business. Given the dynamic nature of this area it will be the locus of an ever-changing set of ethical challenges.

Finally, there will be new developments in the teaching and inculcation of ethics both in academic contexts as well as in businesses. Within business, ethics officers have played an important role in moving beyond compliance responsibilities to those fostering ethical cultures in business. However, their responsibilities remain separated too strongly from the responsibilities of those who are linked with corporate social responsibility. Greater integration of these different functions and responsibilities in business is both desirable and likely. And both need to give greater attention to the ethical dimensions of the products produced, how they are produced, where produced, how marketed, and how recycled.

Within the academic domain there have been important changes in the teaching of business ethics (and CSR). We have long since passed beyond the simple examination of philosophical theories of ethics and their application to business situations. Case studies, forms of casuistry, debates, ethics bowls, practical projects examining real-time ethical situations in business, video creation, and the use of social media are expanding and deepening instruction in business ethics. I assume that there will continue to be an expansion of such innovative and engaging efforts to address business ethics with students. One of the most crucial directions that would benefit business ethics is the examination of mainline theories of finance, economics, strategy, and accounting, as taught in business schools in an effort to uncover their ethical assumptions that many believe inhibit the role and place of business ethics within these academic institutions.

"A business ethics that moves in these directions will be, I think, vastly interesting, rewarding, and potentially more effective than it has hitherto been." (George G. Brenkert)

Peter Pruzan on what is required if business ethics is to have a future

Peter Pruzan asks the fundamental question of how we can ensure that business ethics is to have a future at all. He addresses the need to move beyond purely rational perspectives, and sheds light on how spirituality may enrich business ethics as a field and as a practice.

I have little faith in my ability to see into the future. Therefore, instead of providing a scenario on the future of "business ethics", I will provide some reflections on developments that will be required *now* if the field is to *have* a future – if it is to continue to be regarded as a domain that is worthy of academic recognition and respect that attracts inspired theoreticians and practitioners as well as the eyes and ears of the media and the rest of us.

My personal experience, my readings in the field and the continued stream of scandals one meets virtually every day in the international media have convinced me that the major challenge facing the field of "business ethics" (and of ethics in other fields such as the bioethics, as well as the related fields of "sustainability", "environmental accountability", and "corporate social responsibility") is to help transform the mindset of our leaders to a higher level of conscious awareness. And this is not just in business, but in all fields of organized activity.

By this I do not imply that it is not important for the future welfare of this globe and its inhabitants to improve research in and the teaching of "business ethics" as we know it today. It will continue to be important for organizations of all kinds and at all levels to develop shared and *healthy*[2] perspectives on values, to develop ethical codices and guidelines for employees and suppliers, and to provide training on dealing with moral dilemmas – and other such practical tools and perspectives on ethical challenges in organized commercial activity.

However, and the following is my major concern here, unless we are able to complement the almost exclusively rational approaches to the field (for example analyzing complex decision situations employing consequentialist and deontological approaches) with approaches that emancipate and empower the individual's inner guidance, we will experience widespread and well-deserved skepticism with respect to the raison d'être of the field of "business ethics".

The shift in the consciousness referred to is such that "business ethics" will no longer primarily be regarded as providing limitations on the actions of corporations while they pursue their financial goals, so they are not challenged due to being unethical. Instead, goals and limitations will be reversed – the overall goal will be contributing to the common good – acting so as to contribute to the well-being of the organization's stakeholders – subject to the limitation that the organization is financially healthy and viable.

In order for such a reorientation of goals and means to take place, what I have referred to earlier as a shift in consciousness, will require that decision makers supplement their traditional rational analyses with spiritual-based guidance from their "inner world" when faced with challenging decisions in the external world of business. And a condition for them to be able to do so will be that they actively focus on emancipating and empowering the inherent source of their ethicality. Providing the motivation and the means for so doing will be the main challenge to teachers of a revitalized "business ethics".

It is my strong conviction that unless such a shift takes place we will continue to observe that concern for the common good remains subservient to ego-based aspirations for financial wealth and power. Blatant greed and a lack of sensitivity to the aspirations and needs of the constituencies affected by our organizations' actions will continue to undermine our trust in our business enterprises, their leaders, and the institutional frameworks they function in – and, therefore, in "business ethics".

I realize that an appeal to complement rationality in the theory and practice of "business ethics" with spirituality can and will be seriously challenged in a world where economic rationality dominates the perspectives of both business academicians and practitioners. But unless the field of "business ethics", meaning the academic community and the active practitioners in various corporate, consulting, regulating, and advisory bodies, actively focus on motivating and inspiring the individual decision maker to supplement their economically-based success criteria with perspectives on the "good life" for one self as well as the organization and its sundry stakeholders, "business ethics" will continue to deal more with *un*ethics than with ethics.

At a practical level, perhaps a primary operational challenge is for our theoreticians to inspire the practitioners to expand the standards and measures of performance they use to characterize business decisions, including inner promptings that are not subject to rational analysis. Inspiration can be obtained from such diverse and "unorthodox" sources as "happiness research",[3] research on the interrelationships between management, spirituality, and religion,[4] and research on spiritual-based leadership.[5] The footnotes provide just a few possible sources of inspiration.

But other challenges to the field of "business ethics" will clearly arise from the above focus on a shift in consciousness and the emancipation and empowerment of one's inner guidance. These do not only include developing a widely accepted vocabulary and frame of reference (considered briefly above with respect to "success"), but also to such matters as: Who will be qualified to teach and provide advice on such matters? Who can "teach the teachers" and "advise the advisors"? How to avoid confusing a call for spirituality in business with a return to religious dogma? How to develop decision making environments and tools that facilitate the resolution and dissolution of conflicts in groups where the individual participants each refer to deeply felt convictions that are not in harmony with the convictions of the others, and where each participant strives to experience harmony between their thoughts, words, and deeds?

The challenges are many, as well as potentially hugely rewarding for theoreticians and practitioners of "business ethics" and for the well-being of our societies.

Eleanor O'Higgins on ideals and realities in the future of business ethics

Eleanor O'Higgins takes as point of departure the distinction between the ideals and realities of business ethics – what it is and what it ought to be. She discusses virtues that should be embraced in order for business practices to change fundamentally, and how this relates to the identities of business practitioners.

The question about "the future of business ethics" can be approached from two perspectives. One is the "ideal", that is what *should* the future of business ethics be in practice? The other is the "reality", *what is it likely to be?* If the future state of business ethics resembles business ethics today, then it will certainly not be anywhere near the ideal.

When considering the ideal, it is common wisdom to expect business ethics to feature various characteristics – business practitioners, and especially leaders with a sense of values and principles that balance economic wealth creation with social and environmental sustainability. This can be achieved by mutual respect in human relationships and by recognizing our interdependencies. It also requires actors or agents to appreciate the consequences of their actions on others and take personal responsibility for them.

In pondering questions about consequences and personal responsibility, business practitioners will necessarily consider their own identity, who they are, how they define themselves and the limits of what they would or would not do for the sake of economic success. Business ethics, at its best, should encompass the traditional cardinal virtues of prudence, justice, temperance/discipline, and fortitude. The theological virtue of charity would also not go amiss.

This may be seen in contrast to the situation at present, where mainstream businesses, rather than embracing virtues, may instead be seen to value many of the seven deadly sins, in particular greed and pride/arrogance, largely bolstered by envy. Not only are these not regarded as sins in contemporary business, they are perversely seen as facilitating social welfare, as embodied in economic man.

The definition of business ethics matters here. As long as business ethics is defined as outside the realm of general ethics it is perceived as a special case so that standards of moral right or wrong in our everyday lives – the virtues – are somehow suspended when it comes to business. This boils down to ways of justifying personal selfishness, an atomistic view of human beings where everyone is out for themselves without considering the consequences for others, and a sense of entitlement to materialistic goods without limit. This has enabled business actors to perpetrate harm to society, forcefully manifested in the global economic crisis.

This worldview of business is reinforced by some of those who inhabit the top echelons of business and their self-serving interest in maintaining the current culture. Moreover, research on corporations consistently shows a self-selection bias where many of those who choose and get ahead in business careers possess psychopathic traits like narcissism, egocentricity, cynicism, and ruthlessness. Such traits are exacerbated when individuals reach the top as they abuse their power to ensure that no one else will wrest it from them.

So, if the future "reality" is to be brought closer to the "ideal" in business ethics, it means that we have to change our models of business, its purposes and its mores, to a form that no longer rewards the practice of "sins". Such rewards attracted those who excelled most at unethical behaviour. Instead, business has to be seen as a medium for practicing virtues. This is more easily said than done but, paradoxically, the financial crisis may present an opportunity for change; that is, "You don't ever want a crisis to go to waste; it's an opportunity to do important things that you would otherwise avoid."[6]

Kevin T. Jackson on how future business ethics may transmit authentic human values

The point of departure of Kevin T. Jackson's views on the future of business ethics is that the content of business education conflicts with human nature. He argues that the very salvation of civilization is at stake when it comes to instilling moral and intellectual culture into business, and discusses how we may achieve the goal of transmitting authentic human values through business ethics education.

The job of educating future business leaders is hampered by widespread intellectual weaknesses, not just in business schools but in other institutions of learning too. The weaknesses are simultaneously causes and effects of sundry intellectual vices, methodologies, and ideologies that are hostile to, and incompatible with, a sound understanding of human nature and its relationship to business and economics.

Business ethics needs to somehow absorb the worth and dignity of philosophy – not only for understanding the significance of business as a human endeavor, but also for successfully conducting business. I would exhort business people of all kinds and in all nations to think broader, bigger, and deeper. By emphasizing the significance of philosophy for business and economics, our field of business ethics must be firmly rooted in questions of truth, goodness, and even beauty.

The field of business ethics should foster understanding of the relationship between business and the wider moral and intellectual culture, not for the sake of solving mere intellectual puzzles, but because in a very real way the salvation of civilization is at stake. It is of vital importance that business ethics be equipped to defend the ability of human reason to know the truth. A firm confidence in reason has been an integral part of the western philosophical tradition, but it stands in need of reaffirmation today. At least part of the reason for this is connected with the global financial crisis. Economic crises are accompanied by disturbing perplexity about values, not only financial ones, but moral, epistemological, and aesthetic ones as well. It is this aspect, I believe, that makes such crises particularly disquieting, not just within financial institutions, but also across the wider culture.

I believe that widespread misunderstanding of the relationship between economics (ultimately grounded in truth), ethics (ultimately grounded in goodness), and culture (ultimately grounded in beauty), particularly among those charged with teaching business, weakens the ability of both business schools and business organizations to cultivate and transmit authentic human values. The "ethics" that corporations and business schools try to transmit, when they badly misunderstand the relationship, is "business ethics" only in a superficial sense. All too often it is an agenda for political correctness (motivated from either end of the political spectrum), or it sometimes amounts to "window dressing" to serve the narrow interests of the organization.

So, going forward, business ethics must continuously resist getting captured by corporate interests, political interests, and technocratic interests.

Georges Enderle on different levels and challenges in the future of business ethics

Georges Enderle argues that the future of business needs to encompass multiple levels of analyses in order to address the complex challenges it faces. Furthermore, he argues that business ethics needs to combine descriptive analytical and normative-ethical approaches for a sound treatment of those challenges.

Since the 1980s, business ethics has evolved to a fairly well-established field that covers academic activities such as teaching, training, and research, as well as practical initiatives of businesspeople and organizations, professional associations, governmental agencies, and civil society organizations. Testament to this worldwide evolution in academia is the *Global Survey of Business Ethics in Teaching, Training, and Research* (Rossouw and Stückelberger, 2012). Obviously, the terminology of key concepts varies a great deal according to different languages, cultures, and foci of interest. In English there are particularly three terms that characterize the field or parts of it: business ethics, economic ethics, and corporate responsibility/corporate social responsibility. While the terminology undoubtedly needs to be clarified (see Enderle, 2010), I would like to emphasize some substantive features of "business ethics" (named in short) that are based on the global survey and my recent personal experience in the field.

When dealing with business ethics, one should place it in the key context of "globalization" and strive for "sustainability" as a key goal (see the textbook *Business Ethics* by Crane and Matten, 2010). Taking into account the different foci of interest around the world, business ethics should encompass (at least) three levels of analysis: the micro or individual level, the meso or organizational level, and the macro or systemic level. In order to address complex challenges such as climate change, trust in the financial system, poverty, and wealth creation, it is insufficient and often misleading to limit the analysis on one or two of those three levels. Individual ethics or organizational ethics (for example corporate responsibility) is no substitute for institutional ethics nationally and internationally. Remember that the neglect and ignorance of institutional ethics in the 1920s and 1930s in Continental Europe led to the disaster of the Second World War.

A second lesson from the evolution of business ethics in the global context can be described as "the need to walk on two legs". This means that business ethics would fail if it contented itself with only describing and analyzing the world (for example, why it is "bad") or only postulating and arguing for how the world should be. Rather, business ethics should comprehend both the descriptive-analytical and the normative-ethical dimension of the challenges to be addressed.

A third lesson concerns the relationship between theory and practice. Academia can learn a great deal from good and bad business practices. The development of global standards such as the Global Reporting Initiative, the UN Global Compact, and the UN Guiding Principles for Business and Human Rights is a good example, while corporate scandals like Enron and the global financial crisis have not only shocked the public, they have also shaken up many academics. At the same time, confronted with major challenges, business ethics as an academic endeavor bears a fair share of responsibility to provide guidance for addressing those challenges.

To be more specific, I suggest that three major challenges lie ahead of us, which have emerged with increasing clarity, urgency, and importance, concerning all levels of business, from the individual to the organizational and the systemic level: (1) a rich and comprehensive understanding of wealth creation as the purpose of business and economics; (2) the guarantee of securing all human rights to all people; and (3) the active involvement of the world's religions in meeting the challenges of creating wealth and securing human rights (see Enderle, 2011).

Johannes Brinkmann on criticism, collaboration, and confrontation in the future of business ethics

Johannes Brinkmann emphasizes the multiple roles that business ethics needs to play in the broader field of business. He argues that business ethics should remain a criticism of business, yet develop as a collaborative partner of business, but that it should even – when it is necessary – confront business in order to improve business.
Where should business ethics go?

Our field should continue, elaborate, and deepen our function as a well-founded *criticism* of business practice, business research, and business education, together with (preferably in cooperation with, if necessary in confrontation with) our business school colleagues, but at the same time, we should:

Stay constructive, that is participate actively, together with business practitioners, in drafting and *staging positive examples of and for business practice,* for example in an action research format.

The interesting challenge is of course to weigh and balance our equally important ideals of academic integrity (critique) and practical significance (positive examples) for making the business world a better place.

I am not sure where to start: with business practitioners, with our business school colleagues, or with our business students, and among these groups with the interested and receptive ones, or with the ones with the lowest demand and presumably the highest need for business ethical enlightenment. But we should start somewhere and walk, not only talk.

Josep M. Lozano on personal development in the future of business ethics

Josep M. Lozano's starting point is that business ethics is crucially about personal development, and that this feature of business ethics needs to be emphasized and clarified. He argues that business ethics has the potential to create spaces of freedom wherein individuals can go beyond reproduction of existing cultures, and improve those cultures through criticism and personal development.

Where are we going with Business Ethics?

Business ethics (BE) should be at the hearts of business schools. We have confused offering BE courses with being trained in BE. The subjects are there, but not the BE (or not enough of it, at any rate). It is not just a question of integration and mainstreaming, as we usually say. It is mostly about the fact that the management discourse and its various disciplines have been built self-sufficiently without considering their ethical and value dimensions and, once built, there has been endless debate as to how to incorporate, add, and inoculate them. And, if these dimensions are not intrinsically present from the beginning, they will never be fully present.

BE should focus much more on the processes of learning and personal development. Business schools are full of programs that talk about development (leadership development, management development, and so on) but what needs to be developed is not always obvious. BE is so concerned with the contents that it often only manages – if it does in fact do so – to change the politically correct discourse and increase the complexity of the analysis. But it scarcely accompanies processes of transformation and/or personal development. Not least because the latter does not refer to a subject, but to the educational project of the respective institutions; a topic which, curiously, has rarely impacted upon BE as a discipline. In the current debate on the contribution of business schools, ethics and the need for ethics, or not, as the case may be, comes under discussion. But, we will find little material on BE contributions in connection with business schools' educational models. It seems that BE is considered so important in its own right that it doesn't even bother to reflect on the institution that shapes its context (and, by the way, its credibility).

BE should be a catalyst capable of facilitating business schools, going beyond the reproduction of the dominant culture in relation to vision and business management. Being uncritical reproducers of what is generally accepted serves to train, but not to educate. Generally speaking, a business education has been much more business than education. When I talk about going beyond reinforcing and reproducing what has already been established, and also to question it, I am not referring to techniques, capabilities, and procedures, but to attitudes and mentalities, which is why we are talking about BE in the first place.

BE should help build the personal integration of the education received. It is a common misconception to draw attention over and over again to the risk that what is learned in business schools is a collection of disciplines without anyone taking responsibility for how this accumulation of fragments is personally articulated. It is left to chance, to personal journeys, and to the school of life's hard knocks. And professional ethics has more of a role to play in how each person performs this articulation than the subjects of BE. The quality of business schools in the future will involve more than simply stating that there is a risk of becoming a provider of fragments. And, in this respect, if it is important to address this issue, BE should come up with something appropriate to remedy the situation.

BE should help create spaces of freedom in business schools, even at the price of reducing its own space as BE. By spaces of freedom, I mean that educating the future management world requires voices to be heard in business schools other than those that come from the world of management. We will neither improve management capacity nor will we be able to respond to the future needs of organizations or emerging societies if we understand management education as a closed loop in which management only speaks *of* management *for* management. To better understand what it means to engage in management, business schools should hear other voices from other spheres: culture, spirituality, and politics. BE should be capable of producing spaces of freedom, not of joining in and replicating what goes on in the silos of repetition.

Ove D. Jakobsen on the future of business ethics

Ove D. Jakobsen addresses the interdependence between the economic system and the ecosystems within which economic activities operate. He proposes that a new paradigm is needed, whereby business ethics moves from a materialistic to a spiritual worldview.

Business ethics could not be separated from the question: "Is economics a moral science?" In my opinion, economics is nothing but values. The goal in the economy is to transform nature to commodities that increase the well-being of human beings without disturbing the interplay between different ecosystems. To do this, knowledge is of great importance. Economic activities are, in other words, a combination of natural and cultural values. Production, distribution, consumption, and redistribution (reprocessing of waste) are all necessary functions in the economy. Today the negative unintended consequences of the exploitation of nature and the unfair distribution of wealth are extremely serious. To solve these problems, a deeper change in economics, business administration, and private lifestyles are all necessary. This change has fundamental implications for business ethics.

The present convergence of crises "in money, energy, education, health, water, soil, climate, politics, the environment, and more – is a birth crisis, expelling us from the old world into a new" (Eisenstein, 2011, p. xx). Never in our history have humans faced so many interconnected global crises of our own making. We are now facing an unprecedented challenge which requires us to radically rethink our worldview and to reflect on ecological and societal consequences of how we organize the local, national, and global economy. Creating a viable future is about justice – for humans, for the Earth, and future generations of all species.

Because most scholars today accept that crises are symptoms of a deeper ontological, epistemological, and ethical character, I argue that it is necessary to change from a materialistic to a spiritual worldview. The consequences for business ethics are fundamental. Schopenhauer and Nietzsche's criticism of the modern society represents an interesting introduction to this topic. In addition, Kierkegaard's (2004, 2012) three modes of existence represent an interesting reference to explore the need for a change in business ethics and the driving forces behind the change processes.

In the context of the spiritual worldview the religious mode of existence is of special interest. The reflections on the religious mode of existence represent the interpretative context for development in business ethics. Because spirituality is more closely connected to intuition than rationality, the focus will change towards implications of intuitive ethics. To be conscious that we are integrated parts of the living Gaia leads to a kind of eco-centric moral responsibility with implications for business ethics.

The spiritual worldview represents an interesting contribution to the discussion on how to rethink and develop practical solutions to the most urging challenges in business ethics. Kierkegaard described the progress through the three modes of existence as a gradual appropriation of the individual's existence. "To have an authentic existence one must not be a mere spectator or passenger in life, but seize control of one's own destiny" (Kenny, 1998, p. 300). That is to say, that the process is "an adventure, a risk, a self commitment to an objective uncertainty" (Copleston, 1985, p. 344).

In my opinion the development of business ethics goes along with a deep change from rational self-interest to a kind of intuitive holistic concern. This change process has the potential to unfold solutions to the most urgent challenges, locally, regionally, and globally. Where governments and traditional leaders look at problems from the outside, spirituality comes to understand them intimately, in a holistic perspective from within. The return they seek are not in profits, "but in advances in education, environmental protection, rural development, poverty alleviation, human rights, healthcare, care for the disabled, care for children at risk, and other fields" (Bornstein, 2007, p. 12).

Knut J. Ims and Lars Jacob Tynes Pedersen on character building and the development of moral imagination

In a joint reflection, Knut J. Ims and Lars Jacob Pedersen take a virtue ethical approach and focus on character building and the development of moral imagination to secure ethical behaviour in business and organizational life.

With Hannah Arendt's conception of Adolf Eichmann as a super-efficient manager in mind, empathy and compassion should play an important role in the teaching of business ethics. Adolf Eichmann was one of Adolf Hitler's faithful servants and he was able to plan, organize, and implement a huge project – framed as the project of "Endlosung" that ultimately resulted in sending six million Jews to concentration camps. The term "the banality of evil" was coined as a way of seeing Eichmann as a leader who did not have strong negative feelings for the Jews, but rather saw it as his honor to do a good job – to be an effective manager. Bandura, Caprara, and Zsolnai (2002) write about moral disengagement strategies, practices that form a slippery slope to corporate transgressions. Such mechanisms enable "eminent members of the business community" who are not "dangerous, criminally oriented mavericks" to commit unethical acts without experiencing moral guilt.

How can education contribute to preventing managers from carrying out harmful acts towards other people or other sentient beings? In our view, character building is central, and a main emphasis should be to have a balanced approach between heart and mind. This means to apply at least two perspectives to the same phenomenon – a perspective characterized both by proximity and distance. It means that one should not start with aggregating humans or any other kind of being, one should start with seeing their face. We should think of everybody as a face, and that everybody has a name and address, and thereafter we may proceed towards more distanced, abstract perspectives. This would be in line with Martin Buber's (1958) two ways of living: *I-Thou* before *I-It.* It is also very fruitful to view business through the lens of Ian Mitroff's (1998) four-dimensional perspective: the scientific-technical perspective, the social-interpersonal perspective, the systemic perspective and, not at least, the spiritual-existential perspective. The latter involves posing profound questions related to meaning and purpose in life, and questions about having a "ground (life) project that is worthy and noble".

A main purpose of business ethics should be the development of empathy. Many methodological tools may promote this end, including the use of role play and literature for character development. For more than a decade we have experimented with such methods in the ethics classroom, and students have engaged in and appreciated this approach. By reading classical works such as Henrik Ibsen's *Peer Gynt* and *The Enemy of the People,* students may engage in ethical reflection aimed at opening their minds and hearts. Cases and literature may enable students to develop a moral language that they can master in the environment of business and the culture of particular organizations. This involves understanding that "business ethics embraces much more than simply cultivating the ability to "Just say no" or "Just say yes" to clear-cut alternatives. It includes the discovering, anticipating, encountering, and constructing of moral problems" – many of which are dilemmas that require wisdom in order for the actor to create workable solutions (cf. Ciulla, 1991, p. 212). By exploring the moral shades of business life using economic know-how as well as wisdom, students will be better able to invent sustainable, multicolored solutions. In a larger sense, character building is an attempt to build a link between wisdom and empathy. This should be the main objective of business ethics in the future, in order to stimulate sensitivity, compassion, and wisdom in future business practitioners.

Laszlo Zsolnai and Antonio Tencati on regaining legitimacy and relevance in the future of business ethics

Concluding the dialogue, Laszlo Zsolnai and Antonio Tencati jointly voice the concern that business ethics has lost credibility and relevance, and they ask the question of what is required to regain its legitimacy. They argue that new paradigms and practices are necessary, and highlight four levels on which innovative approaches are necessary.

The recent economic and financial crisis shows that business ethics has lost its credibility and relevance. It became evident that business ethics teaching did not change the general attitude of managers in mainstream business. Ethics and compliance programs were not able to prevent major banks and big corporations from entering into questionable practices and make dirty businesses all over the world.

One explanation of the betrayal of business ethics is that our discipline did not question the underlying models of mainstream business, namely profit maximization and agency theory. Also, business ethics did not have the courage to challenge the institutional structure of contemporary global capitalism. The consequence is that exploitation of our planet's finite sources is continued, the basic needs of society are disregarded, and the interest of future generations is neglected. By its institutionalized greed culture and the "enrich yourself" mentality mainstream, business became a retrograde force in the development of humanity in the twenty-first century.

Therefore, alternative managerial paradigms and business practices are badly needed. This calls for an in-depth reframing of the prevailing assumptions in economics and management based on a negativistic and reductionist view of human nature. If business ethics scholars want to foster careful, collaborative, and responsible attitudes in doing business, they have to foster innovative research at four levels:

- individual (individual level), beyond the unrealistic homo economicus conception;
- firm (micro level), beyond profit maximization and agency theory;
- districts, clusters, industries, and sectors (meso level), beyond the five forces model;
- the economy as a whole (macro level), beyond the infinite growth idea.

These research efforts need to aim at building a new vision of management for really sustainable patterns of development beyond the current competitive, disruptive and, thus, self-defeating approach. This should also have a huge impact on universities and business schools and their role in educating future practitioners.

In conclusion, to gain legitimacy and relevance business ethics should aim to develop a normative but usable theory of ethical management. This theory needs to address systemic and operational issues of managing business in a sustainable, future respecting, and pro-social way. It implies that (i) business activities may not harm nature or allow others to come to harm, (ii) business activities must respect the freedom of future generations, and

(iii) business activities must serve the well-being of society. Ecology, respect for future generations, and serving the well-being of society call for a radical transformation of business. If business ethics cannot contribute to the much-needed transformation of business, it may disappear as a discipline that did not deliver on its promised benefits.

Summary

This chapter has presented views on the future of business ethics that were voiced in the structured dialogue at the closing session of the 7th Trans-Atlantic Business Ethics Conference. The dialogue partners all expressed that ethics should be a vital part of the teaching and practice of management and that, faced with serious crises in business and society in the new millennium, the need for critical and innovative thinking is even more pressing than earlier.

As some of the participants voiced, business ethics has lost credibility and relevance and, therefore, we are in great need of new paradigms and practices. We believe that the central ideas, beliefs and insights expressed by this panel of business ethicists offer new windows into the rich complexity of moral reasoning that is necessary for business ethics to develop and regain its legitimacy and true meaning, and for it to be the important challenger of mainstream business thinking and practice that is necessary.

Notes

1. The following conference attendees also participated in the dialogue, but their contributions are not reproduced here: Thomas Donaldson, Prakash Sethi, Wesley Cragg, Andrew Crane, and Thomas Beschorner.
2. I note that organizations such as the mafia and Hells Angels may excel in developing shared values and norms, with clear guidelines for behaviour and strictly enforced sanctions for behaviour that deviates from the norms – yet most of us would feel

that it would be a serious misnomer to characterize such organizations as "ethical"; the values and norms disregard the values and norms of the many constituencies that are affected by their actions.

3. See for example <http://wwwl.eur.nl/fsw/happiness/>, the website of the *World Database of Happiness: Continuous register of scientific research on subjective appreciation of life* and <http://www.springer.com/social+sciences/well-being/journal/10902>, the website of the *Journal of Happiness Studies* published by Springer.
4. See for example <http://www.tandfonline.com/toc/rmsr20/current>, the website of the *Journal of Management, Spirituality and Religion,* published by Taylor & Francis.
5. See for example <http://www.globaldharma.org/sbl-research.htm>, the website of the Spiritual Based Leadership Research Programme for a knowledge-base of interviews and materials. In 2013, this research program and knowledge base was transferred to the Tyson Center for Faith and Spirituality in the Workplace at the University of Arkansas: <http://tfsw.uark.edu/>.
6. Quote from Rahm Emanuel (Mayor of Chicago) in 2008.

Bibliography

Bandura, A., G.-V. Caprara and L. Zsolnai (2002), "Corporate Transgressions", in *Ethics in the Economy. Handbook of Business Ethics* (ed. L. Zsolnai), Bern: Peter Lang, pp. 151–164.

Bornstein, D. (2007), *How to Change the World – Social Entrepreneurs and the Power of New Ideas,* Oxford, UK: Oxford University Press.

Buber, M. (1958), *I and Thou* (2nd revised edn) (R. G. Smith, Trans.), New York: Charles Scribner's Sons.

Ciulla, J. B. (1991), "Business Ethics as Moral Imagination", in *Business Ethics. The State of the Art* (ed. R. E. Freeman), New York: Oxford University Press,

Copleston, F. C. (1985), *A History of Philosophy* (Book 3), London, UK: Doubleday.

Crane, A. and D. Matten (2010), *Business Ethics. Managing Corporate Citizenship and Sustainability in the Age of Globalization* (3rd edn), Oxford, UK: Oxford University Press.

Eisenstein, C. (2011), *Sacred Economics – Money, Gift and Society in the Age of Transition,* Berkeley, CA: Evolver Editions.

Enderle, G. (2010), "Clarifying the Terms of Business Ethics and Corporate Social Responsibility", *Business Ethics Quarterly,* 20 (4), pp. 730–732.

Enderle, G. (2011), "Three Major Challenges for Business and Economic Ethics in the Next Ten Years: Wealth Creation, Human Rights, and Active Involvement of the World's Religions", *Business and Professional Ethics Journal,* 30 (3–4), pp. 231–252.
Kenny, Anthony (1998), *A Brief History of Western Philosophy,* Oxford, UK: Blackwell Publishers.
Kierkegaard, S. (2004), *Either/Or -A Fragment of Life*, London: Penguin Books.
Kierkegaard, S. (2012), *Fear & Trembling,* New York: Merchant Books.
Mitroff, I. (1998), *Smart Thinking for Crazy Times: The Art of Solving the Right Problems.* San Francisco: Berrett-Koehler Publishers, Inc.
Rossouw, D. and C. Stuckelberger (eds) (2012), *Global Survey of Business Ethics in Teaching, Training, and Research*, Geneva, Switzerland: Globethics.

KNUT J. IMS AND LASZLO ZSOLNAI

Social Innovation and Social Development in Latin America, Egypt, and India

Both creative destruction as well as destructive creation in the economy, and the quality of products and the fairness of the market, need serious reflection (Guiltinan, 2009). There is not always a good match between the products that are produced and marketed and the basic needs of consumers in the market. Some innovations increase the level of "hard" technology and efficiency and facilitate a centralized and globalized economy. The forces of globalization, with all its contradictions and inequalities, have accelerated the growing disparity between rich and poor. Among the manifold consequences of this growing inequality are poverty traps, resulting in poorer health and, ultimately, lower life expectancies in many countries. The real price of financial crises and globalization is not primarily an economic problem, but rather a serious human one with existential consequences that are generally undercommunicated (Stiglitz, 2013). In the same manner, Helena Norberg-Hodge (2000) gives a detailed description of the real costs of modern technical and economic progress.

Our main focus in this chapter is on social innovations that aim to serve that part of the population that is at the bottom of the economic pyramid and cannot afford to pay for the usual products in the free market (see Phills et al., 2008). Inspired by the work of Muhammad Yunus (2010), different types of social innovation in the developing world are explored. While recognizing the importance of technological or financial innovations, we call for serious ethical reflection on the perils of the so-called "devil's doctrine" and the "economic doctrine". The "devil's doctrine" consists of creating and constructing whatever is technically possible regardless of its destructive potential, while the "economic doctrine" is about "selling what can be sold profitably, no matter its substance". We challenge both doctrines by presenting and analyzing successful social innovations. They not only serve the interest of commercial markets, but also advance social development (Ims and Zsolnai, 2014).

The bottom of the pyramid model

The business models of the "bottom [or base] of the pyramid" (BoP) approach (Prahalad, 2004; Hart and London, 2005) receive worldwide recognition today. BoP means developing innovative businesses to serve the largest but poorest socioeconomic group in the world. In global terms, about 4 billion people live on less than $2.50 (US) per day.

The base of the pyramid protocol, developed at Cornell University's Johnson School of Business, is a corporate innovation process for serving communities at the base of the world's income pyramid. With the BoP protocol, companies work to develop new markets and lift up low income communities through the co-creation of innovations and enterprises in the BoP.

By building common ground between companies and communities, the BoP protocol is a model for enterprise and market co-creation. Via the BoP protocol, a company can:

- build a deep understanding of BoP communities' needs and aspirations;
- develop key local partnerships in BoP communities that form the foundation for a sustained business venture;
- incubate and launch businesses that are embedded in the community's aspirations, strengths, and values;
- co-create new markets via new value propositions that neither the company nor its community partners could have created on their own.

The BoP protocol extends the logic of the market and follows profit making as the primary goal of the company while aiming at serving poor communities. In contrast, exemplary social innovation cases do not involve profit as the primary goal, but instead emphasize social, spiritual, and humanitarian goals, such as minimizing suffering and empowering people and communities (Ims and Zsolnai, 2014).

Because of its primary profit orientation and materialistic mindset, the success of BoP models is, in the end, limited. Cornell University's expert, Erik Simanis (2012), argues that there is a fatal flaw in the low-price, low-margin, high-volume strategy that multinationals have been pursuing at the bottom of the pyramid for the past decade. Companies seeking to improve the lives of the world's poor should elevate gross margins far above the company average by pushing down variable costs and boosting the price consumers are willing to pay for a unit of product. They also need to raise the price point for a single transaction.

Not every need of the poor can be transformed into marketable demand and, when it is possible, social, environmental, and cultural externalities may override the material benefits provided by BoP projects. Social development requires going beyond the logic of the market and transcending profit making as the primary goal of business.

The Economy of Communion experiment

We have selected the Economy of Communion (EoC), founded by Chiara Lubich in May 1991 in Sao Paulo Brazil, as our first alternative business model. This model has been analyzed by many scholars; among others, Gold (2003a, 2003b) and Linard (2003). Gold (2003a) uses the concept "economy of sharing", but she is referring to the same phenomenon. Linard describes the EoC as a business paradigm that arises from the spirituality and social praxis of the Focolare Movement, founded by Lubich already in 1943. According to him, this paradigm represents a living "third way" between Marxism and capitalism and is in genuine contrast to many earlier movements of business such as the co-operative or the Quaker business, which have been successful on national as well as on international scales. The problem with those movements, according to Linard, is that "[t]hey have generally merged with the prevailing business ethos to the extent that their outward face to the world is indistinguishable from that of their capitalist competitors" (p. 166). However, according to Linard, the praxis of the Focolare movement is in sharp contrast to those movements. The EoC is not an economic theory; instead, it is characterized by practitioners living an ideal and creating an economic theory. The EoC is no longer a localized phenomenon, but "part of a global praxis – a living 'Third Way' " (ibid.). Within the first decade of its founding, businesses operating within the paradigm spread to 100 countries and attracted some 800 firms, generating almost 100 academic theses.

The Focolare entrepreneurs, workers, directors, consumers, savers, citizens, scholars, and economists are all committed, at various levels, to promoting a practice and an economic culture imprinted on communion, gratuity, and reciprocity. Entrepreneurs are invited to share their profits to sustain the goals of the EoC: that is, reduction of exclusion and its subsequent poverty; diffusion of the culture of giving and of communion; development of businesses and creation of new jobs; business people who can conceive and live their activity as a vocation and service for the common good; aiming at the excluded in every part of the world and every social context (Economy of Communion, 2014).

The sharing of profits in three parts is a practical proposal: (1) one part of the profits would be reinvested in the business in order to develop and create new jobs; (2) the second part would be used to create a new culture, which would inspire women and men capable of incorporating communion into their lives; (3) and the third part would go directly to the poor so as to reinsert them fully into the dynamic of communion and reciprocity (Bruni and Hejj, 2011).

The EoC is not an experience defined by philanthropists or great entrepreneurs who give their goods to the poor without questioning their own lives, without becoming brothers or sisters in equal standing with the "poor" they are helping. The EoC businesses are really an economy of communion even when they have no profits to give, but when they are working they produce a culture of fraternity. Chiara Lubich, the founder, is very explicit on how this redistribution took place in the community of Trent:

> The spirit of unity in charity was the ever-living flame that kept this fraternity alive … it did not require everyone to sell all that they possess, depriving themselves of

> what they could without causing themselves harm ... Everyone brought the extra that they had, above all in money, and committed themselves to giving a monthly sum. The donor and the amount promised remained secret. With the money received, the committee would help, month by month and in secret, those families in the community in need, carrying out this delicate task with the greatest charity and discretion. (Chiara Lubich, "Erano un cuor solo e un anima sola" [They were of one heart and one soul], 1948, cited in Gold, 2003b, p. 148)

Today, business and the economy meet a vital need for communion, happiness, relational goods, and goods of gratuity. By remaining faithful to its vocation day by day, the EoC is increasingly capable of producing these typical "goods". Unity with God and thinking in terms of community are the essence of the Economy of Communion (Bruni and Hejj, 2011). Gold (2003b) emphasizes that this model generates a wider client base and higher loyalty from both employees and clients, especially in an age of crisis. Gold states that an important aspect of the communitarian ethos of the Focolare is the building of relationships based on communion. This ethos may be called a culture of giving, and rests on a different anthropology of the human person than is dominant within economic theory. Thus, the EoC stresses an anthropology that sees human beings as "in search of the other's gaze and in search of communion" (Gold, 2003a, p. 171), not as isolated beings, but as beings who find fulfillment in meaningful relationships with others. As a consequence, barriers of distrust that often exist between businesses and NGOs in formal partnerships have not been a problem within the EoC due to the high level of prior knowledge of the work done by Focolare. Gold stresses that the focus of the Focolare's work is not on poverty alleviation *per se*, but on building relationships based on mutual care and solidarity.

In the early 2010s the Focolare Movement operated in 182 nations with more than 100,000 adherents, and linked to more than 800 companies.[1]

The SEKEM experiment

The organization SEKEM – which stands for "vitality from the sun" in Ancient Egyptian – was founded in 1977 by the Egyptian pharmacologist and social entrepreneur Ibrahim Abouleish in order to bring about cultural renewal in Egypt on a sustainable basis. Located northeast of Cairo, SEKEM includes: biodynamic farms; trading companies for produce and processed foods, herbal teas and beauty products, medicinal herbs and medicines, and organic cotton products; a medical center; a school based on the principles of Waldorf pedagogy[2] open to pupils from any religious or ethnic background; a community school catering specifically to the needs of children from disadvantaged groups; a vocational training center; a college (Mahad Adult Education Training Institute); a research center (SEKEM Academy for Applied Art and Sciences) and the Heliopolis University. It focuses on sustainable development and is working with several partners from Europe in a large consortium. Among its programs there is a basic training course for solar technology instructors. Special importance is given to the interaction between all four sectors of society: the economic, cultural, social, and environmental sectors.

SEKEM's goals are to "restore and maintain the vitality of the soil and food as well as the biodiversity of nature through sustainable, organic agriculture and to support social and cultural development in Egypt" (Abouleish, 2012). By 2005, the organization had established a network of more than 2,000 farmers, and numerous partner organizations in Egypt, and had increasingly extended its experience and knowledge to other countries, including India, Palestine, Senegal, Turkey, and South Africa.

In 1975, Dr Abouleish decided to begin a project of cultural renewal based on a synthesis of Islam and anthroposophy. Two years later, he bought a plot of land in what was, at the time, desert-bordering farmland of the Nile valley. The original goal was to develop the land and improve crop yields using biodynamic methods. A thick border of trees was planted to encircle the 70 hectares of land, and trees were planted along all the roads built; a forest was also planted on part of the land. The Bedouins, who lived nearby, and sometimes on the land, were brought into the project, given work and helped with their living needs. Buildings were constructed using traditional *adobe* – mud brick. The initial farm animals were Egyptian buffalo.

The first large economic venture of the community initiative was production of a medicinal compound, ammoidin, an extract of *Ammi majus*

(Laceflower). The manufacture of herbal teas and a company to market fresh biodynamic produce in Europe followed. The needs of these companies led to many farms throughout Egypt switching to biodynamic methods; the SEKEM organization began an active advisory service to aid these farms in the transition to and the maintenance of biodynamic standards.

Community projects began early in the initiative's history, including a medical clinic using anthroposophical medicine and a Society for Cultural Development sponsoring lectures, concerts, and other cultural activities. In 1987, the center for adult education (Mahad) began its work; children with disabilities were also educated in this center. In 1988, SEKEM opened a kindergarten also for the local Bedouin community. This grew into the SEKEM School, educating children from kindergarten through 12th grade (about 18 years of age) based on Waldorf education (the educational movement of anthroposophy). The school serves Muslim and Christian children alike, encouraging them to live in harmony and have respect for the other's religious practices (SEKEM, 2015; Waldorf Today, 2015). The school also operates a literacy center for children between 10 and 14 years of age. In response to the use of child labor in Egypt, SEKEM founded the project "Chamomile Children", which offers children between 10 and 14 years an education, vocational training, meals, and medical care in conjunction with their work; the children's teachers look after the children throughout the day. There is also a cooperative for employees to organize the social processes (the Cooperative of SEKEM Employees), an independent organization with members from all the businesses and cultural institutions associated with SEKEM.

In 1990, SEKEM founded the Center of Organic Agriculture in Egypt (COAE), an independent organization that inspects organic farms in Egypt, Iran, and Sudan. Faced with pesticide residues in their products that came from aerial spraying on nearby farms, SEKEM took up an initiative to eliminate such spraying in Egypt. As cotton production depended upon the sprays, SEKEM explored organic cotton production on initially small fields. The experiments were successful and yields actually were better than non-organic production achieved. The Egyptian Ministry of Agriculture sponsored further and more extensive tests. Within three years, the ministry agreed that organic pest suppression was superior for cotton farming and began converting the entire area of Egyptian cotton, 4,000 sq. km, to organic methods for controlling pests; the conversion took two years. The conversion resulted in a reduction in the use of synthetic pesticides in Egypt by over 90 percent, and an increase in the average yield of raw cotton of almost 30 percent. SEKEM then created a company, Conytex, to process organic cotton using mechanical rather than chemical methods.

In response to increasing publicity about the novel methods employed by the community in many realms, the association of Muslim sheiks in Egypt gave the community a plaque verifying that SEKEM is an Islamic initiative. This was the result of intensive meetings between SEKEM and Muslim religious leaders. SEKEMs next initiative was the first private pharmaceutical company in Egypt, specializing in medicinal teas. The medical center had now grown large enough to need its own building; concurrently, it considerably expanded its outreach into the Bedouin community, helping establish sanitary facilities and clean water supplies.

In 1997, SEKEM established a vocational training center offering training in metalwork, carpentry, mechanical work, electrical work, tailoring, biodynamic farming, and trading. The vocational center also included an art school. The German Society for Technical Cooperation helped establish this project. In 1999, the SEKEM Academy (now Heliopolis University) opened. Originally a center for agricultural, pharmaceutical, and medical research, the university now also conducts studies and offers training in other areas.

In 2001, a holding company was established to administer the finances of all the SEKEM companies and to oversee developmental projects. The holding company includes a department to help each individual company with its developmental process, and is responsible for the education and training programs for employees.

All SEKEM companies have a policy of ensuring transparency in the production, distribution, and consumption of their goods. They work to ensure fair prices for the farmers supplying them, basing their operations on the principle that the health of the economy depends upon producers, distributors, and consumers cooperating to generate stable businesses.

SEKEM and Dr Abouleish received the Right Livelihood Award in 2003 for integrating commercial success with the promotion of the social and cultural development of society. The organization has been cited as a successful example of social entrepreneurship that has had a significant impact on Egyptian society, both through its influence on the country's agricultural practices and through its educational and cultural institutions. Furthermore, Dr Ibrahim Abouleish and his son, Helmy, were named Social Entrepreneurs of the Year 2003 by the Schwab Foundation for Social Entrepreneurship, and well-known business schools, like IESE in Spain and CIDA City Campus in South Africa, use various case studies (for example, Harvard Business Press) to illustrate this novel approach to sustainable social entrepreneurship. When Dr Abouleish was appointed an "Oslo Business for Peace Honouree" in 2012, he expressed that one of his inspirational sources was the Norwegian philosopher Arne Næss's ideas about "deep ecology".

Yearly seminars that reflect on the aims and goals of SEKEM have evolved the following motto: "We aim towards living together according to social forms that reflect human dignity and further development, striving towards higher ideals. Our main goal is a developmental impulse for people, society and the Earth." The cooperation of economic, social, and cultural activities is stimulated by science, art, and religion, and SEKEM has set itself the following goals for the economic sphere:

- healing the Earth through biodynamic farming;
- developing and producing herbal remedies and any kind of product or service that relates to real consumer requirements and has standards of the highest quality;
- marketing in associative cooperation among farmers, producers, traders, and consumers.

SEKEM aims to advance the individual development of the person through cultural institutions. Free education is the goal of SEKEM's educational institutions for children, adolescents, and adults. SEKEM provides health care and therapies using natural healing remedies. The SEKEM Academy for Applied Arts and Sciences researches and teaches solutions to pertinent questions from all areas of life. Socially, SEKEM furthers a community of people from all over the world who recognize the dignity of the individual, enabling both learning to work and working to learn, and providing equal rights to all.

Aravind Eye Care System experiment

Aravind Eye Care System, founded in Madurai, Tamil Nadu, by Dr G. Venkataswamy (known as "Dr. V"), has implemented an alternative model of health care in India, and Aravind is now one of the largest providers of eye surgeries in the world. Aravind's foundational principles and work ethics set it apart from the usual health organizations run on business lines. The firm is framed by market gurus (Kumar et al., 2000) as a *market driving,* not a market-driven firm. Aravind as a transformational venture goes beyond the practice of social entrepreneurship by adopting and successfully applying the idea of individuals paying for services according to their means. Through its focus on service to the individual as the exclusive motor of any business operation, Aravind's business philosophy moves away from the pure economic logic. Aravind manifests a pattern of entrepreneurial action and a management style that breaks away from profits, markets, products, and consumers. Individuals are no longer a variable that must be adjusted to the imperatives of expansion, viability, or numbers (Virmani and Lepineux, 2014).

Aravind's adoption of spiritual principles of service and compassionate care distinguish it from other health care actors. Its eye camps in villages are one example of how public health interventions can promote solidarity among the population. As a young man, Dr V was a disciple of Sri Aurobindo, an Indian philosopher and saint who lived in service to God and man. The name of the Aravind Eye Care System is inspired by this sage, whose philosophy may be summed up as believing in the interconnectedness of human beings. Aravind has promoted a shift in the social contract with people. As distinct from the traditional view of the social contract in which individuals are isolated from each other, Aravind promotes an interconnected form of social contract, which encourages individuals' involvement, commitment, and responsibility to community welfare. By its elaboration of eye care as public good, not private good, Aravind illustrates how transformational entrepreneurship can contribute to the common good of society (ibid.).

Transformational entrepreneurship is an ongoing process that creates a ripple effect: like a stone thrown into the water, it generates waves from the center, affecting people, institutions, and the state. The necessary transition calls for the development of transformational entrepreneurship ventures, which – like Aravind – can have a multilevel impact and expand the very notion of entrepreneurship. Thirty years after Dr V's modest start of Aravind, the system consists of a network of five regional hospitals performing over 200,000 cataract operations per year, 70 percent of them without cost to the patient (Matalobos et al., 2010). Aravind is not funded through government grants, aid agency donations, or bank loans, since Dr V has chosen to develop an unusually high operational efficiency that, consequently, has extreme cost reductions. Thus, the organization has been supported by an average contribution of $40 (US) from the one-third of its patients who can afford it. Another innovative element is the use of "screening camps". In India, there are large populations of untreated patients who cannot afford to go to the main hospitals. In order to solve this problem, Aravind sets up temporary screening camps in villages. These camps perform classifications of patients and offer free transportation to hospitals for patients who require specialized treatment. In 2006, 1,793 screening camps were established, and more than 2.3 million patients were examined, performing 270,000 sight restoration surgeries. This initiative has been extended to special areas, like school screening camps, and includes innovative technology such as teleophthalmology (ibid., p. 2).

Internationally, Dr V was also one of the founding directors of the Seva Foundation, based in the United States. Seva was created by veterans of the smallpox eradication program, and one of its goals was the elimination of needless blindness, beginning in Nepal and India. Seva's collaboration with Aravind began with small grants to subsidize the cost of cataract surgery, but soon developed into a partnership. Seva eye care workers from Nepal, China, Cambodia, and Tanzania were trained at Aravind. In addition, Aravind sent senior staff to assist in program development, hospital design, and operation training, with no charge by Aravind, to "nurture a growing network of effective eye care centers" (Brilliant and Brilliant, 2007, p. 51).

In 2005, with the encouragement of Aravind, Seva launched the Center for Innovation in Eye Care in Berkeley, California. One of its key activities is the development of the Center for Community Ophthalmology (CCO). These CCOs include at least one institution in every region of the world, drawing strongly on Seva's partnership experience with Aravind. One should also mention that Aravind has shaped international eye care through its emphasis on understanding the human aspect of sight restoration using social science research. Dr V insisted that research should be done to explore why rural patients who could benefit from surgery were not seeking it. One of the findings was that literacy and gender largely determined who got surgery and who stayed blind. "Globally, women carry a greater burden of blindness than men; two out of every three blind are female … not because of genetic propensities, but because, worldwide, utilization of preventive and restorative eye care for females has not been equal to that of their male counterparts. This inequity begins in childhood." (ibid., p. 52). A number of international organizations, like WHO, have now affirmed that the "right to sight" should be equally available to women and men.

Understanding "social innovation"

The three founders of these alternative business models – SEKEM, Economy of Communion, and Aravind Eye Care System – may be regarded as social entrepreneurs. They all experienced a powerful vision or a mission. Mitroff's four perspectives on any problem (Mitroff, 1998) may be a useful framework to analyze the similarities and differences among the three exemplary cases.

The different perspectives in Mitroff's model are scientific/technical, interpersonal/social, systemic, and existential/spiritual (ibid., pp. 58–65). Mitroff emphasizes that the existential/spiritual perspective concerns some of the most basic issues of the human condition: meaning and purpose. Spiritual does not necessarily mean religious, but it means a different perspective than the materialistic one. Mitroff argues that the existential/spiritual perspective is not incorporated in theories of management. When this perspective is not taken into consideration, "[w]e have no comprehensive theories of management!" (ibid., p. 151). The scientific/technical is the ruling paradigm in our culture, emphasizing instrumental rationality that favors scientific/technical solutions. This easily leads to trying to solve most problems with scientific/technical means, even if the problem may be defined as existential or spiritual and therefore in need of different approaches. The scientific/technical paradigm thus easily leads to technocentrism, which may produce powerful and latent side-effects in the short as well as in the long run, causing irreversible threats to the life of plants, animals, and human beings (Ims and Zsolnai, 2006).

Each of these three cases has a fundamental similarity that can only be accounted for by introducing an existential/spiritual perspective. The social entrepreneurs behind the three cases had a *strong vision or mission* that may explain the impetus for their commitment, innovative thinking, and actions. We will not attempt to define spirituality in a scientific way, because we believe that such a definition may inherently miss what is truly spiritual. If we want to keep the notion of spirituality experience based, we have to accept that spirituality is a rich, intercultural, and multilayered concept. As a guideline we used the working definition of the Leuven-based SPES Forum:[3] "spirituality is people's multiform search for a deep meaning of life interconnecting them to all living beings and to 'God' or 'Ultimate Reality'" (SPES Institute, 2015). Most definitions of spirituality share a number of common elements: reconnection to the inner self; a search for universal values that lifts the individual above egocentric strivings; deep empathy with all living beings and, finally, a desire to keep in touch with the source of life (whatever name we give it). In other words, spirituality is a search for inner identity, connectedness, and transcendence (Bouckaert and Zsolnai, 2011).

We think that spirituality melds human beings with something supernatural and that a scientific approach may only be able to measure the effects of spirituality rather than spirituality itself (Newberg and Monti, 2011). As Bruni and Hejj write about the EoC, "[L]ife precedes the theoretical reflection that always accompanies it, because life is denser with truth than theory (which serves life inasmuch as it comes from life and is nourished by it)" (Bruni and Hejj, 2011, p. 381).

All three founders were strongly touched by what they saw when they traveled in the local communities. Abouleish, who was born in Egypt (1937) and educated in Austria, visited Egypt in 1975 and felt deep and poignant grief for the social and economic conditions. He saw destructive pollution of the land due to heavy use of pesticides. Two years later, he founded SEKEM in the desert outside of Cairo. Dr V retired at age 58 from the Government Medical College, Madurai, as the head of the Department of Ophthalmology. However, he was so affected by the suffering of the blind people in his region that he determined to continue his work of organizing and providing quality eye care (Manikutty and Vohra, 2004). Chiara Lubich was a mature woman when she visited Sao Paulo in Brazil and was "confronted with a scenario that powerfully symbolizes the potential contradiction with capitalism: a forest of sky scrapers surrounded by a savannah of slums. The EoC arose as a response to her experience of the suffering of humanity" (Bruni and Hejj, 2011, p. 380). Lubich was deeply moved by the poor neighborhoods where people had no work and no dignity. Within a few days of that trip, the EoC was born. We believe that the success behind all three social innovation models may be explained by the way they integrate *all four perspectives* suggested by Mitroff's framework.

All three founders had strong visions behind their engagement. Originally, Chiara Lubich had such a powerful religious experience that she changed her name from Sylvia to Chiara in honor of Clare of Assisi. This transforming event was dated to December 7, 1943, and it may be said to be the beginning of the Focolare movement. When bombs were falling on her hometown, she communicated one wish to her closest friends. Only one inscription should be carved on her tomb: "And we have believed in love."

In their article "From market driven to market driving", Kumar et al. (2000) use Dr V's enterprise as an exemplary case:

> [What] Aravind Eye Hospital shares with other market driving firms … is the inability of the market driven approach to explain their success … The inspiration for the radical business ideas of these market driving firms came from a visionary such as Dr. Venkataswamy … who saw the world differently and whose vision addressed some deep-seated, latent, or emerging, need of the customer … Ultimately, these firms revolutionized their industries by changing the rules of the game and driving their markets. (Kumar et al., 2000, p. 130)

Concerning the *systemic* perspective, SEKEM is explicitly taking the sustainability of the natural environment into account. This is one of its "raisons d'être" and is one of the strongest markers of its new Heliopolis University for sustainable development and leadership. SEKEM plays a role in transforming the cotton industry of Egypt into organic. It demonstrates that a national environmental revolution is possible when scientific knowledge is used in a pro-nature, innovative way (Abouleish, 2012, pp. 137–143).

Aravind's eye care system has demonstrated that it is self-financed and not dependent upon financial aid or economic support from external sources. The system's efficiency in structures and operations makes it possible to have an economic policy expressed as follows: "We set prices not on our costs, but on who can afford to pay and how much." In this way Aravind cross-subsidizes its different patients to the advantage of the worst-off patients. Aravind is also developing the market and thus generating a demand. In a process of converting a basic need (blindness) into a demand, the model becomes very sound in a business sense. Another core principle in Aravind's business model is excellence in execution of efficiency in providing the treatment, including outpatient services and surgeries. The last core principle is *quality*.

The EoC enterprises also take the *sustainability* perspective into account through its organizing in "business parks", networks, and redistribution of profits through the network. The owners of EoC companies are committed to putting into practice social responsibility by paying fair taxes to the authorities. In that way local development is supported. Probably the most important factor behind the EoC network's success is the empowering element of being a member of the network. When individuals are freed from their poverty traps and regain their dignity, individuals develop hope, inspiration, and energy, which generates their creativity. Such a development changes a person's status from being a victim to external forces to becoming a productive resource contributing to the common good. This personal development is closely linked to *social* perspective in Mitroff's framework, which is emphasized explicitly by all three enterprises.

In the EoC the social dimension is the first and primary principle. The economic exchange of goods plays only a subordinate role. A core principle is acknowledging the dignity of all people in practice by giving them opportunities to become economic players and partners in the EoC network.

In Aravind the organizational vision, to "eliminate needless blindness" as an outreach to people who suffer, and to be able to treat as many patients as possible even if they cannot pay, is a demonstration of interconnectedness and our co-responsibility for each other. In SEKEM, cultural and social events, particularly their morning circles, where all co-workers and leaders meet each other and share their upcoming duties and tasks, are an important part of the daily routine. A number of art activities play an important role in the social life of the SEKEM initiative. In Stockholm in 2003, Dr Abouleish and the SEKEM farm received the Right Livelihood Award. The jury gave this reason for the award: SEKEM is "[a] business model for the twenty-first century in which commercial success is integrated with and promotes the social and cultural development of society through the 'economics of love.'" (Abouleish, 2011, p. 145).

Table 1 summarizes the main characteristics of the alternative social innovation cases presented in this chapter.

What is common in the analyzed cases is the strong intrinsic-spiritual motivation to serve the common good of society and the attempts to understand and measure success in holistic terms that go beyond monetary results

Table 1. Main characteristics of alternative social innovation cases

	Inspiration	Vision	Means	Outcome
Economy of Communion	Christian love, spirituality of unity	Serving the poor and the needy	Building relationships based on a culture of giving, solidarity and sharing	Restoring dignity of persons and communities
SEKEM	Anthroposophy and Islam	Restoring and maintaining the vitality of the soil and the biodiversity of nature	Sustainable, organic agricultural methods	Catalyzing social and cultural development
Aravind	Aurobindo's philosophy of human interconnectedness	Eye care as a public good	Individuals paying for services according to their means	New social contract for the common good

and market outcomes. The primary goal is not profit making, but providing livelihoods and enabling services that enhance people's well-being. We think this kind of motivation may corroborate the concept of embeddedness used by Mark Granovetter (1985) and Johanna Mair and Ignasi Marti (2006), in which they argued that economic functioning is embedded in social and structural relationships that modify economic behaviour. Mair and Marti (2006) state that social entrepreneurship, like entrepreneurship in the business sector, cannot be understood in a purely economic sense, but needs to be examined in light of the social context and the local environment. We have emphasized the spiritual dimension as the major factor in all three cases we have studied. We do not underemphasize the social and local environment, but we believe that our cases demonstrate that we need to take the spiritual dimension into account in order to explain the successes of our sample cases. As stated in the Right Livelihood Award given to Abouleish in 2003, "The SEKEM initiative shows how modern business can combine profitability and engagement in world markets with a humane and spiritual approach to people and respect for the natural environment." (Abouleish, 2011, p. 145).

Concluding remarks: Producing goods that are truly good and providing services that really serve

The market is a powerful institution that provides private goods and services which satisfy the short-term and, to some extent, long-term preferences of people who have enough purchasing power. The strong materialistic bias of the conventional business models, including bottom of the pyramid models, makes it difficult to serve the needs of those stakeholders who have little or no voice in the market system (Zsolnai, 2015). Billions of people in developing countries are unable to pay for products and services that would satisfy their basic needs. The World Bank (2003) states that services that satisfy basic human needs, such as health care and education, are failing poor people – in access, in quality, and in affordability. For an inclusive social development, innovative social enterprise models are needed that transcend the logic of the market in business and the economy and introduce more holistic approaches in order to produce goods that are truly good and provide services that really serve.

Notes

1. See <http://www.focolare.org>.
2. See <http://www.waldorfanswers.com/Waldorf.htm>.
3. See <http://eurospes.org>.

Bibliography

Abouleish, I. (2011), "The economics of love", in C. Egge (ed.), *Levende økonomi. Ledende ¢konomer tenker nytt*, Oslo: Flux forlag.

Abouleish, I. (2012), *SEKEM. En veiviser for det nye Egypt* [A Guide for the New Egypt], Sverre Dahl, trans. (original title: *Die Sekem Vision. – Eine Begegnung von Orient und Okzident verandert Agypten*), Oslo: Antropos.

Bouckaert, L. and L. Zsolnai (2011), "Spirituality and business", in L. Bouckaert and L. Zsolnai (eds), *The Palgrave Handbook of Spirituality and Business*, New York: Palgrave Macmillan, pp. 3–8.

Brilliant, L. and G. Brilliant (2007), "Aravind: partner and social science innovator", *Innovations, Technology, Governance, Globalization*, 2(4), 50–52.

Bruni, L. and T. Hejj (2011), "The economy of communion", in L. Bouckaert and L. Zsolnai (eds), *The Palgrave Handbook of Spirituality and Business*, New York: Palgrave Macmillan, pp. 378–386.

Economy of Communion (2014), *What is the EoC?*, accessed September 6, 2014 at <http://www.eoc-online.org/en/eoc/about-eoc.html>.

Gold, L. (2003a), "Small enterprises at the service of the poor. The economy of sharing network", *International Journal of Entrepreneurial Behaviour and Research*, 9(5), 166–184.

Gold, L. (2003b), "The roots of the Focolare Movement's economic ethic", *Journal of Markets and Morality*, 6(1), 143–159.

Granovetter, M. (1985), "Economic action and social structure. The problem of embeddedness", *American Journal of Sociology*, 91(3), 481–510.
Guiltinan, J. (2009), "Creative destruction and destructive creations: environmental ethics and planned obsolescence", *Journal of Business Ethics*, 89(1), 19–28.
Hart, S. L. and T. London (2005), "Developing native capability: What multinational corporations can learn from the base of the pyramid", *Stanford Social Innovation Review*, 3(2), 28–33.
Ims, K. J. and L. Zsolnai (2006), "Shallow success and deep failure", in L. Zsolnai and K. J. Ims (eds), *Business within Limits. Deep Ecology and Buddhist Economics*, Oxford: Peter Lang, pp. 3–24.
Ims, K. J. and L. Zsolnai (2014), "Ethics of social innovation", *Society and Business Review*, 9(2), 186–194.
Kumar, N., L. Scheer and P. Kotler (2000), "From market driven to market driving", *European Management Journal*, 18(2), 129–142.
Linard, K. T. (2003), "Economy of Communion: systemic factors in the rise of a new entrepreneurship", *Systems Research and Behavioral Science*, 20(2), 163–175.
Mair, J. and I. Marti (2006), "Social entrepreneurship research: a source of explanation, prediction, and delight", *Journal of World Business*, 41(1), 36–44.
Manikutty, S. and N. Vohra (2004), *Aravind Eye Care System: Giving Them the Most Precious Gift (RI)*, Ahmedabad: Indian Institute of Management.
Matalobos, A. D., J. Pons and S. Pahls (2010), *Aravind Eye Health Care Operations*, Madrid: IE Business School.
Mitroff, I. (1998), *Smart Thinking for Crazy Times: The Art of Solving the Right Problems*, San Francisco, CA: Berrett-Koehler.
Newberg, A. B. and D. Monti (2011), "Neuroscience of spirituality", in L. Bouckaert and L. Zsolnai (eds), *The Palgrave Handbook of Spirituality and Business*, New York: Palgrave Macmillan, pp. 26–33.
Norberg-Hodge, H. (2000), *Ancient Futures. Learning from Ladakh* (revised edition of the inspirational classic), London: Rider.
Philis, J. A., K. Deiglmeier and D. T. Miller (2008), "Rediscovering social innovation", *Stanford Social Innovation Review*, 6(4), 34–43.
Prahalad, C. K. (2004), *Fortune at the Bottom of the Pyramid: Eradicating Poverty through Profits*, Upper Saddle River, NJ: Prentice Hall.
SEKEM (2015), Website, accessed May 30, 2015 at <www.sekem.com>.
Simanis, E. (2012), "Reality check at the bottom of the pyramid", *Harvard Business Review*, 90(6), 120–125.
SPES Institute (2015), Outline of *The Spiritual Dimension of Business Ethics and Sustainability Management* by L. Zsolnai, accessed July 26, 2015 at <http://eurospes.org/content/book/spiritual-dimension-business-ethics-and-sustainabiIity-management>.

Stiglitz, J. E. (2013), *The Price of Inequality*, New York: Penguin Books.
Virmani, A. and F. Lepineux (2014), "*Aravind Eye Care Hospital as transformational entrepreneurship: breaking free from economic logic*", paper presented at the conference on "Ethical Leadership: The Indian Way", Bangalore, January 9–10.
Waldorf Today (2015), "Sekem: a flourishing community in Egypt", *Waldorf Today,* accessed May 30, 2015 at <www.waldorftoday.com/2012/05/sekem>.
World Bank (2003), *World Development Report 2004. Making Services Work for Poor People,* Washington, DC: Oxford University Press.
Yunus, M. (2010), *Building Social Business. The New Kind of Capitalism that Serves Humanity's Most Pressing Needs,* New York: Public Affairs.
Zsolnai, L. (2015), "Materialistic versus non-materialistic value-orientation in management", in K. J. Ims and L. J. Pedersen (eds), *Business and the Greater Good – Rethinking Business Ethics in the Age of Economic Crisis*, Cheltenham, UK and Northampton, MA: Edward Elgar Publishing, pp. 107–116.

KNUT J. IMS

Land Ethics in the Context of Economics and Commercialization Effects[1]

Introduction

Mainstream economics and business have supported an all-time high level of materialistic consumption in the West, with destructive impact on different species, the environment, animals, and social justice (Singer, 1995; Ims and Zsolnai, 2006; Steffen et al., 2011; WWF, 2014; Global Footprint Network, 2015). The drive to economic growth and the ideology of more and bigger are better have destroyed the respect for natural limits, frugality, sufficiency, and respect for life. In this paper I will argue that different, deeper, and more complex perspectives on nature can contribute to a necessary transformation of modern economics and business toward a more caring and sustainable form. One main problem is that economic reasoning, which is separated from moral reasoning, has become a main force for rational, instrumental, and narrow self-interested behaviour (Sandel, 2012, 2013).

In the early phases of the development of economics as a discipline economists were interested in agriculture, and one sub-discipline was "Land Economics" (Ely and Wehrwein, 1940, 1984). Ely and Wehrwein argued that the physical reality should be separated from the property relation and should be evaluated on its own premises. If so, the attention would promote a long-run perspective and countervail the devaluation of future generations.

However, Daly and Cobb (1989) state that economics as a discipline has not been connected to the physical world and discusses it as a strong example of what Whitehead (1925, 1967), describes as the fallacy of misplaced concreteness. For economists the concept of land has included the forces of nature as well as the natural environment. In writings from the 1920s, it is clear that a comprehensive literature exists on labor and capital, but not on land as such. Land was exclusively a factor of production, and a property relation similar to other property relations. The main interest was the price of the natural resources and nature was, to a large extent, treated as an unlimited and, therefore, free – zero cost – resource.

On the contrary, the ancient-biblical view of the Earth considers the Earth as a life giver and the source of all that was good. Earth was the production factor par excellence and it was something more, since man belongs to the Earth, honors it, and receives the output with high regard. The Earth includes all the plants and animals.

What implications can the modern negligence of land have? Can we foster a loving relationship towards the Earth and a caring economy? Could business be a solution, rather than a problem in a new life serving paradigm? The paper will introduce Leopold's land ethics and Schweitzer's eco ethics to suggest an alternative way of reasoning.

The fallacy of misplaced concreteness in economics related to land

In the twenty-first century, a new demand for caring for the Earth has been rapidly growing. Climate change, the warming up of the Earth, and more severe hurricanes have stimulated people to reflect upon the western mindset and lifestyle. The impact of business and market behaviour on the environment is devastating. Thousands of species have been exterminated and the exploitation of nature has been disastrous. The material level in most western countries has increased in an unprecedented manner, but not the experiences of a fulfilling life. The material benefits have not been equally shared and the gap between the rich and the poor has increased.

Polanyi (1977) wrote about the great transformation from the feudal to the capitalist market-based society, and argues that the end of this development is that society is nothing more than an "appendage to the market".

The economy is no longer embedded in social interrelationships, but society is embedded in business relations. Sandel (2012) gives a number of examples of the new "market society". It means that the societal context is increasingly subject to the logic of exchange between individuals who are "mutually unconcerned" for one another. According to Ulrich (2002) economic rationality "emancipates" itself from nearly all traditional normative constraints and becomes the dominant societal norm (2013). The dominance of corporations is part of this general pattern. The transformation from an embedded to a disembedded economy reflects the separation of economics from its mother discipline – moral philosophy. Thus, modern mainstream economics has replaced classical political economy with a self-contained neoclassical approach. The instrumental use and exchange of land is an important part of this "development".

According to Sandel (2012) the caring for moral norms is up for sale in the actual "market society". Sandel notes that the more markets extend their reach into traditional "non-economic domains" like education, health, criminal punishment, immigration, and environment protection, the more markets became "entangled in moral". Indeed, Sandel claims that "economists are trafficking in morality". However, there should be moral limits of the market thinking, because markets are eroding moral behaviour. In particular, by allowing and incentivizing certain behaviour with money, interpersonal behaviour will become more and more instrumental, which will have serious effects on how we treat other humans, animals, and nature (Sandel, 2012).

What would be the elements in an alternative framework for economics? With our focus on land, we may problematize the dominant use of abstract thinking and mathematized models. Whitehead's concept of "the fallacy of misplaced concreteness" indicate one huge "error" of modern economics. Whitehead has dealt with the philosophical problem in the modern sciences, (1925, 1967), and has criticized the mainstream thinking of modern economics. When land has been narrowed to property and the buying and selling of land is like ordinary business transactions that takes place in the market, something essential might be missing. Abstractions ease the thinking of soil in a very instrumental way, while the biotic community in itself is so rich that it is doubtful whether the mathematical models are adequate to increase our understanding via the descriptions of the models, as well as the output from the models (see Daly and Cobb, 1989, 1994).

A typical monocultural use of land to serve the international market for agricultural produce typically goes together with heavy use of fertilizing of the land and worse conditions for many members of the local communities. One element of this pattern is the seemingly (but controversial) success of free trade in a globalized world. The economic models used to promote international trade predicted an increase in economic wealth for the exporting as well as the importing countries when the exchange of products was based upon the countries comparative advantages. The application of such models may have contributed to an increased production of materialistic goods in an aggregate sense. However, within each country there has not been equal benefits to all citizens. The distribution of wealth has been biased, and most benefits have been distributed to capital and landowners, not to the laborers, and not to the civil society. It has often contributed to an increase in the socio-economic gap within each country. However, increasing inequality were not components in the macroeconomic models used to promote international trade. The false assumption was that the creation of wealth would benefit all citizens. The sustainability of the land has also been largely ignored.

Within the discipline of economics, there has been a long debate of what role abstract models should play in economics. The result is that economics developed into a theoretical and rigorous social science, with a certain success. The important question is whether the success is shallow. The choice to follow physics implies the choice to formalize and mathematize diverse, dynamic, and complex phenomena. A number of prominent mathematicians have criticized a strict mathematical view of the economy, for example Norbert Weiner: "… the economists have developed the habit of dressing up their rather imprecise ideas in the language of the infinitesimal calculus … Any pretense of applying precise formulae is a sham and a waste of time." (Cited in Daly and Cobb 1989, 1994, p. 31). An alternative way to inform the decision making processes for managers would be to use genuine historical cases as rich and complex sources.

Whitehead (1925, 1967), writes that the problem with economics is that it has been so successfully measured with the standards of the academic world. Smith (1776/1976, 1790) acknowledged as the modern economy's father, was uniting history, moral philosophy, and economy. According to Whitehead (1925, p. 200) the development entailed a "dehumanized industry ... Its methodological procedure is exclusive and intolerant ... It fixes attention on a definite group of abstractions, neglects everything else, and elicits every scrap of information and theory, which is relevant to what it has retained ... But however triumphant, the triumph is within limits. The neglect of these limits leads to disastrous oversights." (1994, pp. 35–36).

Economic science has largely been a deductive science where conclusions are drawn about the real world by deduction from abstractions with little awareness of the danger involved. When economics has developed into an abstract analysis of exchanges on the market, land is sold as any other commodity. However, land is physical space that is decisive for humans as well as other sentient beings. For man, physical space is not only essential in an objective sense, but also in a symbolic, social, and political sense. But land is much more than physical space; nature is probably a better term because it indicates that ecological complexity and the biosphere are central parts of the land concept. In the next section we will consider what is special with land in relationship to other inputs in production processes.

Ancient and modern views of land: Land as space, social and political capital

First, "land is space, on which man has to live and work. The Hebrew word 'maqom' for 'place' means 'standing room'. Nothing can take place – farming, manufacturing, commerce or just living – without space." (Torrey, 1979, p. 1). Torrey asks in a rhetorical way. "And what is war about?" And his answer is: "Who controls space?" The allotment of space is, therefore, an essential question. If there is not justice, it will end in bloodshed. Torrey mention two relevant solutions to land tenure systems: the "Mother Earth-communitarian" solution where Jesus Abbey is one example, and the redistribution of land values through the fiscal system. The latter system will prevent land speculation and create employment, or leave land available for resettlement. Torrey presents the teaching of the Bible with regard to land and describes historical practices.

The law is contained in Leviticus 25, and in particular in verse 23: "The land shall not be sold in perpetuity, for the land is mine:... with me you are but aliens and tenants. Throughout the land that you hold, you shall provide for the redemption of the land."

"But if there are not sufficient means to recover it (a piece of property sold), what was sold shall remain with the purchaser until the year of jubilee; in the jubilee it shall be released, and the property shall be returned." (Verse 28). "I am the Lord your God, who brought you out of the land of Egypt, to give you the land of Canaan, to be your God" (Verse 38).

Torrey writes that the concept of selling land as held in most "civilized" lands today does not exist in the Bible except as a crime. The maximal lease of land was fifty years, but all leases expire in the same year, which is called either the Year of Jubilee, the Year of Liberty, or the Year of the Trumpet.

A precise statement of the Year of Jubilee and the calculus can be found in Leviticus 25, verses 8–13.[2] "You shall count off seven weeks of years, seven times seven years, so that the period of seven weeks of years gives forty-nine years" (verse 8). In the same chapter the Sabbatical Year is introduced as "six years you shall sow your field, and six years you shall prune your vineyard, and gather in their yield, but in the seventh year there shall be a Sabbath of complete rest for the land, a Sabbath for the LORD" (verses 3–4).

But what exactly is understood with the Jewish land concept? In Jewish Scriptures the land is defined in a wide way and includes all the plants and animals. Land is worshipped as all-giving and the giver of life and the source of all good. The people belong to the land, and revere it. According to the Scriptures, there were three factors of production, land, people (Israel), and God (Yahweh). The land was given to the people, and stewardship described the relationship better than ownership. However, there were some selling and buying of the land, in particular in times of crisis. When the urban elite used the land as a monoculture and exported the products, and forced the former farmers to become day laborers, it was regarded as a violation of the covenant.

What is of particular interest is the concept and practice of a jubilee. It meant that land was "reverted" to its original "stewards" (inheritors) (Daly and Cobb, 1989, 1994, p. 103). We may say that this view of the land was connected to agricultural traditions that Jewish people were rooted in. They were nomads, and not dwelling in it.

In this tradition, we also find an interpretation of man as a farmer and servant of the land, and not the dominator beyond other living beings. In Genesis 2:3 we find that humans are not created in the image of God, but out of arable soil, the soil cultivated in the growing of grains, vineyards, and orchards. One implication is that the human being is not distinguished from other forms of life but, identified with them. Plants (Gen. 2:9) and animals (2:19) are both produced by God out of the same arable soil from which the first human was made. Thus, animals share the agricultural identity of humans. The human being and animals alike are called "living beings" (Gen. 2:7, 19) an identity that English translators have been unable to accept. It means that the Yahwist view the human vocation as one of dependence and service (Hiebert, 2009, pp. 681–683).

That land should be treated differently from other property is not only an ancient Judo-Christian tradition, land as property is a subject, which is discussed in a profound manner among influential philosophers like John Stuart Mill in his *Principles of Political Economy*. According to Mill "The Grounds of property in land are different from those of property in moveables", (Mill, 1965, pp. 226–34) since "The essential principle of property being to assure to all persons what they have produced by their labour … this principle cannot apply to what is not the produce of labour, the raw material of the Earth. If the land derived its productive power wholly from nature,… it would be the height of injustice, to let the gift of nature be engrossed by (a few) individuals." Mill also discusses the limitations of the property in land and uses the case of the owners of Irish estates, who "do nothing for the land but drain it of its produce". "That the greatest 'burthen on land' is the landlords. Returning nothing to the soil, they consume its whole produce, minus the potatoes strictly necessary to keep the inhabitants from dying of famine" (Mill, 1965, p. 229). Mill states that "No man made the land. It is the original inheritance of the whole species." "When private property in land is not expedient, it is unjust" (Mill, 1965, p. 230).

Mill made a distinction between moveable and land. While "moveable goods" do not prevent others for acquiring it by the same means, as regards land, the owner will keep others out of the enjoyment of it. One example used by Mill is an owner who tries "to shut up a part of the Highland and exclude the rest of mankind from many square miles of mountain scenery" .... "This is an abuse, because it exceeds the legitimate bounds for the right of landed property." Such a man is morally bound "to make his interest and pleasure consistent with the public good". When land is not cultivated, "no good reason can in general be given for its being private property at all" (Mill, 1965, p. 232).

Mill set the stage for Henry George (1839–1897), another critical voice of the contemporary system with landowners. George advocated that those who occupy land should pay rent to the community for the use of land or the natural resources beneath, on, or above land. This "Georgist Paradigm" (Harrison, 1994) argues for a rehabilitation of the importance of land in the economy and society to create a more just society. Assuming that wages and profits are the rewards for the input of labor and capital, the rent is the reward to land. The rent will be "the measure of the beneficial use of particular sites, or natural resources, which have no cost of production, and which are literally provided free by nature" (Harrison, 1994, p. 174). The problem was that politicians supported by economists assimilated "land" into "capital" and treated nature as just another marketable commodity. Thus, owners of urban land were free to exercise their power and, according to Harrison (op. cit.), land speculation is the primary cause of the collapse of the urban economy. During the transformation from feudal to industrial society, the landlord class was protected, while many of the landless were exploited and fell into endless poverty. As regards the environmental issue, Harrison claims that "The depletion of natural resources and the despoliation of nature is due to a single reason: The failure properly to measure the rental value of all of nature's resources, and to make the users pay the community for the benefits they receive" (Harrison, 1994, p. 189). The logic is that if the users had been obliged to pay rent, they would have nurtured nature, and not exploited her.

In their influential book *Land Economics*, Ely and Wehrwein (1964) give a substantial perspective of land in a broad physical, institutional, and economic perspective. They assert that the "physical facts are the framework within which the economics of land operates". However, this framework is not neutral, but has an effect on the physical conditions of man himself. Taking a historical perspective, Ely and Wehrwein state that the Earth has been the inert and man the active factor in the partnership between Earth and man. They compared the crop yield when the Native American *Indians* "owned" America, and after the industrial revolution. Before the industrial revolution, the land of the USA supported only a few million Indians, but afterwards the land supported more than 200 million.[3] But this seeming success was not without obvious costs. According to Ely and Wehrwein (1964), the production of goods depends upon the intelligence, inventiveness, culture, and the stage of civilization of a people (op. cit. p. 25), but the "industrial man" becomes a cog in a "machine civilization". Forests were removed and fertilizers increased the produce of the land, and the result was that the balance of nature was disturbed. In contrast, "primitive man" only scratched the surface of the Earth (Ely and Wehrwein, 1964).

In *Land Economics* the soil was described as "the final synthetic expression of all these factors"; vegetation, climate, parent rock, and slope working together in the production of the actual soil (Ely and Wehrwein, 1964, pp. 35–36). Ely and Wehrwein expand the meaning of soil, writing that soil is much more than ingredients that nourished the plants. Social and political privileges of land ownership have been very important (op. cit. p. 99). In the western world, only landowners were allowed to vote and be members of parliament. Such extra-economic factors are more important than the purely economic factors in determining land values (op. cit. p. 101). The political and social significance of landed property has resulted in most nations limiting the possession of the soil by foreigners. It may also explain why for instance, mining corporations often owns all the land and houses occupied by their employees. It gives the corporations a certain power over their business interests. There are also social psychological reasons that certain foreign farmers in US had a hereditary passion for landed property. They "clings to the soil in spite of better economic opportunities elsewhere" (op. cit.). In fact, there is a higher percentage of farm ownership among foreign-born whites in the US. Ely and Wehrwein (1964) believe this "intense desire for land" is due to the fact that the foreigners could not satisfy their wishes in their native land. This tendency may express the old view that "The true saving bank is the soil."

Ely and Wehrwein (1964) recall also the social and political view of the American president, Thomas Jefferson (1743–1826), who worked on the Declaration of Independence. Jefferson stated that equality needs widespread land ownership, which is necessary to maintain one's independence. In general, private ownership is seen as a stabilizing force. There is no agrarian unrest when the attachment to the land is strong (op. cit. p. 102).

Land ethics and reverence for life

Modern economics has reduced land to a commodity where rent is the decisive element (Gaffney and Harrison, 1994). In order to maximize profit, large-scale agriculture has been using huge machines, artificially fertilizing the soil, and trying to kill insects with chemicals means such as DDT.[4] The care for land is challenged since the extant use of chemicals leads to many non-intended effects in the long run (see Carson, 1962/2002; Schumacher, 1999). To reestablish land as living biota, we need a land ethic. One of the most famous drafts of a land ethic is sketched by Leopold (1966). According to Leopold an ecological ethic is a limitation on freedom of action in the struggle for existence. A land ethic should reflect an ecological conscience, based upon a conviction of individual responsibility for the health of the land. "Health is the capacity of the land for self-renewal." "Conservation is our effort to understand and preserve this capacity" (Leopold, 1966, p. 236).

Leopold's (1966) concept of land embraces much more than soil. Leopold states that there is no ethics dealing with man's relation to land and to the animals and plants that grow upon it. Land is still property. The land-relation is still strictly economic, entailing privileges, but not obligations (op. cit. p. 218). Therefore, according to Leopold, land ethics is an ecological necessity, and implies guidance for meeting ecological situations "so new or intricate, that the path of social expediency is not discernible to the average individual". One of the challenges is to move from animal instinct towards "community instinct".

Leopold observes different ways to perceive and treat land. One way is to "regard the land as silt, and its function as commodity-production". The other way is to "regard the land as a biota". Within the field of forestry, land as "silt" represents agronomic interests, being content with growing trees, which can "produce" cellulose, which has high market price, feeling no inhibition against violence. On the contrary, land as biota goes beyond economics and involves a richer and more complex perspective. Leopold gives a precise and detailed account of this "biota" perspective (Leopold, 1966, p. 237). As part of this image, Leopold operates with different roles of man, "land as slave and servant" versus "land the collective organism", which influence the way land is treated (Leopold, 1966). Even if we can increase the crop from depleted soils by using imported fertilizing, the food-value is not necessarily improved. The "products we make of fertile soil may be qualitatively as well as quantitatively superior" (Leopold, 1966 p. 238).

The heart of Leopold's land ethics is the conservation of land: "Decent land use should not only be seen as an economic problem". It should also be a question of what is ethically and aesthetically right. "A thing is right when it tends to preserve the integrity, stability, and beauty of the biotic community. It is wrong when it tends otherwise" (1966, p. 240).

The *concept of community* is essential in the Leopoldian view. The boundaries of the community are extended to include, soils, waters, plants, and animals, or collectively, the land. "A land ethics cannot prevent the management of these 'resources' but it does affirm their right to continued existence, and, at least in spots, their continued existence in a natural state" (op. cit. p. 219).

Leopold's land ethics changes the role of *Homo sapiens* from conqueror of the land-community to member and citizen of it. It implies respect for his fellow-members, and respect for the community as such (1966, p. 220). Leopold states that hopefully, we have "learned that the conqueror role is self-defeating". The conqueror believes he knows, but does not actually know the essential things that he controls. The biotic processes are so complex that no one is able to understand it completely (op. cit. p. 220).

The concept of land as a biotic pyramid is another important concept in Leopold's view. Leopold emphasizes that a mental image of land as a biotic community is central (op. cit. p. 230): "since we can be ethical only in relation to something we can see, feel, understand, love or otherwise have faith in". Leopold argues that this mental image of a pyramid is better than the usual slogan of "the balance of nature" (p. 230). The Leopoldian view is based upon evolution regarding food chains as steadily increasing their complexity, and the biota becoming more elaborated and diversified (op. cit. p. 231). In Leopold's view, land is a "fountain of energy flowing through a circuit of soils, plants, and animals". "Food chains are the living channels, which conduct energy upward; death and decay return it to the soil. The circuit is not closed: some energy is dissipated in decay, some is added by absorption from the air, some is stored in soils, peats, and long-lived forests, but it is a sustained circuit …" (Leopold, 1966, p. 231).

Birds as well as predatory mammals are members of the biotic community and should have a biotic right to continue to live, regardless of economic profit (Leopold, p. 226). Lack of economic value is sometimes characteristic not only of species or groups, but of entire biotic communities: Marshes, bogs, dues, and "deserts" are examples (op. cit. p. 228). "No special interest has the right to exterminate them for the sake of a benefit, real or fancied, to itself."

Leopold land ethics can be complemented by the views of Næss and Schweitzer. Leopold's writings were well known by Arne Næss, who formulated and structured "Deep ecology", a radical alternative to the mainstream consumer-oriented lifestyle (1989). Næss believed that philosophy and activism could help chart a way out of the ecological crisis, and invites people to work out their own "ecosophy", that is a philosophy of ecological harmony. This kind of *sophia* or wisdom is openly normative and contains norms, value priorities, and hypotheses.

"Self-realization for all beings" is the hypernorm in deep ecology, and the platform is summed up in eight points. The two first points are directly referring to land in a broad sense: (1) the well-being and flourishing of human and nonhuman life on Earth have value in themselves independent of their usefulness for human purposes; and (2) richness and diversity of life forms are values in themselves.

The central motto of deep ecology is "to live a rich life with simple means" or "simple in means, but rich in ends". The main suggestion is to produce and consume less, to trade lighter and wiser, and to celebrate the virtues of slowness and smallness (Næss, 1989).

Schweitzer's (1998) eco ethics is centered on the view of *reverence for all life*. Schweitzer found the elementary and universal concept of his ethical philosophy during a journey up the river from his hospital in Lambarene in a canoe: "[W]e were making our way through a herd of hippopotamuses, there flashed upon my mind, unforeseen and unsought, the phrase 'reverence for life' … Now I had found my way to the principle in which affirmation of the world and ethics are joined together!" (Schweitzer, 1998, p. 155). Schweitzer summed up his ecological perception with the phrase, *I am life, that wills to live, in the midst of life that wills to live.* It means that there is an individual will to live shared by all other living beings. In this view, a human being is part of a relational web of all living beings. It also means that nature is a relational good.

Schweitzer's philosophy is an experience-based and action-oriented philosophy. In this way it has similarities with Arne Næss's ecosophy and Gandhi's unconditional commitment to nonviolence.

Discussion

Pope Francis published *Care for Our Common Home* as an encyclical letter in 2015. The letter concerns ecological and economic issues and explains the cause of the environmental problems as social and economic problems. Nature should not be viewed only as a source of profit, and it is a wrong to go for quick and easy profit. According to Francis, it is important to reject the beliefs that humans are given dominance over the Earth, because every creature has its own value, beside the immense dignity of each person. The natural environment is a collective good, and the patrimony of all humanity. Thus, everyone has to be responsible for it.

When land is degraded to a commodity that can be bought and sold as any other commodity on the market a number of consequences arise.

Commodification and commercialization effects change human behaviour, and might result in expected as well as unexpected outcomes. The most obvious outcome is reduced respect for the "object". Any intrinsic value will be diluted, and important moral norms and principles may be violated, which may increase the misuse of land in many senses. One actual example is rural areas consisting of fruitful soil, but which are located close to urban centers. There is a strong push by profit-minded persons to buy such land and transform it into factories, office buildings, shopping malls,[5] and parking lots. This push is perfectly logical from an economic point of view. Whenever there is a buyer that has purchase power to buy a property in order to "allocate" the property to a "better" "more efficient" use, it is not easy to stop the deal. The forces of the market, often understood as a "natural force" represent the highest moral principle in the "market society", which is often the winner while the fruitful land is the loser. The Food and Agricultural Organization (FAO) of the United Nations estimates that, in the future, cropland will continue to be lost to industrial and urban development. This will lead to less land used for food production. An important "economic law" is that the greater the shortage of a commodity, the higher price, which will mean less food and higher prices, which will have grave consequences in many developing countries where food shortage is a problem. Poverty in itself leads to bad treatment of land because the users, by reason of need, will act with a very short time horizon. Furthermore, conversion of forests to cultivation results in a loss of biodiversity and increased soil erosion.

How can we avoid the tendency to transform land into "urban development"? One answer will be to pay high respect to nature. We might be inspired by the Judo-Christian tradition where land was not tradeable, or by Leopold's suggested land ethics where land is a very complex living biota, where "decent land use" should be seen as going far beyond economics and should be a question of what is ethically and aesthetically right.

Leopold's "in spots" concept plays a large role in Norway, for the time being, where special areas for wolves have been established. Local farmers regard wolves as a great threat to their livestock in the same "in spots" because some of these wolves have attacked their sheep, sometimes killing a great number in one attack. According to Norwegian law, a certain number of families of wolves are allowed to live in the same habitat as the farmers. When the wolf establishes his family outside the "in spot", the alpha wolf and his family are shot. This is a conservationist action and does not represent a view that animals have intrinsic value.

The essence of the Leopoldian land ethics is that the integrity, stability, and beauty of the biotic community should be preserved as having intrinsic value. Using the precautionary principle, strongly proclaimed by Hans Jonas (1984), this multidimensional perspective has many consequences. It means that we should be reluctant to intervene into any biota unless we have an extended timeframe, and expand the biota into the biosphere. For Jonas, the object of responsibility is the changeable and the perishable, where caution should be the core of moral action. Jonas's moral imperative is "Act so that the effects of your action are not destructive of the future possibility of the permanence of genuine human life" (Jonas, 1984, p. 11). Towards land, humans have an elementary nonreciprocal responsibility and duty. The responsibility stems from our power to act, and the future of mankind cannot be separated from the future of nature. According to Jonas, the care should not only be for mankind, but as solidarity with the organic world. To care for nature is a necessary condition of man's own caring. The interest of man coincides, beyond all material needs, with that of life (1984, p. 136). There is an "overall symbiotic balance, and a common destiny of man and nature" (op. cit.).

Commodification involves a kind of distancing and separation between man and the object. We found that in the Judo-Christian tradition man, land and Yahweh were forming a trinity – a holy unit. When man is separated from land, nature will be regarded as something outside, which will lead to an alienation process. Vetlesen (2015) writes that the more we are alienated from nature, the more we are driven into regression and fascination with technology, which leads to a spiral of destruction. Even if the rhetoric is shifting from "resource management" to "ecosystem services", "the commodification of nature is prevalent" (Vetlesen, 2015, pp. 2–3).

Concerning commodification, there are some strong examples of violating "land" or even the better term, nature. One example given by Sandel (2012) is the Inuit's selling their right to shoot a walrus to "trophy hunters". A number of problems are involved. For trophy hunters hunting is a hobby and entertainment, while the animals have little chance to avoid a deadly shot or serious harms due to modern weapon technology. There might also be unequal bargaining conditions between Inuits and rich people, which could undermine meaningful consent. When one objectifies what one sells, landed property or animals, one fails to respect the dignity of the living beings, and erodes the respective norms. A market practice involves a degradation that can result when turning "all human relationships into transactions and treat all good things in life as if they were commodities" (Sandel, 2013, p. 125). Even if the markets improve efficiency, markets may corrupt or crowd out nonmarket norms of moral importance. One suggestion to stop such commodification of land and animals is to promote Schweitzer's eco ethics' "reverence for life". With such an ethic, there will be no acceptance to sell or buy the right to kill walrus or any other animals for entertainment purpose. As Schweitzer wrote: "We must reach the point that killing for sport will be felt as a disgrace to our civilization." (Schweitzer, 2017, p. 39).

Another example of misplaced use of the market mechanism to protect nature is incentivizing the right to pollute. The Kyoto Protocol on climate change gives countries a choice. A country can reduce their own greenhouse gas emissions or pay another country to reduce theirs. The rational of the Kyoto arrangement is that the market mechanism can be used to lessen the global pollution when buyers and sellers of carbon can trade quotas. A country that pollutes more than their national quota is permitted to buy quotas from a country that has a surplus of "carbon" capacity. This market-based logic to pollution is described as "Tradable Pollution Permits". However, what attitude toward the land do we want to promote? When countries buy the right to pollute, it "would be like letting people pay to litter" (Sandel, 2012, p. 73). Sandel's argument is that we should try to strengthen, not weaken, the moral stigma attached to pollution of the land. The current arrangement between countries means that rich countries can buy their way out of the duty to reduce their own emissions, and thereby undermine the sense of shared sacrifice that is necessary to reduce the pollution of land. One essential aspect of this phenomenon is that global warming is a cumulative harm. It does not matter from which places or who emits carbon from the planet to the sky, "But it does matter morally and politically" (Sandel, 2012, p. 76). The message is that it is ok to use nature as a dumping ground for those who can afford it. Sandel concludes that we have to develop new environmental ethics, and these ethics should affirm moral norms as going beyond economics.

There is an urgent need for ecological consciousness as a cultural norm among people and businesspeople in particular. However, ecological conscience is not necessarily connected to "more quantities of education". For example, Leopold (1966) is worried that the quality of education is diluted, and that higher education seems deliberately to avoid ecological concepts. A better understanding of ecology may "benefit from a great diversity of courses that does not have the label ecology, but geography, botany, agronomy, history, or economics". A central question is how to foster ecological consciousness, and whether formal business education might play a positive or a negative role. The company Patagonia, owned and led by Yvon Chouinard, demonstrates that business can be part of the solution (Chouinard, 2006). One step in the right direction is to take land ethics seriously. A land ethics will imply a number of obligations, but "obligations have no meaning without conscience". Leopold (1966) holds that it is not the "volume" but the content of education that is the key. Looking at different professional fields, in order to gain external legitimation, we find that most of them have taken the same path as economics, became abstract and emphasized explicit knowledge in the tradition of episteme. Thus they are like economists, easy victims of the fallacy of misplaced concreteness because of their way of selecting, articulating, and organizing their knowledge (Whitehead, 1925/1967).

Leopold was educated as a forester and is famous for his dramatic change towards the intrinsic worth of nature when he was hunting in the wilderness; he had shot into a "package of wolfs" and witnessed an old wolf dying. The change took place when Leopold's and the dying wolfs eyes met. The meeting with the dying wolf turned out to become a life-changing experience for Leopold. The context of the meeting was the wilderness.

Does it mean that economists should realize excursions not only to Wall Street and city banks, but also to Antarctica or real Patagonia in the South of Argentina, or to Svalbard close to the North pole to experience wild nature? Should students in economics learn to use a kayak, a surfboard, or climb a mountain wall? According to Leopold, we need presence in the nature and a certain skill to deal with it to see the value of it.

However, there are some obvious problems with such a strategy. Leopold warns against mass tourism, and Vetlesen (2015) warns about the climate change.[6] When the ice and snow have melted, the urban cities have grown, and children use more of their time before the screens (gaming or sending likes to each other), children and adults will not have the chance to stay in nature. Probably, they will also not take the time necessary to travel to the spots, which may be preserved, because it may be seen as waste of time. The natural state will be the end of nature, which will be replaced by the "second nature", dominated by technological tools where human behave and communicate with other human beings and robots via screens (McKibben, 2003/1989).

Zsolnai et al. (2016) posed similar questions concerning incorporating "nature as primordial stakeholder into the functioning of human organizations, including businesses, public administration and civil society organizations". They stated that "the crucial question concerns how to develop the ecological sensitivity and responsiveness of people at different levels of organizations, and how then to translate the emerging ecological consciousness into effective and caring organizational practices through which organizations develop a culture of Earth Citizenship".

The encyclical letter by Pope Francis (2015) proposes the application of an integral ecology where fundamental values are the common good and justice between generations. In order to care for future generations, people have to change their lifestyle in the direction of self-restraint and frugality. Thus, an education that creates "ecological citizenship" is required, together with a new, universal solidarity. There is an intimate relationship between the poor and the fragility of the planet (Pope Francis, 2015).

In the book *Caring economics*, researchers dialogue on how to develop and train altruism and compassion (Singer and Ricard, 2015). Some practical applications in terms of leadership practice are expressed by George (2015), referring to a study at Harvard. The data shows that only one out of three trust their leaders, their values, and their ethics. It is a very challenging task to stimulate compassion through a top-down approach.

Economics is a core component in any business education. Do we have to redefine economics? The subtitle of Daly and Cobb's (1989/1994) book is *Redirecting the Economy toward Community, the Environment and a Sustainable Future*. This implies a mindshift in economic thinking. As Ghoshal (2005) rightly states, bad management theories may destroy management practice. The emphasis should be on community and land, not on individual preferences. It means that the main ideological idea of self-interest should be undermined to the advantage of the centrality of a community, which is not an easy task, but would have a number of benefits (Singer, 2009, pp. 73–78). In our Hellenistic tradition it was a matter of fact that human fulfillment could only be reached within a community. Human is a social being and can realize himself in a polity (Aristotle, 1985). This view is also held by Etzioni (1988) who emphasizes that the economy should be a subsystem of society. Etzioni's "I am We" paradigm is a good step in this direction. McKibben's (2007) concept *Deep Economy* is relevant, because community should play an important part of the solution, and economics should be subsumed to important norms in society – and not the opposite. One of the tasks is to re-embed the economy into society again. This position is in accordance with Polanyi's (1977) substantive definition of economy where he distinguishes between formal economizing, which refers to making the best of one's means, while substantive economizing deals with man's dependence of nature and his fellows for his livelihood. One of the implications will be that frugality, not efficiency, will be the primordial purpose of resource-use in economics (Bouckaert et al., 2008). It also means that the long-standing presumption that "more" equals "better" should be actively attacked.

A shift in mindset needs a different structure of organizing knowledge than through an extensive use of abstract mathematics. Instead, utopias, historical cases, and concrete experiences of nature might play important roles in education of economists. Opening up economics towards a political view, as presented by Mill (1965) under the label "political economy" will represent one way. Daly's (1980) *Economics as a Life Science* was the inspirational source to ecological economics. The central concept is life process, which makes the subject matter of biology and economics into one and the same. The photosynthesis is carried out by green plants, which draw their inputs form the physical environment of air, soil, water, and sunlight (1980). According to Boulding, the throughput should be regarded as something to be minimized – not maximized (see Daly, 1980, p. 258).

Vetlesen (2015) asks us to question the ruling ideology in our western world. The acceptance of limits is a precondition for a necessary reorientation, which demands emotional maturity. Vetlesen also opposes the key statement in the 1987 United Nations requested Brundtland Commission report in 1987. "The ideology of managerial, technology optimistic 'resourceism' on display here qualifies as a grand, and to this day immensely influential, exercise in denial, conveniently avoiding the growth – and profit-based capitalist roots of the problem …" (Vetlesen, 2015, p. 12).

Central in economic theory is economic growth. The ideology is that when making a bigger "materialistic" cake to divide, everybody can get a bigger piece, and everybody will be satisfied. The problem is that larger production means more use of non-renewable resources, which are finite, and even if the resources are renewable, the generation, the transportation, the use of the resource typically lead to depletion, pollution, and waste.

Zsolnai et al. (2016) proclaim that the size or scale of the economy relative to Earth's biophysical systems must be explicitly addressed. The new geological age, Anthropocene, requires an ecological transition, and put forward the following points: (i) business should not transgress its fair Earth share; (ii) business should contribute to the preservation and regeneration of life forms; (iii) business should not violate the freedom of future generations, and; (iv) business should contribute to the flourishing of people and their communities. In order to satisfy the above requirements innovative business paradigms are needed.

We noted that in the Judo-Christian tradition land was not tradable in the long run. Land was highly respected as part of a holy unity. The extent of this praxis, and how long it has lasted, is not documented. However, we know that the idea of treating land as holy has been a strong impulse for community buildings through our history. One example is the Koinonia farm[7] in a small village in Georgia, USA. There is anecdotal evidence that they have been able to recreate the fruitfulness of land by giving great respect to the land by humans deeply inspired by Christianity.

According to Daly and Cobb (1989) the problem is that "modern economics is not materialistic, but idealistic". Thus, it has ignored the physical basis of human existence, the "irreducible stubborn facts of life". However, it is time to revisit the field and regard "the physical basis of human existence" in a more holistic sense, where the material as well as the spiritual basis are included. Such a view is in contrast to the mechanistic worldview that has dominated the modern sciences since the enlightenment. The ancient view of land, and examples concretized by the Judo-Christian tradition, Torrey (1979) and SEKEM in Egypt (Abouleish 2012) illustrate that conventional market thinking about land/nature are challenged.

Alternative models of land use and property relations exist. The attitude of being either a player on the market, with land as an object of exchange depending upon supply and demand in the market, or a steward of land, have essential practical and symbolic consequences. The stewardship model can draw inspiration from the ancient Judo-Christian tradition. Torrey put it eloquently in this way: "The two facts of life are land and labour or space and man …". This alternative model was revitalized by Pope Francis's recent encyclical letter (2015) where the Pope proclaimed the need for an integrative ecology. Thus, there is need for a new vision, which requires an ecological comprehension of land.

Sandel (2012, 2013) discusses the prevalent tendency to commodificate a number of new activities and their implications. Sandel's reasoning illuminates some of the devastating consequences of the commodification of land. The buying of the right to shooting walrus only to be used as trophy is a glaring example of violence towards innocent living beings, and a blatant example of that markets do not hinder perverse desires to be satisfied. Buying and selling of "Tradable Pollution Permits" between countries, are also eroding ecological moral norms. Sandel asks the question: Whether to commodify a good, we must consider more than efficiency and fairness (2013, p. 130). One problem is that market norms crowd out nonmarket moral norms. Through a number of cases across a diversity of domains, Sandel (2012, 2013) shows that the crowding out of norms represents a loss worth caring about. The conclusion is to take "commercialization effects" seriously, even if mainstream economics have ignored these effects, and even made the word "incentivization" one of the newspeak common words in the twenty-first century. The common often hidden assumption has been that the commercialization process does not affect the product. This is a mistaken assumption "looming large" in the rising "economic imperialism", and the attempts to extend economic analysis into domains of social, political (Hirsch, 1976), and natural life. A transactional view implies a degradation of moral values as dignity and the intrinsic values of nature – and such degradation ultimately leads to more exploitation and violence to land. The second hidden assumption is that economists assume that virtues are a scarce resource to be economized. This view assumes that the more we act out of altruistic or love motives, the more those virtues will be depleted. It follows then, according to mainstream economists, that we should save those important virtues to our family. However, according to Aristotle (1985), virtues are to be cultivated in practice. Without any use or cultivation, virtues will be depleted. Use it or lose it (Sandel 2012, 2013).

Inequality is another reason why commodification should be bothering (Harrison, 1994). When more of our lives are "up for sale", persons with little money will lose many essential opportunities. On the opposite, when you are affluent you can buy access to prestigious universities, political influence, good medical care, and finally buy land for financial speculation, and not for cultivation.

Schumacher's *Small is Beautiful* (1999) demonstrates that the scale of business is important. DDT was very useful and saved millions of lives each year, and improved the quality of life of millions living in malarial regions. The serious problem arose when agribusiness started using DDT on a large scale to improve crop yields (Lovelock, 2006, p. 108). Lovelock warns that Rachel Carson's (1962/2002) book *Silent Spring* did not come only from the poisoning by pesticides, but because there are so many humans that would like to live a first-world lifestyle, which means to displace "our partners on the planet, the other forms of life". Reducing emissions of greenhouses gases is only part of the picture, "we have also to stop using the land surface as if it was ours alone. It is not: it belongs to the community of ecosystems that serve all life …" (Lovelock, 2006, p. 109).

Lovelock's statement is in line with Leopold's promotion of the biotic community, and the danger of "economism", the tendency to reduce all values to economic values. One false assumption is that the economic parts of the biotic community will function without the uneconomic parts. Such assumptions may have the net result that we will have less soil, fewer healthy forests, and many floods (Leopold, 1966, p. 230).

We have criticized a one-sided use of highly abstract models in economics and suggested to promote historical and authentic cases that are relatively easy to share, grasp and to be inspired by. Using a narrative, case-based logic, we can explore the philosophy and the practice of progressive business companies (O'Higgins and Zsolnai 2018). Examples like Patagonia (2016) fits well as a future-oriented role model for business.

Openness for nature, being in nature, and the beauty of wilderness in particular is regarded by Leopold as a fundamental human faculty to be developed. It means that we might regard nature as a relational good. Bruni and Zamagni (2007) and Bruni (2012) develop the concept of relational good in economics, which may be seen as a contrast to positional goods that foster relationships of competition, envy, and distrust. A relational good implies that economic objects are no longer reduced to their exchange value, but acquire a relational and intrinsic value. It means a transition from an instrumental to a relational approach. With regard to land, it means to reconsider land from viewing it exclusively as a means to satisfying our needs and desires to acquiring an inner experience of reverence for life. Bouckaert (2017) suggests that a transformation of capitalism into a relational economy should go together with frugality as *art de vivre*, "which implies low material consumption and a simple lifestyle to open the mind for nonmaterial, social, cultural, and relational goods" (op. cit. p. 26). Bouckaert argues that the principle of frugality could be regarded as an economic concretization of Schweitzer's reverence for life since frugality implies an open attitude of "reverence for life", and a contemplative attitude towards our environment. It means to learn to enjoy life in a more detached manner by consuming less. "If things are perceived as gratuitous gifts and as unique expressions of the will to live, then we may retain an awareness of their nonmarketable and nonexchangeable value" (Bouckaert, 2017, p. 26). When land such as rainforests, climate conditions, and living beings are regarded as relational goods with a high degree of vulnerability, those goods are dependent on limited and responsible use by all stakeholders. Therefore, when economists speak of growth, they should advocate a relational growth economy. A relevant illustration can be found in the index of GNH (Gross National Happiness index, 2015), developed in Bhutan, in which there are indicators that measure the capacity to grow in interconnectedness.

Leopold's view is that ethics and aesthetics should be equally important as economics. In his Sand County ALMANAC he discusses different components in relation to wilderness. One of the components is perception. According to Leopold, the development of the perceptive faculty of people is the only true development. We should build receptivity into people's mind, because perception entails no consumption and no dilution of any resource (Leopold, 1966, p. 265). According to Leopold, there are some who can live without wild things, and some who cannot. "For us (who cannot) the opportunity to see geese is more important than television, and the chance to find a pasque-flower is a right as inalienable as free speech" (Leopold, 1966, p. ix).

Notes

1. I am grateful for comments from anonymous reviewers, from the opportunity to present some of the ideas on the "New Economy-old traditions: Caring entrepreneurship" conference, September 2017, Tel-Aviv University, and from the presentation on "Our Sacred Earth" conference August 28–September 1, 2018, Lucca, Italy, and detailed comments from Peter Pruzan on an earlier draft.
2. Leviticus 25 (verses 8–13) corresponds to the Norwegian official Bible with "the third Mosebok", Chapter 25).
3. In 2017 there might live more than 330 million people in USA.
4. DDT is not any longer permitted to be used in the western world, but may still be used in some places in some countries.
5. In 2015 IKEA was able to buy fruitful and productive land in a central part of Norway in order to build a new outlet – a mega department store – for their products.
6. Vetlesen is a very conscious traveler who give an unusually high priority to travelling by train.
7. <https://en.wikipedia.org/wiki/Koinonia_Partners>.

Bibliography

Abouleish, I. (2012) SEKEM. *En veiviser for det nye Egypt* (A Guide for the New Egypt) Sverre Dahl, trans. (original title: Die Sekem Vision – Eine Begegnung von Orient und Okzident verandert Egypten), Oslo, Antropos.

Aristotle (1985) *Nichomachean Ethics.* Translated by Terence Irwin. Hacket Publishing, Indianapolis.

Bouckaert, L., Opdebeeck, H. and Zsolnai, L. (eds) (2008) *Frugality: Rebalancing Material and Spiritual Values in Economic Life.* Peter Lang Publishing, Oxford.

Bouckart, L. (2017) "Authenticity and Sustainability: The Search for a Reliable Earth Spirituality", in Jakobsen, O. and Zsolnai, L. (eds), *Integral Ecology and Sustainable Business.* Emerald Publishing, Bingley, pp 15–29.

Bruni, L. (2012) *The Wound and the Blessing. Economics, Relationships and Happiness.* New City Press, New York.

Bruni, L. and Zamagni, S. (2007) *The Civil Economy. Efficiency, Equity, Public Happiness.* Peter Lang, Oxford.

Carson, R. (1962/2002) *The Silent Spring.* A Mariner Book, Houghton Mifflin Company, Boston.

Chouinard, Y. (2006) *Let My People Go Surfing. The Education of a Reluctant Businessman.* Penguin Books, New York.

Daly, H. (1980) *Economics, Ecology, Ethics: Essays toward a Steady State Economy.* W. H. Freeman and Company, San Francisco.

Daly, H. (2008) "Frugality First", in Bouckaert, L., Opdebeeck, H. and Zsolnai, L. (eds), *Frugality: Rebalancing Material and Spiritual Values in Economic Life.* Peter Lang Publishing, Oxford, pp. 207–226.

Daly, H. and Cobb, J. B. (1989/1994) *For the Common Good. Redirecting the Economy toward Community, the Environment and a Sustainable Future.* Second edition updated and Expanded. Beacon Press, Boston.

Ely, R. T. and Wehrwein, G. S. (1940, 1984) *Land Economics.* Madison: University of Wisconsin.

Etzioni, A. (1988) *The Moral Dimension. Toward a New Economics.* The Free Press, A Division of Macmillan, New York.

Gaffney, M. and Harrison, F. (1994) *The Corruption of Economics.* Shepheard-Walwyn Ltd, London.

George, W. (2015) "Compassionate Leadership", in Singer, T. and Ricard, M. (eds), *Caring Economics. Conversations on Altruism and Compassion, between Scientists, Economists, and the Dalai Lama.* Picador, New York.

Ghoshal, S. (2005) "Bad Management Theories Are Destroying Good Management Practices", *Academy of Management Learning and Education,* Vol. 4, No. 1: pp. 75–91.

Global Footprint Network (2015) <http://www.footprintnetwork.org/en/index.php/GFN/>

Gross National Happiness Index (2015) *The Centre for Bhutan Studies & GNH Research*. Thimpu, Bhutan.

Harrison, F. (1994) "The Georgist Paradigm", in Gaffney, M. and Harrison, F. (eds), *The Corruption of Economics*. Shepheard-Walwyn Ltd, London.

Hiebert, T. (2009) "The Human Vocation: Origins and Transformations in Christian Traditions", in Minch, M. and Weigel, C. (eds), *Living Ethics: An Introduction*, Wadsworthy Cengage Learning, Belmont, CA.

Hirsch, F. (1976) *The Social Limits to Growth*. Cambridge University Press, Cambridge.

Ims, K. and Zsolnai, L. (2006) "Shallow Success and Deep Failure", in Zsolnai, L. and Ims, K. (eds), *Business Within Limits. Deep Ecology and Buddhist Economics*. Peter Lang, Oxford.

Jonas, H. (1984) *The Imperative of Responsibility: In Search of an Ethics for the Technological Age*. University of Chicago Press, Chicago & London.

Leopold, A. (1966) *A Sand County ALMANAC with Other Essays on Conservation*. Oxford University Press, New York.

Lovelock, J. (2006) *The Revenge of Gaia. Earth's Climate in Crisis and the Fate of Humanity*, Basic Books. A member of the Perseus Books Group, New York.

McKibben, B. (2003/1989) *The End of Nature*. New York, Random House.

McKibbon, B. (2007) *Deep Economy. Economics as if the World Mattered*. Oneworld Publications, Oxford.

Mill, J. S. (1965) *Principles of Political Economy with Some of Their Applications to Social Philosophy*, Books I–II University of Toronto Press, London, Routledge & Kegan Paul. (First version was completed in 1847, posthumous 8th edition in 1878 and 9th in 1886).

Næss, A. (1989) *Ecology, Community and Lifestyle: Outline of an Ecosophy*. Cambridge, Cambridge University.

O'Higgins, E. and Zsolnai, L. (2018) *Progressive Business Models. Creating Sustainable and Pro-social Enterprise*. Palgrave Macmillan, Palgrave Studies in Sustainable Business in Association with Future Earth, Cham, Switzerland.

Patagonia (2016) <https://www.patagonia.com/international/>

Polanyi, K. (1977) *The Livelihood of Man*. Academic Press, New York.

Pope Francis (2015) *Praised Be: On the Care of Our Common Home*. Encyclical Letter LAUDATIO SI" of the Holy Father Francis. The Vatican City.

Sandel, M. J. (2012) *What Money Can't Buy*. Penguin Books, London.

Sandel, M. J. (2013) "Market Reasoning as Moral Reasoning: Why Economists Should Re-engage with Political Philosophy", *Journal of Economic Perspectives*, Vol. 27, No. 4 (Fall): pp. 121–140.

Schumacher, E. F. (1999) *Small Is Beautiful: Economics As If People Mattered: 25 Years Later … With Commentaries.* Hartley & Marks Publishers, Vancouver.

Schweitzer, A. (1998) *Out of My Life and Thought: An Autobiography.* John Hopkins University Press, Baltimore, MD.

Schweitzer, A. (2017). *Reverence for Life: The Words of Albert Schweitzer.* Maurice Bassett, Anna Maria, FL.

Singer, P. (1995) *Animal Liberation.* Pimlico, London.

Singer, P. (2009) *The Life You Can Save.* Random House, New York.

Singer, T. and Ricard, M. (2015) *Caring Economics. Conversations on Altruism and Compassion, between Scientists, Economists, and the Dalai Lama.* Picador, New York.

Smith, A. (1776, 1976) *An Inquiry into the Nature and Causes of The Wealth of Nations* (ed., E. Cannan), Two Volumes in One. The University of Chicago Press, Chicago.

Steffen, W. et al. (2011) "The Anthropocene: From Global Change to Planetary Stewardship", *AMBIO,* Vol. 40: pp. 739–761.

The Green Bible, (1989/2008) *New Revised Standard Version.* HarperCollins Publishers, New York.

Torrey, A. (1979) "The Land & Biblical Economics", *Land & Liberty*, July–August: pp. 60–63.

Ulrich, P. (2013) "Integrative Business Ethics", in Zsolnai, L. (ed.), *Handbook of Business Ethics. Ethics in the New Economy.* Peter Lang, Oxford.

Vetlesen, A. J. (2015) *The Denial of Nature. Environmental Philosophy in the Era of Global Capitalism.* Routledge, New York.

Whitehead, A. N. (1925/1967) *Science and the Modern World.* The Free Press, New York.

WWF (World Wide Fund for Nature). (2014) *Living Planet Report 2014: Species and Spaces, People and Places.* WWF International, Gland. <http://wwf.panda.org/about_our_earth/all_publications/living_planet_report/>

Zsolnai, L., Ims, K. and Tencati, A. (2016) Business Ethics in the Antrophocene. Paper presented on *The Ninth TransAtlantic Business Ethics Conference (TABEC) "Justification and Application: Bridging the Gap in Business Ethics"* September 28–October 1. University of St Gallen, Switzerland.

KNUT J. IMS

Caring Entrepreneurship and Ecological Conscience: The Case of Patagonia Inc.

Entrepreneurship is critical for increasing the capacity for innovation and responsiveness of business, and the entrepreneurial spirit may manifest itself in number of ways. The opening up of new markets, the creation of new products and methods of production and management, and even the establishment of new businesses and new organizational forms are the potential results of entrepreneurial activities. However, not all activities and all businesses promote good values and are dedicated to solving the most serious of society's problems. One of the biggest challenges today is how to make modern business part of the solution to climate change, which is the major problem for the planet and, in particular, for future generations. Climate change is a human-made "mess" caused by the massive and unprecedented violation and exploitation of the Earth.

The economic and psychological theory of entrepreneurship can be used to gain insight into and under what circumstances individuals become entrepreneurs, and what characterizes the development and character of entrepreneurs that genuinely care about nature as a primordial stakeholder. This paper presents the case of Yvon Chouinard, founder of Patagonia, a clothing company. Chouinard has succeeded in creating innovative solutions and has implemented a number of pro-social and pro-environmental activities that can inspire other business leaders, organizations, and businesses to engage in more sustainable practices.

This paper gives an account of the background to Chouinard's ecological consciousness that makes him and his business a role model for other companies. This paper also reflects upon what would be a more fitting education for business leaders at a time when management theories and the field of economics are dominated by abstract theories whereby narrow self-interest and a mechanical worldview support subjecting nature to rent-seeking behaviour. Such an approach does not diminish the violent treatment of nature and overconsumption in the western world.

Introduction

Entrepreneurship has been a topic of discussion and inquiry among economists and business scholars for decades, but there is no consensus about what constitutes entrepreneurship, nor a cogent theory of the entrepreneur. Schumpeter (see Kurz, 2012) described the entrepreneur as a heroic innovator who discovers new combinations and initiates discontinuous change. Harper (2003), who delves into the psychological, cultural, political, and institutional context focuses on the phenomenon of entrepreneurial alertness and its determinants.

While Weber (1930) argued that the protestant ethic set the stage for the industrial revolution on the European continent, his hypothesis has not been conclusively validated. However, his seminal work gives many clues to understanding entrepreneurship and industrial capitalism (Kalberg, 2002). Western culture, of which individualism is a crucial element, provided a climate that was very suitable for the emergence of modern entrepreneurship. The Calvinist notion of demonstrating one's faith through the performance of good work stimulated believers to choose business as an occupation for achieving their life's goals.

However, there is a more developed literature that explores non-conformists as successful entrepreneurs. One interesting study involves a major religious maverick in Norway, *Hans Nielsen Hauge* (1771–1824), a lay preacher who was engaged in a great number of entrepreneurial initiatives, and who was able to collect a large group of followers from around Norway during a relatively short period. Hauge is known for his life-changing conversion experience at the age of 25, following which "Hauge was launched into a whirlwind of religious and social activism that continued until his final arrest in 1804" (Dalgaard and Supphellen, 2011; Breistein, 1955).[1]

The case of Hauge demonstrates how one single person under the right circumstances can drive fundamental change at a societal level. Hauge led a spiritual awakening in Norwegian society and succeeded in starting a number of enterprises which employed many thousands of his followers – children, women, poor peasants, and unemployed people. He disseminated about 250,000 pieces of writing among the largely illiterate population of 900,000 and gave followers a feeling of self-worth and pride, which became an important component for the Norwegian liberation process of the nineteenth century (Breistein, 1955).

The unit of analysis in this paper is the individual entrepreneur, not the entrepreneurial team or firm. We note that economics, when speaking of entrepreneurs, typically considers not the individual but the function of entrepreneurship; a special device that is introduced to simplify such frameworks (Harper, 2003: 8).

Zsolnai, Ims and Tencati (2016) proclaim that the size or scale of the economy relative to the Earth's biophysical systems must be explicitly addressed. Our new geological age, called the "Anthropocene", requires an ecological transition whose implications for business are as follows: Business should … (i) not transgress its fair Earth share; (ii) contribute to the preservation and regeneration of life forms; (iii) not violate the freedom of future generations; but, (iv) contribute to the flourishing of people and their communities. In order to satisfy the above requirements, innovative new business models are needed.

The "more is more" and "bigger is better" ideology has destroyed the respect for natural limits and the idea of sufficiency. By transgressing ecological and ethical limits, business activities have become the primary driver of the materialism of our societies and the joyless behaviour of people. This is an outcome of adhering to a mainstream economic approach which is premised upon rational, instrumental, and self-interested behaviour. However, business is an existential enterprise because its decisions and policies greatly influence the fate and survival of nature, society, and future generations.

One interesting question is whether entrepreneurs are significantly different to non- entrepreneurs across situations. In particular, there is one variable that has received attention in the literature about the psychology of entrepreneurship. This is the individual perception of control and self-efficacy (Boyd and Vozikis, 1994). Self-efficacy is defined as the individual's subjective expectations about the extent to which one can produce effects/outcomes and exert power over what happens in one's life. It thus concerns beliefs about the locus of control and self-efficacy (or expectations of competence) (Bandura, 1977: 193). Thus, it involves two types of expectations about the self: the confidence that a given behaviour will lead to a certain outcome (an internal locus of control), and a conviction that one is able to produce the required outcome. Harper's central thesis is that when "an agent's locus of control is internal and he or she has the knowledge and capabilities to carry out the tasks, the more acute and sustained will be his or her alertness to opportunities" (Harper, 2003: 37).

This paper focuses on *Yvon Chouinard,* an ecologically conscious entrepreneur. Chouinard became a self-made company owner in adult life and has created a private, profitable apparel company based upon manufacturing, repairing, reselling, and recycling products. The company, operating under the global brand name Patagonia, provides a lifetime guarantee for all of its wares. If a product can no longer be repaired, Patagonia will recycle it and reimburse the customer (Hoang 2017). The business idea is that products should last a lifetime. In a sensational full-page advertisement on Black Friday, 2011, the company proclaimed an Anti-Growth Strategy. The advertisement in the *Times* read "Don't Buy This Jacket". The text in the advert reported on the environmental cost of the company's top-selling sweater and recommended that buyers should think twice before buying it, or other products. Paradoxically, the advertisement boosted sales significantly the following year (see MacKinnon, 2015). In 2016, Patagonia made more than $710 million (US) in sales, and each year 1 percent of sales revenue is donated to hundreds of environmental organizations. Patagonia has undertaken a number of novel ecological initiatives that have inspired other companies. Their financial philosophy is to be a product-driven company, and the bottom line is the amount of goods that the business has created over the year. Making a profit is not a goal, since the company is inspired by the Zen saying that profits are created "when you do everything else right" (Chouinard, 2006, 2016: 151). Naomi Klein, who has written several books revealing some of the inconvenient truths about capitalism (2014), and who explicitly does not endorse multinational corporations – even "green" ones like Patagonia – writes that "Yvon Chouinard's experiment … is like nothing that has come before, and therefore deserves our closest attention" (Chouinard, 2016: viii–ix).

Methodology

This paper uses the case study method as a research and learning tool. The conventional wisdom about case study research should be updated, because the scientific status of this approach is underrated (Flyvberg, 2006). In business ethics it is of paramount importance that paradigmatic cases of business (in the sense that "they shine") are selected and studied. The case-study approach is in agreement with the Kuhnian insight that a scientific discipline without a large number of thoroughly executed case studies is a discipline that lacks the systematic production of exemplars.

Business ethics as a discipline would be meager and ineffective without including enlightening exemplars that inspire, and can be imitated. Such exemplars can also negate the false assumptions that good managers ought to be opportunistic. One important problem with such assumptions is that they often turn out to be self fulfilling (Ghoshal, 2005). Therefore, the appropriate strategy is to identify an enlightening case that shows that it is possible for business to be part of the solution to climate change, and is not a big contributor to environmental pollution and other "ills". The question of which case to select among the companies that are caring in an ecological sense is not straightforward. In an earlier study, the present author selected SEKEM and its founder Ibrahim Abouleish (2012) as a paradigmatic case (Ims and Zsolnai, 2015). This time I have selected a different case to provide deeper insight and to generate more knowledge about successful entrepreneurs.

Flyvberg (2006) notes that it is not possible to use fixed criteria when selecting a paradigmatic case, but experts can use their intuition and their experience based upon thousands of other cases to make an informed selection. As Dreyfus and Dreyfus (1986) write, it is only the novice who needs strict rules as criteria. Based upon expectations about the information content associated with Patagonia and its entrepreneurial leader, Yvon Chouinard, it appears to be an interesting case to study.

Case studies also often contain substantial narrative elements. A good narrative is often a story with surprises, paradoxes, and complexities. These may make the narrative difficult to summarize in the form of general propositions and theories, but a thick narrative should be understood as a "rich problematic" (Flyvberg, 2006: 237).

By selecting Patagonia and its founder and leader for study, I hope to increase our understanding of ecologically based businesses in general, and enrich our concepts about entrepreneurship. Also, the selection demonstrates that extreme positive cases exist and that such businesses are not condemned to be losers in a globally competitive market. More specifically, my unit of analysis is the individual person, Yvon Chouinard, his entrepreneurial character, his action, and the social context in which he operates.

We claim that Patagonia is a paradigmatic example of how to combine business activities and caring for the Earth. The case illuminates how the company's founder and leader Yvon Chouinard was able to integrate an ecological conscience into his company and business in general.

There is a stream of research about social entrepreneurship, which is partly inspired by Gilligan's ethics of care (for example, Andre and Pache, 2016; Dey and Steyaert, 2016), which involves an investigation of the different forms that the ethical practice of social entrepreneurship can take. The example of Patagonia is related in many ways as it also demonstrates caring for employees as well as clients. However, the main principle of the responsibility of Patagonia is care for the Earth (Jonas, 1984).

My research is mainly based upon Chouinard's autobiography (Chouinard, 2006, 2016), other texts (Chouinard et al., 2011; Mason, 2017), interviews with Chouinard, official documents from the homepage of his company (<https://www.patagonia.com/home/>), and some internet sources (MacKinnon, 2015).

The founder of Patagonia: From loner to rebel and eco-oriented businessman

Modern business has been very destructive towards land and ecology (Carson, 1962/2002; Georgescu-Roegen, 1971; Ims and Zsolnai, 2006; Zsolnai et al., 2016; WWF, 2014; Chouinard, 2016; Hayha et al., 2016). However, there are "business" entrepreneurs that have based their operations on a philosophy of doing minimal harm to nature. One of these is Doug Tomkins, who founded the company "The North Face". Inspired by Norwegian philosopher Arne Næss, Tomkins established the *Foundation for Deep Ecology* (FDE) in 1990. FDE is an institution that is intended to act as "a voice for wild nature" and to "support efforts to protect wilderness and wildlife, promote ecological agriculture, and oppose the use of destruction of mega-technologies that are accelerating the extinction crisis" (<http://www.deepecology.org>; accessed August 17, 2018). Tomkins spent much time in wild nature with his good friend Yvon Chouinard, and each developed a business based upon their special interests in nature and their special competences, in spite of their lack of formal business education.

Patagonia Inc. is a clothing company founded by Yvon Chouinard. As a young child Chouinard wanted to be a fur trapper – not a businessman. He grew up as a French Canadian and moved to California with his family when he was 7 years old. Having come from a French-speaking Catholic school, Chouinard was challenged when starting public school, as he was the smallest child in the class and, moreover, was unable to speak English. In addition, he had to "defend [him]self because [he] 'had a girl's name' ". In his own words, "I ran away" (Chouinard, 2016: 4). Soon he was transferred to a parochial school where he received help from the nuns who worked there. "My report card from that year showed Ds in every subject" (Chouinard, 2016: 7). Language and cultural differences made him a loner, and he spent most of his time off by himself. He often biked many miles to reach a lake where he fished. Later, he spent every day after school "gigging frogs, trapping crawdads, and hunting cottontails with my bow and arrow" (op. cit. p. 7). He describes his worst experience as being a student in high school. “I had pimples, I couldn’t dance, and I had no interest in any of the subjects except for the shop classes … I was always in detention” (Op. cit. p. 7). However, he excelled in athletics, but not while people were watching. One of his main interests was training hawks and falcons for hunting. The challenge was to find a hawk nest, and then capture a young hawk to train. Chouinard states that this was the “most formative time of my life” (op. cit. p. 7). When Chouinard was 15 years old, his greatest challenge was “to trap a wild goshawk, stay up all night with her until the bird finally developed enough trust to fall asleep on his fist, and then train the proud bird using only positive reinforcement; well, the Zen master would have to ask, ‘Just who is getting trained here?’ ” (op. cit. p. 7).

In order to catch a hawk he had to climb mountains. He wanted to make his own climbing hardware since he had started to climb the big walls in Yosemite, and that required hundreds of piton placements. He was an autodidact and taught himself blacksmithing (2016: 10). He developed his own pitons, which were stronger than imported pitons and could be used several times over, by trial and error. Friends of Chouinard also wanted some, so he started selling them for one-and-a-half dollars each. At that time European pitons cost only 20 cents, but Chouinard’s were better, and were required for climbing the difficult mountain walls. In 1957, 18-year-old Chouinard borrowed 825 dollars and 35 cents from his parents and drove to Los Angeles to ask the Aluminum Company of America to make him a special machine that he could use to forge his self-designed pitons (Chouinard, 2016: 13). Now he only needed to work for some months during the winter producing his pitons, and could spend the rest of the year climbing, either in Yosemite, Wyoming, Canada, or in the Alps. He was able to support himself by selling his self-made equipment from his car and lived on “fifty cents to a dollar a day” (Chouinard, 2016: 13). Chouinard writes that he “slept two hundred days a year or more in my old army-surplus sleeping bag. I didn’t buy a tent until I was almost forty ….” (Op. cit. 13). He and his friends were rebels against consumer culture. Politicians and businessmen were “greaseballs, and corporations were the source of all evil. The natural world was our home” (Chouinard, 2016: 13). “We were like the wild species living on the edge of an ecosystem – adaptable, resilient, and tough” (Chouinard, 2016: 13).

How did Patagonia start? On a trip in the mid-1960s, Chouinard bought a rugby shirt. When back home he used it when climbing because it was made out of tough material. Everybody said, "Wow, where'd you get that cool shirt?" He started importing a few pieces of colorful sportswear from England and was able to sell them straightaway. Then the snowball slowly gained momentum. He started making shorts, and ... finally, he understood that he had been pulled into the clothes making business (Hanh, 2007: 202). The next important step was to make clothes out of synthetic material. "You could fall into a river in the winter, take it off and shake it and all the water would run out, and put it back on and it would save your life." The clothes were functional and looked good too. "Suddenly our business took off ..." (Hanh, 2007: 203).

In the mid-1980s Patagonia's revenue was $20 million (US), and by the mid-1990s it reached $100 million (US). By 2006, revenue was more than $200 million (US), and by 2013 it was $600 million (US), and the company was employing 2,000 people. *Fortune* and *Working Mother* magazines named Patagonia as one of the 100 best companies in USA to work for.[2] Patagonia has now donated more than $22 million (US) since 1985. One of its initiatives was in support of the health of the oceans. In *Let My People Go Surfing* (Chouinard, 2006, 2016), Chouinard speaks of how Buddhism and mindfulness enhanced his business. He said that the business world was "the perfect place I found to apply Zen Philosophy ...", and that compassion and mindfulness could make his businesses a pleasure "to ourselves and a gift to our employees and the world" (Hanh, 2007: 200).

The company is based in Ventura, California, and is a member of several environmental movements and a certified B Corporation. This means that the company is obliged to contribute to public-benefit concerns in parallel with making a profit. In 1989, recession hit, but the company survived and learned a lot through the crisis. With "ten of our most important people we took a walkabout" in the real country of Patagonia in South America, and asked ourselves "Why are we in business? What are our values?" Through this, the following values were expressed. Quality: making the very best climbing equipment in the world. Making functional, hard-wearing, yet good-looking clothing.

The next value was "flexi time", which led to the following company policy: "You go surfing when the surf comes up" (Hahn, 2007: 205). The third value involved blurring the distinction between work, play, and family. Accordingly, Patagonia started a child care center (one of the first on-site corporate child care centers in America). Eighty percent of the employees were women. The fourth value was continuing to hire friends. No-one in the company had an MBA, but everybody had a degree in anthropology, biology, or sociology. Chouinard himself has a degree from a high school in auto mechanics: "Instead of hiring people who studied business in school, we'd much rather hire passionate people who'd be interesting to go to dinner with and who did the sports that we were making stuff for …" (Hahn, 2007: 206).

These values were written down and turned into a philosophy of doing business (Chouinard, 2016: 63–64) The mission statement became: "Make the best quality product. Cause no unnecessary harm. Use business to inspire and implement solutions to the environmental crisis." The company is committed to sustainable "natural growth" and to selling their products to people that need them (Hanh, 2007: 209). Chouinard then spoke with each individual employee to "instill in my company, at a critical time, lessons that I had already learned as an individual and as a climber, surfer, kayaker, and fly fisherman. … You have to be true to yourself; you have to know your strengths and limitations and live within your means. It was time to apply a bit of Zen philosophy to our business" (Chouinard, 2016: 66). Chouinard says: "I learned that 'profits happen when you do everything right'" (Hahn, 2007: 209).

The company then developed an environmental philosophy consisting of five principles: 1. lead an examined life; 2. clean up your own act; 3. do what you can do; 4. support civil democracy; 5. be a role model (Hanh, 2007: 209–18). Step five implies influencing other companies and other people, because one cannot save the world by oneself. "The proudest thing I've done is I've started an organization called *One Percent for the Planet* which is now an alliance with 224 other companies that are all pledged to give one percent of their sales to environmental causes" (Hanh, 2007: 217)

The company acknowledges that Patagonia is a producer and thus is part of the ecological problem (Lovelock, 2006). Patagonia makes products which use fossil fuels, builds factories that use water and other resources, creates waste, and emits carbon into the air. It ships products around the world in boxes and plastic bags, and consumes electricity – using renewable as well as non-renewable resources. Employees drive cars and ride on airplanes. "As individuals, we consume products … probably more than we need" (<https://www.patagonia.com/home/2016>).

The challenge is, therefore, to focus on what Patagonia can do to reduce, neutralize, or even reverse the root causes of climate change. Accordingly, the horizon of responsibility includes:

1. Reducing the environmental impact of Patagonia and its supply chain
2. Supporting grassroots activists by paying an Earth Tax
3. Using Patagonia's voice to advocate for systemic change
4. Empowering customers by making quality products that can be repaired
5. Supporting regenerative practices in ranching and agriculture
6. Envisioning a new approach to business

Point 1) relates to measuring Patagonia's carbon footprint. In 2015 the estimated emissions of Patagonia's global operations were 3,617 metric tons of CO_2. When opening new stores, Patagonia prefers to improve existing buildings. Green building projects are part of its strategy. A transportation program "Drive – Less" has been implemented for employees. The program uses monetary incentives to promote riding bikes, using a carpool, or taking public transportation. A measurable result of the program is that in the first year (in 2011), 900 employees collectively drive 690,000 fewer miles and reducing CO_2 emissions by 500,000 pounds.

The Chemical and Environmental Impacts Program is a supply chain initiative that supports managing chemicals in a more careful way. The program covers all areas of environmental systems including waste, water use, and energy use. Another supply chain initiative involved the company becoming a founding member of the "Fair Labor Association" (FLA) in 1996. In 2007, the program "The Footprint Chronicles" traced the social and environmental impacts of Patagonia's products.

Patagonia implemented a human-trafficking detection tool in 2011. Audits of suppliers revealed that labor brokers were charging migrant workers from Asian countries up to 7,000 dollars to get a job in the Taiwanese factory mills that supply Patagonia.

Points 2) and 3) involve supporting grassroots activists by paying an Earth Tax. This form of funding started in 1985. Patagonia now gives 1 percent of its sales revenue to the preservation and restoration of the natural environment. In 2015, the amount given to grassroots environmental groups was $70 million (US). Patagonia's employees can work up to 320 hours for environmental groups while continuing to receive their full salary from Patagonia. Another program helps oppose the construction of new dams and supports the transition towards lower-impact energy and water sources that cause no harm.

Point 4) involves making products that are durable using raw materials that "cause less environmental harm than their counterparts". A related motto is "Repair is a radical act". In one of Patagonia's adverts, the text reads: "To make one of our best-selling jackets requires 135 litres of water, so by repairing it you will save many natural resources, Co2 emissions and waste. Don't buy what you don't need. Think twice before you buy anything." In 2015, an internal scorecard was created to rate the quality of Patagonia products on a scale from 1 to 10 (1 being the worst, 10 the best). Quality is defined and measured by product durability, reparability, multi-functionality, non-obsolescence, and lack of environmental harm (see also Chouinard et al. 2011). A checklist of criteria for Patagonia's designers to consider is listed in Chouinard (2016); namely: "Is it functional? Is it multifunctional? Is it durable? Is it repairable? Does it fit our customer? Is it as simple as possible? Is the product line simple? Is it an innovation or an invention? Is it a global design? Is it easy to care for and clean? Does it have any added value? Is it authentic? Is it beautiful? Are we just chasing fashion? Are we designing for our core customer? Does it cause any unnecessary harm?"

Point 5) concerns the environmental program "Worn Wear". The message is "invest in quality and repair when things break and celebrate the clothing that travels with us through life." Patagonia has the largest garment repair center in the USA, located in California.

As regards Point 6), an internal investment fund has been established to help "like-minded responsible start-up companies bring about positive benefit to the environment". The purpose is to inspire and use business to help mitigate the environmental crisis. Another initiative is being guided by The Stockholm Resilience Center concept of "planetary boundaries" (Steffen et al., 2015).

Discussion

Why did Chouinard become a successful entrepreneur? What were the decisive factors that drove the founder of Patagonia – character, context, or lucky circumstances? The hypothesis that was put forward by Max Weber concerning the cultural individualist orientation of entrepreneurs seems to be supported. It is evident that the individualistic culture in the USA involves a strong entrepreneurial spirit. Beliefs in the "self-made man" and the American Dream (the idea that anybody has a fair shot at becoming a millionaire) creates a cultural context that stimulates entrepreneurial activity. Individuals that succeed are characterized by strong independence and a drive for achievement. These personal characteristics of successful entrepreneurs are supported by a British study (see Cox and Jennings, 1995). Chouinard clearly fits the classification of an "elite independent entrepreneur".

Reading Chouinard's (2006, 2016) autobiography, it becomes clear that Bandura's claims about the importance of the locus of control and self-efficacy make sense. As Harper (2003: 40–43) writes, "inner-directed people are more likely to be able to recombine and synthesize elements into new forms, to be independent in their thinking and to resist coercion and pressures to conform." When Chouinard was only 7 years old, language and cultural differences due to his family's relocation made him into a loner. His business had a very simple, natural start. Chouinard needed better tools for his mountain climbing, and knew that he was a competent handyman. At the same time Chouinard was fascinated by hawks, which did not have to climb, but could naturally and elegantly fly above his head while he had to climb carefully step by step, often with much effort to reach the summit. After teaching himself blacksmithing, by trial and error, he was able to make his own supreme quality pitons. His friends observed his innovative and robust tools, and Chouinard realized that he had a market. Chouinard's self-efficacy is convincingly documented by the fact that, when he was a young man, to construct a machine for his self-design products, he borrowed a small amount of money and embarked on a long journey to meet personally with the managers of a big corporation. Later, Chouinard came into the sportswear industry almost by accident. A rock-climbing friend of Chouinard noticed the fancy but strong rugby shirt that Chouinard had bought on a visit to Ireland. Chouinard responded by ordering more shirts, and the business adventure took off. Via successive innovations, he established his company in the sportswear industry and named it Patagonia (named after the southern part of South America). The company grew and went through downturns, but demonstrated an ability to grow in a relatively stable and sustainable way.

If we investigate the tendency of non-conformists to be over-represented among entrepreneurs, we find convincing evidence that supports the findings of Dahlgaard and Supphellen (2011), who studied Hans Nilsen Hauge. Chouinard admits that he and a small group of mountain climbers "were rebels from consumer culture". Politicians and "businessmen" were "greaseballs", and "corporations were the source of all evil" (Chouinard, 2016: 13). As a self-declared Buddhist, he also fitted well into the non-conformist camp.

Reading Chouinard's (2016) autobiography, it becomes clear that his general view of business people is not positive, but at the same time he acknowledges that business can produce goods that are required and contribute to the common good. He writes:

> I've been a business man for almost sixty years .... I've never respected the profession. It's business that has to take the majority of the blame for being the enemy of nature, for destroying native cultures, for taking from the poor and giving to the rich, and for poising the Earth with the effluent from its factories. Yet business can produce food, cure disease, control population, employ people, and generally enrich

> our lives. And it can do these good things and make a profit without losing its soul ... (Chouinard 2016: 1)

According to Chouinard, "business as usual" is a ruthless enterprise. But business has the potential to do good, and good deeds are not empty words, according to Chouinard, who through Patagonia has implemented a number of pro-social and pro-environmental activities that make his company an industry leader. We might recall that Patagonia supports grassroots activists by paying an Earth Tax, an idea which has been imitated by a number of other companies. The company acts as a role model by paying 1 percent of its sales revenue to support the preservation and restoration of the natural environment. Paying 1 percent of sales revenue is much more demanding than paying 1 percent of company profit. Also, the amount given to grassroots environmental groups, $79 million (US) since 1985, (Chouinard, 2016: 70) is extraordinary. Another distinctive fact is the "human resource" policy of allowing employees to work eight weeks a year for environmental groups while receiving full payment from Patagonia.

Zsolnai, Ims and Tencati (2016) pose some fundamental questions regarding how "nature as a primordial stakeholder" can be incorporated into the functioning of human organizations, including businesses, public administration, and civil society organizations. They state that "the crucial question concerns how to develop the ecological sensitivity and responsiveness of people at different levels of organizations, and how then to translate the emerging ecological consciousness into effective and caring organizational practices through which organizations develop a culture of Earth Citizenship".

Following up on these questions, we may ask from where does an ecological conscience and ecological sensitivity come? Can having an ecological conscience be taught in business schools as a part of formal education? In particular, is the development of an ecological conscience promoted in business schools? Chouinard did not have a business education, and Tomkins, the entrepreneur behind the giant land conservation project, did not either. This might indicate that a formal business education is not necessary for becoming a successful eco-minded entrepreneur and businessperson.

As a child, Chouinard used nature as a protective and fascinating free "second home". As a young adult he spent months in wild nature every year. Chouinard writes that "the natural world was our home … we were like the wild species living on the edge of an ecosystem – adaptable, resilient, and tough" (Chouinard, 2016: 13).

The ecological consciousness of Chouinard and his friend, Tomkins, was stimulated throughout their lives when they actively spent time in wild nature. It is interesting to note that other well-known eco-friendly persons have had similar experiences.

Aldo Leopold (1966), educated as a forester, is famous for having a dramatic change in attitude towards the intrinsic worth of nature following a hunt in the wilderness. With his rifle he shot at a "pack of wolves" and witnessed an old wolf dying. The critical change took place when Leopold's eyes met those of the dying wolf. The meeting with the dying wolf turned out to be a life-changing experience for Leopold. The context of the meeting was the wilderness.

Arne Næss, founder of "Deep Ecology" (Næss, 1989), spent extended time in nature. He was employed as a professor by the University of Oslo, but he left the university and the capital whenever he had an opportunity to travel (by train) to his small, simple, and robust cabin high up in the Norwegian mountains. One estimate is that Næss spent a total of ten years of his life in the mountains. Næss was a professional philosopher, but for him deep ecology had more of an ontological than a philosophical basis. For Næss, deep ecology was based upon a conviction about the intensive interdependence of man and nature, not philosophical reflection about it. The central requirement for Næss was being in nature, not thinking about nature. The tentative conclusion is that ecological sensitivity and consciousness is independent of the type of education, but may be reinforced by spending a lot of time in nature.

Does this mean that students of economics and business should take excursions not only to Wall Street and Citibank, but also to Antarctica or Patagonia in the South of Argentina, or to Svalbard close to the North Pole to experience wild nature? Should students of economics learn to use a kayak, a surfboard, or climb mountains? According to Leopold (1966), we need to have a presence in nature and have certain skills for dealing with it to see its value. Chouinard's skill as a rock climber, kayaker, and surfer was at a very high level. The same could be said about Arne Næss as a rock climber.

Chouinard acknowledges that his practice of Buddhism was a fruitful way to do business. Through his life he has repeatedly shown that his business is not based upon greed. His personal life is very frugal (Bouckaert et al., 2008; Daly, 2008). In one interview, he speaks about himself as a "dirtbag" because he lives in the same jacket that he has worn for decades. Surprisingly, he warns buyers about his products: "Don't buy this product – unless you really need it." Vetlesen (2015) asks us to question the prevailing ideology in our western world. He argues that the acceptance of limits is a precondition for a necessary reorientation, which also demands emotional maturity. It seems clear that Chouinard has reached emotional maturity.

A central assumption of economic theory is the need for economic growth. The ideology is that if we make a bigger cake to share, everybody will get a larger piece and everybody will be satisfied. The problem is that larger-scale production means more use of non-renewable finite resources, and even if such resources were renewable, their generation, transportation, and use will typically lead to depletion, pollution, and waste (Georgescu-Roegen, 1971). In Patagonia's marketing some of those ecological footprints are revealed, such as when the company explains how much water is required to produce one particular jacket (Global Footprint Network, 2015). This is relevant information for producers as well as customers, and this kind of information will lead to customers being more cautious about their purchasing behaviour.

Whenever business students now learn about land ethics, this is a step in the right direction. But intellectual learning is not enough, according to Leopold (1966). Land ethics imply a number of obligations, but "obligations have no meaning without conscience" (Leopold, 1966: 225). In order to take land ethics seriously, an ecological conscience is a prerequisite.

Capra and Jakobsen (2017) claim the need for general ecoliteracy, which involves an understanding of how ecosystems works and how one can build a sustainable human society in which humans are understood as members of a community of communities. Ecoliteracy builds upon the science of systems and ecology and tries to foster learning processes that foster a deep appreciation of nature and the human role in it. Systems thinking recognizes the world as an integrated whole, where the links between different parts are more crucial than the single parts in isolation.

Arne Næss (1989: 159) is critical of traditional formal education and writes that we know that formal education is no longer a powerful agent of change. Rather than "green education" he suggests a "naturalist" orientation in its original, deep, romantic sense. "We [should] engage not teachers as we know them today but instead people who have internalized deep ecological norms, even if a small minority, and make them more central in … communities" (1989: 159–160). Such "naturalists", through their very existence, can make people aware of things they had never thought of before.

Concluding remarks

We have described the childhood and young adult life of the entrepreneur Chouinard in particular. Even though his first years must have very tough, he successfully overcame adversity, and later showed evidence of extreme resilience as a business entrepreneur, with an ability to bounce back from very difficult financial situations. Such characteristics place him in the category of "elite entrepreneurs", described by Cox and Jennings (1995). As a young boy, Chouinard did not believe that he could achieve the American dream. Nevertheless, he did because he had some remarkable skills and an ability to transform ideas into reality, demonstrating an unusual alertness to opportunities. Chouinard did not leap forward, but took small steps that finally turned into a long successful journey as a business leader. He may be considered a real entrepreneur, and is responsible for a number of innovations, of which many can be characterized as eco-oriented. Perhaps the most remarkable of these is Chouinard's high level of ecological sensitivity, and his extraordinary generosity towards groups and organizations that are pro-environmental. In this respect he has become involved in a number of initiatives that other organizations have also imitated.

In this account we have also introduced Chouinard's good friend, Doug Tomkins, the entrepreneur who bought huge areas of land in Chile and Argentina in order to restore them. In an earlier study, Ims and Zsolnai (2016) described the social entrepreneur, the Egyptian pharmacologist Ibrahim Abouleish, who founded the organization SEKEM (which stands for "vitality from the sun" in Ancient Egyptian) in 1977. SEKEM's goals are to "restore and maintain the vitality of the soil and food as well as the biodiversity of nature through sustainable, organic agriculture and to support social and cultural development in Egypt" (Abouleish, 2012). SEKEM has stimulated the moral impulse in Egypt through their manifold forms of work based around a biodynamic farm in the middle of a desert northeast of Cairo. The ideology of SEKEM is to award special importance to the interaction between economic, cultural, social, and environmental sectors.

Earlier, we asked the question "from where does ethical sensitivity come?" If we only focus on Chouinard and Tomkins, one simple answer would be "from spending much time in wild nature, and having a high level of skill at dealing with wild nature". However, Abouleish (2012) does not fit this pattern. The impression we receive from the autobiography of Abouleish is that he was enlightened by ideas about anthroposophy rooted in the philosophy and pedagogy of Rudolf Steiner, not by spending time in nature to the same extent as Chouinard and Tomkins. We have also discussed Arne Næss because he was a founder of the Deep Ecology movement, but Næss was a professional academic, not a business entrepreneur.

In the introduction, I mentioned the Norwegian entrepreneur Hans Nielsen Hauge, who went through a short but a hyperactive period of activity before he was jailed. There exist many written sources by Hauge. He was a maverick, but was also able to start a great number of enterprises within the framework of a religious movement which have made a lasting impression on the Norwegian entrepreneurial and religious spirit. What can we learn from these entrepreneurs, individually and as a group?

Is it plausible that they all would emphasize Leopold's (1966) call for the promotion of the biotic community, which means a love of and respect for land as a community of interdependent organisms inhabiting the same environment or habitat and interacting with each other? Together with the abiotic factors and physical landscape, these elements comprise an ecosystem. It also means looking upon biota as a pyramid consisting of layers and food chains. Moreover, we might wonder if they all are aware of the danger of "economism"; the tendency to reduce all values to economic value (Daly and Cobb, 1989/1994). One false assumption is that the economic parts of the biotic community will function without the uneconomic parts. Such assumptions may have the net result that we will have less soil, fewer healthy forests, and many floods (Leopold, 1966: 230).

As a critique of the one-sided use of highly abstract models in economics (Daly and Cobb, 1989/1994), this paper has described historical, authentic cases of caring entrepreneurs that are relatively easy to share as narratives, and are in themselves rich sources of learning about a concrete, context-dependent reality. We have in particular explored the philosophy and the practices of the entrepreneur Chouinard and his business company, Patagonia. We conclude that Patagonia embodies a future-oriented role model for business. The Buddhist-inspired entrepreneur Chouinard shows a remarkable ability to innovate, as well as an extreme pro nature attitude enabled by Patagonia. Chouinard had a lifelong friendship with the forest conservationist, Tomkins, and they spent much time together in wild nature. I do not know whether Chouinard knows the Egyptian founder of SEKEM, Abouleish, who found inspiration in different sources. One common denominator is that none of these remarkable business entrepreneurs had a business education, but all of them had a strong ecological consciousness, which is a precondition for caring for the Earth.

Acknowledgements

The author is grateful for comments received from anonymous reviewers, for the opportunity to present the ideas contained in this paper at the conference, "New Economy, Old Traditions: Caring Entrepreneurship" in Tel-Aviv between September 4–6, 2017, and for detailed comments by Peter Pruzan, Ora Setter, Bram Timmerman, and Laszlo Zsolnai on an earlier draft.

Notes

1. The real reason that Hauge was arrested was probably because the King of Denmark and Norway feared that Hauge would lead a revolution "from beneath" – one supported by peasants and poor land laborers – that would ultimately destroy the power of the ruling, privileged class (Breistein 1955).
2. See <http://archive.fortune.com/magazines/fortune/fortune_archive/2007/04/02/8403423/index3.htm> accessed August 17, 2018.

Bibliography

Abouleish, I. (2012). SEKEM. *En veiviser for det nye Egypt* (A Guide for the New Egypt) Sverre Dahl, trans. (Original title: Die Sekem Vision – Eine Begegnung von Orient und Okzident verandert Egypten), Oslo: Antropos.

Andre, K., & Pache, A. C. (2016). From Caring Entrepreneur to Caring Enterprise: Addressing the Ethical Challenges of Scaling Up Social Enterprises. *Journal of Business Ethics,* 133, 659–675.

Bandura, A. (1977). Self-Efficacy: Toward a Unifying Theory of Behavioural Change. *Psychological Review*, 84(2), 191–215.

Bouckaert, L., Opdebeeck, H., & Zsolnai, L. (eds) (2008). *Frugality: Rebalancing Material and Spiritual Values in Economic Life,* Oxford: Peter Lang Publishing.

Boyd, N. G., & Vozikis, G. S. (1994). The Influence of Self-Efficacy on the Development of Entrepreneurial Intentions and Actions. *Entrepreneurship: Theory and Practice,* 18(4), 63–77.

Breistein, D. (1955). *Hans Nielsen Hauge. Kjøpmand i Bergen.* (*Hans Nielsen Hauge. Merchant in Bergen*). Bergen: John Griegs Forlag.

Capra, F., & Jakobsen, O. D. (2017). A Conceptual Framework for Ecological Economics Based on Systemic Principles of Life. *International Journal of Social Economics*, 44(6), 831–844.

Carson, R. (1962/2002). *Silent Spring.* Boston: Houghton Mifflin.

Chouinard, Y. (2006). *Let My People Go Surfing. The Education of a Reluctant Businessman.* New York: Penguin Books.

Chouinard, Y. (2016). *Let My People Go Surfing. The Education of a Reluctant Businessman*, Foreword by Naomi Klein, 2. Edition. New York: Penguin Books.

Chouinard, Y., Ellison, J., & Ridgeway (October 2011). The Sustainable Economy. *Harvard Business Review*, 89(10), 52–62.

Cox, C., & Jennings, R. (1995). The Foundations of Success: The Development and Characteristics of British Entrepreneurs and Intrapreneurs. *Leadership & Organization Development Journal*, 16(7), 4–9.

Dalgaard, B. R., & Supphellen, M. (2011 March). Entrepreneurship in Norway's Economic and Religious Nineteenth-century Transformation. *Scandinavian Economic History Review*, 59(1), 48–66.

Daly, H. (2008). "Frugality First", in Bouckaert, L., Opdebeeck, H., & Zsolnai, L. (eds) *Frugality: Rebalancing Material and Spiritual Values in Economic Life.* Oxford: Peter Lang Publishing, pp. 207–226.

Daly, H., & Cobb, J. B. (1989/1994). *For the Common Good. Redirecting the Economy toward Community, the Environment and a Sustainable Future* (2nd edn, Updated and Expanded). Boston: Beacon Press.

Dey, P., & Steyaert, C. (2016). Rethinking the Space of Ethics in Social Entrepreneurship: Power, Subjectivity, and Practices of Freedom. *Journal of Business Ethics*, 133, 627–641.

Dreyfus, H., & Dreyfus, S. (1986). *Mind over Machine: The Power of Human Intuition and Expertise in the Era of Computer*. New York: Free Press.

Flyvberg, B. (April 2006). Five Misunderstandings About Case – Study Research. *Qualitative Inquiry*, 12(2), 219–245.

Georgescu-Roegen, N. (1971). *The Entropy Law and the Economic Process.* Cambridge, MA: Harvard University Press.

Ghoshal, S. (2005). Bad Management Theories Are Destroying Good Management Practices. *Academy of Management Learning and Education*, 4(1), 75–91.

Global Footprint Network (2015). <http://www.footprintnetwork.org/en/index.php/GFN/>.

Hanh, T. N. (2007). "Work and Pleasure: The Example of Patagonia", in *The Art of Power.* New York: HarperOne, pp. 199–218.

Harper, D. A. (2003). *Foundations of Entrepreneurship and Economic Development.* London: Routledge.

Hayha, T., Lucas, P. L., van Vuuren, D. P., Cornell, S. E., & Hoff, H. (2016). From Planetary Boundaries to National Fair Shares of the Global Safe Operating Space – How Can the Scales Be Bridged? *Global Environmental Change,* 40, 60–72.

Hoang, L. (January 16, 2017). Patagonia's Circular Economy Strategy in *News & Analysis,* January 16, 2017 <https://www.businessoffashion.com/articles/news-analysis/how-patagonia-transformed-the-circular-economy>

Ims, K., & Zsolnai, L. (2006). "Shallow Success and Deep Failure", in Zsolnai, L., & Ims, K. (eds) *Business Within Limits. Deep Ecology and Buddhist Economics.* Oxford: Peter Lang.

Ims, K., & Zsolnai, L. (2015). "Social Innovation and Social Development in Latin America, Egypt, and India", in Enderle, G. & Murphy, P. E. (eds) *Ethical Innovation in Business and the Economy*. Studies in Transatlantic Business Ethics. Cheltenham, UK: Edward Elgar.

Jonas, H. (1984). *The Imperative of Responsibility: In Search of an Ethics for the Technological Age*. Chicago & London: University of Chicago Press.

Kalberg, S. (2002). *The Protestant Ethic and the Spirit of Capitalism*. The Expanded 1920 Version. Authorized by Max Weber for Publication in Book Form. New Introduction and Translation by Kalberg, S. Los Angeles, CA: Roxbury Publishing Company.

Klein, N. (2014). *This Changes Everything. Capitalism vs. the Climate*. New York: Simon & Schuster Paperbacks.

Kurz, H. D. (2012). Schumpeter's New Combinations: Revisiting His Theorie der wirtschaftlichen Entwicklung on the Occasion of Its Centenary. *Journal of Evolutionary Economics*, 22, 871–899.

Leopold, A. (1966). *A Sand County ALMANAC with Other Essays on Conservation*. New York: Oxford University Press.

Lovelock, J. (2006). *The Revenge of Gaia. Earth's Climate in Crisis and the Fate of Humanity*. New York: Basic Books. A Member of the Perseus Books Group.

MacKinnon, J. B. (2015). Patagonia's Anti-Growth Strategy. *The New Yorker.* May 21/<https://www.newyorker.com/business/currency/patagonias-anti-growth-strategy>

Mason, C. (2017, Summer). Patagonia's Journey into a New Regenerative Performance Approach. *People + Strategy*, 40(3), 30–34.

Næss, A. (1989). *Ecology, Community and Lifestyle: Outline of an Ecosophy*. Translated and revised by D. Rothenberg. Cambridge: Cambridge University Press.

Patagonia (2016). <http://www.patagonia.com/international>

Steffen, W., Richardson, K., Rockström, J., Cornell, S. E., Fetzer, I., Bennett, E. M., Biggs, R., Carpenter, S. R., de Vries, W., de Wit, C. A., Folke, C., Gerten, D., Heinke, J., Mace, G. M., Persson, L. M., Ramanathan, V., Reyers, B., & Sörlin, S. (2015). Planetary Boundaries: Guiding Human Development on a Changing Planet. *Science*, 346(6223), 736, 1259855–1–10.

Vetlesen, A. J. (2015). *The Denial of Nature. Environmental Philosophy in the Era of Global Capitalism*. London and New York: Routledge.

Weber, M. (1930). *The Protestant Ethic and the Spirit of Capitalism.* Translated by T. Parsons. London: G. Allen & Unwin.

WWF (World Wide Fund for Nature). (2014). *Living Planet Report 2014: Species and Spaces, People and Places*. WWF International, Gland. <http://wwf.panda.org/about_our_earth/all_publications/living_planet_report/>

Zsolnai, L., Ims, K., & Tencati, A. (2016). *Business Ethics in the Anthropocene*. Paper presented at The Ninth TransAtlantic Business Ethics Conference (TABEC) "Justification and Application: Bridging the Gap in Business Ethics" September 28–October 1, University of St Gallen, Switzerland.

KNUT J. IMS

The Computer and Science as Bearers of Culture: From Intuition to Calculation

Introduction

Computers are associated with multiple types of phenomena – from star wars and arms races to different profit-enhancing endeavors, such as automation and rationalization. Information technology has gained widespread use throughout the world. The computer is a well-known and accepted tool in education, research, and business.

Man is becoming increasingly involved with these "thinking machines". The computer's ability to think instrumentally and logically is being strengthened and developed. In our culture, the computer seems to be nothing less than a hallmark of rationality and soundness.

However, that would suggest that the computer also features other less mentioned aspects – aspects that have a direct bearing on our social lives. In earlier periods, machines and tools modified *our self-perception*. The computer seems to be our time's "defining culture", that is technology that involves a link between science and culture and is always available to serve as a suggestive metaphor. What are the cultural implications of the rapidly growing use of computers? Throughout the centuries, intuition seems to have become increasingly overshadowed by instrumental thinking, calculation, and measurement. As an advanced expression of modern high technology, the computer seems to be amplifying these tendencies. Where is this leading us? Are we becoming more "unidimensional" and mechanized? Are we becoming more metaphysically oriented? Studies (Turkle, 1984; Weizenbaum, 1976) indicate that the presence of computers can lead to both tendencies, but the first-mentioned force seems to be predominant and the most widely recognized. This easily gives rise to a perception of "mechanized man". We also find features of the "anthropomorphization of the machine". In fact, are metaphors for humans and human thinking shaped by the computer. Does that mean *machines and human beings are becoming more alike*? These questions are just as challenging for humanists as they are for technologists, but seem to receive less attention. Focus on computers' instrumental function has apparently taken precedence over the cultural aspect. It may therefore be appropriate to begin exploring the relationship between technology and man's social life. Questions about the consequences of automation, explanations behind computer widows, hackers, and a specialist computer language are cultural elements that incite fear in some people. But perhaps these are not the essential questions? The question of how the computer will be developed in the future may not be the most interesting in the post-computer era either; for example, whether it will be virtually able to replicate human intelligence. The crucial question may be *what we, as human beings, will become?*

This article does not aim to answer the question of man's future evolution, but is of relevance. While working on this article, we found a relationship between scientific discoveries, applied technology, and our perception of the world, indicating that technological advancements have driven change in human beings throughout history, the most recent being the computer. I focus on the subjective and "irrational" aspects of the computer. Special attention is devoted to artificial intelligence, which is a conflict-laden melting point between science and technology, and possibly the area where man's natural limits are the most challenged. With that in mind, it is important to attempt to shed light on the context that led to the first mechanization of mind and society, as a forerunner of the building of machines. A cultural-historical light is, therefore, cast on the period before, during, and after the Renaissance.

The main notion in the following is that human thinking and our self-image has changed in tandem with technological developments, which can shed light on the traits of our society. The theme can also be motivated instrumentally.[1] A more holistic reflection concerning the use and misuse of the computer is required. The best technology and best expert knowledge are of little use if social elements are disregarded. Postmodern society seems to be missing a historical perspective, and so we tend to set aside traditions that facilitate insight into our way of thinking, our moral assessments, and even our analytical categories. Our language, technology, and culture are ultimately at the root of our shared understanding of reality. In that respect, this article can also be considered as an element of the modern view of science as economics. The article is a modest attempt to emulate Odd Langholm's masterly stroke of combining culture, science and economics.

The article is structured as follows: A historical perspective is provided to evoke the "leap in consciousness" associated with the Renaissance. I then explore the second main theme which relates to the "anthropomorphization of the computer", followed by the third theme, the "mechanization of man". I then look at some questions concerning the relationship between human warmth and high tech. Finally, I look at various aspects of research in artificial intelligence. The question is whether intuition as a human talent is superior to the logical machine. Intuition cannot easily be coded, but is it a sign of human beings' foremost talents and hence their dignity?[2]

Technology and human beings – A brief overview of selected literature

As I assume that some of the cultural and historical sources below are not that widely known, I will comment on some of the main sources I have used. I will not comment in particular on the standard sources from the North Atlantic artificial intelligence community, many of them from the Massachusetts Institute of Technology (MIT), nor on a few popular scientific works (some bestsellers), novels, and mystical literature. The latter sources are the most speculative, although they have made a clear contribution to intellectual history, which has provoked relevant issues.

Frude (1893) in *The Intimate Machine* and the anthropologist Sherry Turkle (1894) describe cultural dimensions of the relationship between humans and machines. Frude cites animism as a factor behind man's more intimate relationship with computers and robots, that is man's ability to attribute a soul to nature and artefacts, while Turkle analyses the computer's ability to stimulate reflection about metaphysical questions. Turkle gives an account of her experience after having lived in and studied the "hacker community" at MIT for many years.

There are books dealing with the history of today's advanced computers. Many of them tend to focus on technological aspects such as vacuum switches, transistors, bits and bytes. Technological developments are rarely explored in a broader cultural-historical context. One exception is Lewis Mumford (1934), who makes an immense historical contribution with regard to western civilization and technology. As Mumford is key in the following, his view of history deserves some attention. Mumford invokes the human force behind technology and cultural developments (1934, 1967, and 1970). He posits (1970, p. 415) that technology is the "formative part" of human culture as a whole. Dreams, desires, impulses, and religious motives have profoundly altered technics at every level of development. This influence comes from human beings' subconscious, not from the practical needs of daily life. According to Mumford, technology has played an active part in human beings' process of self-transformation. In his two-volume work *The Myth of the Machine* he firmly defends the subjective element of technological development. One of his original contributions is to *deal with technics as an integral part of human beings' higher culture*. Mumford does not consider scientific discoveries and technological innovation as the only purpose of human existence. Instead, he sees life itself as a primary phenomenon (Weizenbaum, 1976). Creativity, rather than the act of conquering nature, is seen as the ultimate criterion for man's biological and cultural success. His harsh criticism of Marshall McLuhan, whom he believes has misunderstood representations of both man and technology, is of particular interest. Mumford attacks McLuhan's representation of the "electronic hut" – which consists of the entire world being gathered into one unit through the electronic network. Mumford claims that dialogue between human beings in different cultures is a very difficult process, and that the key to success is *local* electronic contact between people who share the same language, culture, and conceptual definitions. J. David Bolter (1984) provides a more recent contribution to knowledge and understanding of the origin of western culture. He compares the modern "electronic man" to Goethe's Faust in the 1700s and the natural human being in Antiquity.

His conclusion is that "Turing's Man" has lifted North Atlantic thinking up to a higher level. We have now been pushed to the extreme when it comes to drawing the line between the natural and the artificial.

A retrospective

A historical retrospective can place ideas and theories into a broader context and bring about new perspectives. What is it that has ushered us into the comprehensive mechanization of man and led us into the techno-economic paradigm characterizing the West today? Which aspect of the West's cultural history has been decisive for representations of man as a machine? A glance at modern history suggests that man's self-image has changed over a long history. A turning point occurred during the Renaissance, which merits closer examination.

From speculation to testing

During the Renaissance, man's status in the universe was radically altered. Scholastic theology was challenged and rejected (see for example Næss, 1967, 1968; Polyani, 1966, p. 63). The feudal system was disintegrating. Perhaps religious detachment was of key significance, as it led to a break between theology and science. Science shifts from being in a state of fusion with theology, to separating from it and following its own path. How does this influence man's self-image? Let's start with the religious dimension.

We seem to be facing a paradox. Man's status in the universe is downgraded, but the result is an enhancement of man's self-image. Man's privileged relationship to God is shaken. From being the centre of creation, the laws of Copernicus reduce us to inhabitants on a small cloud of dust in a vast space.[3] Such thoughts were unthinkable, while a century later they were no less than a given. What caused this a leap in thought?

People of the Antiquity had an intimate relationship with nature, which they admired. God was nature's creator and master. Consciousness of imagery and myths was presumably highly developed. Mythical thinking is concrete, emotional, and highly subjective. According to Lewis Mumford, each culture harbours its own dreams. The Christian was a resident of a "fabulous" heavenly world. God, saints, devils, demons, angels, archangels, cherubims, and seraphims, along with principalities, powers, and authorities, were part of this celestial world. Man also had a soul that was separate from life in the material world. Emile Male posits that during the Middle Ages the *idea* of an object shaped by a human being for himself was more real than "the actual" object, which is why there was no concept called science. Examining artefacts was seen as meaningless, while nature was studied to discover eternal truths expressed by God through creation in its totality (Mumford, 1934, p. 29). The external world had no conceptual niche in the mind during the Middle Ages. Natural facts were insignificant compared with the purpose and holy order that Jesus Christ and his Church had established. The visible world was only a guarantee and a symbol of the eternal world in which people were waiting to take part. Mumford poses the question: To what extent could the mind apprehend scientific measurement and observation when the mystical numbers three, four, seven, nine, and twelve filled all relationships with conceptual imagery? (Op. cit. p. 29).

However, thinking gradually overtook the scene and suppressed imagery and myths in our consciousness (Jung, 1958). While the artisan's interest in nature first was somewhat perplexed, it continued to grow, shifting outwards towards external nature. The shift from being oriented towards heaven to becoming earthbound was a dramatic attitudinal change. Existing knowledge, and hence the worldview, had to be broken down. In this sense, it was a process that could operate more easily in the naïve mind than in the erudite mind. The most important difference in human thinking before and after the Renaissance can be illustrated by the formulation *from existing to thinking about something.*

A shift in consciousness was a precondition for this leap in thought. Consciousness of the ego and the self increased, and the dualism between the self and the outside world became prominent during the Renaissance.

The world becomes an object outside the self, stimulating an urge to explore. Experience and experiment are exalted, and "natural" explanations become increasingly observable and verifiable. The result is that human beings develop a sense of *conscious trust in their own experiences*.

Where do machines come in? Which forces facilitated their introduction and which hurdles did machine builders face? Strangely, agriculture was the last area where machines made their entry. The focus was on functions aimed at maintaining and cultivating life. Machines first made their entry into areas where the body was – out of habit – most shamefully treated, namely in the monasteries, in the mines, and on the battlefield (Mumford, 1934, p. 36). Orthodox believers in the Middle Ages despised their bodies and were prepared to hurt them. Instead of disliking the machine, which could mimic the body and that act, machines were welcomed.

The vast array of technical improvements from the 1500s onwards rested on a dissociation between the living and the mechanical (Mumford, 1934, p. 32). The main difficulty lay in *the propensity towards animistic thinking*.[4] As long as all objects, animate or inanimate, were perceived as a house for the spirit, as long as a tree or a ship was expected to act like a living creature, it was nearly impossible to isolate the special function one wished to present as a mechanical sequence. The process of abstraction was difficult and threatening when the world in its essence was animistic and "external forces" menaced. The only way of escape was to discipline the self or to conquer other people. The Church disciplined the self – but not as severely in the case of the peasant, as in monasteries and universities.

The church regarded the world as an independent and orderly part of reality. Descartes exploited this dichotomy. He described in a purely mechanical fashion the physical world of nature, but deliberately omitted the soul. The soul was the church's domain. Science could continue to explore nature – provided it did not infringe upon the church's territory. It may not be a coincidence that serious seventeenth century scientists, such as Galileo, Descartes, Leibniz, Newton, and Pascal, were persons of worship.

The mechanical clock was the first truly autonomous machine built by man. Mechanical clocks had already been constructed in the 1200s, and were first used to regulate life in the monastery. During the 1400s, the number of clocks increased and they eventually became ubiquitous outside the monasteries. The clock made it possible to divide time into hours, minutes, and seconds. The world could now be seen in a new way, with daily life regulated according to the clock – not at one's own discretion. *Direct experience is renounced.* The clock created an impulse that alienated man from nature, placing what can be measured and weighed at the centre. Modern natural sciences were born with Galileo (1564–1642). Galileo's advice to natural sciences is: "Measure what is measurable and make measurable what is not so" (Ranheimsæter, 1970, p. 54). The new "Machine Age" is ushered in. Nature is left standing as a complicated, lifeless machine. In the 1700s, the ideal of mechanical regularity and mechanical perfection becomes an integral part of all human activity, from the drilling of soldiers to seed drilling, from double bookkeeping to establishing study routines in school. There is no room for man in this new configuration. Man has been reduced to a standardized servomechanism.

Did specific traits of western culture prepare the way for the cultural leap in consciousness? According to Mumford, the proliferation of monasteries was a significant precondition.

"What better preparation could a whole culture have for such an effort than the spread of the monastic system dedicated to living a humble and self-abnegating life under a strict rule? Here in the monastery was a relatively non-animistic, non-organic world … One of the first experimental scientists, Roger Bacon, was a monk. The monks stood high in the roll of mechanics and inventors. The spiritual routine of the monastery, if it did not positively favor the machine, at least nullified many of the influences that worked against it" (1934, pp. 34–35).

There seems to be little doubt that the west's brotherhood of faith crafted more productive and complex machines than the Buddhists' prayer wheels.

What shapes science? What are the conditions for scientific development?

In every period a set of materials, techniques, and tools are not only employed as aids, but also contribute to shaping human beings' cultural and philosophical views. This is also referred to as the "defining technology". Let's take a closer look at the defining technologies of different periods.

The spindle, the clock, and the steam engine have influenced human thought in earlier times. The spindle was a defining technology of Antiquity. Plato's universe was divine and alive and compared to a spinning shaft with eight concentric whorls to ensure its stability (Bolter, 1984, p. 21). For the Greeks and Romans, the materials were clay, wool, and wood, and the instruments were spindles and the potter's wheel. Human beings became fascinated with rotary movements, with their picture of a rotating universe, in which the stars, sun, and moon moved in endless circles around the world. According to Bolter (1984), the literature of that time was characterized by an imaginative reaction to technology.

In the 1600s, this universe is replaced by a divine and gigantic clock. Mumford (1934, p. 14) writes that we cannot overestimate the importance of the weight-driven clock, the key machine of the modern industrial age. At each phase of its development, the clock is both a fact and a symbol of the machine. The clock, as a triumph of mechanization, offered a powerful technical metaphor. The world was now seen as a whole based on the analogy suggested by the weight-driven clock. The clock metaphor was not only used to describe the macrocosm, but also plants, animals, and the human body. The value of the metaphor lies in the clock's cyclical movements, regularity, and independence from "outside" forces. The clock represented a precise and ordered cosmos and fostered the abstraction and quantification of the experience of time. Great scientists such as Descartes and Leibniz could now explain all nature – except the mind – based on mechanical movements. Leibniz even discussed mental clocks.

Dynamic technology started in the Middle Ages, using water and wind; for example, water-powered sawmills and windmills. They existed before the Renaissance, but were made more efficient during the Renaissance. The interest in technical innovation intensified. The steam engine became the defining technology by the end of the 1700s. It combined two qualities that were separate earlier, one in the clock and the other in the water wheel. To some extent, the machine was a *self-controlled instrument* transforming heat energy into an exploitable force. Steel and iron were materials that were hard and durable, and could be used to make machine equipment that could make other machines. The idea of machines creating machines entailed a new level of technological abstraction.

During the 1800s, the world and man tended to be seen as possessing many of the attributes of a heat machine. How did that shape the view of humanity? Bolter (1984) writes that these attributes are not particularly evocative. They suggested that the universe is an imperfect steam engine that gradually recedes, disintegrates, and wastes away as useless heat. The brutality of this new metaphor reflected the harsh conditions prevailing during the industrial revolution. During the 1800s, one considered and studied the living organism as a warm engine that burnt fat, proteins, and glucose, transforming them into water, carbon dioxide, and urea (Wiener, 1948).

According to Bertalanffy (1968), this development is the result of a "de-anthropomorphization" of the worldview. Von Bertalanffy points to this as one of the most significant traits of modern science, which he describes as a process of increasingly rapid "de-anthropomorphization". Erroneous sources due to human experience are progressively eliminated. Physics began with sensorial experience of the eye and ear, leading to optics and acoustics, and finally direct experience, being replaced by self-learning algorithms. Consequently, that which is specific to human experience has been progressively erased. Knowledge that is "obscure, reduced and uniform" is readily deemed as being the only valid knowledge. Modern science's ideas must be clearly delimited, rely on principles, and be logically consistent (Bråten, 1983, pp. 128–135). With the introduction of the computer, the rationalist view of society and man was reinforced. *Rationality is distorted* into becoming synonymous with logic (Weizenbaum, 1976, pp. 13–14 and Vickers 1973). Max Horkheimer also highlighted this development. In 1947, he described how instrumental thinking was increasingly dominating nature and man: "Concepts have been reduced to summaries of the characteristics that several specimens have in common. By denoting similarity, concepts eliminate the bother of enumerating qualities and [are] thus thought of as mere abbreviations of the items to which they refer. Any use transcending auxiliary, technical summarization of factual data has been eliminated as a last trace of superstition. Concepts have become streamlined, rationalized, labor-saving devices … thinking itself (has) been reduced to the level of industrial processes … in short, made part and parcel of production" (Weizenbaum, 1976, p. 249).

Von Bertalanffy (1968, p. 199) claims that the diversity of concepts and approaches to social sciences ("the Sciences of Man") would be healthy were it not for one disturbing fact. All the theories share a common picture of man. This picture has its origin in the physical technological universe and is taken for granted by otherwise contrary theories such as behaviourism and computer models of cognitive processes, and it is demonstrably false. It represents the *robot model* of human behaviour.

Mumford accuses the paradox of an apparent elimination of subjectivity from the predominant world picture. He sees the desire for perfection and the impulse to go beyond also within technological fields, and argues: "Strangely, the very existence of these pre-conscious sources of technology have been ignored on the supposition that science and technics have no subjective attachments. Nothing could be further from the truth" (1970, p. 419).

And what has become of man? Mumford describes man as having become a banished person, out of sight and therefore out of mind, in exile, and likened to a starving prisoner in a concentration camp built by man himself (1970, p. 419).

Summary

Karl Popper sums it up as technology and natural sciences having shaped a new perception of reality through history, culminating in a deification of the machine and measurable facts. Belief in God is gone, while the life of the soul is filled with new secularized religions linked to nations, social classes, existentialism, positivism and behaviourism (cf. Popper, 1981, pp. 248–249).

Technology has provided useful tools, but it has also changed our representations. The telescope enabled the discovery of new stars. Jet aircraft allowed us to cross a continent in the space of hours. This has led to new concepts of time, distance, human cooperation, and communication. Technology catalyses changes – not only in *the way we work and what we do, but the way we think and ultimately in what a human being is*. The computer or modern information technology, also referred to as "thinking machines", is perhaps one of the key drivers behind changes in man's self-consciousness, consciousness of others, and relation to the world.

The relationship between man and the computer

The computer is an instrument that expands human capacity and broadens man's range of abilities. Electronic text processing, CAD (Computer Assisted Design), and telecommunications change our perceptions of cooperation. For example, an author can receive same-day feedback irrespective of location as long as he is online. The disabled benefit from new aids, the blind use reading machines, and machines help the deaf learn to talk.

The computer is *more than an instrument*. The presence of computers penetrates the way we think about ourselves (Nordenstam, 1983). According to Turkle (1984), the computer has two essential features, one of them resembling an analytical engine. This is the well-known computer metaphor. It includes the computer's ability to make logical, numerical calculations with unique speed and precision. Nevertheless, the computer is also an *evocative object*. This is the second feature of the computer. The computer *tends to stimulate reflection about fundamental questions* (Schank, 1984, p. 27). Adults begin to toy with the idea that the human brain is a computer program following a set of specified procedures. According to Turkle, the computer also has a chameleon-like property. When you program it, the computer becomes your creation – your own micro world. We create private worlds, and through them we can expose and explore ourselves – we create a mirror. For specialists the machine becomes a faithful friend that is difficult to push away. For "computer spectators" the machine arouses fear. Whether the relationship with computers is one of ecstasy or phobia, the presence of computers leads to human position-taking and hence creates a relationship between man and machine. Let's explore some of these types of relationships.

Anthropomorphization of the computer

In 1980, Johannes Lunde wrote a feature article in the Norwegian daily newspaper, *Aftenposten*, entitled: Do electronic brains need a heart? He starts by writing the following:

> A passenger who cannot bear travelling backwards on the train, asked for a seat in the travel direction when ordering a ticket. The train company employee replied that seats are computer-assigned, so we just have to take what we get. (Aftenposten, March 11, 1980)

The example above contains an element of anthropomorphism, that is the attribution of human traits to machines. Using the computer as a self-managed unit capable of taking action on its own is something most of us have experienced in different practical situations. In some situations, it may come in handy. The train company employee can blame the machine, if it might reinforce a positive self-image. The employee is also in a position to sympathize with the passenger who must sit opposite to the travel direction for eight hours. In the literature on computers there are many examples of this form of anthropomorphization of the machine. Some have more dramatic consequences than the example above. A classic example is related to the artificial expert system called ELIZA (Weizenbaum, 1976).

Joseph Weizenbaum, professor at the Massachusetts Institute of Technology, built and tested between 1964 and 1966 a computer program (DOCTOR) that could "converse" in English. The program, which simulated a psychotherapist, had been given a number of simple messages to respond to what the patient said by posing a number of specific questions and some more general questions. The client had to type his questions on a keyboard that was connected to the computer. The computer "answered" in the form of a machine-written paper printout.

First, the client who talked with the computer became rapidly and deeply emotionally involved in the conversation with the computer. *The clients humanized the computer*. They conversed with the computer as if it were a person one could talk with about the most intimate questions.

Weizenbaum was aware of the fact that human beings form emotional bonds with different machines such as musical instruments, cars, and boats, but was surprised at how rapidly they were deceived by the computer:

> What I had not realized is that extremely short exposures to a relatively simple computer program could induce powerful delusional thinking in quite normal people. (1976, p. 7)

Second, Weizenbaum was appalled by the reaction of several professional psychotherapists. A number of practicing psychiatrists truly believed that the computer program could be developed to perform a wholly automated psychotherapy. Computer-administered psychotherapy was proposed should there be a shortage of therapists, for example in mental hospitals and psychiatric centres.

> Because of the time-sharing capabilities of modern and future computers, several hundred patients an hour could be handled by a computer system designed for this purpose. The human therapist, involved in the design and operation of this system, would not be replaced, but would become a much more efficient man since his efforts would no longer be limited to the one-to-one patient-therapist ratio as now exists. (Colby et al., 1966, quoted in Weizenbaum, 1976)

Here Colby paints a rather sterile and rigid picture of a therapist, describing him as an "information-processor" that is dependent on following certain rules – not as an engaged human preoccupied with healing another human being.

> The patient is, in other words, an object different from the desired object. The therapist's task is to detect the difference, using difference-detecting operators, and then to reduce it, using difference-reducing operators, and so on. (Weizenbaum, 1976, p. 181)

Third, the reaction showed that even well-educated persons are capable of assigning highly exaggerated attributions to a technology they do not understand. Weizenbaum frames the problem as follows:

> This reaction to ELIZA showed me more vividly than anything I had seen hitherto the enormously exaggerated attributions an even well-educated audience is capable of making, even strives to make, to a technology it does not understand. (Op. cit. p. 7)

Other expressions of the relationship between man and machine also exist.

The computer as a reflective medium

Hackers illustrate another aspect of our relationship to machines. A hacker is a person that lives his/her life in close proximity their computer, with the computer becoming a new reflection of herself/himself – a reflective medium (Turkle, 1984).

A teenager describes it as follows:

> When you program a computer there is a little piece of your mind, and now it's a little piece of the computer's mind. And now you can see it. (Turkle, 1984, p. 155)

With the computer we can create a micro world similar to the one children create with drawing and dolls. A computer program is a reflection of the programmer's consciousness. A person who has written a program and begins to work with it is offered an opportunity to see himself in a different way. Turkle refers to this as creating a "second Self". The computer becomes a means to reflection.

But hackers are not the only ones who can enjoy gazing at their own reflection in their "electronic image". Many people have a blurry self-understanding, and that insecurity induces us reflect on who we are. Mirrors play an important role in human development. They allow us to see ourselves from the outside and objectivize aspects of ourselves that we only sensed in an obscure and ambiguous manner. With the computer, we can learn about our "*second self*".

A system for the interactive use of the computer as an instrument for human and professional advancement was developed and described by Keen and Tschudi (1985). The danger lies in human beings not evolving, but ending up trapped in the machine. The computer provides reactive and interactive company, but lacks the reciprocity and complexity inherent in human relations (Turkle, 1984, p. 19).

The computer as an expressive medium

In his study of children and young people's relationship to the computer, Turkle finds considerable individual differences, partly reflecting how we use the computer in different ways to express ourselves, our longings, desires, fears, and anxiety. The projections reveal something about the child's thinking, feelings, and organization of experience, reflecting an image of who the child is or perhaps wishes to become. In that sense, the computer can be a projection of the "self" and a reflection of consciousness.

In terms of learning, human beings go through *three stages* in their relationship with the computer, according to Turkle. The first stage is metaphysical. We question to what extent machines think, feel, and are alive. The second stage is proficiency, where mastering technology is key. The third stage involves personal identity, that is the metaphysical self.

Mechanization of man

We have seen several examples above of the computer thought of in human, mental terms, that is the anthropomorphization of the machine (Bråten, 1983). In contrast, human beings can consider themselves and others as machines, which can be called a "mechanization of man and consciousness". Such mechanization can be expressed in various forms; for example, linguistic forms and metaphors derived from the computer's mode of operation.

The concept of the machine

Franz Reuleaux posited that machines are a combination of resistant components, each specialized in function, operating under human control to exploit energy and perform work (Mumford, 1967, p. 191). This definition also allows Mumford to include "collective machines", which require the mechanization of human beings. According to Mumford (1967), this development preceded the mechanization of work instruments. These collective machines are called "invisible machines". If employed in collective actions aimed at power and destruction, they are called "military machines". If all the components – economic, military, political, and ruling authorities – are included, Mumford refers to them as a "megamachine". When collective machines were first acknowledged, the mechanism rapidly spread, not only used in self-defence, but by being imposed by kings acting as gods. Mumford argues that the secret to mechanical control is to have only one single mind with a well-defined goal at the top of an organization, and a method for sending messages through a series of intermediate officials until they reach the smallest unit. Both an exact reproduction of the message and absolute fealty are essential, the result being a mechanized group that can be manipulated by command.

The mechanization of human beings did not, therefore, come with the computer age. One can trace it back to the Pyramid Age in Egypt. But in our own culture hardly anyone has described the tendency more clearly than the Frenchman La Mettrie in his book *L'homme Machine* in 1748. He formulated and published the philosophical case of considering man exclusively as an ingenious machine. This prompted strong reactions. It was also the period where romanticism flourished. The romantic took a negative view of science with its cold and calculating methods. Mary Shelley was a leading romanticist who objected to rationalism. Motivated by a fear of, and as a warning against, innovative science, Mary Shelley wrote a novel about Frankenstein in 1819. While monsters were formerly made by gods, Mary Shelley's novel about Frankenstein is an image of a humanlike creature made by human beings.

The self-concept and the computer

Psychology and related disciplines have developed concepts to help grasp human beings' experiences, behaviour, "self" and the human organism, for example concepts such as the world of experience, cognitive structure, stimulus-response, learning, schemata, and so on. Turkle (1984) invokes young people who are not familiar with these theories. Stimulated by the interaction with the computer, they acquire a vocabulary for use in their self-understanding. The computer resembles the human mind to the extent that that it can form analogies between the self and programs. Some use explicit computational metaphors, while others employ the expression "clear buffers" when they have to change a topic of conversation. Some use debugging to describe having been through psychotherapy. The mind as a program is likely the most pervasive and influential metaphor. In this context, human consciousness is to be understood as a specific set of procedures that perform a sequence of actions. For example, we use expressions such as having to "reprogram ourselves" to deal with new situations.

These young "button-pushers" are unconcerned about whether the computer will be one day emulate human beings, but realise that human memory is similar to a computer's memory. Turkle's conclusion is that the computer affects the manner in which people think – especially about themselves – regardless of whether we ever manage to create artificial intelligence that can think like a person (Turkle, 1984, p. 162).

The mechanization of man does not only involve linguistic processes about ourselves. Lunde (1980) offers the example of the train company employee and the unfortunate passenger to shed light on *the forces that surround, and are partly fostered by, technology*. From this perspective, Lunde attempted to demythologize the computer. It is the human brain that creates the computer and the human brain that controls it. But what are the forces that drive human beings and what misuses of the computer might that entail?

The reason behind the misuse does not lie in the computer itself, but in profit and efficiency maximization requirements – which the computer provides new scope for accommodating. Limitations required by human and social considerations are not moderated. Those who control the computer have, as Lunde argues, only one objective: optimal profitability and technical efficiency, whether the computer is used in the state railway system, elderly care, the tax system, or in manufacturing.

Lunde describes an elementary *tension* between economics and well-being, efficiency, and humanity, and argues that *an equilibrium ratio inevitably emerges*. But the computer user has not yet experienced this "anthropomorphization process". According to Lunde, the computer is still so new, it does not yet lend itself to public critique, so unilaterally and uncontrollably mastered by the technological ideology.

Lunde (1980) argues that to the extent human considerations can be expressed quantitatively, the computer can react in a human manner. But since all things cannot be quantified, a remnant in need of individual human assessment will subsist. Lunde suggests that this remnant must be integrated into the system. Intuition-driven efficiency thinking does not alone suffice, but social consciousness must also play a part.

Lunde further argues that an advanced technique has a dangerous power over specialist mentality. Moreover, it is easy to believe that non-economic, efficiency- limiting considerations are unprofessional and unfounded. Apart from Lunde, other researchers also see a risk that all new techniques can take on absurd forms of use if their development is not controlled. We are easily impressed by what is new. According to the German pastor, Dietrich Bonhoeffer, the true problem is not the evil of the few, but the willingness of the masses to be impressed by displays of power and to march in lockstep. The question is where responsibility should be placed. It cannot be placed with an organization, but only with the individual.

In the context of teaching and learning, Russel Ackoff (1974, especially at p. 74) is preoccupied with the *destruction of the concept of self*. He sees a great danger associated with machine teaching in our time's "machine age" – because machines do not have a self. He asks us to reflect on what happens when the teacher has been trained to act as a machine. Mechanization of man easily leads to man being objectified and results in human contact only, producing a feeling of being an object in a world of objects. Talking to a machine might then be preferable to being talked to by an expert in a white coat? The Norwegian scientist Stein Bråten (1983) interprets the reaction to the computer program ELIZA in that way. Bråten argues that if we more readily confide in the computer program, it is because we are afraid of being reprimanded when opening up to other human beings. Using a model of consciousness built on self-reflective dialogues, Bråten explains the client's openness to the machine.

> Rather than perceiving the conversation with ELIZA as a conversation with someone other than a person, it is possible to see this computer program simply as a means of stimulating the person's conversation with himself – about himself. The feeling of understanding comes from the person's self-reflective dialog with himself – through the cautiously repetitive and questioning computer program. It is not a given that a feeling of being understood will transpire in a conversation with a person in a white coat with analytical aptitudes. (Bråten, 1983, p. 24)

Bråten is more hesitant regarding human objectification than the quest to preserve our human identity through artefacts, such as the computer. Contrary to being considered an object, one can see a meeting with oneself or with another human being as a Me-You dialogue, featuring closeness, empathy, and human warmth. Bråten claims that the computer, as an "it", can in certain cases give rise to a genuine me-you-dialogue within and between people "where their humanity [is] thrown into relief against the backdrop of the machine" (op. cit. p. 24). It should also be noted that ELIZA simulated a Rogerian psychotherapy model, where the client knows best and where the client is the one who takes the initiative and sets the premises for the conversation.

But what happens when it becomes clear to the client that the machine actually does not understand his/her problems, but just reformulates the client's own words? Well, a person that has attempted to converse with the machine can tells us the following.[5] When I began to suspect that the machine did not understand, I start taunting it. The machine's wordplay was then unmasked, and I became aggressive. Unlike an engaged fellow human, the machine does not offer any help to channel the aggression onto a constructive path. Whether or not the machine, which has not lived, experienced, fought, loved and longed *can ever understand a human being* in a deeper sense is a fundamental question (Weizenbaum, 1976, Chapter 7, especially pp. 200–201).

Human warmth versus the computer

In a previous section, we referred to Johannes Lunde (1980) and his postulation that in the longer run an equilibrium ratio between efficiency and humanity necessarily emerges. The same is posited by John Naisbitt (1984) in his book *Megatrends.*[6] Naisbitt places the two concepts of "High Tech" and "High Touch" in opposition to each other with many examples of how high tech has nearly pressed forth a stronger need for human contact and warmth. His model indicates that *high technology, somewhat further ahead, will require a human warmth component.* One example is the rising technification of hospitals, which has led to rising numbers of home births, home care, and demands that the same person/nurse follow the patient during a stay in an institution. Life extension technology reflects a growing desire for a dignified death, and "quality of death" has become a frequently discussed concept. The introduction of television did render man passive, but created countercultures in the form of group therapy, movements, and study circles. People still went to the cinema and there was a renewed awareness of the need to experience movies together with others. The model illustrates that there is a need for human closeness and warmth through, for example, conversation with others. Fløystad believes that man has a need to create and recreate values through active participation in a community (Fløystad, 1987, p. 6).

Lysgaard's book *Arbeiderkollektivet* (The Workers Collective) describes similar reactions against the technical-economic forces. Lysgaard's (1961) model explicitly involves a business organization. The existence of a basic tension in the workplace is the central postulate. This tension promotes the formation of an improvised collective among the workers on the floor with a view to easing tensions. Traditionally, tensions tend to exist at the worker-manager level. Lysgaard provides a more nuanced picture using the concept of the technical-economic system (T/E). This is the primary, planned system underlying the other systems in a firm. The system is summarized in three words:

- Insatiable: Increased productivity, not productivity, is the goal. Man is naturally limited in this regard.
- One-dimensional: The T/E system divides tasks into an increasing number of parts so work becomes specialized. Yet, man is multidimensional by nature.
- Inexorable: It is not in the T/E system's interest to keep members that are unable to fulfil requirements. The employee is dependent on work

to earn a living, and hence seeks security. If we compare the requirements of the T/E system and essential features of man, we find the following conflicting traits:

- insatiable – limited.
- one-dimensional – multidimensional.
- inexorable – security-seeking.

This engenders tension, and workers respond by forming a "workers collective".

Ims (1976) suggests that if tension rises above a certain level, workers go on strike. Naisbitt's conclusion is that if the technological input is amplified beyond a certain level, a *technological backlash* may occur.

An empirical study of a large US company (Zuboff, 1983) surprisingly found that during the transition to electronic mail, communication and contact points did not decline, but increased. The difference was that the new form of contact became a type of electronic contact. Social talk had earlier been controlled by protocol in accordance with one's place in the hierarchy. Now operators "on the floor" began to send electronic messages to the director. The new form of contact is, however, a "pseudo-contact" because contact is limited to "look (and hear), but don't touch" (Bråten, 1983).

The findings above relate to the relationship between man and high tech in general, but how does it specifically relate to the use of computer? There are many anecdotes about hackers and obsessive-compulsive programmers (Weizenbaum, 1976), about asocial button pushers who replace human contact with machine contact.

Their human contact is virtually marginal. Both Turkle (1984) and others, for example Vallee (1984), report that hackers do form invisible colleges, that is social groups that communicate via the electronic network. This electronic communication takes place through bulletin boards – generally at night. Hackers rarely meet face to face and they are identified by aliases, so-called handles. Information is the good traded. A "hacker", called "the cracker" (Landreth, 1985), describes it as follows:

> "In most cases, "trade information" means "I'll give you what you need now, and next time you get something good, let me know." (1985, pp. 15–16)

The information traded is generally about how to access different computer systems, whether it be the computer of a phone company, a bank, or NASA. What makes the contact special is that it is channeled through a network. It can be electronically spontaneous, but does not involve physical contact.

Human knowledge superior to artificial intelligence?

The theme of man and artefacts is in one sense a history of *self-imitation*. Man has made statues of himself as art, represented himself on cave walls, on the screen, and more recently on video and TV. Is it an endless ego trip? We are preoccupied with ourselves, but we are also preoccupied with the representations we make of ourselves. As Pamela McCorduck observes, we are ten times more fascinated by clockwork imitations than by real people performing the same task (1979, pp. 4–5). Like others, Mumford was of the view that attributing greater importance to the artefact than to man reduces the mystery, transforming it into absurdity.

Attempts to make artificial people started early, already with the Greek gods in the myths of the statue Galatea who came to life as solace for Pygmalion. The myth of Pandora's box is from the same period. In a broader context, it offers us insight into man's curiosity and the possible consequences of blindly following that curiosity. Curiosity killed the cat. Pandora was the creature who, commissioned by Zeus, was given the mission to punish human beings because they had accepted the fire Prometheus had given them as a gift. Pandora brings back a box but is forbidden to open it. But curiosity gets the better of her – she opens the box and all the evil of the world is released. One can say that Prometheus and Pandora mysteriously embody two facets of human knowledge – lightness and darkness, evil and the gift. This battle between advocates and opponents of new knowledge and technology is one of the links in the chain that characterize our history. Perhaps the conflict reaches its climax in the case of artificial intelligence.

In the 1950s, 60s, 70s, and 80s, there was an intense debate between "pro-technologists" and philosophers, who refuted the machine's ability ever to possess intelligence on a par with that of human beings. What was at the core of the conflict? Why was it not possible to solve the conflict by arriving at an unambiguous answer accepted by all parties?

Dreyfus and Dreyfus (1986) support the idea that expert knowledge – the highest level of human knowledge – cannot be coded and stored in a computer. In their book *Mind over Machine: The Power of Human Intuition and Expertise in the Era of the Computer*, they convincingly advance the notion that computers used as logical machines will always will be inferior to human beings as a source of insight and reflection. They constructed a model comprising five stages of skills acquisition; novice, competence, proficiency, expertise, and mastery. Albert Einstein draws a similar conclusion:

> To these elementary laws there leads no logical path, but only intuition, supported by being sympathetically in touch with experience. (cited in Dreyfus and Dreyfus, 1986, p. 41)

Dreyfus and Dreyfus show that the two final stages are characterized by holistic acknowledgement and swift, subconscious, and often intuitive actions that go far beyond what today's most advanced technology can offer. Dreyfus and Dreyfus also distinguish between "know how" and "know that". While "know how" is a skill or "silent knowledge" that can hardly be articulated, "know that" is lexical knowledge that is easily formulated and coded.

The authors describe the differences between the properties of man and the machine. The machine is rigorous, reliable, and durable, traits that are superior to those of man. Human beings, on the other hand, show flexibility, exercise judgment, and use intuition, traits that resist decomposition of specification and deduction.

Hubert L. Dreyfus, one of the Dreyfus brothers, has always been considered a controversial author within the area or artificial intelligence. He openly confronted leaders in the field, namely the trio Allen Newell, J.C. Shaw, and winner of the Nobel prize in economic sciences, H.A. Simon. They published works at the end of the 1950s fostering enthusiastic expectations that machines would, in a foreseeable future, be able to emulate human beings (see also Newell et al., 1963a, 1963b). One of the starting shots for the debate was triggered by Simon in 1958 where he claimed that their computer program (General Problem Solver) was ushering in the era of artificial intelligence:

"It is not my aim to surprise or shock you … But the simplest way I can summarize is to say that there are now in the world machines that think, that learn and that create. Moreover, their ability to do these things is going to increase rapidly until – in a visible future – the range of problems they can handle will be coextensive with the range to which human mind has been applied." (Simon and Newell, 1958).

These expectations were harshly criticized by Hubert L. Dreyfus (1979) who compared their research to alchemy.[7] Accordingly, the partisans of artificial intelligence were associated with the chemical science of the Middle Ages and speculative philosophy. As for Dreyfus, he maintained that he was not first and foremost attacking their technical effort, but the philosophical assumptions they applied. Dreyfus believed the assumptions were erroneous, partly because they made an excessive distinction between mind and body. Dreyfus submitted an alternative basis for scientific theory (phenomenology), strongly inspired by Wittgenstein's later works (Wittgenstein, 1958; see also Baker and Hacker, 1980). Nevertheless, Dreyfus stood alone in his critique. Somewhat later, Joseph Weizenbaum (1976) offered an in-depth critique of the «pro-technologists». Like with Dreyfus's critique, it was difficult to dismiss Weizenbaum's critique, because Weizenbaum was a renowned scientist at MIT who had just become known for developing the artificial expert system ELIZA.

But why was the disagreement so vehement? Newell and Simon and the other pro-technologists' academic pride was offended when they were compared to scientists of the Middle Ages. But Weizenbaum (1976) also claimed that progress achieved within the field was not primarily due to research, but to technology, and the innovations were attributed to the hackers in the MIT community. They were described as "…bright young men of dishevelled appearance, often with sunken glowing eyes, can be seen sitting at computer consoles, their arms tensed and waiting to fire their fingers, already poised to strike, at the buttons and keys on which their attention seems to be as riveted as a gamblers on the rolling dice … They work until they nearly drop, twenty, thirty hours at a time. Their food, if they arrange it, is brought to them, coffee, cokes, sandwiches. If possible, they sleep on cots near the computer. But only for a few hours – then back to the console or the printouts. Their rumpled clothes, their unwashed and unshaven faces, and their uncombed hair all testify that they are oblivious to their bodies and to the world in which they move. They exist, at least when so engaged, only through and for the computers. These are computers bums, compulsive programmers. They are an international phenomenon" (1976, p. 116).

The bitter controversy seems inextricable because it is rooted in philosophical assumptions about method and the researcher's neutrality. Dreyfus argues that the researcher himself is a part of the method. Dreyfus (1979) posits that all human beings, including the scientist, perceive things based on their own intentions, but therein lies a source of methodological error in all research. This theoretical postulation is to some extent supported in interviews with the relevant scholars. Pamela McCorduck (1979) offers insight into the human element, just as Simon and Dreyfus do. McCorduck found that both Dreyfus and Simon, since the very beginning of their careers, appeared to have taken a stance on artificial intelligence – whether that was possible or not! Simon thought it was feasible, while Dreyfus reacted very aggressively towards anyone of that opinion. McCorduck advances the following: "When I finally met and spoke with Professor Dreyfus, an idea occurred to me that will please neither him nor the field he so relishes attacking, and it is that they (Simon) have more in common than they think. As surely as Herbert A. Simon was firmly convinced from the beginning that a scientific explanation exists for what we call mind, Hubert Dreyfus was from the beginning equally convinced that one does not" (1979, p. 203).

McCorduck further argues:

> Predisposition, or world view – call it what you will – have more to do with opinions on this scientific question than evidence. I've been amused to discover the same thing within myself. (1979, p. 205)

There are, in summary, some crucial differences and similarities between Simon and Dreyfus. The phenomenological viewpoint supported by

Dreyfus places emphasis on the body's role in unifying and organizing our experience, the situation's role as to setting the background, so that behaviour can follow certain patterns without following rules, and finally how human beings' subjective goals and needs influence how the situation is organized. The similarity between Simon and Dreyfus is that they do not feel tied to the Cartesian worldview. Descartes had to draw a distinction between body and soul. For him, drawing that demarcation line was a historical necessity to enable him to practice his science. But perhaps we scientists should attempt to reunite mind and body as the phenomenologists suggest?

"The electronic man" – metaphors and fears

Why are we afraid of the mechanization of man? Before exploring that question, let's look at the linguistic expressions or metaphors used to describe that reality.

The image of man that Newell, Shaw, and Simon built on was man as "an information processor with metaphors from the computer's functioning". Bolter (1984) also denotes this image of man as «The Turing Man» after Alan M. Turing, the mathematician who first described the symbolic characteristics and theoretical possibilities of logical machines. In 1936, Turing demonstrated how one could in principle construct the "universal machine". Through the promise (or threat) of replacing man, the computer entails a new definition of nature and man. Man becomes an "information processor" and nature becomes "information to be processed" (Bolter, 1984, p. 13). This electronic machine image of human thinking was attacked by quite a few academics from different angles, for example the psychologists George A. Miller (1982) and Ulrich Neisser (1976). Miller maintains that the computer's manner of functioning hardly seems to have any more relevance to the way our brain works than a wheel is illustrative of how man walks. Neisser claims that the information processing model is rigid and neglects emotional aspects that are also important in terms of understanding thought. Richard Boland (1987) provides a more recent and meaningful critique of Simon's models of human thinking. Like Churchman (1968) and Enid Mumford (1987), Boland also laments the absence of any ethical reflections on the part of the system designer.

Describing the computer in human terms; for example, the logical human is not the same as describing the computer using technological terms. Metaphors appear to have a limited place in modern social science. Nevertheless, both Aristotle and Immanuel Kant were aware of the importance of metaphors. According to Aristotle, each metaphor brings to light an intuitive perception of similarities and difference and is for that reason a mark of genius, the ultimate element. According to Kant, the act of speaking in a metaphorical language, that is to speak using analogies, is the only way thinking can manifest itself. What does it signify when the metaphor "man as an information processor" is widely used? Lakoff and Johnson (1984) posit that "we live via" the metaphors we employ. Since the essence of a metaphor is to experience and understand a "thing" via another thing, metaphors such as "man is a machine" or "the heart is a pump" participate in structuring perceptions of the subject who observes because the metaphor focuses on and conceals aspects of that which is observed. Valle and Eckartsberg (1981) give a broad overview of the metaphors used for human consciousness. They maintain that physics has undergone various conceptual revolutions in the twentieth century. This has clearly revealed the limitations of the mechanical world perspective that has left its mark on our perceptions. They claim that new science should push in the direction of a more organic and ecological worldview. The universe is no longer seen as a machine composed of countless separate objects.

But why do we fear this human being? Is it because we live with images of people passed down from earlier cultural periods and that are threatened by "The Turing Man"? Is it because the electronic man seems superficial and devoid of history? Are we still attracted to Goethe's Faustian figure featuring a quest for depth and recognition? Is there still a part of us that lives in the same cultural world as described by the Greek philosophers? Does the use of logical machines and the strongly propelled and extensive mechanization of mind and society mean that we are not following the virtues of Aristotle and Socrates? We are familiar with many of them. Choose the mean and avoid the extremes on either side and know yourself. Aristotle advocates a balanced life where the goal must not been seen independently of the means. Socrates says "know yourself", that is tread cautiously to avoid moving beyond the limits of life. Have Socrates and Aristotle captured the essential conditions of a good life? According to Bolter (1984), computer culture also consists of such a peculiar combination of Greek and western European qualities, but the image of the electronic man has moved North Atlantic culture a step forward.

But perhaps the Greeks did not lead us only forward? Did Plato and Socrates, as founders of critical thinking, aim at greater objectivity only contributed to the good? Dreyfus and Dreyfus (1986) raise that question. For example, in Nietzsche's eyes Socrates was not fully part of our cultural heritage. Nietzsche held the view that since Socrates had lost the ability to trust intuition, Socrates was "*the first to degenerate our culture*". Aristotle also opposed Plato and Socrates' rational thinking. According to Aristotle, human beings need *wisdom or judgment* to apply rules to concrete situations. Experiences of daily life are important, but cannot always be made explicit. Aristotle defined man as *zóion logon echon*, which means equipped with logos. In Greek logos could mean understanding situations in their totality and logical thought. When logos was translated into Latin as ratio, its meaning was substantially diluted (Dreyfus and Dreyfus, 1986, pp. 202–203).

Where can we find correctives to the prevailing and one-sided "calculating" rationality and see the limits to the dominance of computer metaphors? Maybe we should seek insight from more intuitive cultures such as Japan, and perhaps we must explore ideas that existed *before Plato and Socrates* to find a more holistic image of man? Here lies an important mission for science in the time ahead.

Conclusion

The computer is symptomatic of the more fundamental characteristics of our civilization. The computer has amplified forces that have driven man towards a steadily more mechanical self-image and rationalistic view of society. Different types of instruments enhance our body, muscles, and senses. They have enabled new forms of interaction. But there are also machines that engender a more radical change in our perception of reality. The clock was one of these autonomous machines that altered reality. It weakened the version of the former reality because the direct experiences that made up the former perception of the world were denied.

Human beings are influenced by their relationship with technology. It is far from certain that more human warmth will be called as high tech spreads, as Naisbitt postulated in his model about the balance between "High Tech" and "High Touch". Naisbitt's model is based on historical data and has provided insight into man's reactions against technology in the 1960s, 70s, and 80s. However, does it provide any indication of peoples' reactions in the 1990s? What will be the outcome if developments result in even more mechanized minds?

Are there no differences in the characteristics of man and the machine? There are models of expert systems and expert knowledge where the genuinely human is prominent. *The ultimate form of expert knowledge is intuition. Non-logical, non-calculable and non-codifiable information represents the highest level of human knowledge.* This means that the use of technology is subject to an upper limit. When it comes to exploiting the comparative advantages of man and machine, the "natural" boundary for a good division of labor between man and machine may be well within this upper limit. In that respect, it is not only a question of what computers *can* do, but what they *should* do. According to Weizenbaum (1976), an alternative to increased use of machines is restructuring, and even the elimination of tasks. But this requires thinking. Our culture seems to contain components of what economists call the "Pig Principle": if something is good, more of it is even better. One can ask oneself if the computer has preserved many social and administrative institutions that otherwise could have been changed?

Technological development has forged the world into one unit. However, as did Weizenbaum (1976), we can ask whether technological advancement has also closed many doors that were previously open. Have certain forms of behaviour that were possible earlier become impossible? Is this an irreversible development? One cannot concluded with certainty that this is irreversible, but several doors appear to be closed. The computer's ability to manipulate inconceivable amounts of data has resulted in a new problem of understanding. Understanding the results of computer processing has become a sort of closed door.

The new technology has made our society vulnerable. We have been relegated to technology and to ourselves. We have been given a new common destiny. According to Arthur Koestler (1979) a new era started on August 6, 1945. Before that day, man had lived with the prospect of dying as an individual. Since the day the first atomic bomb fell on Hiroshima, human beings must live with the prospect of being eradicated as a species. We need new concepts and a framework of understanding that raise awareness about this common destiny. It may seem as if a contrast exists between the unique technological achievements of the West and our incompetence in dealing with social conditions.

At the dawn of the 1990s, computers are spreading through the West, from the workplace to institutions of learning. Machines and instruments have altered the way we think of ourselves. Intuition and "the act of being" have steadily given way to instrumental thinking. The danger lies in instrumental thinking becoming absolute. There are two risks here. On the one hand, intuitive thinking – being man's highest form of knowledge as well as a prerequisite for creativity and innovation – is a form of "primitive" knowledge that must be preserved. On the other hand, humanity is put at risk if human beings are solely treated as a "thing" or "it", and as a means to something else. The most interesting question arising in the computer age may be how man's self-image can develop in the context of increasing contact with and the anthropomorphization of thinking machines?

This question is inevitable since computers are changing the way we live and *how we see the world*. We should seek to increase our awareness of the linguistic expressions that we use and "live through". The computer as a conceptual lens for much of what we see is one-sided and restrictive. We need new metaphors to refine the perception of what it means to be a human being. As many of us tend to find inspiration and resonance in images of "that which is human", which has been passed down from the Faustian tradition in the 1700s and Greek philosophy before the Renaissance, might we just have a basis for innovation? The computer, as a calculator, should be supplemented by other metaphors. Schank (1984) proposes metaphors such as the computer as a partner, teacher, and servant. But how does anthropomorphization of the machine impact our relationship to fellow human beings? Maybe we should do what Bolter (1984) recommends – refrain from using expressions such as artificial intelligence and, instead, call it synthetic intelligence? We would then be highlighting the instrumental function of the computer and maintain focus on a fruitful cooperation between human and machine attributes.

In western culture, theology has lost its authority, and we are left with a scientific worldview. Our culture and our science have been "stripped" in a process of progressive de-anthropomorphization. Even though the scientific methods reflect an increasingly greater distance to man's subjective life, his instincts, feelings, and intuition, scientific knowledge only mirrors some facets of reality, and each facet only holds a relative truth. As such, it is suggestive of the limitation as well as *the dignity of human knowledge* (Bertalanffy, 1968).

As a reaction against the uncontrolled subjectivism in earlier worldviews, western culture may have gone to the opposite extreme. While excessive authority was given to fantasy in the past without any correctives, modern science may have overcorrected this source of error and perhaps produced an equally false perception, namely that the organization of physical and bodily activities can develop in a world devoid of a mind? Have we held on to the Cartesian worldview where the mind was the church's domain and the body a subject of scientific research? The human mind is much too central for it not to be subject to scientification. Moreover, Descartes's dichotomy appears to be an unfruitful and possibly misleading one! Against that backdrop, there seems to be a need to *attempt to reconcile intuition and calculation, the subjective and "objective" aspects of human experience.*[8]

Notes

1. It should be noted that scientific research has been displaced from emphasis on cultural argument towards economic arguments, partly to push the evolution of science in a more positivistic direction.

2. Polanyi (1966) indicates the difficulty to code intuition in his eminent work *on tacit* knowledge, which in many ways might be regarded as intuition. Polanyi gives emphasis to the importance of the body concerning the perception of external objects. Such a view is in accordance with the so-called "expert paradox", which means that an expert in a specific field is not able to articulate or "decode" his or her expertise. In a comprehensive study Bastick (1982) develops a theory on intuition, and puts forward a number of interesting hypotheses.
3. Polanyi (1962) maintains that Copernicus' reason to change the paradigm was the greater intellectual satisfaction the new one created. Thereby he gives preference to the human joy generated by abstract theory. The price is to negate the sensual evidences.
4. One may note that Giddens (1979) argues that there is a contradictory relationship between human beings and nature in all societies. He writes: "Human beings exist in contradictory relation to nature because they are in and of nature, as corporal beings existing in material environments, and yet at the same time they are set off against nature, as having a second nature of their own, irreducible to physical objects or events." (p. 161).
5. I build upon my own, personal communications with a friend of mine who attempted to talk with ELIZA.
6. See also Mumford (1970, p. 341, graphical section no. 2, photo 27) on the counterculture's rituals, who writes: "The depressing monotony of megatechnic society, with its standardized environment, its standardized foods, [...] its standardized daily routines, produces a counter-drive in overstimulation and over-excitement in order to achieve a simulation of life."
7. In December 1965 Hubert L. Dreyfus published the article *Alchemi and Artificial Intelligence*, The RAND Corporation Paper P-3244. Since the article was published as a RAND report, it was regarded as a highly reputable source. This made the "pro-technologists" furious since they maintained that Dreyfus unfairly utilized the situation since he only had a three-month consultancy assignment with RAND, see for example Dreyfus and Dreyfus (1986).
8. I would like to thank the following persons: Leif B. Methlie for giving me good advice, comments, and literature; Gunnar E. Christensen, Rigmor E. Grimsø, Olav Kvitastein, Knut Lie, and Jon Segtnan have given me useful comments; and John A. Fundingsrud and Finn Tschudi who have been active conversation partners. Mostly, I would like to give thanks to Odd Langholm who, through his unique scientific contributions, has been an inspiring example. Finally, without Helle and Biancha Snellingens meticulous effort and expert competence, this article would never have been translated into English thirty-two years after its original publication. It was a true pleasure to cooperate with them.

Bibliography

Ackoff, R. L. (1974) *Redesigning the Future. A System Approach to Societal Problems*, John Wiley & Sons, New York.

Arendt, H. (1978) *The Life of The Mind*, Harcourt Brace Jovanovich, Publishers, New York.

Baker, G. P. & P. M. S. Hacker (1980) *Wittgenstein: Meaning and Understanding, Essays on the Philosophical Investigations*, Vol. 1, The University of Chicago Press, Chicago. (Paperback ed. 1985).

Bastick, T. (1982) *Intuition: How We Think and Act*, John Wiley & Sons, Chichester.

Bertalanffy, L. (1968) *General System Theory. Foundations, Development, Applications*, Penguin Books Ltd, Harmondsworth.

Boland, R. J. (1987) The In-Formation of Information Systems, in *Critical Issues in Information Systems Research*, R. J. Boland Jr & R. A. Hirscheim (eds), John Wiley & Sons Ltd, New York.

Bolter, J. David (1984) *Turings Man*, Penguin Books Ltd, Harmondsworth.

Bråten, S. (1983) *Dialogens vilkår i datasamfunnet. Essays om modellmonopol og meningshorisont i organisasjons- og informasjonssammenheng*, Universitetsforlaget, Oslo.

Churchman, C. W. (1968) *Challenge to Reason*, McGraw-Hill Book Company, New York.

Dreyfus, H. L. (1979) *What Computers Can't Do: The Limits of Artificial Intelligence*, revised edition, Harper Colophon Books, Harper & Row, Publishers, New York. (The original edition was published in 1972. This revised edition has a long introduction where Dreyfus sums up important aspects of the development from 1970 to 1979 within the domain of artificial intelligence).

Dreyfus, H. L. & S. E. Dreyfus (1986) *Mind over Machine: The Power of Human Intuition and Expertise in the Era of the Computer*, The Free Press, A Division of Macmillan, Inc. New York.

Fløystad, G. (1987) Etikk og helhetstenkning i næringslivet. *Norsk Harvard*, No 2, 2. kvartal

Frude, N. (1983) *The Intimate Machine. Close Encounters with Computers and Robots.* A Mentor Book, New American Library, New York.

Giddens, A. (1979) *Central Problems in Social Theory: Action, Structure and Contradiction in Social Analysis*, University of California Press, Berkeley and Los Angeles.

Ims, K. J. (1976) Streik som prosess: Bedriftens ytre og indre miljø som forklaringsfaktorer på utbrudd av streik – i historisk lys. Fra diagnose til tanker om et sunnere arbeidsliv. Stensil, Norges Handelshøyskole.

Ims, K. J. (1987) *Leder i dialog: En studie av informasjonsøkning med metoder for personlig utvikling*, Universitetsforlaget, Bergen.

Jung, C. G. (1923/1958) *Psychological Types* (Rev. edn), Pantheon Books, New York.

Keen, T. & F. Tschudi (1985) *Repertory grid data utilizing feedback mechanisms for elicitation & subsequent analysis*, manuscript. University of Oslo.

Koestler, A. (1979) *Janus: A Summing Up*, Pan Books, London.

Lakoff, G. & M. Johnson (1984) *Metaphors We Live By,* University of Chicago Press, Chicago.

Landreth, B. (1985) *Out of the Inner Circle: A Hackers Guide to Computer Security*, Microsoft Press, Washington.

Lunde, J. (1980) Trenger elektroniske hjerner et hjerte? *Aftenposten 11.3.1980.* (Do electronic brains need a heart?) See also Lunde, J. (1982) *Etikk i økonomikk: Artikler og undersøkelser*, Norges Handelshøyskole.

Lysgaard, S. (1961) *Arbeiderkollektivet: En sosiologisk undersøkelse av en norsk industribedrift*, Gyldendal, Oslo.

Miller, G. A. (1982) *New Scientist.*

McCorduck, P. (1979) *Machines Who Think. A Personal Inquiry into the History and Prospects of Artificial Intelligence.* W. H. Freeman and Company, San Francisco.

Mumford, E. (1987) Managerial Expert Systems and Organizational Change: Some Critical Research Issues, in *Critical Issues in Information Systems Research*, R. J. Boland Jr & R. A. Hirscheim (eds), John Wiley & Sons Ltd, New York.

Mumford, L. (1934) *Technics and Civilization*, Harcourt Brace Jovanovich, Inc., New York.

Mumford, L. (1967) *The Myth of the Machine: Technics and Human Development*, Harcourt Brace Jovanovich, Inc., New York.

Mumford, L. (1970) *The Myth of the Machine: The Pentagon of Power*, Harcourt Brace Jovanovich, Inc., New York.

Næss, A. (1967) *Filosofiens historie: Fra oldtid til renessanse*, Universitetsforlaget, Oslo.

Næss, A. (1968) *Filosofiens historie: Fra renessanse til vår tid*, Universitetsforlaget, Oslo.

Naisbitt, J. (1984) *Megatrends: Ten New Directions Transforming Our Lives*, Warner Books, New York.

Neisser, U. (1976) *Cognition and Reality*, W. H. Freeman & Co, San Francisco.

Newell, A., J. C. Shaw & H. A. Simon (1963a) Chess-Playing Programs and the Problem of Complexity, in *Computers and Thought*, E. A. Feigenbaum & J. Feldman (eds), McGraw-Hill, New York.

Newell, A., J. C. Shaw & H. A. Simon (1963b) Empirical Explorations with the Logic Theory Machine: «A Case study in Heuristics», in *Computers and Thought*, E. A. Feigenbaum & J. Feldman (eds), McGraw-Hill, New York.

Nordenstam, T. (1983) Et pragmatisk perspektiv på datautvecklingen i: *Datautveclingens filosofi* B. Gøranzon (ed.) Carlsson & Jønsson Bokførlag AB, Stocholm.

Peirce, C. S. (1972) *Charles Sanders Peirce:* Utvalg og innledning ved Ingemund Gullvåg, Pax Forlag A/S, Oslo.

Polanyi, M. (1962) *Personal Knowledge: Towards a Post-Critical Philosophy*, The University of Chicago Press, Chicago.

Polanyi, M. (1966) *The Tacit Dimension*, Routledge & Kegan Paul, London.

Popper, K. (1981) *Fornuft og rimelighet som tenkemåte. Utvalgte essays*, Dreyer, Oslo.

Ranheimsæter, H. (1970) *Europas idehistorie: Grunnlinjene*, H. Aschehoug & Co., Oslo.

Schank, R. (1984) *The Cognitive Computer: On Language, Learning and Artificial Intelligence*, Addison-Wesley Publishing Company, Inc., Reading, MA.

Simon, H. A. (1978) *The Sciences of The Artificial*, MIT Press, Cambridge, MA.

Simon, H. A. & A. Newell (1958) Heuristic Problem Solving: The Next Advance in Operations Research, *Operations Research*, Vol. 6 (January-February), pp. 1–10.

Turkle, S. (1984) *The Second Self: Computers and the Human Spirit*, A Touchstone Book, Simon & Schuster, Inc., New York.

Valle, R. S. & R. Eckartsberg (1981) *The Metaphors of Consciousness*, Plenum, New York.

Vallee, J. (1984) *The Network Revolution: Confessions of a Computer Scientist*, Penguin Books, Berkeley.

Vickers, G. (1973) *Making Institutions Work*, Associated Business Programmes, London.

Weizenbaum, J. (1976) *Computer Power and Human Reason: From Judgment to Calculation,* W. H. Freeman and Company, San Francisco.

Wiener, N. (1948) *Cybernetics*, John Wiley and Sons, New York.

Wiener, N. (1954) *The Human Use of Human Beings: Cybernetics and Society,* Doubleday Anchor Books, Doubleday & Company, Inc. Garden City, New York.

Wittgenstein L. (1958) *Philosophical Investigations*, G. E. M. Anscombe & R. Rhees (eds), 2nd edn, Blackwell, Oxford.

Zuboff, S. (1983) New Worlds of Computer Mediated Work, *Harvard Business Review,* September–October, pp. 142–152.

KNUT J. IMS

Mismeetings: An Essay about Money, Urban Life, and the Ethics of Proximity[1]

Introduction

As I walk around the modern city of Oslo, I ask myself: what kind of a person do I really want to be? Once I was walking down the main street, Karl Johan, when I saw a man lying on the pavement. Many people passed – hurrying on without paying attention to him. I was tempted to do the same, but an inner voice said: «Knut, you have to stop and check, the man may be ill, he may need help!» So I made up my mind, walked over to him, and tried to wake him. After a while, the man opened his eyes, looked at me and said: «give me 10 kroner!»

Afterwards I had mixed feelings. I was not quite sure whether I had done the right thing. Would it have been better just to let him sleep? I really do not know.

I find it problematic walking in cities meeting hundreds and hundreds of people. Looking into other peoples' faces, I see loneliness and a longing for contact, but nobody takes the responsibility to stop and say: «Hey, how are you? Why don't we have a nice chat?»

I am trapped in the civic, cold, modern urban culture, where *strangers* are the new category of people. People who grow up in the countryside do not feel well among modern scyscrapers of glass and steel and streets overcrowded by people. Should we take the uniform blasé attitude towards everybody and not discriminate between persons? Should we only pass by, letting the old and the young, the homeless, cripples, and outlaws be categories of people in which nobody has an individual face? Is it right for me to say that I am no more responsible than all the others walking through the streets? Can we preserve our autonomy and integrity by using strategies of mismeetings, exchanging some factual viewpoints about the weather, and then disappearing – unattached, looking back at the happening as an unimportant episode?

I feel a need to reflect upon what kind of a person I really should strive to be and how I should act in a modern city, given that I want to preserve my moral consciousness and feelings of sympathy towards other human beings. To assist me in my effort to understand city life, I will in particular use the work of the social scientist Georg Simmel, whose writings have received new attention recently (cf. Simmel, 1991). As a contrast to this, ideas from the direction called ethics of proximity will be introduced. In particular, I will draw upon the philosopher Emmanuel Levinas, Zigmund Bauman, and to a certain degree the Dane, Knut E. Løgstrup. In Simmel's *grande ouvre*, *The Philosophy of Money*, there is a kind of moral consciousness that is seldom found in newer scientific works. He struggles to preserve autonomy and individuality in the face of overwhelming social forces in the cities.

Sennet (1969) says: «the field of urban studies is now plagued by a kind of superstitious belief in scientific purity of research, as though, like Faust, if we had perfect and pure knowledge the world around us would suddenly change … The emotional perspective of the essayists in this book (Simmel, Weber, Spengler), … for all the generality of their ideas, are probably the only guides to the conduct of modern urban life on a moral plane …. that are relevant to renewing a sense of freedom and vitality in our cities» (p. 19).

Ethics of proximity – relevant for management?

The accumulation and use of knowledge since the Renaissance has been enormous, and the technological progress probably has no counterpart in the history of human beings. Technology can be put into good or bad use. The Holocaust may be viewed as a hypermodern project in which scientific management, together with advanced and sophisticated technology, were put into effective use. In an ethical sense, the Holocaust and Auschwitz were also basically *moral projects,* and Adolf Eichmann acted according to (positive) law, but not to legitimate law. The genocide committed by the Nazis was a kind of *enlightened barbarity* (Bauman, 1989). Fifty years after the Second World War we are witnessing the same kind of brutality and ethnic purging in the former Yugoslavia. How can these evil acts continue in modern Europe, where the humanistic tradition is long and well established? As C.P. Snow pointed out several decades ago, there is a gulf between technological and humanistic cultures. Therefore, one challenge is to bridge the gap between these cultures. Unfortunately, I have to ignore this subject in this essay. However, even within the humanistic tradition, we have to adapt to the new *postmodern* situation. It is in this light the ethics of proximity is to be seen. What, then, is the advantage of knowledge when it is misused – used against people instead of for the common good – when technological knowledge is used to exploit the third world, and even to build and test nuclear weapons in 1995? The lessons drawn from the Holocaust are the dangers of socialization and blind obedience. Critics have to stand up and voice their opinions and see responsibility as a personal project.

Managers have a particularly important responsibility because they have much power. They have the capacity to make things happen through the use of powerful tools and many other people. Directly or indirectly, they are in a position to influence what is right or wrong, good or evil. As Harvey (1988) argues under the heading «Eichmann in the Organization», it requires true moral resolve and clear thinking to recognize a *tiny atrocity* and say no. Cruel acts may well start from seemingly banal suggestions (Arendt, 1976). Tolerating small atrocities today may lead to the toleration of more evils tomorrow and, finally, we close our eyes in front of any evil. In everyday mundane behaviour life is lived at a safe distance from the extreme acts like those seen during wartime. Harvey sees the solution for the moral agent to act out of moral principle, to respond bravely to small atrocities, because they will inevitably catch us in a vicious circle at a later point of time. Bauman (1993) takes an even more radical point of view as he underlines that moral *practice* needs a horizon that establishes standards that cannot be reached. Self-assurance and self-righteousness are not compatible with morality. My point in this essay is to make managers and students of management more aware of the ethical dimension. We all need to investigate our personal moral base and evaluate whether we serve the humanity or not, and we should remind ourselves that organizations are never goals in themselves, but means to achieve a greater end.

A huge body of management books is available, and some of them point to the need for learning social knowledge. Leavitt (1975) and Bennis and Nanus (1985), for example, regard the mastering of interpersonal relationships as the essence of management, and Bennis (1989) sees integrity as the basis for trust. Shein (1992) states that the creation and management of culture are crucial for managers. Thus, it is important to understand the individual human being, the social dimension. Authors like Giddens (1979, 1984) try to make a synthesis of social knowledge theories, and there are indications that an increasing number of students of management pay attention to these structural theories that combine objective and subjective approaches. It is a great challenge to develop better empirical understanding of real behaviour. The intellectual, analytic tradition in which managers are seen as purposeful agents that maximize shareholders' profit is still dominant. Technocratic consciousness, calculated interests, and the urge of efficiency are also elements of this pattern.

In this essay I want to put forward viewpoints that may seem quite contrary to prevailing lines of thought within management. As Davis (1971) proclaims in his article, with the conspicuous title *That's Interesting*, the purpose of *good theories* is *to attack the taken-for-granted world of their audiences – the more strongly held assumptions the theories negate, the better.* The ethicist Emmanuel Levinas (1985, 1987) has a radical way to portrait responsibility. In seeing responsibility as more important than knowledge, he opposes the mainstream intellectual tradition in our culture. Levinas may be categorized as a representative of the ethics of proximity, where the other person' s face and feelings are central (Vetlesen and Nortvedt, 1994). Levinas' ethics is a *Humanism of the Other.*

The ethics of proximity has clear consequences for all organizational members. Proximity is the realm of intimacy and morality, while distance is the realm of estrangement and law. More sensitivity will strengthen the ties among coworkers. A close relationship is a fruitful point of departure for personal growth and development because such a relationship gives both persons an opportunity to become confirmed (Ohnstad, 1993). As Simmel argues, all relations with others are mere stations along the road by which the ego arrives at its self. It is through relationships with others that I, as a lonely monad, am freed. Hjorth (1993) argues that close relationships are the most important single aspect for being healthy.

Levinas' ethics is first and foremost dyadic, that is an ethics of the relations between any two persons. When a third person enters the dyadic relationship, the issue of justice becomes significant. And justice is exercised through institutions. But they must be checked by interpersonal relations. In the following I will not say much about justice, which is covered in depth by Rawls (1971) and Eckhoff (1971). For an interesting sociological analysis of the triad, see Wolff (1950, pp. 145–180).

Mismeetings

Urban survival needs a capacity for the fleeting, restricted relationship. Thus urban life is characterized by mismeetings. People respond politely on the surface. But on the deeper level, after the exchange of agreeable phrases, they leave each other unattached (Bauman, 1993). Typically, individuals are replaceable and disposable in the urban crowd.

There are different strategies for these mismeetings.

> Rule one: Prevent physical space from being social.

Social space involves engagement, interaction, and commitment (cf. Lofland, 1973, pp. 51–57). Thus, the strategy is to place the stranger in the sphere of non-engagement of emotional void. Let neither sympathy nor hostility intrude. Avoid the risk of sympathy. The main defense against all strangers is detachment (cf. Bauman, 1993).

On the surface, take an indifferent attitude. Avoid eye contact. See while pretending not to. Disguise observation as indifference.

> Rule two: Regard the meetings as episodes or intervals with no history, which leaves everything as it was.

This means fragmentation, where life is split into episodes which are self-enclosed. Experience from one meeting is not significant for the next meeting. My life can be lived in "fragments", and I can play different roles in different settings.

> Rule three: Do not obtain knowledge about others and deny them knowledge of yourself.

Let the Other stay in the background. Don't give her a face. Let her be a featureless and an empty shell. You probably will never meet the Other again, so don't care about her (cf. Bauman, 1993).

> Rule four: Do not seek any confirmation and do not receive it.

Avoid the risk of engaging in a dialogue and avoid the risk of losing face and feeling shame. Do not expect reciprocity. You shouldn't believe that the stranger wants *to* meet you and, in any case, do not be personal so that your emotions may be triggered. It is always better to operate intellectually, being impersonal, than to touch on subjects that are tied to the heart. The weather is always an interesting topic.

> Rule five: Cultivate intellectualism.

Measure, weigh, and calculate everything and always try to be as exact as possible. Let economic rationality pervade all spheres of life, including social life. Avoid the risk of being emotional and impulsive.

> Rule six: Cultivate money.

Give special attention to money, because the mathematical character of money gives the relationships of the different elements of life strong precision and unambiguousness. Try to reduce any qualitative value to quantitative. Gauge values in terms of money. Then you can reach the ideal of numerical calculability.

> Rule seven: Try to transform the relationship into a purely commercial one.

In the market the actors are faceless and concerned of quantity only – not unique subject bound qualitative values. The Other person should be completely indifferent to us. He has "lost his face" and we need not care about him.

> Rule eight: Do not make any financial transactions with either friends or enemies.

A personal relationship may intervene between seller and buyer and may spoil the neutral and purely commercial relationship. Also avoid entering into transactions with enemies because hostile intentions may intervene. You should meet the Other with a detached, analytic, blasé attitude.

Shame – a forgotten word in the vocabulary of management?

At the time when I was living outside Cleveland, I often met strangers who greeted me: «hey, how are you doing?» Because this approach was unexpected for me as a Scandinavian, it made me a little confused. However, I interpreted this as an invitation to have a conversation, and began stottering about my mental and physical state, while the stranger just moved on. Left back talking aloud to myself, I felt shame. After several similar shameful events I understood that this phrase was not an inquiry, but only another way of saying: «I have no time to speak to you. I am in a hurry.» The lesson is this:

> Others do not want to hear about your indigestion;
>
> "how are you?" is a greeting, not a question.

Mismeetings may easily make a person who does not master the urban strategies a fool – producing feelings of shame – an intensive displeasure we experience when a pursued reciprocity remains unrealized. This is the approach taken by Ikonen and Rechardt (1993), who define shame as a reaction to the absence of approved reciprocity. The urge for reciprocity occurs between the self and the others as an aspiration to find a reciprocal other person, but also «as an aspiration to become whole and expand»

(p. 103). When this expectation is unfulfilled, the person is paralyzed and brought to a standstill, and wants «to sink underground». The expression of shame is a downturned head and gaze. Shame is *tied to the whole self.* The avoidance of shame prevents a person from perceiving reality. He/she denies reality and stops thinking. Another consequence is a collapse of self-esteem, which may result in the giving up of one's identity and the seeking of symbiotic relationship (1993, p. 102).

One of the best ways of reducing shame is using humor. If a person is able to laugh at his own shortcomings and failures and understand that, all things considered, they do not represent our whole self, the feeling of shame will not go so deep. The trick is to regard the shortcomings as being caused only by a part of the self or by «a momentary self»: «what a fool I was at that moment!» (Ikonen and Rechardt, 1993, p. 113).

Meetings

We have seen strategies for mismeetings. Now it is time to shift the perspective and ask: what is the condition for good meetings – meetings that lift us up, inspire us to deeds, and energize us?

Let us highlight some principles of the ethics of Levinas.

> Principle one: Our responsibility for the Other is infinite.

For Levinas the responsibility to respond is an infinite responsibility. Related to dialogue, the ethical relationship does not manifest itself so much in the content, as in the demand for response.

> Principle two: Responsibility is asymmetric.

Nobody else can take my responsibility for you, but you are not responsible for me.

> Principle three: Responsibility is laid upon me.

Responsibility is not taken, it is not object for choice, it is not an option.

> Principle four: Responsibility individualizes me.

In the relationship between you and me we are both seen as unique. We do not have any stand-ins and cannot be substituted.

> Principle five: Categorization violates reality and reduces the Other to the same as others.

We reduce the Other person to an instance within a category.

> Principle six: The ethical demand is silent.

We have to listen in order to act.

> Principle seven: Ethics is before ontology. "First philosophy is ethics."

Morality is autonomous and is not to be justified by anything different.

Comments to principle one

What does it mean to have infinite responsibility for other human beings? It means that my charge as a caring person cannot be described or thematized through language. And the charge laid on me goes even further. It goes beyond my instincts and consciousness. Because, says Levinas, «no one is good voluntarily». It is the demand made by the face of the Other in his nakedness which forces me to this unlimited care for the Other. The charge is also before and beyond my commitment. Otherwise it would have been a question of whether the Other "deserved" the fulfilment of my obligation. However, the relationship to the Other's face give me no choice. And it is obvious that this responsibility is beyond contractual responsibility, which assumes an exchange of services in which my obligations as a partner are strictly confined. Why does the charge escape thematization? Because responsibility for the Other cannot be reduced to a principle or a formula, that is responsibility is before knowledge and ontology, that is before descriptions of what is.

These short introductory remarks to Levinas' ethics will be developed further through the next comments.

Comments to principle two

Why is responsibility asymmetric? According to Levinas, «I meet you and you are privileged. It is a motion from you to me. You command me and I am subjected to you. I become your hostage.»

Levinas distinguishes between «being with» the Other, which is symmetrical, and *being for*. The mode of being for the Other precludes not only solitude, but also indifference. And, morally speaking, indifference and ethical neutrality are the same. Levinas says that «I am for the Other, whether the Other is for me or not; his being for me is his problem, because my being-for-the-Other includes respect for the Other's freedom and autonomy. He may do what he wants. I do not demand to be repaid in the «you-for-me» sense» (cf. Bauman, 1993, pp. 49–50).

Morality does not calculate on reciprocity. It is not a rational stance to take, nor a means to an end. Thus, morality is contrary to a business transaction, in which we expect a balanced and fair repayment of what we offer. So, what is wrong with calculability? Because it assumes that thinking precedes doing and that ethics is the outcome of rational decision making. In contrast moral call is non-rational and deeply personal because it appeals to my responsibility.

But morality is also solitude. As a social person I am *with* the Other and as Maurice Blanchot has written: «everyone here has his own prison, but in that prison each person is free». As a moral person I am *for* the Other. And being for the Other means that I do not let tools of thought like concepts, rules, or standards play any role. These «tools» make an exit to the moral impulse, to the moral urge. As Bauman points out: even though the moral act starts in solitude, the ethical person wins himself during this process of reaching out and "reaching beyond the solitude in the act of self-sacrifice which is both the hub and the expression of 'being *for*' " (p. 61). Because we are moral persons, we *are* society. «Society is not a constraint, not even an opportunity. *It* is us.» (Etzioni, 1988).

It is you who individualize me through the meeting. As Bauman (1993) states, duties tend to make humans alike; responsibility is what makes them individuals (p. 54).

Comments to principle three

Responsibility is laid upon me. I give myself to the Other as a hostage. I take the responsibility as if it was there «already» and «always». This responsibility is *unconditional.* It does not depend upon the quality of the Other. Meeting the Other as a face, means that the face is an authority because I am willing to listen to the Other. The face is naked and appears absolutely defenseless. It is without a mask, without any covering. It is precisely this weakness that makes my relation to the Other a moral act. It is my presence that makes the Other strong and makes the silence speak (cf. Bauman, 1993, pp. 73–74).

Responsibility is before (elder than) freedom because responsibility is elder than cognition (intellect). This means that practice has priority over theoretical knowledge which leads to abstraction and distancing. Usually in ethical literature we postulate *autonomy* as a premise for moral behaviour. Autonomy is tied to freedom of choice. According to Levinas, responsibility is not asked for and I have no choice because of *heteronomy.* From Dostoyevsky, Levinas has borrowed the saying: «we are all responsible, but I am more responsible than all the Others».

Comments to principle four

Responsibility individualizes me. I am constituted as a human being by taking responsibility and by my responsibility I simultaneously constitute the Other. It is through the meeting with the Other that I am liberated from my enveloped reality. The Other disturbs me and destroys my centering about myself. Facing the other, I stretch out and then liberate me from myself. Then I sober up and I am not I any longer, but «I am *for*». And being for the Other is the birth of my own self. Then I find myself unique, irreplaceable, and not part of a category. According to Levinas the self is born out of *union* – through stretching out *towards* the Other.

In Levinas' (1985) own words:

> To be human means to live as if one were not a being among beings My responsibility is untransferable, no one could replace me the very identity of the human I starting from responsibility ... (which) is what incumbent on me exclusively, and what, humanly, I cannot refuse. This charge is a supreme dignity of the unique. (pp. 100–101)

This means that responsibility emancipates me from living in fragments. It means that I accept the ambivalence which constitutes the moral self and I negate the comforting, orderly, and logical foundation that stems from standards, rules, and law. As Bauman puts it, «I am free as far as I am a hostage» (1993, p. 78).

Comments to principle five

Why is categorization regarded as an «evil»? The man who was sleeping on the pavement; was he «a homeless » person? A person that has less than ordinary dignity and moral status? Bureaucrats are often blamed for seeing and treating any person as member of a category – not a face and, in fact, this is the main idea of bureaucracy; impersonality of interpersonal relations and a system of procedures that secure a fair treatment of the clients. Media tries to give people a face. But how does the audience react? Is a 5-year-old girl suffering in a cruel war on the television screen seen as a face? Or is she just an instance within a category? I assume that most television watchers, in order to survive, have a tendency to reduce the Other to a member of a category in order to bring their emotional situation under control. This reminds me of the lady who was asked «how are you today?» and replied: «quite well, thank you, but I'm worried about Blake in «Dynasty» (the «TV soap »).

Categorization assumes an essence of things and persons. But how can we have any essence if we are unique as individual persons? The danger is to regard persons in a technocratic way, to look upon an individual as a thing, and not being sensitive to his or her individuality. We would expect that moral philosophy would not abstract itself from individuality and uniqueness among people.

Comments to principle six

The ethical demand is unspoken – silent. The Other is a face, not a force. The demand is radical and it is precisely the silence that makes the demand radical. It is radical because the Other has no right himself to make the demand and only I am able to discover what will serve the Other best. Because the demand is vague and confused, barely audible, I have to make my own interpretations and, as with all interpretations, I can never be sure that mine is the correct one. Because the demand is radical, I can never know whether I have acted correctly, and I can never know if I have done enough for the Other. No rules or conventions help me as a moral person.

> [I]t is the moral *anxiety* that provides the only substance the moral self could ever have. What makes the moral self is the urge to do, not the knowledge of what is to be done … The moral self is a self always haunted by the suspicion that it is not moral enough. (Bauman, 1993, p. 80)

Comment to principle seven

Ethics as first principle, that is ethics as the carrying idea and perspective on the world is a core viewpoint in Levinas' work. To an academic this is a provocative question: is there anything before ontology? In an ontological sense, nothing is before being, but for Levinas morality is «before» being in the sense of being «better».

Ethics is superior to everything. To define ethics through a denotative definition is always a reduction of what ethics might be because *ethics* escapes *every concept of knowledge.*

In Levinas' (1985, p. 8) own words, «ethics cuts across ontology» … «It is radically and irreducibly – » otherwise than being, or beyond essence.» In order to illuminate the meaning of this we may draw the line of history back to Socrates who taught us the method of categorization. Things are in the same category if and only if they have certain properties in common. These properties are necessary and sufficient conditions for defining the category. Categorization means a way of thinking that violates the richness of individual phenomena in real life. The categorical way of thinking makes what is different the same. It is held to be a fruitful scientific procedure because it involves simplification and a reduction of the multidimensional reality. Methodologically it gives us the ability to manipulate reality, to control and dominate nature. It is a platform for model thinking. The categorical way of thinking also puts the thinker at a distance from the phenomenon itself. This makes it easier to regard and treat persons as objects to be manipulated.

Levinas breaks the assumption that knowledge is essential, that there is an order of things or an essence of things. In the Socratic tradition the knowledge of the essence, essential knowledge, precedes the existential. We have to understand this essence before we act. We have to know before we commit ourselves to belief and action. Before we can do any good, we must first know it. *To know the good* is *to do the good.*

In contrast to this knowledge approach, Levinas maintains that the ethical question is one of obligation and responsibility – not a question of « what is,» that is ontology. Ontology obscures the ethical question. The essence *of ethics* is *not to have an* essence. Its « identity » is not to have an identity. Ethics is precisely ethics by disturbing the complacency of being (Levinas, 1985, p. 10). Ethics is beyond thematization, representation, and the reasoning of essential thinking. The question of essence, « What is ethics » excludes the force of the ethical question. It collapses the « what ought to be » of ethics into the « what is » of ontology (1985, p. 8). *Ethics* is *better than being* – and nobody can tell in advance what this « better » implies in a concrete situation. Perhaps the best illustration of this idea is to watch Mother Theresa in her work in Calcutta. She takes care of the people she meets, without asking questions.

Levinas' ethics are humanistic. They oppose the academic emphasis on conscious, intentional thinking. According to Levinas the question of justice establishes the human: a person has capacity to fear injustice more than death, to prefer to suffer rather than to commit injustice … (cf. Levinas, 1985, p. 85). Socrates is a primary example of this preference to suffer. He emptied his cup of poison because he would not be blamed for deceiving the youth. Thus, it is impossible to encircle what ethics is. Ethics is compassion. It is the Other person's face that demands us. It is impossible to reduce the ethical relationship to a formula, a principle, a rule. It is a question of responsibility.

A juxtaposition of viewpoints

Levinas postulates the ethics of heteronomy, which may be summarized in «I am the hostage of the Other», «the Other decides over me », and «responsibility cannot be excluded», that is I can never be free from the Other. *Is Levinas' ethics a return to a slave morality?* To be the Other's hostage, is that a feasible moral standard in 1995? Is not the hostage role precisely what women's liberation has opposed? From a psychological perspective we may well ask if the caregiver is a sick, neurotic person that has unfulfilled demands and uses an unlimited caring for the Other to abreact from his insane self instead of working with his own development in order to become a balanced, mature personality? Or do we see an example par excellence of the fact that *realisation through the Other* is *the only way to self-actualization?*

We must ask ourselves: does Levinas' theory work? Does *this subjection in front of the Other guarantee a good meeting?* Or could it be that a person who wants to go through fire and water for the Other really is overwhelming and threatening? Could this attitude of infinite responsibility in effect create *distance and detachment* rather than a warm and personal relationship? Do we have *a primary case of «willing what cannot be willed»*? Several authors point to the fact that some *very important social and mental states can only be realized as by-products of actions undertaken for other ends.* Martin Buber holds that persons become themselves and reach self-actualization not by aiming directly at these goals but as a by-product of dialogue (Friedman, 1992). Jon Elster (1983, p. 43) argues that natural behaviour, spontaneousness, faith, self-respect, or love never can be brought about intentionally or intelligently «because the very attempt to do so precludes the state one is trying to bring about» (Elster, 1983, p. 56). Elster distinguishes between intellectual fallacy and moral fallacy of by-products. Intellectual fallacy is the belief that we intentionally and instrumentally have been able to produce the state which is present. Moral fallacy is the attempt to reach states which are useful or desirable while the attempt to realize them is bound to fail. They are both fallacies because we cannot do it, even though we very much would like to. Apparently, this knowledge is a frustrating fact.

Are we in fact unable to reach these important states, or should we merely sit down, do nothing, but hope that a passive approach may lead to the end? Aristotle may give us a hint. He gives the following recipe on how to become virtuous: *act as if you already are virtuous.* The hint is *to make an arrangement* that put yourself in a situation where the wanted state *may be* reached.

So, *how do we obtain a good meeting?* First, we have to define a good meeting. In general it depends upon the situation, demands, and interests of the Other. Martin Buber says that «every living situation has, like a new-born child, a new face that has never been seen before and will never come again. It demands of you a reaction that cannot be prepared beforehand. It demands nothing of what is past. It demands presence, responsibility; it demands you» (cited in Friedman, 1992, p. 28).

A good meeting is

- to meet the Other as a partner and not an object,
- to experience the Other's side of the relationship without losing one's own,
- to bring oneself as a whole person, (an authentic personality) accepting the Other as the person he or she is in their Otherness (Friedman, 1992, p. 41), confirm the Other as he or she is, not *trying to alter the Other*,
- to give the Other deep respect and positive regard,
- to touch upon topics which the Other finds essential in his or her life,
- to be in complete openness (Løgstrup, 1972, 1991).

Face and dialogue are linked together because the face speaks. The authentic relationship with the Other person is dialogue. Most of us know from personal experience that it is difficult to be silent in other persons' presence. Therefore, it is necessary to speak of something. This is to respond to the Other.

Martin Buber's concept *essential relation* between persons refers to the fact that the Other becomes present in the depths of one's substance. The two persons participate in each other's lives. They exist in the sphere of the "between", and they exist fully in the life of dialogue.

Perhaps the most striking aspect with Levinas' ethics is the apparent *negation of knowledge as a base for morality.* How could it be so? In the introduction it was stated that knowledge was a precondition for the Holocaust. Thus, pure knowledge is not enough. In Schumacher's (1974) analysis of different resources, *education,* not money or physical resources, is seen as the most important resource. Schumacher states that it is not a «lack of know-how that causes the despair not only of religious thinkers like Kierkegaard but also of leading mathematicians and scientists like Russel and Hoyle. We know how to do many things, but do we know *what* to do?» (p. 84). Schumacher admits that science cannot produce ideas by which we could live. The task of education is «first and foremost, the transmission of ideas of value of what to do with our lives» (1974, p. 80). Schumacher ends with admitting that we suffer from a metaphysical disease, and the cure must, therefore, be metaphysical.

As far as I can see, Levinas negates such a viewpoint because he does not trust any metaphysical base. Metaphysics is onto-theo-logy and thus based upon myths that can never be proved or disproved. As a consequence, the core of Levinas' moral is based upon *the Other's face. Access to the face is straightaway ethical.* A face cannot be reduced to features. «A face is without defense. The skin of the face is that which stays most naked.» Therefore, one often tries to mask the face by taking on a countenance. «The face of the Other is destitute, he or she needs help and I am called.»

For Levinas onto-theo-logy has been the wrong standard of ethics, because ontology is an inferior standard. Ontology tells us «what is», but ethics is a question of what we *ought to do.* Levinas negates the Socratic position that to know the good, is to do the good, that is the strong tie between thinking and behaviour. Levinas postulates that ethics is before ontology because *ethics* is *better than ontology*. Ethics is the force to intervene in a reality that should be different.

This view may be supported by the special power of ethics, which is demonstrated by the story of how Mahatma Gandhi, as leader of the Indian liberation movement, met the British colonial administrators. Mahatma was given the opportunity to speak out before a big group of administrators and his speech was short, brave, and simple:

> «Leave India.»

The power of ethics based upon the Face is not the usual power. It is of another type – different. It does not oppose power with more power – with more guns or a bigger army. It opposes power with sincerity and responsibility. The British used violence against the non-violent Indians and finally the British faced their own inhumanity. In the proximity of face-to-face, the British became aware of the moral force. Thus, moral force is not stronger than totalizing powers, but it is different – it is better.

How can Levinas' heroic ethical theory be adapted in a modern urban culture? This is a serious dilemma, and it is optimistic to believe that we may find an acceptable answer. The question is: *who shall be given a Face?* A serious problem is that we may become desensitized in the city and may as a matter of fact protect ourselves against any face. In a big city I am drowning in hundreds of faces in such a way that I need *to create distance* between myself and the Other. And by creating distance to the Other I become alienated from myself at the same time. In addition to the distance I intentionally create there are a number of other forces that work in the direction of detachment.

It is a challenge to inquire into these *distancing forces* in the modern urban culture and hopefully better understand the practicability of Levinas' theory. Interestingly, on the other side of this distancing effect we suspect personal freedom – *the freedom to be let alone,* the most important single aspect of privacy. Levinas' «the Other decides over me» stands in sharp contrast to the new relation between freedom and dependence, which is the essence of city life. When I am in a big city nobody knows me. I am *free from the tyranny of intimacy*. I am *free from* community and *free to be let alone* (cf. Sennett, 1978, 1992). I am an anonymous being that may hide in the crowd. There I can do whatever I want, without risking any feeling of shame. This may give a restrained person emancipating feelings – a phenomenon paraphrased as: the longer away from home, the worse my morality. The same type of phenomenon occurs when I as a passenger in an airplane discover that I am willing to confess relatively intimate relationships to the Other passenger sitting next to me. I imagine that there is no risk in such a confession, because I expect never to meet the fellow passenger again. The meeting is merely an episode for me as well as for the Other.

Still another distancing factor is *intellectualism,* which is to a considerable extent supported by the strong forces of economic rationality and instrumental thinking that pervades modern culture. From a historical point of view, the money economy and the dominance of the intellect are intrinsically connected (Simmel, 1990). Thus the culture characterized by intellectualism is a precondition for a new kind of freedom from the Other which is manifested in the detached, blasé attitude in which everything is made the same. As a consequence, it may be of interest to investigate more thoroughly the specific elements which influence social relationships in a money culture.

In his work, *The Philosophy of Money*, Simmel (1990) argues that money plays a very important role in modern city life, which drives a new kind of instrumentality – a calculating ends-means rationality. The transformation from barter to a money economy meant a «development from a life of unity to a life of fragmentation». The money economy created organizations of a purely technical character. The medieval guild included the entire person in a living community in a social, religious, political, as well as occupational respect. By contrast, the main link to the new organization is money. We see it clearly demonstrated by the joint-stock company. In most cases, this type of company is not influenced by any particular shareholder. «The money has slipped like an insulating layer between the objective totality of the association and the subjective totality of the personality» (Simmel, 1991, pp. 18–19).

With money in our pockets we are free .... (George Simmel)

Within the former barter economy, we had to bring with us in a physical sense what we wanted to exchange, and usually there was a negotiation, a kind of dialogue between the buyer and seller. Buying and selling was a kind of social ritual, in which it was important to find out what kind of person the interlocutor was. A good dialogue was equivalent to a good bargain. In the money economy I typically interact with different individuals each time. I become independent of any particular person or supplier. In short, I become self-reliant. This gives me a personal independence and strong individuality. It gives me a kind of freedom, but at the same time it means estrangement, loneliness, and a flat and dull life. In Simmers own saying: this freedom means a «vapidity of life and a loosening of its substance!» Therefore, the transformation from barter to a money economy has *substantial costs in a cultural and interpersonal sense.* My personal independence, my self-reliance, *alienates* me from other people. When I enter into short relationships I do not care about the persons' individuality. The Others' personal characters are of no interest to me in the supermarket. What counts is the relative prices and how much I can afford to pay. The mental distance between me and the Other is particularly visible in this context due to the bodily proximity and narrowness of space. The more people around me, the stronger the feelings of being alone and lost. So this concept of freedom is not reflected in my emotional life as comfort.

I have argued that money produces a previously unknown impersonality in relationships on the market. Money mediates the relationship and money is perfectly objective. So the money economy is supported by two opposing streams; (i) independence of the person, that is the autonomy of the individual character, and (ii) levelling and equalization.

Money is «common» because it is the equivalent for anything and everything. This «tragedy of levelling» process pulls the highest down to the level of the lowest and converts the most heterogenous elements into money. The salient question turns out to be more money – or less. This nourishes the metropolitan blasé attitude, in which the question of what something is worth is displaced by *how much.* Acquisition of money becomes the proximate goal for most peoples' lives and the general psychological effect of the means which colonize the ends is manifested. «Money is the most obvious instance of means becoming and end. As means money reveals its true character – it becomes useless as soon as life is concentrated on it alone – it is only the bridge to definitive values, and one cannot live on a bridge» (Simmel, 1991, p. 25).

How shall we go about living in the town? Comparing Levinas' ethics with the urban culture a tension concerning responsibility is obvious. Being in the crowd, who shall be given a face? There is no definite answer, only a foundation wall of uncertainty based upon a listening to silent demands.

Another approach to the question is indicated by the modern feministic literature which discusses the tension between separation and connection for female adolescents. How is it possible to negotiate between the contradictory strivings to be separated and yet connected (Gilligan, 1982; Gilligan et al., 1990)? Simmel emphasizes *the inner circle* as an opportunity to choose a community, a group of people that we feel «belongs to us». Within the small circle we can obtain confirmation – the Other can serve as a mirror for us, and we have the valuable opportunity to express ourselves. As far as I can see there is a considerable difference between the inner circle formed in the city and the circle that characterize a rural culture, the small village. In the city we may voluntarily choose the people with whom we want to interact. However, the same freedom of choice does not exist in a village. Nevertheless, as for the monopoly and tyranny of meaning – the phenomenon of possessing the only correct opinion and, consequently, the suppression of the opinions of the Other – we cannot a priori expect any differences between the circles. We expect more heterogeneity in the village circle, and a stronger community based upon professional or ideological beliefs in the city circle. However, the strength of the exclusive control exerted by any one of the group members is left open. We may say that any inner circle may intimidate the personal inner freedom and, by enlarging the circle, the outer freedom will increase.

An important attribute of freedom is the ability to choose. Thus, in the mainstream of ethical thought, freedom is the basic precondition for morality. There is no morality without freedom. Only a free man or woman can be made morally responsible for his or her actions. This viewpoint can be traced back to Aristotle, who discussed *degrees of freedom* (Wheelwright, 1951). On the contrary, Levinas postulates that ethics is before freedom; first ethics and then freedom. In Levinas' terminology,

I am subjected to the Other. As far as I can see Levinas bases his theory on the point of view that humans are social beings – that man and woman *can only exist fully in relationships with others.* Thus, in a way he is saying what Aristotle told us 2,500 years ago, «man is born for citizenship, that is phusei politikon».

However, I think Levinas, more than Aristotle, emphasizes the importance of responsibility contra being a «free», autonomous individual. My interpretation of Levinas is that it is meaningless to speak about freedom if you do not take responsibility. *Freedom from community* is *a delusion.* Freedom preconditions relationships, that is community. As a person I am constituted by taking responsibility, that is living in relationships.

The final question we have to pose is: do we have to understand Levinas in order to understand his work? In short – is understanding of another author necessary, or do I believe in semantic autonomy on the basis of which I can banish the author? The early Max Weber wrote that in order to understand somebody we need to *relive* the experience of the other – getting inside his or her head. This is impossible in the case of Levinas and, even if possible, I would hesitate to go through all the suffering of this man. The entire family of Levinas was extinguished during the Holocaust. Thus, the subjective thoughts of Levinas are inaccessible. Nevertheless some data about his life might illuminate his thoughts. Levinas was born in Lithuania in 1906. He studied philosophy in Strasbourg between 1923 and 1930, and stayed in Fribourg from 1928 to 1929, close to Husserl and Heidegger, and he became naturalized French in 1930. Levinas was appointed professor of philosophy at the University of Sorbonne in 1973.

If, on the other hand, Levinas is *an essential thinker* in Heidegger's (1987) sense – a genius like Nietzsche – Levinas is «the founder of that which never becomes visible in images, which can never be historiologically related or technologically calculated … (p 4).»

Levinas might be an essential thinker in Heidegger' s sense. Whether he is or not, is not decisive for me, because I do not attempt to reconstruct what might have been in Levinas' mind when he was writing. However, I take the privilege to be inspired by his texts and let his texts speak to me. I use the text freely from my point of view based upon my interests for the time being. (For an interesting discussion of the validity in interpretation se f.ex. Hirsch, 1967). My viewpoint does not mean that I see Levinas' work independent of his life. I am not indifferent to Levinas biography. On the contrary I am informed by it.

This liberating way of using a text should not be particularly bold in the case of Levinas since he often makes use of a poetic language, which in its essence stimulates new perceptual bridges and breakdowns. As Nussbaum (1990) puts it in *Love's Knowledge*, there are many dimensions which cannot be adequately conveyed through a strict academic language. Therefore, the use of narratives and rich metaphors creates a contextual knowledge that is a necessary complement to a strict academic language. Connected to this I find Levinas' study of Heidegger important. Heidegger's transformation of the word ontology to onto-theo-logy is an illustrative example of Heidegger's attempt to emancipate from the ordinary use of language in order to present a new and fresh view of reality. Heidegger is the primary example of a person who attempts to think without recourse to metaphysics, and to seek a language free from historical and metaphysical bias. In this critique of ordinary language and metaphysics Levinas succeeds Heidegger.

Seeing Levinas' work as a countervailing force to the main streams of thought in the age of modernity may give his texts a sense of meaning. Leading modernistic thinkers like Thomas Hobbes (egoistical hedonism) and John Locke (a modified form of self-preservation) saw individuals related to society at large in an external and instrumental way. Social institutions exist in order to preserve and protect the self-interests of individuals. Accordingly, individuals have no obligations towards each other, except following the law. The ethics of proximity forwarded by Levinas is clearly an antithesis to these ideas. And in the history of thought we often find a dialectical tension between thesis and antithesis. Against this background, Levinas' focus on the Other's face and the unlimited responsibility for the Other is a radical step. And I interpret Levinas as saying this about human nature: the existence of human beings is based upon taking responsibility for the Other. I am freed from myself and lifted out of my lonely shell by caring for the Other.

Conclusion

A perfectly (moral) life is not absent from moral conflict. Each time I walk the streets in a city, I I am uncertain as to what actions I should take, and if I «intervene» actively, I do not know if I have done enough. And I do not have the means to find out what the correct action really was. There is a demand, but the demand is vague, unclear, opaque. Thus, it follows from the ethics of proximity that *I am moving on a wall of uncertainty.*

It is easy to furnish citizens with laws, rules, and codes of ethics, but do these external symbols change behaviour? Do they really have any effect? Do they mean anything to the persons they refer to? Morality goes deeper than external rules which may or may not be used. What is even worse is that remedies in the form of external rules as well as academic, categorical concepts *may* come *in-between* me and the Other. Thus, the all too common remark: «why talk to me? I am only doing my job!»

In the same way as modernity was the age of ethics, postmodernity is the age of morality. In correspondence with Levinas' point of view we may say that morality is a natural destroyer of any order. Adolph Eichmann's «I only follow the law» is a core example of allegiance to the external law that never can be any guarantee of morality. This was the message of the Holocaust, where evil actions were carried out in correspondence with law, order, and Nazi- morality. Law and order are based upon categorical thinking and the danger is to put what is different into a category. Therefore, we may well ask whether categorical thinking necessarily is evil? To think in «rights», that is human rights, represents an universalistic kind of ethics. The gains are that all moral persons have the same rights. The homeless in my example is, therefore, not inferior to the King of Norway in dignity and moral status. In this sense categorization is the base of an universalistic morality. In practice we categorize all the time. Thus, the pitfall is not categorization as such. It is the overuse and misuse of categories that we should be aware of.

Who is *allowed to become a face? The* first step towards the good is to listen to the unspoken demand, and take and perform responsibility. However, there is no final answer on the delicate question: how should

I use my time and resources to do good? In postmodern society the moral self is constituted in contradictions and in a constant dissatisfaction. *My humanity resides not in security, but in ambivalence.* I cannot see faces all the time. But it makes a difference whether I sharpen my awareness or not.

I cannot take part in meetings all the time. However, it is a challenge to become aware of the tendency to perform mismeetings and try to transform some mismeetings into meetings. It is a question of frame and a question of dialogue. Instrumental meetings are problematic. I know that closing my fists too hard, leads to a loosening of the grip. We have the options – *being aside, being with, and being for* the Other and we may choose between a technocratic, blasé attitude of attachment and a human attitude of sensitivity! Hopefully I have documented the importance of uncertainty in social life and fundamental value of meeting the Other. Seeing the Others' face is a precondition for human existence.

Note

1. I have benefited from dialogues with Hans Chr. Holmsen, Lars Chr. Opdal, and Jorun Ur, and I am grateful for having participated in a course in the Norwegian program of ethics «*The Other in Ethical Discourse*», March 20–24, 1995, where Zigmundt Bauman, Arne Johan Vetlesen, Svend Andersen, and Asbjørn Årnes were among the contributors.

Bibliography

Arendt, H. (1976), *Eichmann in Jerusalem.* Viking, New York.

Bauman, Z. (1989), *Modernity and the Holocaust.* Polity Press, Cambridge.

Bauman, Z. (1993), *Postmodern Ethics.* Blackwell, Oxford.

Bennis, W. (1989), *On Becoming a Leader.* Addison-Wesley Publishing Co, Inc. Reading, MA.

Bennis, W. & B. Nanus (1985), *Leaders: The Strategies for Taking Charge.* Harper & Row, New York.

Davis, M. S. (1971), That's Interesting! Towards a Phenomenology of Sociology and a Sociology of Phenomenology. *Philosophy of the Social Sciences,* no. 1 pp. 309–344.

Eckhoff, T. (1971), *Rettferdighet.* Universitetsforlaget, Oslo.

Elster, J. (1983), *Sour Grapes. Studies in the Subversion of Rationality.* Cambridge University Press, New York.

Etzioni, A. (1988), *The Moral Dimension: Towards a New Economics.* The Free Press, New York.

Friedman, M. (1992), *Dialogue and the Human Image. Beyond Humanistic Psychology.* Sage Publications, London.

Frisby, D. (1992), *Simmel and Since. Essays on Georg Simmel's Social Theory.* Routledge, London.

Giddens, A. (1979), *Central Problems in Social Theory. Action, Structure and Contradiction in Social Analysis.* University of California Press, Berkeley.

Giddens, A. (1984), *The Constitution of Society. Outline of the Theory of Structuration.* Polity Press, Cambridge.

Gilligan, C. (1982), *In a Different Voice. Psychological Theory and Woman's Development.* Harvard University Press, Cambridge, MA.

Gilligan, C., N. P. Lyons, & T. J. Hanmer (1990), *Making Connections. The Relational Worlds of Adolescent Girls at Emma Willard School.* Harvard University Press, Cambridge, MA.

Harvey, D. (1989), *The Condition of Postmodernity. An Enquiry into the Origins of Cultural Change.* Basil Blackwell, Cambridge, MA.

Harvey, J. B. (1988), *The Abilene Paradox and Other Meditations on Management.* Lexington Books; D. C. Heath and Company, Lexington, MA.

Heidegger, M. (1987), *Nietzsche, The Will to Power as Knowledge and as Metaphysics (Vol. III), Nihilism* (Vol. IV), Harper San Francisco (Harper-Collins Paperback 1991).

Hirsch, E. D. (1967), *Validity in Interpretation.* Yale University, New Haven, CT.

Hjorth, P. F. (1994), Helse for alle! Foredrag og artikler 1974–93, *Utredningsrapport nr. U1 – 1994.* Avdeling for samfunnsmedisin . Seksjon for helsetjenesteforskning.

Ikonen, P. & E. Rechardt (1993), The Origin of Shame and its Vicissitudes. *The Scandinavian Psychoanalytic Review* 16, 100–124.

Kemp, P. (1992), *Levinas.* Forlaget ANIS, Danmark.

Levinas, E. (1985), *Ethics and Infinity* (*Conversations with* Philippe Nemo. Translated by Richard A. Cohen). Duquesne University Press, Pittsburgh.

Levinas, E. (1987), *Collected Philosophical Papers.* Nijhoff, Dordrecht.

Leavitt, H. J. (1975), Beyond the Analytic Manager. *California Management Review,* Vol. 17, No. 3 and 4 (Spring), pp. 11–21.

Lofland, L. H. (1973), *A World of Strangers. Order and Action in Urban Public Space.* Basic Books, Inc., Publishers, New York.
Løgstrup, K. E. (1972), *Norm og spontanitet. Etik og politik mellem teknokrati og dilettantokrati,* Gyldendal, Copenhagen.
Løgstrup, K. E. (1991), *Den etiske ford ring,* 2.utg. Gyldendal, Copenhagen.
Nussbaum, M. C. (1990), *Love's Knowledge. Essays on Philosophy and Literature.* Oxford University Press, New York.
Ohnstad, A. (1993), *Den gode samtalen.* Det Norske Samlaget, Oslo.
Rawls, J. (1971), *Theory of Justice.* Harvard University Press, Cambridge, MA.
Schein, E. H. (1992), *Organizational Culture and Leadership.* 2nd edn. Jossey-Bass Publishers, San Francisco.
Schumacher, E. F. (1974), *Small Is Beautiful. A Study of Economics as If People Mattered.* Abacus, London.
Sennett, R. (1969), Introduction, in *Classic Essays on The Culture of Cities.* Prentice-Hall, Inc., Englewood Cliffs, NJ.
Sennett, R. (1978), *The Fall of Public Ma.* Faber and Faber, London.
Sennett, R. (1992), *Intimitetstyrra nniet.* J. W. Cappelens Forlag A.S., Oslo.
Simmel, G. (1969), The Metropolis and Mental Life, in *Classic Essays on The Culture of Cities*, R. Sennett (ed.). Prentice-Hall, Inc,. Englewood Cliffs, NJ.
Simmel, G. (1990), *The Philosophy of Money.* Routledge, London.
Simmel, G. (1991), Money in the Modern Culture, in *Theory, Culture & Society.* SAGE, London, Newbury Park, Vol. 8, pp. 17–31.
Vetlesen, A. J. & P. Nortvedt (1994), *Følelser og moral.* Ad Notam, Gyldendal, Oslo.
Wheelwright, P. (1951), *Aristotle:* Seven of the most important books of Aristotle, The Odyssey Press, New York.
Wolff, K. H. (1950), *The Sociology of Georg Simmel,* Translated, edited and introduced by K. H. Wolff. The Free Press, New York.

PART II

KNUT J. IMS AND LASZLO ZSOLNAI

Shallow Success and Deep Failure

In our modern society we tend to favor and celebrate short-term success, pseudo-solutions, and window-dressing activities, at the peril of ignoring long-term consequences. An obsessive hunt for short-term gains, often concretized as profit, produces detrimental effects for all life conditions in the long run. In the functioning of today's corporations, and in whole societies as well, we find many activities that result in grave failures rather than the creation of real solutions to pressing problems.

The problem

"*A pyrrhic victory*" is a nice metaphor for a seeming success that, all things considered, finally costs too much. The term can be traced back to King Pyrrhus, who won a battle against the Romans in 279 BC, but the victory came at too great a cost. It would have been better had the King and his army avoided the battle. During our modern history we find too many similar examples, and one recurring question is how to avoid painful and destructive failures. Can we, under certain conditions, be better able to know in advance what to do and what to avoid and restrain, and thereby not waste resources or harm man and nature? There are many examples of the strong forces in organizations that work in the direction of treating symptoms rather than the underlying problems (Cyert and March, 1963). One interesting concept formulated by Zappfe (1996) is the distinction between a real solution and a surrogate solution. It is typical to run away from problems (and responsibility) and stick to a surrogate or pseudo solution.

The terms "shallow" and "deep" as used here demonstrate the inspiration we received from *Arne Næss* and his work on *ecosophy*. By shallow, Næss means a reformist approach to the relationship between man and nature. The shallow approach represents a conventional anthropocentric perspective in which the central idea is to secure the health and affluence of people. In contrast, the deep approach favors a relational, total field image, looking at organisms as knots in intrinsic relation.

In his opus magnum *On the tragic* (in Norwegian, 1996), *Zappfe* discusses the concerns of the individual entities and a scale of abilities. Concerned individuals (sentient beings) attempt to realize their potential by using their abilities. The abilities may be inadequate (deficiency) or adequate (sufficiency). An interesting issue arises when there is a surplus of ability related to the demands of the situation. Such a surplus of abilities may have destructive consequences. Zappfe cites an example of this in the case of a huge red deer (megaceros euryceros) with antlers that grew excessively from generation to generation through natural selection. At first the huge antlers were an important means for the deer's survival because they could be used for protection against other animals, but eventually the antlers hooked into bushes and trees and became a heavy burden to carry. The antlers gradually became disproportionate to the deer's body. The *strength* became *transformed* into a serious *weakness*. The long-term effect was that this species finally became extinct because of one-sided growth along only one dimension.

This case captures the essence of the tragedy of imbalance of abilities. Deer do not have a choice, but people do.

The fallacy of misplaced techno-centrism

The choice of interpretation determines how we respond to a certain event. A deeper awareness of and competence in interpreting events, formulating problems, and implementing solutions are essential. A theoretical framework by Mitroff (1998) illustrates a systematic way of looking at different kinds of problems and the consequences it has on which solutions we prefer. Mitroff's framework will be used as a point of departure in order to arrive at different solutions to important problems. The logic behind the framework is that, although generating knowledge to reach a deeper understanding is a detour and an indirect way to action, deeper insight is essential to finding appropriate solutions for urgent problems.

One fascinating concept used by Mitroff (1998) is the so-called E3 (Type III error) that relates to the problem formulation process. E3 means *solving the wrong problem precisely.* We need critical thinking to solve the right problem. Even an approximate solution to the right problem is better than an elegant solution to the wrong one. Mitroff asks us to scrutinize the assumptions we hold about the stakeholders, and the importance of picking the right stakeholders, which generally means expanding the set of stakeholders. To manage the important problems in a fruitful way, system thinking is indispensable if E3 is to be avoided.

Mitroff (1998, p. 59) distinguishes four perspectives one may take on any problem. The perspectives are scientific/technical, interpersonal/social, existential/spiritual, and systemic (see Figure 1).

The scientific, technical way of thinking is the dominant perspective in our western culture, which favors technical or economic solutions. This may easily lead to the so-called "*pig principle*": one dimensional growth that, in the long run, will change and pervert the quality of the dimension.

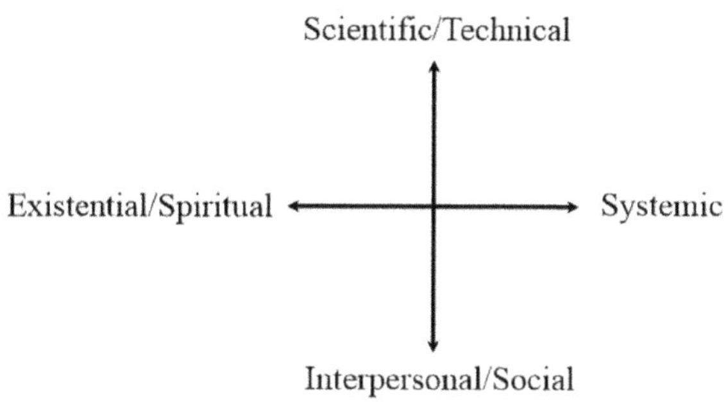

Figure 1. Four perspectives on any problem. *Source:* Mitroff (1998).

The more, the bigger, and the faster – the better it is. This is a profound assumption. This is frequently a serious fallacy, because usually we do not need more of the same, but more of the right things.

We need a blend of efficiency and effectiveness. One example of connecting a technical solution to an existential problem is to see uncertainty in the organization as a bureaucratic problem efficiently solved with more rules. This is a quick and easy solution, which may cause serious dysfunctional effects in the situation because it may stifle ethical reflection and contribute to a reification of rules. Such rules might also create distance between individuals when proximity would have been the better choice. Only under conditions of proximity, and when encountering the other's face, might the moral competencies of empathy and compassion develop as moral impulses capable of discerning what is good or evil (Bauman, 1993).

The Mitroff framework is illustrated through examples inspired by Beck's interesting work on *Risk Society* (1986). We may view the careless use of pesticides in the west and, in particular, the Third World as a technical solution to a non-technical problem. We call this the *fallacy* of *misplaced techno-centrism*. The crops and associated profits were expected to increase by the use of new toxic chemicals that killed the plants' enemies. So the solution was to use DDT and other chemicals to increase the harvest. In the short run this was an obvious and tangible success, because the artificial fertilizers and chemicals increased the yields per hectare.

In the long run, powerful and latent side effects became visible. The fertility of the soil declined, many plant and animal species disappeared, and the lead content in the milk of nursing mothers increased. These secondary effects undermined the natural basis of agricultural production and created an ecological crisis. The secondary effects struck back like a boomerang. The pesticides used in the Third World returned to the industrialized countries in the imported fruit, tea leaves, and cocoa beans. The lesson to be learned was that the problem of unsatisfactory crop yields cannot be solved by artificial, technical means.

The problem might be better defined as a systemic one, with important second- and third-order effects. Some of the pesticides may produce irreversible threats to the life of plants, animals, and human beings. It is better to think of the use of pesticides in a systemic way. They intervene in an ecological system in which everything is interdependent. This perspective encourages us to be much more reflective and cautious with the use of new and artificial materials that, in the long run, may endanger all life and civilization. As Næss suggests, "If we hurt nature, we hurt ourselves" – because nature and man are one: a unity. In this sense the use of pesticides, while having a systemic character, also has an existential effect on peoples' lives. It might even threaten the existence of all life on Earth.

Although pollution is one of the primary outcomes of advanced chemical plants, these hazardous industries have even more serious effects on the social conditions of people. Beck (1986) maintains that pollution has a democratic effect, because it affects everybody, even the polluter, in the form of a boomerang effect. But the hazardous effects do not affect the perpetrator and the victims equally. It is a known fact that the poor have a "systematic attraction" to extreme risk. Unemployed people, the poorest of the poor, settle close to the smokestacks of the chemical factories in the industrialized centers of the Third World. Poor people have a higher acceptance rate of extreme risk. Who were living close to Bhopal in India when the Union Carbide plant had a catastrophic outburst of toxic gas? The poor. In a social perspective, poor people are leaving their local communities and their children to make money where they can. They move to the cities that gradually develop around chemical plants, which initially were localized far from established cities because of their risk potential. "People are living too close to the plant … We built [it] in 1968 and no people were here. Now the people come closer and closer and we are bothered about this. The waste air made some people sick …" (Weir, 1987). The social and existential conditions for the poor migrants living too close to the chemical plants of transnational corporations are often miserable.

If the problem of unsatisfied employees is defined as a technical problem, paying higher salaries might solve it. The underlying assumption here is that the most important incentives for man are external, and money is the primary motivational factor. This assumption corresponds well with the view of man as an empty shell equipped with rational preferences. There are, however, several examples demonstrating that the use of monetary incentives can press out (the crowding-out effect) intrinsic motivation and lead to reduced activity (the hidden cost of reward [Frey, 1997]).

The Danish philosopher, Knud Løgstrup, is inspired by Kierkegaard, but strongly argues against some of his writings. Løgstrup cites the phenomenological conditions of man as consisting of compassion, trust, love (in the meaning of eros), the propensity for speech to openness and, finally, fidelity (Løgstrup, 1971). According to Løgstrup, we human beings are in a profound sense interdependent. This implies that in any relationship we are vulnerable, due to the exertions of the other's power. Løgstrup writes that we put ourselves into the hands and power of other persons, and in any life encounter the self is confirmed, increased, or reduced. Any encounter has effects on our self-image and our self-respect. Thus, our selves are always vulnerable to others. According to Løgstrup, trust is the way of living with anxiety.

Sartre focuses on the other's gaze, and he illustrates in a dramatic manner that a person is not an object but a relation. One of his main postulates is that any human is free, because in every moment we can step back *(se recule)* and reflect upon ourselves and, as a result, we are able to transcend ourselves. Freedom implies responsibility, even if most people, according to Sartre, try to avoid freedom as well as responsibility. Emmanuel Levinas (1985, see also Kemp, 1992) assumes a non-symmetrical responsibility for the *Other*. This asymmetric relationship belongs to the basic constitutions of the responsible subject. Levinas postulates that a person has a responsible relationship toward all human beings; the self of a person is born out of a union with other persons, stretching itself towards the other – living for the other, not only with the other.

Arne Næss (1989) develops the concept of self further. He assumes an ontology that implies that human beings are one with all nature, and that self-realization cannot be attained without taking a much deeper and broader perspective. Through identification with others, you arrive at self-realizing consciousness, and your own self-realization depends on whether other beings have attained self-realization. As a consequence, non-violence as a general principle becomes essential. Næss's deep ecology is highly inspired by Buddhism, wherein the self does not have absolute boundaries. Every thing has a living, flowing connection with everything else and there are no isolated entities. To ignore our dependency on our close partnerships with other forms of life has contributed to the creation of a master–slave role, which leads to the alienation of man from himself.

According to Zappfe (1996, p.43), under certain conditions abilities involve an inclination to manifestation, converting capabilities into action. Human capabilities might also be detrimental to life in their maximal manifestation. A surplus in technical efficiency (material culture) is one example. The problem may arise when all material needs are satisfied, but the production of means does not stop. The capabilities get out of control and the tools – the means – become predominant. It is no longer the need that cries for the appropriate means, but the means that cry for a need. Supply exceeds the demand. The means become the goals and shadow for the purpose for which they should be used. With Zappfe, we may say that technical insight and competence is exploited for a destructive purpose.

It is an existential fact that we have to live and choose within a world in which we do not know the consequences of our actions. Because our mega-technological means have become so powerful, we have to stop and reflect upon the situation of the Earth (Jonas 1984). Technology is double faced and can lead to either good or evil. And even its good consequences have the potential of becoming perverted due to excessive growth along one or a few dimensions. "Thus, moral responsibility demands that we take into consideration the welfare of those who, without being consulted, will later be affected by what we are doing now. Without our choosing it, responsibility becomes our lot due to the sheer extent of the power we exercise daily ...." (Jonas, 1996, p. 99). Even with the best intentions, we may face long-term effects that are detrimental for future generations and all life on Earth. We need to be aware of this new magnitude of responsibility that we bear.

We need scientific and technological knowledge, but we also need a better understanding of the existential conditions of human beings to avoid the fallacy of defining most problems as technical economical-scientific, and solving them in purely technical ways. We should gain a better understanding of self-realization and what self-realization means in the perspective of deep ecology and sustainability.

Toward real solutions

Here we summarize the main themes and messages of the book, showing what deep ecology and Buddhist economics can offer for solving the most pressing problems of business, economic, and social life in our contemporary world.

Self-realization

Arne Næss points out that *self-realization* is the realization of the deeper and broader self. It is identification with or seeing something of yourself in others. Your feelings are somehow adapted to those you identify with. This identification can extend to your friends, your neighbors, and your country. It can even extend to the whole of humanity. You can also identify yourself with animals, plants and other natural elements. Through identification with others you find self-realization.

Self-digging means not developing the depth and broadness of the self but, instead, "digging" down to ego-fulfillment as opposed to actualizing oneself through self-realization. You develop a cult of your own self-seeking. You have a reputation, a social image that enhances the feeling that you are somebody who does not need anybody else. Ultimately, you cut off identification with others to such an extent that you are left alone.

Diversity in every aspect of our existence should be a norm, whether it be biodiversity, cultural diversity, or economic diversity. Diversity of ideas is also important. If we were to think that there is only one correct idea, one absolute truth, one right way to sustainability, then we might end up creating a form of eco-fascism. It is only through multiplicity, plurality, diversity, and inclusivity that we can find self-realization. There is no one final definition of self-realization. We all find our own meaning in this word. It is through practice that we find realization. Just as we all have our own bodies, we all have our own "realization".

Nel Hoftsra and *Aloy Soppe* note that deep ecologists go further than the more pragmatic environmental reformers and revisionists, who stress conservation and efficient use of resources, and promote a more effective use of nature. They criticize the theoretical assumptions of modern economics and meet the weaknesses and contradictions in the "dominant social paradigm" of capitalism. Deep ecology can only exist if people act according to nature rather than making nature subordinate to human beings.

A *life according to nature* implies several directives: (i) to discern and respect "reason" in all things and life (system-thinking), (ii) to discern and respect "law" in all things and life (cyclical thinking), and (iii) to discern and respect the "divine" in all things and life (soul and humility).

Richard Welford stresses that there are many intrinsically valuable aspects of our lives that could be developed: love and appreciation of nature for its own sake; appreciation of aesthetics, honesty, and integrity; self-respect and respect for others; humility, generosity, understanding, and knowledge. An emphasis on economic growth and materialism has often led to little space for the development of these aspects of human existence.

We must recognize that self-esteem should come from who we are inside. Not external factors, but factors within ourselves provide meaning, security, value, and self-worth. None of this can be based on the human ego, but can only developed through a process of humility and inner development.

Greed

Welford states that the *ecological crisis* in which we find ourselves is a manifestation of a mindset that fails to recognize interconnectedness and complexity. Human beings, through their materialism, consumption, and greed, are directly responsible for most of the unsustainable practices that we see. If there is to be a real move toward sustainable development, it will not be enough to rely on businesses, governments, and other institutions. Change will have to occur within people, and especially within those of us who live in the West.

In western economies people seem to believe that the pursuit of *money* and *material possessions* is the path toward greater happiness, satisfaction, and contentment. But in many cases, no matter how wealthy people become or how high their standard of living, they often die without realizing contentment. Indeed, the greedier we become, the less likely we are to find happiness and contentment.

Nel Hofstra and *Aloy Soppe* note that in our contemporary world the primary function of money and capital (that is its role as unit of account and unit of exchange) threatens to become subordinate to the *creation of financial wealth as a goal in itself.* A high interest rate increases pressure for economic growth in order to make the direct investment profitable. This accelerates pressure on future environmental resources and employee performance. A greedy and sometimes "violent" approach to future growth generates uncontrolled expectations and volatility of market prices.

Economism

Ove Jacobsen and *Stig Ingebrigsten* point out that, throughout the last few decades, an increasing number of societal sectors have been subjected to privatization and the market economy in the quest to reach goals connected to economic efficiency. This trend is characterized as "economism". *Economism* means that the economic value system plays a dominating role in society, ignoring or reducing other values.

The economic invasion of the lifeworld not only leads to an impairment of culture, it also undermines creativity and trust, two important pillars of the market economy. When economics invades culture, instrumental rationality replaces communicative rationality, functionality represses intentionality, atomistic competition destroys integrated cooperation, consumption becomes more important than cultivation, and utility (converted to a monetary scale) replaces value pluralism.

Zsolt Boda refers to *Karl Polanyi,* who, in his seminal work, *The Great Transformation,* argues that the market economy requires a market society where social interactions and activities, such as labor and human relations, the cultivation of land, the management of natural resources, and even the evolution of culture are coordinated by the logic of the market. Polanyi argues that the social and environmental consequences of this process are dramatic, because a single logic rules over all others.

Economic theory

John Gowdy points out that at the core of traditional economic theory is a concept of human nature that is devoid of social context, mechanically rational, and driven by an insatiable appetite for material possessions. The only outlet for human creativity and self-realization is the consumption of market goods. Responding to the desires of such a creature *(Homo economicus)* in order to maximize profits is the only ethical responsibility of business.

Research in behavioural economics and game theory has already shown that the self-centered model of *Homo economicus* is a poor characterization of human behaviour. The evolutionary feasibility of strong reciprocity, that is, cooperative behaviour not based on reciprocal altruism is demonstrated. Also, experimental results and case studies show that sharing is a part of "normal" human behaviour.

Relational ontology

Julie Nelson notes that deep ecology and Buddhism share a thoroughly *relational ontology*. In both deep ecology and Buddhist philosophy what really *is* are relations and processes. Things exist in a state of dependence on the relations that constitute them.

The diversity and elaboration of these relations and processes has value. The intrinsic worth of relationality, and the responsiveness of humans to this worth through gratitude, compassion, and care form the basis for ethics, which permeates the ground of being.

Modern *western thought* is based on a *substantivist ontology*. Relations are seen as secondary – as simply the way that pre-existing "stuff" is arranged in patterns with, or bounces against, other "stuff." Since the rise of modern science, the physical world has been conceived of as a sort of ethically neutral clockwork, driven by the "laws" of physics. Some modern thinkers try to take a thoroughly reductionistic approach, seeing all issues of ethics, aesthetics, and emotions as simply the epiphenomena of indifferent processes of evolution.

The insights of relationality extend to big, human-made, and materially oriented institutions, such as corporations and economies. Recognition of symmetric mutuality opens our thinking to ways in which co-workers might treat one another with respect. The recognition of asymmetric mutuality further opens up the possibility of thinking about relations of respect among people with different levels of power and different roles. Not all workers in an enterprise have equal abilities in leadership, inventiveness, or finance. Enterprises can be structured in ways that take advantage of people's different qualities of power, while still retaining a fundamental attitude of mutuality.

Buddhist view of the world

Richard Welford stresses that *Buddhist philosophy* turns the whole western mindset upside down. In contrast to the anthropocentric worldview commonly characterized by western culture, the Buddhist cosmology has the entire universe at its center. Human beings are humble in the totality, and are just grains of sand in the vast, limitless ocean of space.

The Four Noble Truths of Buddhism address the dynamics of human suffering and give us a starting point for moving beyond the barrier of overconsumption. We can see the *Four Noble Truths* as typifying the problems in the achievement of sustainable development:

1. *Life is suffering.* This has to be comprehended. With increasing secularism and dissociation from nature and the environment, and rising levels of expectations inside and outside work, people are becoming less satisfied with life and the lifestyles they adopt.
2. *The cause of suffering is desire.* Unchecked desire has to be abandoned. Heightened levels of dissatisfaction have implications for consumerism. First, there is the erroneous perception that

purchasing goods is going to make us happy, and second, because we are increasingly dissatisfied and thus unhappy or stressed, we are unable to deal with the changes needed.

3. *The cessation of suffering is the cessation of desire.* This has to be realized. By becoming aware that there is a root to the general societal malaise of avoiding environmental and social responsibilities, we know that there is a way of stopping such complacency to begin a path to sustainability.
4. *The path to the cessation of desire is practice.* To stop doing what makes us dissatisfied, we have to realize the cause of that dissatisfaction and keep trying to behave in a more sustainable manner. Buddhism shows us that this is difficult and requires ongoing commitment and practice.

Buddhism points out that even if one attains what one desires in the short run, greater desires always emerge. The *ego mindset* cannot be fulfilled, and its greed for more satisfaction and recognition becomes the source of its own destruction. This becomes a source of suffering because the human spirit becomes captured by the avaricious mind.

Happiness might be seen as the ratio between wealth and desire. The capitalist system has been successful in increasing levels of wealth and thus increasing happiness to some extent. But through the epiphenomenal increase in people's desires for more, stoking up avarice and greed, capitalism has not produced great increases in happiness overall. The Buddhist approach suggests that people reduce their desires so that even the maintenance of current wealth increases happiness. As people become detached from desires, their levels of happiness will increase. This may result in a reduced demand for consumer goods. It is an opportunity for businesses to concentrate less on providing non-essential consumer goods and more on providing essential goods and services to developing countries, introducing technologies that will remediate environmental damage and serve the poor and needy. Indeed, business will then become less exploitative and damaging, and more worthwhile and productive.

Buddhist economics

Richard Welford states that Buddhism can be the source of greater individual contentment and satisfaction and that this is more consistent with protecting nature and caring for the environment. There is a need to move away from mass-consumption economies toward a more environmentally restorative one.

While western economics emphasizes self-interest and material development, Buddhist economics stresses interconnectedness and "inner development". It would also place an emphasis on culturally appropriate economic approaches. A Buddhist approach involves an emphasis on a much more sustainable development, where both human beings and living creatures can realize their potential, and where inner development and economic development are compatible, all in the context of a just society and healthy ecosystem.

Buddhist economics sees little problem with activities that are beneficial to oneself, to one's business, and to one's country, but only in circumstances of *non-harmfulness* to others. Establishing mutually beneficial transactions rather than exploitative ones is important. One distinguishing feature of Buddhism is that its adherents have never engaged in a religious war. Its emphasis on *peace* and non-harm needs to be translated into modern economics. Non-harm means respecting all human beings and all other creatures, and developing a sense of respect for all life.

An economics based on respect would certainly help to reverse the mounting ecological crisis. Economics should be based on notions of fair "give and take". If we are going to take something from the environment, we must be prepared to ensure that it can be replaced in one way or another. We need to develop a *restorative economy* where whatever damage is done to the environment is either restored or fully compensated for.

Buddhist economics is based on a concept known as the "*middle way*". Moderation reflects the knowledge that "things" and consumption are not the root of happiness and contentment. The search for true happiness is not a material activity, but a spiritual one. Reaching the goals of happiness and contentment requires that we all spend less time consuming and more time contemplating – less time following prescribed roles and more time being creative.

Traditionally, businesses have considered natural resources as freely available for their use in satisfying the desires human beings have acquired. But a Buddhist business considers the stock of natural resources as potentially available for use by all generations, therefore that stock has to be very carefully protected and used only when necessary. When we use natural resources we are borrowing from future generations and the cosmos, and they are not legitimately available for exploitation.

Material flow analyses

Peter Daniels emphasizes that "material flow analysis" covers several distinctive approaches. But they are directed toward the systematic physical measurement of the magnitude and "location" of material (and energy) flows. The flows cover environmentally significant materials as they move through the socioeconomic metabolism of human geographic systems. The main assumption is that the *impact* of *human economic activity* is a *physical phenomenon* caused by the magnitude, concentration, dissipation, and transformation of flows of material and energy.

Material flow analysis, as the systematic analysis of the physical environmental consequences of livelihood activities, is a way to identify how and where the socio-economy can be transformed into a more sustainable form. It serves as an important methodology for making *eco-restructuring strategies* the best prospect for operationalizing sustainable development.

The most important link between Buddhist economics and material flow analysis is that both are predicated on reducing the metabolism of human economies. The Buddhist view calls for minimum environmental intervention and human-induced material flows and transformations. An economic system founded on this caution would be supportive of technological and consumption related savings of material and energy, recycling of products and waste (including closed-loop, integrated production systems), and the reduced toxicity and physical flows of pollution.

Physical growth in intervention will create imbalance and ecosystem instability. Flow accounting based on metabolism reduction and Buddhist economics are both built on comparable natural law principles that "scientifically" prescribe minimum disruption and violence, and livelihood and consumption patterns that are in harmony and balance with the external world.

Dialogue between culture and economics

Ove Jacobsen and *Stig Ingebrigsten* stress the importance of *dialogues* based on the principles of *discursive rationality* as a tool to combine the cultural lifeworld with the economic system without undermining either of them.

Dialogues involving all relevant stakeholders represent a force for changing the direction of development and domesticating the undesirable effects of market forces. Meanings are created, knowledge is developed, and learning takes place in the dialogue among different stakeholders. Dialogue-based relationships among the actors might contribute to strengthening values like sympathy and empathy, and to using learning and innovation as productive forces.

In the process of coordinating the interplay among the different stakeholders, a new institutional setting is proposed – "*the communicative arena*". In the communicative arena, the agents concord their "strong evaluations" through processes based on communicative rationality and coordinate their economic actions to bring about common ends. Sustainable development presupposes an arena where economic and cultural agents coordinate and evaluate their different values and norms through communicative processes.

Environmental commons

Zsolt Boda argues that we need *complex institutional arrangements* to secure the *global environmental commons.* Privatizing them is not a good solution. It pretends to provide incentives for protecting the commons, but just the opposite happens: it voids the concept of the commons of its inherent ethical content.

Business must acknowledge and respect the inherent value of the commons. This claim implies the following three correlative *duties:*

i. Business shall refrain from appropriating common resources.
ii. Business shall take part in the governance of the commons and contribute to effective regimes to govern and manage them.
iii. Business shall contribute financially to the maintenance of the commons.

Theory of the firm

Julie Nelson argues that limiting possible relationships to either arms length contracts or hierarchical control rules out the idea that values, group identity, mutuality, non-hierarchical structures, or ethics could play a role within and among contemporary business organizations. Yet the evidence on employee behaviour suggests otherwise. Real humans do not simply leave their needs for social relations, their values, their loyalties, and their creativity at the workplace door. Many managers and researchers in organizational behaviour share the insight that people work better when they are supported, empowered, and allowed to draw on their own creativity, rather than when they are consistently treated as potential shirkers who have to be brought under control.

Buddhist thinking does not prescribe replacement of for-profit businesses with systems of small-scale and cooperative enterprises as the cure for economic suffering. The ethical merit of organizations cannot be prejudged on the basis of size alone, or by the purposes written on their articles of incorporation. Organizations must be evaluated by what they do. Small, purportedly "loving" families are too often the sites of domestic violence. Small non-profit hospitals too often exploit their own workers for the sake of keeping costs in line. Large, for-profit corporations have at times taken actions that show that they can be good workplaces and responsible members of social and environmental communities – when given a chance and especially when encouraged in these directions by consumers, shareholders, and political activism.

Personal responsibility

Knut Ims points out that *business* as a human and existential enterprise is a *personal* and *social activity,* and it consists of webs of interpersonal relationships and widespread cooperation. Emotions are a central feature of a person's character. It is the emotions and personal values that constitute persons as extended selves, which enable them to identify with a larger community. *Personal responsibility* is not reducible to social, professional, or organizational roles. Its essence is to respond to different stakeholders through an authentic and active dialogue.

Ecology of spirit

Mike Bell stresses that we should see organizations in the context of the living universe. The "Ecology of Spirit" refers to a complex set of relationships and systems, infused with an inner life-force (or spirit), that links the land and its creatures to individuals, people, communities, organizations, and to the entire universe.

If the land is living and we are living, then our organizations must also be living. They have unique spirits that are linked to our culture. Our organizations can only be as healthy as our culture; and our culture can only be as healthy as our organizations. As members of organizations, we are called to represent and serve our people and to care for and nurture the spirit of our organizations. We must find a way of realigning our primary relationships and maintaining a balance of spirit. Our primary relationships are our relationships with these: (i) the spirit of the land, (ii) our own inner spirit, (iii) the spirit of our human community, our families, relatives, and the people who are closest to us, and (iv) the spirit within our organizations and workplaces.

Ethical business

Laszlo Zsolnai states that modern business activities considerably affect the fate and survival of natural ecosystems and the life conditions of present and future generations. For this reason business has a one-way, non-reciprocal duty to care for the beings affected by its functioning. To become truly ethical, businesses should be transformed into *sustainable, pro-social*, and *future-enhancing* enterprises. Doing ethical business is not a luxury for advanced societies. It is a requirement for modern-day business to survive in a world of large-scale ecological disruption and social disintegration.

About the future

Richard Welford insists that Buddhist economics is founded on *changes* at the *individual level,* which will eventually filter through to institutional change, and later impact on the activities of governments and businesses. This is promising because, rather than waiting for radical change through the democratic process, each of us can begin that change. In developing a spiritual attitude of caring and compassion we can improve our own lives, the lives of others, and the planet as a whole.

Bibliography

Bauman, Z. (1993): *Postmodern Ethics.* Oxford: Blackwell.

Beck, U. (1986): *Risk Society. Towards a New Modernity.* London: Sage Publications.

Cyert, R. and J. March (1963): *A Behavioural Theory of the Firm.* New York: Englewood Cliffs.

Frey, B. S. (1997): *Not just for the Money. An Economic Theory of Personal Motivation.* Cheltenham, UK: Edward Elgar Publishing.

Jonas, H. (1984): *The Imperative of Responsibility. In Search of an Ethics for the Technological Age.* Chicago: University of Chicago Press.

Jonas, H. (1996): *Mortality and Morality. A search for the Good after Auschwitz.* Evanston: Northwestern University Press.
Kemp, P. (1992): *Levinas.* Denmark: Forlaget ANIS.
Levinas, E. (1985): *Ethics and Infinity* (Conversations with Philippe Nemo. Translated by R. A. Cohen). Pittsburgh: Duquesne University Press.
Løgstrup, K. E. (1971): *The Ethical Demand.* Philadelphia: Fortress Press (Originally 1956).
Mitroff, I. (1998): *Smart Thinking for Crazy Times. The Art of Solving the Right Problems.* San Francisco: Berrett-Koehler Publishers.
Næss A. (1989): *Ecology, Community and Lifestyle. Outline of an Ecosophy* (Trans. and ed., D. Rothenberg). Cambridge: Cambridge University Press.
Perrow, C. (1984): *Normal Accidents. Living with High-Risk Technologies.* New York: Basic Books, Inc.
Reinert, E. S. (ed.) (2001): *Evolutionary Economics and Income Inequality.* Cheltenham: Edward, Elgar.
Weir, D. (1987): *The Bhopal Syndrome. Pesticides, Environment and Health.* Center for Investigative Reporting. San Francisco: Sierra Club Books, 1988.
Zapffe, P. W. (1996): *On the Tragic* (in Norwegian: *Om det Tragiske).* Oslo: Pax Forlag AS.

KNUT J. IMS

Deep Ecology

Deep ecology is an alternative philosophy called ecophilosophy, as well as a campaign platform. As ecophilosophy, deep ecology is a fundamental approach to environmental problems and focuses on the underlying causes, the roots of the problems. It assumes a relational, total field perspective that fits into a holistic, non-reductionistic, non-anthropocentric worldview. It is a contrast to shallow ecology, which represents the technocratic attitude to pollution and resource depletion – using rules like "the polluter pays", assuming that by treating the symptoms through technological quick fixes the ecological footprints will be lower. According to deep ecology, one has to change the basic ideological structure, which ultimately means changing ourselves.

Deep ecology is highly relevant because global reports (WWF International, 2008; Millennium Ecosystem Assessment, 2005) describe a serious bio(logical) diversity crisis. An estimated loss of different extinct species every year might be in the range of 10,000 to 100,000 species. Human activities are almost wholly responsible for this loss, and the loss of species and habitats occurs at an ever-increasing rate. According to deep ecology, humans are an integral part of nature, thus the human fate is tightly linked with biodiversity, and the loss will weaken the ability of the living systems and the web of life on which we depend to survive. The United Nations proclaimed the year 2010 as the International Year of Biodiversity.

A much more debated problem of our planet is climate change, which in large part can be explained by a continuous rise in greenhouse gas emissions (in particular CO_2). Since 1961 the ecological footprint has more than doubled. In 2005 the ecological footprint exceeded biocapacity by 30 percent, which means that the world's resources will be used faster than they can be renewed.

Inspirational sources – The deep ecological tree

There are many supporters of deep ecology. The Norwegian philosopher Arne Næss (1912–2009) is acknowledged as a pioneer who formulated some of the core concepts. A metaphor used by Næss to suggest the inspirational sources behind deep ecology is "the deep ecological tree" (see Hegdal and Olsen, 2001), a tree with long and strong roots, and different branches, consisting of ideas from Hinduism, Confucianism, and Buddhism on the one hand, and Aristotle, Heidegger, and Spinoza on the other. Thus, the ideas of deep ecology are not new, but ancient. One of the inspirational sources is Mahatma Gandhi's concept of oneness, which means that everything is interrelated. Gandhi postulates a holistic worldview which naturally leads to nonviolence.

In his book on Baruch de Spinoza, Næss (1999b) writes that many of his intellectual friends have great respect for Spinoza, but wonder how it is possible for anybody to take Spinoza's metaphysics seriously today? Spinoza is an ontologist – not an ethicist. One of Spinoza's core concepts is the substance, and this substance is causasui: its own cause. If there is a God, God has to be in the substance, and God is the substance, and the substance cannot be separated from nature. This teaching represents monism and a kind of panpsychism, which involves the notion that all living entities are internally related to one another. It supports the fundamental idea of the oneness of all life in spite of diversity. People and all sentient beings are related because every species and every being is part of the same substance, and they have an equal right to life-unfolding. Any increase in the level of perfection involves a deep joy *a la hilaritas*. And a perfection cannot develop far without sharing joys and sorrows with others, which follows from our ability to identify with others and, thereby, become close to all kinds of life (Næss, 1999b).

Spinoza is also a key inspirational source for the importance given to spontaneous experiences in deep ecology. Gestalt psychology – which values the perception of gestalts or a network of relationships – also supports the belief that human beings are able to feel a strong sense of wide identification and a deep sense of empathy with other sentient beings, and nature in general. Those identification processes lead to a natural inclination to protect non-human life.

Self-realization as the hypernorm

Self-realization is an essential norm in deep ecology, and it must be interpreted as self-realization for all beings (Næss and Rothenberg, 1989; Næss, 1999a). The meaning of self-realization within deep ecology extends the usual concept of realizing oneself in western society, where self-realization typically is regarded as an ego trip; an individual's effort to satisfy his or her own wishes. Deep ecological thinking involves a redefinition of the human concept of self, and opens up for the possibility that all sentient beings are ecological selves, or *eco-selves*. And, as Gregory Bateson argues, the unit of survival is not the organism alone, but the "organism plus environment".

Self-realization in the wide sense assumes that human beings have the ability to identify with other sentient beings, and to develop a transpersonal self. There are empirical as well as intuitive forms of knowledge that support the possibility of a transpersonal self. The thesis has as a logical consequence; if we damage nature, we hurt our selves. Næss does not "difficultivate" the concepts of the self – focusing instead on the human ability of identification with other beings.

Live a rich life with simple means

One central motto of deep ecology is "to live a rich life with simple means", or "simple in means, but rich in ends". The required change of the ideological structure is that life quality is more important than economic welfare, which is typically measured by Gross National Product (GNP). Although GNP is by far the most used measure to compare economic development between countries, it is not appropriate from an environmental perspective, because a big GNP often involves a high degree of pollution. So, ironically, GNP may be called Gross National Pollution. A central theme in deep ecology is to produce and consume less – tread lighter and wiser on Earth, celebrating the virtues of slowness and smallness in an age of speed and scale.

The Four Levels framework

It may be useful to make a distinction between four levels of deep ecological thinking (see Drengson and Inoue, 1995). The most profound level is the metaphysical level, which refers to the ultimate principles a person holds. The next level is the platform level, which might be seen as the core level and which unites different kinds of groups – from radical ecocentrists, groups from the peace movement, ecofeminists, direct action groups, as well as religious groups. Level three is the policy level, and level four is the practical action level, where different activities might range from "ecotage" to the support of politically oppressed people.

This framework admits a great diversity at level one, the level of ultimate philosophies. Thus deep ecologists do not have to subscribe to the same ultimate ecological philosophy to work cooperatively. As Næss claims: the front is very long – and each person may contribute on his own premises. To indicate this long front of level one we can compare the premises of Gandhi and Næss. Metaphorically, Gandhi looked upon everybody as drops of water and writes: "This ocean is composed of drops of water; each drop is an entity and yet it is a part of the whole; the one and the many. In this ocean we are little drops." For Næss, this is not an appropriate image, and he states that "I would rather look upon myself as a little tree in a large forest."

One idea behind the four-level framework is to stimulate continuous back and forth processes between the levels, thereby assuring that our fundamental premises and practice are in harmony. The framework challenges us to pose deep questions in exploring the ultimate premises and norms. When a more or less articulated position on the first level is appears, we can move to the other levels.

The core level – the platform

The platform level is summed up in eight points:

1. The well-being and flourishing of human and nonhuman life on Earth have value in themselves – independent of the usefulness for human purposes.
2. Richness and diversity of life forms are values in themselves.
3. Humans have no right to reduce this richness and diversity except to satisfy vital human needs.
4. We need a substantial decrease of human population.
5. The present interference with the non-human world is excessive.
6. Policies which affect basic economic, technological, and ideological structures must be changed.
7. The ideological change is mainly that of appreciating life quality (dwelling in situations of inherent value) rather than adhering to an increasingly higher standard of living. There will be a profound difference between big and great.
8. Those who support these points have an obligation to directly or indirectly attempt to implement the necessary changes in a non-violent way.

Personal ecosophy and personal responsibility

Næss invites all people to work out their own subjective ecosophy, by which he means a philosophy of ecological harmony. This kind of *sophia* or wisdom is openly normative, and contains norms, value priorities, and hypothesis. Wisdom is policy-wisdom, prescriptions, and hypothesis, not only scientific description and prediction.

Næss walked his talk and participated directly in society's value struggles. His behaviour is in accordance with point 8 at the platform level and, as a good philosopher, Næss always argued for his practical position. He admits to having learned much from ecology as a science. But there are limits to science. Even if deep ecology relies on the science of ecology (in particular acknowledge the value of unity, symbiosis, and diversity), ecologism, the view that takes ecology as the ultimate science and final authority, involves over-generalizing and universalizing ecological concepts. Ecology cannot be a substitute for philosophical analysis. Any change has political implications and, in Næss's own words, "we need to fight against depolitization". The science of ecology, concerned as it is with facts and logic, cannot answer the questions on how we should live a good life.

From point 8 at the platform level, it follows that supporters of the movement have the responsibility to act in harmony with all the guiding points in the platform. There are many ways to work for a change towards basic deep ecological values. Næss's life give many concrete examples of how to act in a responsible way, and he was deeply influenced by Gandhi's nonviolent strategies. Gandhi's lifelong practice in nonviolence against rulers with powerful means was remarkably effective. One primary example was the Salt March in 1930 – a campaign that started with only a few people. The concept behind the campaign was the Satyagraha (truth-force) – a nonviolent protest which openly violated British law. The campaign ended in a large scale of acts of civil disobedience in which millions of Indians participated. Gandhi was arrested and put in jail, but the event received extensive news coverage worldwide. More than 80,000 Indians were jailed, but the campaign had a decisive effect on changing world attitudes – and even British attitudes – toward Indian independence. The Salt March demonstrated the impact of the wise and courageous use of civil disobedience acts.

Gandhi's practice taught that nonviolent actions require more courage than violent actions and, if nonviolence shall have the intended power, it needs creativity as well as personal suffering. Leaders need extensive training as well as education. The formula for Gandhi consisted of three factors: fearlessness not fear, self-sustainability not dependency, and identity not emptiness and alienation. The protesters have to conquer their own fear of suffering, and even death; they should not depend upon the suppresser in any way and, finally, they should have a core that they believe in within themselves – an inner flame, a truth. One has to strengthen oneself, since the power over oneself is a prerequisite for power over another. Any serious conflict should be transformed to a higher level, involving transforming the opponents into partners. The strategy entails breaking cooperation with the other as suppressor, while at the same time developing a good relationship with the other as a human being. The action should not be posed as a threat against the suppressor as a person, only against the system. Nonviolence as *ahimsa* (Hindu) means doing positive acts of love toward the suppressor. The ethics behind is that the means are essential, and they can never be compromised for a noble purpose. Gandhi's soft law involved always acting in a way that realizes the good, the nonviolence in everybody, including in yourself. Your actions should mirror the society you want to obtain, not the eternal spiral of violence (See Gandhi, 1999, in particular "the introductory essay" written by Johan Galtung).

Those thoughts had a major influence on Næss's philosophy and participation in nonviolent actions to protect important wilderness values. When Næss acted against the regulation of a beautiful waterfall, Bardøla, in 1970, he tied himself with 300 other demonstrators in protest against the building of a dam and the subsequent destruction of the waterfall. The actions were not popular in Norway because the government enjoyed a high level of legitimacy in the postwar era. But the protesters had a major political significance in the Norwegian society because they had increased the level of awareness about environmental issues, and contributed to undermine the high position the experts enjoyed. In an action in 1980 to preserve the river of Alta in the northern part of Norway, which partly passed alongside the land of the Sami population, we witnessed five young men of the Sami population who participated in a starvation strike clearly inspired by Gandhi's ideas.

Eco-topia? Real places or romantic nonsense for urban dreamers?

Abu Dhabi is a city state in the United Arabic Emirates where there is a plan to construct a modern ecocity – Masdar City. Is it a real ecotopia? The purpose is to develop a genuinely green city with zero carbon emissions and with no garbage. The hypermodern technology used in United Arabic Emirates is almost the opposite of deep ecological thinking, because it requires a massive use of resources and is based upon the belief that high-tech strategies necessarily represent sustainable solutions to environmental problems. However, high-tech, high-cost, complex hardware and infrastructure lead to a high degree of specialization, elitist expertise, complexity, and vulnerability, and is aligned with a shallow technocratic paradigm. In contrast to Abu Dhabi, close to the city of Bergen in Norway, we find the "Bergen Ecological Village", using permaculture as one of the key concepts. "Perma" is the first part of the word permanence, and the permaculture philosophy holds that the activities should typically have a sustainable character – an indefinite horizon. Other interesting examples are that the activities typically should have a sustainable character – an indefinite horizon. Other interesting examples of practices that accord with deep ecology are movements like Cittaslow, Slow food, Slow travel, Slowdown etc. A common denominator of such movements is the mindset that "less is more" and downshifting means an improvement in life quality.

Is it possible to create a political and economic system with a green responsible policy? Since trees and plants store and absorb carbon, forest planting and protection of rain forest will be part of the solution. However, the possibility of green practice is negated by the centralistic technical-industrial way of production. We need a movement away from the economic-growth paradigm and a culture dominated by competition toward one focused on quality of life, where self-realization for all livings is at the core (Ims and Jakobsen, 2008). We will need to explore which system will work best, but we already know that the role of the economy will be reduced as compared with the prevalent thinking in the western world. We know that the economy will have to adapt to local conditions and local cultures and be based upon economic needs – not demands orchestrated by huge marketing departments. We may develop a model that transcends the difference between socialism and capitalism and hints at a deep ecological community using two simple dimensions, the vertical/horizontal and the collective/individualistic (Galtung, 1999). The first dimension concerns the degree of egalitarian order, whereas the second dimension concerns the degree of individual freedom and self-fulfilment. This model results in four types of societies (see Table 1).

To satisfy the criteria for deep ecology we need a combination of the "horizontal and individual dimension", or what amounts to a postrevolutionary society. Such a society is still unrealized in practice. However, within such a society we may seek maximal self-realization for all beings, in combination with minimal differences along the vertical axes, thus eliminating big bureaucratic institutions. It might be worthwhile to research eastern traditions and cultures to study relevant concepts for a new economy. Many interesting concepts do exist, for example "sufficiency economy" (Chaisumritchoke, 2007) and "mindful economy" (Magnuson, 2007), and Buddhist economics where simplicity and nonviolence, want negation, and purification of human character are central values, and where the main task of economizing is to provide peace and permanence (Zsolani, 2007; Zsolnai and Ims, 2006).

We may look upon deep ecology as a kind of strategy toward creating a sustainable future. Such a strategy will place an emphasis on culturally appropriate soft and intermediate technologies and make it essential for both human beings and all living creatures to realize their potential within a healthy ecosystem. The eight points from the platform level may inspire

Table 1. Four types of societies

	Vertical	**Horizontal**
Collective	Conservative	Revolutionary
Individual	Liberal	Postrevolutionary

the formulation of some of the essential ideological issues in our time. How can we care for future generations, determine what humans' place in nature is, and arrive at technology and consumptions patterns that are sustainable?

Bibliography

Bateson, G. (1972). *Steps to an Ecology of Mind.* New York: Ballantine Books.

Chaisumritchoke, S. T. (2007). Sufficiency Economy, The King's Philosophy: An Application of Buddhist Economics to Develop Thai Local Pharmaceutical Industries for Sustainable Well-being, *Society and Economy*, Vol. 29, No. 2, August, pp. 235–252.

Drengson, A. and Inoue, Y. (1995). *The Deep Ecology Movement: An Introductory Anthology.* Berkeley, CA: North Atlantic Publishers.

Galtung, J. (1999). "*Pluralism and the Future of Human Society*", and "*Two Ways of Being Western: Some Similarities between Liberalism and Marxism.*" Refers to two papers by Johan Galtung: The two papers can be ordered from the Peace Research Institute in Oslo.

Gandhi, M. (1999). *Vi er alle søsken.* Med et innledende essay av Johan Galtung, Oslo, Bokklubben Dagens Bøker, (*We are all Brothers,* with an introductory essay by Johan Galtung in the Norwegian translation).

Hegdal, O. A., and Strand Olsen, T. (2001). *Jeg, Arne Næss. Et tegnet liv.* Oslo: Kagge forlag. (*I, Arne Næss. A Drawn Life*).

Ims, K. J., and Jakobsen, O. (2008). Quality of Life – The Golden mean between Materialistic Consumerism and Spiritual Asceticism, in Zsolnai, L. (ed.) *Europe-Asia Dialogue on Business Spirituality, European SPES Cahiers.* Antwerp: Garant.

Magnuson, J. C. (2007). Pathways to a Mindful Economy, *Society and Economy, Journal of the Corvinus University of Budapest*, Vol. 29, No. 2, August, pp. 253–284.

Millennium Ecosystem Assessment. (2005). *Ecosystems and Human Well-being: Synthesis.* (Washington, DC: Island Press. Available at <http://www.millenniumassessment.org/documents/document.356.aspx.pdf>.

Næss, A. (1999a). *Økologi, samfunn og livsstil. Utkast til en økosofi.* (Med innledende essay av Thomas Hylland Eriksen, 5 omarbeidet utgave) Oslo: Bokklubben

Dagens Bøker. (*Ecology, Society and Lifestyle. A draft to an Ecosophia. With an introductory essay by Thomas Hylland Eriksen. 5. reworked edition*).

Næss, A. (1999b). *Det frie menneske. En innføring i Spinozas filosofi.* Oslo: Kagge forlag. (The Liberated Human Being. An Introduction into the Philosophy of Spinoza).

Næss, A., and Rothenberg, D. (1989). *Ecology, Community and Lifestyle. Arne Næss* (Trans. and ed., David Rothenberg). Cambridge: Cambridge University Press.

WWF International. (2008). *Living Planet Report 2008*, <http://www.panda.org/about_our_earth/all_publications/living_planet_report/>

Zsolnai, L. (2007). Western Economic versus Buddhist Economics, *Society and Economy, Journal of the Corvinus University of Budapest*, Vol. 29, No. 2, August pp. 145–153.

Zsolnai, L., and Ims, K. J. (eds) (2006). *Business within Limits. Deep Ecology and Buddhist Economics.* Peter Lang: Oxford.

KNUT J. IMS AND OVE D. JAKOBSEN

Deep Authenticity: An Essential Phenomenon in the Web of Life

Introduction

The concept of authenticity has received an increasing amount of attention recently. Taylor (1991) criticizes the increased degree of individualization in modern society, the "malaise of modernity". He claims that we have the presence of an individual in a deeper sense, a "self" that has ties to communities as well as a "horizon of meaning". For Taylor authenticity is the courage to be true to oneself within a horizon of important values, while traditional concepts of individuals are one-dimensional and instrumentalist and do not emphasize the importance of social and dialogical relations with others. Authenticity is often connected to the search for individual "wellbeing", "enjoyment of life", and "quality of life". People living in harmony with their own values and principles and experiencing meaning and purpose in their lives are supposed to be authentic. The fallacy is that individuals are not open to the horizon of significance. One consequence is loss of meaning. We suggest a conception of authenticity that interprets man as a part of a web of life.

Interesting contributions to a new conception of man are offered by Richard Niebuhr (Kaiser, 1995), who makes a phenomenological analysis of man's moral existence. Niebuhr's man is relational, and the self exists in a triadic relation to itself, to society, and to nature. And the appropriate action is a fitting one within the context (Kaiser, 1995). This view has important consequences. It follows from here that the question "Who am I?" is more important than "What is real?" Niebuhr argues that we should reflect upon the former question because self-knowledge is an important avenue to a responsible life. And Niebuhr states that when we ask about responsibility, we ask about ourselves!

In another strand of literature, the "deep ecology movement", the main norm is self-realization for all sentient beings (Næss, 1989). This means that all sentient beings have the right to self-unfolding on their own premises. The mature deep-ecological man has developed an ecological self, a self which is not impersonal but transpersonal and has the ability to identify with all other sentient beings. The ecological self negates the minimal self, which is self-centered, self-obsessed, attached to hedonistic pleasures and, as a consequence, is alienated from itself, from other beings, and nature. One consequence of being an ecological self is that if you hurt nature, you hurt your own interests and, in a deep sense, you hurt yourself! (Næss, 1989; Sessions, 1995; Drengson and Inoue, 1995).

In accordance with deep ecology-thinking, we argue that the interpretation of authenticity should be expanded or "deepened" to encompass a holistic rather than individualistic context. This means that the individual's authenticity is richer and more meaningful because of its connection to the web of life. In other words, how we choose to live our lives is dependent upon our connections within the web of life. Both society and nature will have an impact on human authenticity, on both the individual and the collective levels.

Our argument is based on the presupposition that understanding is always guided by the interpretative framework used, or our background assumptions. According to Rose our interpretative framework is a continually evolving construction that functions to "highlight or bring into focus certain features of experience, moving them from the background to the foreground of experimental awareness" (Rose, 2002, p. 11). This indicates that understanding starts from a set of ontological and epistemological presuppositions, which can be aggregated up to a worldview. In accordance with this view, we contend that authenticity depends on the interpreter's worldview. In the following paragraphs we will state some reasons for these assumptions and conclusions. More precisely we will draw a demarcation line between shallow and deep authenticity.

We will define shallow authenticity as a mode of being based upon a mechanic worldview, indicating that people might be authentic independent of their relations to other living organisms. Shallow authenticity is a characteristic of an individualistic mindset. In contrast, we define deep authenticity as based upon an organic holistic worldview, indicating that authenticity is a relational concept. Deep authenticity is a characteristic of the individual strongly embedded in society and nature. To be deeply authentic implies being an integrated part of social and natural systems. In other words, deep authenticity is based upon organic interdependence as described in Næss's "Ecosophy" (Næss, 1989; Drengson and Inoue, 1995) and in Whitehead's "Philosophy of Organism" (Whitehead, 1925/1967, 1927/1985, 1933/1967).

From the statement that everything is intensively interconnected, it follows that a precondition for deep authenticity is cooperation between interrelated, self-realizing subjects. A deeply authentic person is simultaneously free and committed to society and nature. Shallow authenticity is possible within an atomized market based on a mechanic worldview where all entities are isolated atoms with a natural inclination to compete against each other. Deep authenticity is inseparably connected to communicative coexistence between human beings and nature. We will illustrate the discussion with some practical examples. First, we will examine and elaborate organic and mechanic worldviews as concepts in some detail. Next, we will discuss shallow and deep authenticity respectively. The meaning of authenticity depends on the worldview being discussed.

Organic and mechanic worldviews

Referring to Capra's (1982) definition, an organic worldview refers to a reality composed of living entities having a high degree of "nonlinear" interconnectedness. This means that individuals and communities simultaneously create and make each other and require each other. Thurow goes into more detail when he argues that societies are not merely statistical aggregations of individuals engaged in voluntary exchange, but something much more subtle and complicated. "A group or community cannot be understood if the unit of analysis is the individual taken by himself. A society is clearly something greater than the sum of its parts" (Daly and Cobb Jr, 1994, p. 7). According to Whitehead a society is self-sustaining and has its own reason for being. A society is more than a set of entities to which the same class-name applies. "The self-identity of a society is founded upon the self-identity of its defining characteristics, and upon the mutual immanence of its occasions (...) and the creative advance into the future" (Whitehead, 1933/1967, p. 204). Therefore, it is important to ask questions about patterns, organization, rhythm, and flow. We have contrasted some essential differences between organic and mechanic worldviews in Table 1.

We argue that organic thinking is based on the concept that culture is a collective phenomenon rather than a sum of individuals exclusively. Within this complex and dynamic framework, individual behaviour is both multifaceted and context-dependent; hence, accepting the organic worldview has far-reaching consequences for the interpretation of the individual as a self-realizing person in society.

In addition to holistic integration the organic worldview is based upon the concept of inherent values. The motivation for introducing values and purposes into nature is the acknowledgement that materialism cannot give an adequate account of the emergence and meaning of life. In accordance with this reasoning we argue that we have to rethink the status of life in nature and accept that integrated organisms have inherent value. The term "life" refers to the appreciation of things like "self-enjoyment", "freedom",

Table 1. Mechanic and organic worldviews

Mechanic worldview	**Organic worldview**
Linear connectedness	Non-linear interconnectedness
Instrumental values	Inherent values
Deterministic	Co-creation
Physical laws	Creativity
Dualism	Holism
Isolated ego	Extended self
Purposelessness	Purposefulness

"creativity", "purpose", and "subjectivity", derived from the past and aimed at the future. The idea of including human nature as an element in ecosystems results in the notion that variables such as value and freedom can no longer be excluded from the descriptions of nature.

Within the organic worldview *life* and *mind* are interwoven with matter and motion. It is the essence of life that it exists for its own sake, as an intrinsic reaping of value. The point is that we can understand neither physical nature nor life unless we fuse them together as essential factors in the composition of the whole universe. Nothing in nature could be what it is, except as an integrated ingredient in nature as a dynamic whole.

In contrast to the organic worldview, mechanism presupposes that physical matter is identical with reality, and that everything can be explained in terms of imposed physical laws. From this perspective, the social sciences are characterized by the idea that bits of matter are isolated individuals (atomism), related to one another purely externally. "The material universe, including the human organism, was a machine that could in principle be understood completely by analyzing it in terms of its smallest parts" (Capra, 1995, p. 21). Through natural laws, society represents no real unity in itself. Society is nothing more than a mere mechanism based on the interplay among egocentric individuals seeking their own ends ("the economic man").

As a consequence, the explanations based upon the mechanic worldview claim that every biological or social event is a pattern of non-biological occurrences. This formulation cannot be interpreted as an assertion that all organisms are like machines. To avoid some of these problems we are using the term mechanism more broadly than machine. In other words, mechanism is a worldview claiming that physical matter is reality, complete and total. Everything in the universe can be explained in terms of physical laws.

The great forces of nature, such as gravitation, were entirely determined by the configurations of masses. Accepting that the whole universe is completely causal and deterministic has serious consequences concerning the possibilities of human creativity, freedom, and self-fulfillment. Interpreted from a mechanic worldview, "all that happened had a definite cause and gave rise to a definite effect, and … the future of any part of the system could – in principle – be predicted with absolute certainty if its state at any time was known in detail" (Capra, 1997, p. 120). A consequence of this logic is that "dead" nature can provide no reasons, and it aims at nothing.

Today, the mechanical worldview still forms the basis of many scientific disciplines, including mainstream economics. In lifeless nature, problems can be solved within the framework of physical laws. It is, at least in principle, possible to arrive at unambiguous solutions to these kinds of convergent problems. Schumacher (1977) argues that convergent problems have nothing to do with self-consciousness or life functions. In accordance with Georgescu-Roegen (1971) and Daly and Cobb Jr (1994) we contend that this assertion still holds validity for modern mainstream economics.

The mechanic worldview is useful for describing and dealing with delimited physical phenomena encountered in our daily environment. We have argued (Ims and Jakobsen, 2006) that the ontological presuppositions are important for understanding appropriate market behaviour, as indicated in Table 2. We maintain that a mechanic worldview implies competition among actors in the market, as exhibited in cell 1 in Table 2. However, we should be aware of the problems connected to using abstractions based on the limited worldview of mechanism. According to Whitehead we tend to forget that the mechanic worldview is an abstraction and, even worse, we tend to mistake the abstraction for the concrete actuality ("the fallacy of misplaced concreteness"). When emotions and values are missing, we lose the connectedness to living society (cell 3 in Table 2). Also, cell 2 shows a constellation between the mechanic worldview and cooperation, which may be exemplified with different types of collusion. Collusion is an illegal activity because it may lead to price-fixing or the carving up of markets between different companies. In cells 3 and 4 we assume an organic worldview, and in these constellations the partners are perceived as

Table 2. Principles of coordination, in the context of different worldviews

	Mechanic	**Organic**
Competition	1. Efficiency	3. Disintegration
Cooperation	2. Collusion	4. Responsible co-creation

integrated, through dialogical process, and they share common long-term values and interests. The organic paradigm presupposes cooperation (cell 4), which leads to responsible co-creation. The organic worldview involves cooperative communication processes among the partners as being essential, inherent parts of the market (network economy). We claim that market behaviour based on competition will often lead to disintegration and egocentric behaviour. Within an organic worldview the use of one-sided power is destructive (cell 3).

Awareness of the ontological and ethical preconditions can help us to see the limitations of the mechanic worldview. To grasp the whole human being, as an individual and as a member of society, it is necessary to expand the perspective. The organic worldview is more fundamental, as the cosmos is seen as interrelated and connected manifestations of one inseparable reality – always in motion, alive, organic, spiritual, and material at the same time (see cell 4 in Table 2).

Reflections on authenticity

For Taylor authenticity is a kind of self-fulfillment that is inseparable from a horizon of important values. And according to Taylor a self has an identity defined in terms of certain essential strong evaluations that provide the horizon. And humans have the capacity for second-order desires, that is, strong evaluations of desires, and they are responsible for their evaluations. However, the deepest evaluations are the most hidden and most subject to interplay with illusion. Thus, the struggle for self-interpretation is not trivial. Taylor uses the term "deep reflection" as a kind of radical evaluation, as "a reflection which engages the self most wholly and deeply" (1985, p. 42).

The concept "authenticity" has been described and discussed within philosophical existentialism for centuries. Soren Kierkegaard (1843/1978) strongly maintains that we are able to choose ourselves, and we have the ability to move from living a purely aesthetical life, in the sense of being an ego, to a stage of being that is called ethical. The aesthetic man lives in dread and despair because he cannot sense that he is meant for something higher. In contrast, the ethical man truly chooses himself. Individuals who do not live authentically often lose the meaning of life and can be hit by chronic anxiety, boredom, or despair. In our modern societies, people avoiding the responsibility of living authentically, in the worst case, end up anesthetizing themselves with alcohol or other drugs. For most people it is important to live as authentically as possible; however, in our globalized, complex world it is not easy to achieve this goal. In the western world the advertising pressure oriented towards individuals, which is designed to delude the populace into believing that self-realization depends upon consumption habits, is very strong. The intriguing question is, what does it mean to live an authentic life? Some people choose to appreciate more of their own current approach to life, while others might choose to introduce new activities and lifestyles.

From an existentialistic point of view authentic persons must see themselves as fundamentally free, thereby acknowledging responsibility for their actions and lives. The only way to experience freedom is to regard and treat other people as free, since the acceptance of other people's authenticity is a condition of oneself being authentic. Some existentialists argue that authenticity is based upon intimate personal relations – love and friendship. To elaborate the concept of "self-understanding" within an organic worldview, we distinguish between the following dimensions: "inner" and "outer" experience, and "me" and "other" beings, as shown in Table 3.

According to the organic worldview we are integrated parts of the web of life, embedded in human societies and nature. To understand our selves in this perspective requires a multidimensional perspective: First (cell 1), I have to know myself. This is the old Socratic imperative: "Know Thyself!" It is a profound point of departure. Schumacher frames it thusly: I have to

Table 3. Reflected self-understanding (inspired by Schumacher, 1977)

	Inner experience	**Outer experience**
Me	1. My inner world	3. Me in the eyes of others
Other beings	2. Other beings' inner world	4. The world in the eyes of other beings

"feel what I feel like". And the preconditions for understanding my inner self are self-awareness and self-confidence. "Without self-awareness, the exploration and study of the inner man, that is, one's interior world is completely impossible" (Schumacher, 1977, p. 66). Without self-awareness and self-confidence we think and act mechanically, like a programmed computer. The result is that we have no freedom to act in accordance with our own ideas and intentions.

Second (cell 2), I have to be sensitive to the inner worlds of other people and "feel what you feel like" (empathy). We believe that we may know something about the inner life of animals, but still more of what goes on inside other human beings. We do not believe that we are able to know much of a plant's inner life, and we know nothing about inanimate matter. Human society depends on "our ability to understand others and their ability to understand us" (Schumacher, 1977, p. 81). According to a number of wise men in different traditions – Buddhists as well as Christians – it is impossible to understand others without first understanding ourselves.

Third (cell 3), in order to know myself it is necessary to know how I am known by others. Schumacher argues that the only source for knowing myself is my own inner feelings, and that I tend to understand myself as the center of the universe. Having the ability "to see ourselves as others see us, will help us to see our contradictions" (Schumacher, 1977, p. 97). Contradictions are not manifestations of error; instead, they are manifestations of truth. Opposites are a part of real life and we have to learn to keep opposites in mind simultaneously. The best way to gather knowledge about myself is by putting myself in other people's situations. This is possible because we are reflective social beings, living with others.

Fourth (cell 4), according to Schumacher this "is the real homeland of every kind of behaviourism: only strictly observable behaviour is of interest" (Schumacher, 1977, p. 100). Many people believe that this is the only field in which true knowledge can be obtained. The strange thing is that we in cell 4 have most knowledge concerning inanimate matter and least about the inner life of human beings. In cell 2 it is the other way round, we can know most about the higher levels of life and least about inanimate matter. In cell 4 we can observe movement and other kinds of changes in matter, but we cannot observe meaning, purpose, feelings, or values through our physical senses. In terms of maturity we might say that "the more mature the object of study, the less mature the science studying it" (Schumacher, 1977, p. 103). We argue that authenticity presupposes knowledge based upon all the four different cells.

Charles Taylor (1991, 1989) supports our reasoning by arguing that self-creation must be added to self-realization. Creativity is always reflexive and is exercised over and with respect to the self. Since the self is social, creativity is transactional and multidimensional. Therefore, creativity is both self-creative and co-creative. Taylor points out both the good and damaging aspects of the modern development of an authentic self, and he mentions the importance of some moral measurement system anchored in strong evaluations. To become authentic, we have to take part in cooperative dialogues where we exchange our ideas with others and construct our values and beliefs in the course of communicative processes. In other words, to discover and be aware of our authenticity we have to converse with other people. Dialogue is essential if we want to develop our own authenticity and, according to Taylor, authenticity is essential if we want to live a better life. This emphasis on dialogue is in accordance with Niebuhr, who argues that our thinking is an inner conversation where we deal with other concrete, particular selves (Kaiser, 1995).

Concluding comments

To penetrate authenticity we find it illuminating to draw a demarcation line between "shallow" and "deep". By using these concepts, we indicate our inspiration from "the deep ecology movement" (Næss, 1989; Drengson and Inoue, 1995; Witoszek and Brennan, 1999). The difference between shallow and deep ecology concerns the willingness to go to the roots of problems and ask critical questions about oneself as well as modern society with its prevailing institutions. The danger is to confuse the "self" with the "egocentric self". Deep authenticity depends on our ability to flip to an organic worldview where awareness of an "eco-centric self" becomes possible. We try to visualize the radical and necessary change in mindset in Table 4.

We argue that the organic worldview is more suitable than the mechanic for interpreting authenticity. Hidden and unintended consequences are very often connected to advanced technologies. The result is that what we do today will have consequences on many levels for ourselves, for others, for nature, and for future generations. Once we discover that we as individual persons are always relating to other human beings, society, and nature we have to clarify our positions in the web of life. A consequence of ecocentered awareness is that we become more responsible in the social and ecological worlds. To become more responsive openness and dialogue are vital. We need courage to change our interpretative map to discover the interconnectedness in the social and environmental realities. The concept of deep authenticity is essential for grasping these new perspectives. In the context of deep authenticity we are free, and we are aware of the responsibility that is laid upon us. Freedom is a burden because we are called upon to act in response to the environment. But the single individual has to face the situation and, if necessary, break with the dominating norms. In our western world the external pressure to become wealthy and outwardly successful is strong for each individual. To handle the individual, social, and environmental challenges we are facing today we have to transition away from the idea of "economic man" to that of "ecological man".

To be inauthentic is to choose the easiest way and thereby follow the avenue to self-deception – to circumvent and avoid facing the real problems in the world. Even ignoring inconvenient truths, crucial facts about the state of the world hoping that the problems will in the end be resolved by someone else.

Table 4. Shallow and deep authenticity

	Mechanic	**Organic**
Egocentric	1. Shallow Authenticity	3.
Eco-centric	2.	4. Deep Authenticity

In the perspective of marketing and management literature, assumptions about an organic worldview lead to new perspectives for all stakeholders and the relations between them. Responsibility towards all living beings illuminates the inherent value in the web of life. One important consequence is the shift from maximizing shareholder values to increasing the enjoyment of life for all stakeholders in the context of a long-term perspective.

Bibliography

Capra, Fritjof (1982): *The Turning Point – Science, Society and the Rising Culture.* London: Flamingo.

Capra, Fritjof (1995): "Deep Ecology – A New Paradigm", in George Sessions (ed.) *Deep Ecology for the 21st Century.* London: Shambhala.

Capra, Fritjof, (1997): *The Web of Life – A New Synthesis of Mind and Matter.* London: Flamingo.

Daly, Herman and John B. Cobb Jr (1994): *For the Common Good – Redirecting the Economy toward Community, the Environment, and a Sustainable Future.* Boston, MA: Beacon Press.

Drengson, Alan and Yuichi Inoue (1995): *The Deep Ecology Movement – An Introductory Anthology.* Berkeley, CA: North Atlantic Books.

Georgescu-Roegen, Nicholas (1971): *The Entropy Law and the Economic Process.* Cambridge, MA: Harvard University Press.

Ims, K. and O. Jakobsen (2006): Cooperation and competition in the context of organic and mechanic worldview: A theoretical and case-based discussion, *Journal of Business Ethics* (66: 19–32).

Kaiser, R. M. (1995): *Roots of Relational Ethics: Responsibility in Origin and Maturity in H. Richard Niebuhr.* Atlanta, GA: Scholars Press.

Kierkegaard, S. (1843/1978): *Enten – Eller*. Samlede Værker. Bind 2, (*Either – Or*. Collected Work, Vol. 2). Copenhagen: Gyldendal.

Næss, Arne (1989): *Ecology, Community and Lifestyle: Outline of an Ecosophy.* Translated and revised by D. Rothenberg. Cambridge: Cambridge University Press.

Rose, Philip (2002): *On Whitehead.* Belmont, CA: Wadsworth Philosophers Series.

Schumacher, E. F. (1977): *A Guide for the Perplexed.* New York: Harper & Row.

Sessions, George (ed.) (1995): *Deep Ecology for the 21st Century – Readings on the Philosophy and Practice of the New Environmentalism.* London: Shambhala.
Taylor, C. (1985): *Human Agency and Language: Philosophical Papers.* Cambridge: Cambridge University Press.
Taylor, C. (1989): *Sources of the Self: The Making of the Modern Identity.* Cambridge: Cambridge University Press.
Taylor, C. (1991): *The Ethics of Authenticity.* Cambridge, MA: Harvard University Press.
Whitehead, A. N. (1925/1967): *Science and the Modern World.* New York: The Free Press.
Whitehead, A. N. (1927/1985): *Process and Reality.* New York: The Free Press.
Whitehead, A. N. (1933/1967): *Adventures of Ideas.* New York: The Free Press.
Witoszek, Nina and Brennan Andrew (eds) (1999): *Philosophical Dialogues – Arne Næss and the Progress of Ecophilosophy.* Oxford: Rowman & Littlefield Publiners Inc.

KNUT J. IMS

Deep Ecology and Personal Responsibility

Inspired by Arne Næss's writings on ecology, this paper explores his viewpoint about the necessity of a deep ecological approach and the need to take a personal ecosophical position. According to Næss, deep ecology is proceeding in two directions; it is trying to bring about a change and it is trying to develop an alternative philosophy called "ecophilosophy". While ecophilosophy is a descriptive science without fundamental postulates about values, ecosophy includes the development and eliciting of norms. Arne Næss invites everybody to work out their own individual ecosophy, by which he means a philosophy of ecological harmony. This kind of *sophia* or wisdom is normative and contains value priorities as well as norms.

Concerning contributing to practical change, Arne Næss walked his talk and participated directly in society's value struggles. He learned from ecology as a science, but as a philosopher he was very aware of the limits of science and understood that ecology and ecologism could never substitute for philosophical analysis. Any change has political implications and Næss actively tried to fight against the "depolitization" of ecological science from the 1970s onwards. Science in itself cannot provide the ultimate answers to ethical questions about how we should live. In the same manner, we should be aware of the danger of ecologism, and we should also see economism as a threat to our western society.

In the spirit of Arne Næss this paper attempts to broaden our understanding of a deep ecological approach and emphasizes the need to develop a personal world view. Today there is an urgent need to create an alternative pathway for the western world to develop along (see for example, Zsolnai, 2011). *The Living Planet Report* (2008) presents strong evidence that the exploitation of natural resources and the level of consumption in western countries have led to a situation of huge ecological overshoot. This is one reason why the thinking promoted by deep ecology has garnered many supporters over the last decades. There are now green movements, foundations in different parts of the world, educational programs, discussion groups, and even a church of deep ecology.

Deep ecology and its inspirational sources

Deep ecology was not invented by the Norwegian philosopher Arne Næss. As Glasser writes, Næss has given old ideas "a name, structure, and a philosopher's insight to a minority tradition, a quiescent utopian social movement that has been with humans since at least Buddha, Chuang Tzu, Asoka, Saint Francis of Assisi, Thoreau, John Muir, and Aldo Leopold" (Glasser, 2011, p. 58).

Arne Næss himself referred to some of the inspirational sources behind deep ecology by using the metaphor "the deep ecological tree", by which he meant a tree with long and strong roots and different branches consisting of ideas from Hinduism, Confucius, and Buddha on the one hand, and Aristotle, Heidegger, and Spinoza on the other.

In accordance with a central Aristotelian view, every living creature has a *telos,* a goal, and can be expected to realize itself and to flourish. Gandhi, by giving concrete examples of how to live a simple, but a rich and non-violent life, was a least significant source of inspiration for Arne Næss. Gandhi's metaphysics represents the concept of oneness. It implies that everything is interrelated and that only a holistic worldview is adequate. Such a worldview naturally leads to the undertaking of practical action on a non-violent basis.

Spinoza (1632–1677) is another major source of inspiration for Arne Næss. In his book on Spinoza, Arne Næss (1999b) writes that many of his intellectual friends have great respect for Spinoza, but wonder how it is possible for anybody to take Spinoza's metaphysics seriously today. Spinoza claims that we have an ability to identify with others and thereby come close to all kinds of life, assuming *biophilia* to be a central human attribute. Contemporary ecologist Wilson writes that biophilia is "the inborn affinity human beings have for other forms of life, an affinity evoked, according to circumstance, by pleasure, or a sense of security, or awe, or even fascination blended with revulsion" (Wilson, 1994, p. 360; cited in Glasser, 2011, p. 53).

As a philosopher, Arne Næss was always asking fundamental questions like "how can we deal with the essential ideological struggle of our time?", "how can we care for future generations?", "what is the place of humans in nature?", and "which technology and consumption pattern is appropriate?" It was in order to get to the roots of such problems that deep ecology was named and structured.

Shallow versus deep ecology

In the early 1970s Arne Næss criticized dominant environmental thinking for only being concerned with short-term problems and measures. What we need, Næss said, is a *long-term, deeper approach*. Næss argued that we need to distinguish between deep and shallow approaches to environmental problems. *Shallow ecology* represents a technocratic attitude to pollution control and resource depletion by supporting rules like "the polluter pays" and assuming that symptoms can be treated through technological fixes. It follows that ecological footprints will be smaller and thus, after some time, there will be no serious ecological problems at all. Such a shallow approach assumes an anthropocentric worldview in which human affluence is vital. As a consequence, levels and patterns of production and consumption are not challenged.

In contrast, *deep ecology* assumes a relational, total field perspective that fits into a non-reductionist, non-anthropocentric worldview. It focuses on the underlying causes; *the roots of the problems*. Accordingly, the cure is more radical. The prescribed medicine is to change the basic ideological structure, which ultimately means changing how we as humans regard ourselves. Thus, deep ecology redefines the very notion of self as a subject and opens it up to transformation into an eco-self.

To sum up, deep ecology is both a philosophical approach and a campaigning platform. It is a process of reflection leading to action.

Arne Næss – the man and philosopher

Arne Næss was born in 1912 and studied in Oslo, Paris, and Vienna. At the age of 24 he took a doctoral degree and at 28 became the first full professor of philosophy at the University of Oslo. For 30 years he served as chair of the philosophy department in Oslo. At the age of 58 he retired and became an environmental activist and was directly involved in many demonstrations aimed at protecting the wilderness, rivers and waterfalls in Norway. At 92 years of age Arne Næss was involved in a demonstration in front of the Norwegian parliament.

Uebel (2011) emphasizes that Arne Næss was "a kind of metaphysician" who did not exclude "conceptual poetry" from metaphysics. Næss leaped from logical empircism in his early career to ecophilosophy as a mature man. As Uebl argues, Næss wanted philosophy to be open to the "questions of life". In Næss's own words, he was "looking for wisdom through the study known as ecophilosophy". Næss did not stop with philosophy, but strived for ecosophy– "a total view inspired in part by the science of ecology and the activities of the deep ecological movement". Ubels conclusion is that Næss remained remarkably consistent throughout his philosophical career. He succeeded in carrying out his early plan to develop an *a-metaphysical* philosophy. This attitude accords with the approach of Wittgenstein: philosophy should not be regarded as a system of distinguished propositions, but primarily as the process of clarifying meaning (Uebel, 2011, p. 101).

In his work on ecosophy, Næss emphasizes that we need to undertake normative evaluations and that objective science cannot provide us with the principles for action. "Without an ecosophy, ecology can provide no principles for acting, no motive for political and individual efforts" (Næss, 1989, p. 41). Also, experts need norms in order to act. For Næss this awareness led to the need to "take greater account of the pragmatic dimension of his philosophy than is common for academics. Exercising powers of persuasion over more than one's colleagues (…) became an essential part of his task" (Uebel, 2011, p. 79).

In Næss's many works it is not difficult to see the underlying reasoning that explains his behaviour as an activist engaged in civil disobedience. For

Næss it was clear that one should not reduce himself or herself to only being an instrument for other people's projects. Accepting that we daily decide between conflicting considerations and interests, Næss emphasized that working for a "more ecologically responsible personal and societal lifestyle is not only the ecologist's job. Nor the philosophers. We should all do it together, as generalists rather than as specialists. We all have a certain amount of practice in choosing courses of action." Also, Næss stresses that "to act in an organization merely as a functionary is to renounce human dignity in favor of the anonymity of the automaton" (Næss, 1989, pp. 44–45).

Ecosophy – the wisdom of living systems

Ecophilosophy may be defined as "the utilization of basic concepts from the science of ecology, such as complexity, diversity and symbiosis to clarify the places of our species within nature through the process of working out a total view" (Næss, 1989, p. 3). The central ecological principles are *unity* and *diversity*. This means that differences in perspectives should not be lost in unity. The environmental movement is stronger if a concise set of "principles are derived from a variety of world-views". This position reflects a pluralist view.

The word *ecosophy* comes from the Greek terms *oikos* and *sofia* which mean "house" and "wisdom", respectively. In ecology, "eco" has a much wider meaning than housekeeping and can be interpreted as caring for the Earth (Næss, 1999a, pp. 18–20). Ecosophy is a personal philosophy; a reasoning process that may lead to something like the platform of deep ecology. Given that it is a personal philosophy, there may be many other ecosophies (A, B, C, D, … X). Arne Næss's own system of reasoning is called ecosophy T. The T symbol stands for his mountain hut, Tvergastein, but it is the personal nature of the system that is the main point (Næss, 1989, p. 4). Næss acknowledges that every person may have a different, distinct, and valid ecosophy. An underlying assumption is that, as people, we pretend to act and decide on the basis of having a total overview.

The four-level system

To better understand the deep ecological worldview, it may be fruitful to consider the four levels it is said to be composed of. The most basic level is the metaphysical level. The second is the platform level and the third level relates to the policy level. The fourth level refers to level of political action (Table 1).

Table 1 is a simplified form of Næss's Apron Diagram (See Drengson and Inoue, 1995; Næss, 1995).

The logic behind the framework is that there should be a continuous process of going back and forth between the four levels to keep our understanding and practice in harmony.

On *level one,* posing deep questions and exploring ultimate premises and norms is a primary task. Having established a more or less articulate position on the first level we may move to the lower levels. This framework admits a great diversity of ultimate philosophies. We do not have to subscribe to the same ultimate ecological philosophy in order to work cooperatively. The frontier is long and each person may develop his or her own premises.

Næss invites everybody to work out his or her *individual ecosophy*, which is openly normative and contains norms, value priorities, and hypotheses.

Næss's view begins with one basic norm: *self-realization*! This is understood as "self-realization for all beings!" The self that should be realized for humans is not the ego self but the larger ecological self.

Table 1. Four levels of questioning and articulation

Level I	Ultimate premises	Taoism, Christianity, ecosophy T, etc.
Level II	Platform principles movement	Peace movement, deep ecology movement, social justice movement, etc.
Level III	Policies	A, B, C, etc.
Level IV	Practical actions	W, X, Y, etc.

Næss focuses on the human ability to identify with a larger sense of self. Humans naturally have this capacity. This can be observed as a cross-cultural phenomenon. Supporting the principles of the peace movement is also a part of Næss's philosophy. But Næss stresses that social justice is not enough. We have to produce and consume less, and tread more lightly and more wisely on the Earth. His motto is "simple in means and rich in ends". This speaks for quality of life instead of standard of living, and celebrates the virtues of slowness and smallness in an age of speed and scale.

To sum up, Næss's holistic worldview negates the dominant metaphysics of our age which sees humans as essentially different from the rest of nature. Næss claims that humanity is inseparable from nature: *If we hurt nature, we hurt ourselves.*

Taking a closer look at level two, the platform level, we can explore what unites different individuals and groups in the deep ecology movement. The platform is typically summed up using eight points. This paper lists the eight points next and elaborates on some of the points that may be more controversial and difficult to understand.

1. The well-being and flourishing of human and nonhuman life on Earth have value in themselves – independent of their utility to humans.
2. Richness and diversity of life forms are also values in themselves.
3. Humans have no right to reduce this richness and diversity, except to satisfy vital human needs.
4. We need a substantial decrease in human population.
5. The present interference with the nonhuman world is excessive.
6. Policies which affect basic economic, technological, and ideological structures must be changed.
7. The ideological change is mainly that of appreciating life quality (dwelling in situations of inherent value) rather than insisting on an increasingly higher standard of living. There will be a profound awareness of the difference between big and great.
8. Those that support these points have an obligation to directly or indirectly try to implement the necessary changes in a nonviolent way.

Another great environmentalist, *Aldo Leopold,* in his book *A Sand County Almanac* writes about an experience with a wolf that triggered a total re-orientation of his life and, finally, the development of an eco-centric land ethic (Leopold, 1966).

The moral of Leopold's story corresponds nicely with Næss's writings, and the story goes like this: The US government had developed a rational, scientific policy for eradicating wolves from the entire US, and Leopold was appointed to implement this policy. One morning, he was out with some friends on a walk in the mountains. Being hunters, they carried their rifles with them in case they got a chance to kill some wolves. Around lunchtime they sat down on a cliff. Soon they saw a group of a wolves. They took up their rifles and began to shoot excitedly into the group, but with little accuracy.

> In a second we were pumping lead into the pack, but with more excitement than accuracy: how to aim a steep downhill shot is always confusing. When our rifles were empty, the old wolf was down, and a pup was dragging a leg into impassable slide-rocks.
>
> We reached the old wolf in time to watch a fierce green fire dying in her eyes. I realized then, and have known ever since, that there was something new to me in those yes something known only to her and to the mountain (...); I thought that because fewer wolves meant more deer, that no wolves would mean hunter's paradise. But after seeing the green fire die, I sensed that neither the wolf nor the mountain agreed with such a view. (Leopold, 1966, p. 130)

Two decades later, a masterpiece of a book on *land ethics* by Leopold was published posthumously which stated that: "A thing is right when it tends to preserve the integrity, stability, and beauty of the biotic community. It is wrong when it tends otherwise" (Leopold, 1966, p. 240).

Supporters of deep ecology stress the importance of such spontaneous experience (seeing the fierce green fire dying in the wolf's eyes). A key aspect of these experiences is the perception of gestalt networks of relationships. In such gestalt we feel a strong sense of *wide identification,* an increased sense of empathy and a natural inclination to protect non-human life.

Point seven underscores the fact that we have to stop the thoughtless consumption of material goods, which have had a devastating effect on the environment and do not lead to increased happiness. "More is always better" is a tragic fallacy believed by much of the western world (Zsolnai, 2011). This point involves the need to shift from a welfare approach (measured by economic growth in GDP) to a quality of life and happiness approach (for a review of this literature, see Ims, 2013). Often, more is less and less is more.

Barbara McClintock, who won a Nobel Prize for her work in biology, writes that "one must have the time to look, the patience to hear what the material has to say to you, the openness to let it come to you. Above all one must have a feeling for the organism" (see Hallen, 1995, p. 207). Mclintock says: "no two plants are exactly alike. They are all different (...) I know them intimately and I find it a great pleasure to know them" (Keller, 1983 cited in Hallen, 1995, p. 207). Hallen emphasizes that this is an example of embracing the world, which is quite different from desiring to conquer it. While Francis Bacon expressed the desire to dominate, manipulate, and control, or to "put nature on the rack and torture her", Hallen and McClintock's love of plants represents an alternative tradition of science which we need to understand and emulate. This view stands in strong contrast to a science based on detachment and the division between subject and object, between self and the other.

Human happiness does not arise from consuming the maximum amount of material goods. We can learn from Buddhist economics that consumption should be guided by the "Middle Way". "The effective path lies between the extremes of hedonistic self-indulgence and sensual pleasure, and self-mortification or asceticism" (see Daniels, 2011, p. 46).

A crucial question to ask is this: how do the supporters of these eight points contribute to the necessary and radical changes? Transition strategies are vital for reducing our ecological impact. In the field of business, organizations like *The Natural Step* are working within such a framework.

So what are the main critiques of these ideas? Criticism has emerged concerning androcentrism (or patriarchal values), population control, and accusations of misanthropy when animals and humans are presumed to have an equal footing (see Zimmerman, 1995).

Concerning the critique of androcentrism, feminists may maintain that the central dichotomy is not the "human-nonhuman" relationship, but the "male–female" one. The dichotomy between humans and nonhumans results in a lack of attention being given to such issues as the domination of humans by other humans and the exploitation of women by men. The deep ecology movement should give priority to deeply questioning the validity of socio-cultural institutions that are based on traditional patriarchal values.

Strategies for changes

There are many ways to work for change. We can:

- Live according to ecological ideals.
- Encourage compromise between the present and the ideal state, nurturing the coming into existence of a realistic transition path to a new society.
- Try to change the system directly, for example via political action and acts of eco-sabotage.

Arne Næss wrote about the important balance between actively participating in direct actions and working more from day-to-day with less spectacular means. A main aspect of any action is to attract the attention of the public. While an action is part of a campaign and a campaign is part of a movement like deep ecology, it is necessary to adhere to strict nonviolent actions. The possibility of success is highly dependent on the level of nonviolence in the actions, campaign, and movements. In this context it is essential to contact the opponent before a direct action. Næss refers to the Alta (trying save the river) demonstrations (1980–1981) in arctic Norway, where the powerful opponents were continually contacted for at least nine years of the campaign, and during the direct actions the opponents were treated with coffee and invited to discuss the action in order to avoid misunderstanding. Næss had learnt from the Gandhian approach to "maximize contact with your opponent!" This strategy fits well with being a philosopher who always has to argue for his positions, acknowledging the *limits of science.* Any change has political implications and in Næss's words, "we need to fight against depolitization". Governments often hire experts from natural sciences like physics, chemistry, and mineralogy to avoid questions of how to change our societies in the direction of sane ecopolitics. Ecology as a science, concerned with facts and logic alone, cannot answer ethical questions about how we should live. We have to go to ethics, the practical branch of philosophy, to get answers to such questions.

Ingebrigtsen and Jakobsen (2004, pp. 50–52) argue that value pluralism is important and that it is crucial to see different sectors living in reciprocal tension with each other in order not to be colonized by economic models, economic concepts, and fixed interpretations. They differentiate between economy, nature, and culture, seeing these sectors as being involved in a simultaneous complex interplay, although they acknowledge that the sectors have characteristic traits that make them genuinely different. One implication of this is that the interplay between economy and culture has to be based on their characteristic value differences, and that economic values should not be mixed with or displace cultural values. The same relationship is important for the interplay between economy and nature.

From an ecosophical perspective it is important that neither economy nor ecology is accepted as the dominant view. Economy and ecology should be the *means* within an ecosophical framework. In the following illustration such a framework is presented as a hierarchy (Figure 1).

In the spirit of deep ecology, humans are equal in value to other sentient beings. However, this position does not imply that we as humans can drop our burden of responsibility. As human beings we have immense

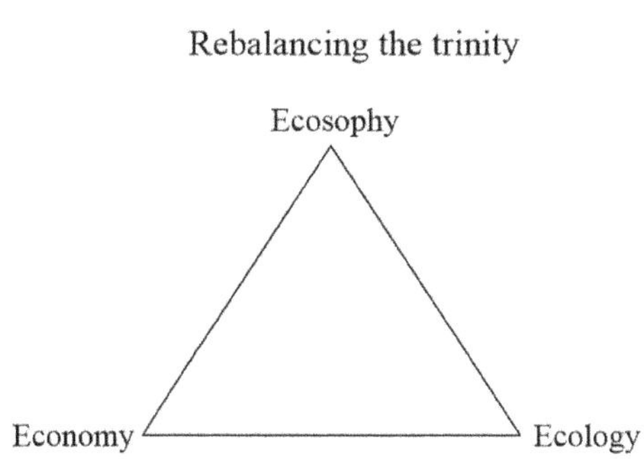

Figure 1. The hierarchy of ecosophy, economy, and ecology.

power in the world and therefore a far-reaching responsibility (*Noblesse oblige*!). The existing pattern of production and consumption is disastrous and unsustainable. We need a radical change of mindset. As Glasser maintains, "Nature can-and will continue to be viewed in instrumental terms as a resource and waste sink, but its use must become framed by the larger context of nature as a partner, friend, inspiration, mentor, and final arbiter" (Glasser, 2011, p. 55). We therefore need to ask a deep question: what is good for the community of life on Earth?

The story of Leopold's shooting of the wolves shows that there is *hope* for a change in mindset. Sometimes radical changes occur sooner than anybody thinks possible. In Norway there are strong instrumental and institutional forces that try to reduce political issues to problems to be solved by experts. Vital issues are often not debated in an open and democratic manner. This practice implies that specialists follow established norms and do not take ecological issues seriously. Often the prevalent norms satisfy rather narrow economic conditions as if the cheapest solution were always the best. Experts from different fields can make important contributions as regards clarifying the means that are appropriate for achieving a given goal. However, economists and ecologists should be engaged and take explicit positions about controversial issues in the society, not let professional politicians debate the fate of society alone. Experts are not better equipped than non-experts to decide which values and norms should be part of the value system of society. Experts should participate in the process of deciding what goals and priorities local communities have, as well as society as a whole.

Ecosophy is a personal position, a personal total worldview. Everybody should work out a worldview and act in accordance with it. It is not possible to articulate a complete "total view", but by clarifying fundamental norms and values one will be better able to see and respond to the challenges of life in the ecosphere. As Arne Næss wrote, "the ecopolitical frontier is immensely long ..." (Næss, 1989, pp. 161–162). It follows that there is work for everyone to change the system by using his or her own capabilities.

Bibliography

Daniels, P. 2011. Buddhism and sustainable consumption. In *Ethical principles and economic transformation – A Buddhist approach.* Vol. 33, ed. L. Zsolnai, 35–60. Heidelberg: Springer, Series. Issues in Business Ethics.

Drengson, A., and Y. Inoue. 1995. Introduction. In *The deep ecology movement: An introductory anthology,* ed. A. Drengson and Y. Inoue, xvii–xxviii. Berkeley: North Atlantic Books.

Glasser, H. 2011. Næss's deep ecology: Implications for the human prospect and challenges for the future. *Inquiry* 54 (1): 52–77.

Hallen, P. 1995. Making peace with nature: Why ecology needs feminism. In *The deep ecology movement, an introductory anthology,* ed. A. Drengson and Y. Inoue, 198–218. Berkeley: North Atlantic Books.

Ims, K. 2013. From welfare to well-being and happiness. In *Handbook of business ethics. Ethics in the new economy,* ed. L. Zsolnai, 217–247. Oxford: Peter Lang Academic Publishers.

Ingebrigtsen, S., and O. Jakobsen. 2004. *Økonomi, natur og kultur.* Oslo: Abstrakt forlag.

Leopold, A. 1966. *A Sand County almanac. With other essays on conservation from Round River.* New York: Oxford University Press.

Næss, A. 1989. *Ecology, community and lifestyle. Outline of an ecosophy.* (Trans. and revised by D. Rothenberg). Cambridge: Cambridge University Press.

Næss, A. 1995. The apron diagram. In *The deep ecology movement: An introductory anthology,* ed. A. Drengson and Y. Inoue, 10–12. Berkeley: North Atlantic Books.

Næss, A. 1999a. *Økologi, samfunn og livsstil,* 5th revised edn. Oslo: Boklubbens Dagens Bøker.

Næss, A. 1999b. *Det frie menneske. En innføring i Spinozas filosofi.* Oslo: Kagge Forlag.

Uebel, T. 2011. A kind of metaphysician: Arne Næss from logical empiricism to ecophilosophy. *Inquiry* 54 (1): 78–109.

Zimmerman, M. E. 1995. Feminism, deep ecology, and environmental ethics. In *The deep ecology movement, an introductory anthology,* ed. A. Drengson and Y. Inoue, 169–197. Berkeley: North Atlantic Books.

Zsolnai, L. 2011. Why Buddhist economics? In *Ethical principles and economic transformation – A Buddhist approach.* Vol. 33, ed. L. Zsolnai, 3–20. Heidelberg: Springer, Series. Issues in Business Ethics.

KNUT J. IMS AND OVE D. JAKOBSEN

Peace in an Organic Worldview

Introduction

Peace economics concentrates on a reduction of conflict and on peace enabling structures in the economic sphere. To be more specific, peace economics, inspired by Johan Galtung's work (2007, 2012a, 2012b) and Santa Barbara et. al. (2009) focuses on reducing the conflict between the economy and the social and natural environment in order to increase the quality of life and wellness for all living beings. In this article we discuss to what extent authenticity and cooperation may contribute to reduce the number and levels of conflicts and promote peaceful, non-violent, and sustainable relations between the economy and the social and natural environment.

We will discuss peace economics on two levels: on the micro level related to the individual economic actor, and on the macro level related to the systemic interplay between economic actors. On the micro level, we will make a distinction between shallow and deep authenticity, and on the macro level we will distinguish between competition and cooperation. Our central argument is that a change from shallow authenticity and competition towards deep authenticity and cooperation presupposes a paradigmatic shift from a mechanical to an organic worldview.

In order to elaborate on these issues we have structured the article in the following way. First, we give a brief description of relevant literature to put our topic into the context. Second, we present elements in Galtung's peace theory, which is summed up in a "peace formula". Third, we present and discuss some defining characteristics of mechanic and organic worldviews. Fourth, we introduce shallow and deep authenticity as relevant concepts in peace economics. Fifth, we focus on the differences between competition and cooperation in relation to peace economics. Finally, we discuss the four possible connections between the two dimensions. We conclude that peace is most likely in the combination between deep authenticity and cooperation, and a war is characterized by the combination between shallow authenticity and competition.

The metapsychological roots of war and peace

The purpose of the conference "Business for Peace – Strategies for Hope" is commemorating the centennial of the First World War. What was the cause of the destructive world war in which twelve million people were killed? Was the war an accident, designed, inevitable, or planned? What should be learned from the First World War to avoid new wars? The immediate reaction to the assassination plot of Archduke Franz Ferdinand and his wife Sophie on June 28, 1914 in Sarajevo was that Austria-Hungary put forward a number of claims to Serbia. The Serbian government would not accept these demands and war broke out. Some political leaders hoped for a local, short lasting war. However the chains of the events lead to a global war that lasted four years.

It is naive to believe that the real cause is to be found in the assassination. According to Eric Fromm:

> The brutal and massive violence of the "Great War" was due to a mixture of economic interests, ambition, and vanity on the part of the leaders, and a good deal of stupid blundering on all sides. But once the war had broken out it became a "religious" phenomenon. The state, the nation, national honor, became the idols, and both sides voluntarily sacrificed their children to these idols. (Fromm, 1973, p. 206)

Webel (2007) argues that modern mainstream psychology has not been geared to understand and solve the most important conflicts of all – the conflicts that lead to war. However, in early psychoanalysis we find some exceptions. In 1914, Freud tries to understand new dimensions of human aggression based upon his own theory (Fromm, 1973), and a famous exchange of letters between Albert Einstein and Sigmund Freud took place in 1932 (see Einstein and Freud, 1933, p. 193): "Warum Krieg? Ein Briefwechel".[1]

"Metapsychology" is one term used by Freud who wrote several essays just after the start of the First World War. Here Freud develops his tripartite "structural theory" in which the ego, the superego, and the unconscious are in continuous interaction. This is Freud's theoretical effort to provide a portrait of the dynamics of emotional life. Webel (2007) applies and revises Freud's theory in the field of political psychology of peace and conflict formation. He claims that there are three dialectical, dynamic "spheres" of greater or lesser peace: (i) the "inner" peace that is the mental and emotional life of the individual defined as "unconscious, pre-conscious and conscious thoughts, impulses, needs, desires, and perception"; (ii) the "outer" economic arena that is macro-economic and political forces; and (iii) interpersonal peace that is the field of human interaction in everyday life and work. States of inner peace are characterized by low degrees of inner conflict, but even a psychologically sane person has difficulty maintaining his/her harmony in a pathogenic environment. The interpersonal sphere mediates inner and outer peace. Being at peace is possible, but difficult in an environment in which cultures are desperately poor and war-ridden. These spheres of inner, outer, and intersubjective peace are never static and always in interaction (Webel, 2007, p. 11).

Webel's beliefs seem to be in accordance with the Dalai Lamas' writings: "Peace must first be developed within an individual. And I believe that love, compassion, and altruism are the fundamental basis for peace .... Once these qualities are developed within an individual, he or she is then able to create an atmosphere of peace and harmony. This atmosphere can be expanded and extended from the individual to his family, from the family to the community and eventually to the whole world." (Dalai Lama in *Thich Nhat Hanh*, 1991, p. vii, as cited in Webel and Galtung, 2007, p. 3).

In several books, Fromm (1955, 1956, 1960, 1973) develops and revises Freud's original ideas in the direction of what he calls a humanistic psycho analysis, emphasizing that progress can only occur when changes are made simultaneously in the economic, socio-political, and cultural spheres. One concept developed by Fromm is *narcissistic character*, described as the opposite pole to love, reason, and objectivity. Fromm writes that "the insane person is the one who has completely failed to establish any kind of union" (Fromm, 1955, p. 30). The necessity to unite with other living beings, to be related to them, is an imperative need on the fulfilment of which man's sanity depends. For the narcissistically involved person there is only one reality, his own thoughts and processes, his own feelings and needs. The world outside is not experienced or perceived objectively, that is as existing in its own terms. The insane person has lost contact with the world and has withdrawn into himself.

"Alienation" is another important concept used by Fromm, defined as the mode of experience in which the person experiences himself as an alien, estranged from himself: "He does not experience himself as the center of his world. The alienated person is out of touch with himself as he is out of touch with any other person." A related concept used by Fromm (1955) is *identity*. Fromm emphasizes that the development of western culture in the direction of individuality has its problems. The hope was to make the individual politically and economically free, but this hope was not always fulfilled. The original clan identity had disappeared, and to feel "I" in the sense that a person was the center and active subject of his own power was problematic. According to Fromm, "many substitutes for a truly individual sense of identity were sought for" ("I am a businessman"), but not found. And the "need to feel a sense of identity stems from the very condition of human existence, and it is the source of the most intense striving" (Fromm, 1955, p. 63).

When no identity is found, one attitude towards oneself is described as a *marketing orientation*. This is a phenomenon in which man experiences himself as a thing and aims to sell himself successfully on the market, competing with other individuals (Fromm, 1955, pp. 141–142). According to Fromm, such a man does not experience himself as a man with love, fear, or convictions, but as an abstraction alienated from his real nature, which fulfils certain functions in the social and economic system. His sense of value depends on his success. Also, the alienated personality will lose a sense of dignity.

Fromm supports his thesis drawing on Ibsen's dramatic poem "Peer Gynt" (Fromm, 1955, p. 186). Ibsen made the loss of self the main theme of his criticism of modern man (see also Ims and Zsolnai, 2010). Peer Gynt is a modern capitalist who chases after profit and material gain, and finally discovers that he has lost his self and that he is like an onion with layer after layer but without a core. Losing himself, he also loses his identity. Then he is attempting to acquire a secondary sense of self by being successful and useful – a saleable commodity fitting well into one of the current patterns.

In his comprehensive opus *The Anatomy of Human Destructiveness,* Fromm (1973) analyzes in depth the character of Adolf Hitler, categorizing him as a "highly narcissistic, withdrawn person for whom phantasy is more real than reality" (p. 427). Hitler failed the *Realschule* and was a drop out. Many young students are drop outs, but the problem with Hitler was that "he was not only not interested in his school subjects, he was also uninterested in *everything*" (Fromm, 1973, p. 431) – perhaps with the exception of his dog later in his life. Fromm emphasizes that the drop out was felt as a personal failure for the young Hitler. But Hitler did not change his behaviour. Instead of "taking a step toward reality he withdrew more into his phantasy world and away from closer contact with people" (Fromm, p. 425).

Another central concept used by Fromm to explain Hitler's character is *necrophilia,* an attraction to death and to everything dead. Necrophilia can be contrasted to biophilia, which means a passionate love of life and all that is alive, whether it concerns a person, a plant, or a social group. In contrast to Freud, Fromm argues that biophilia is the normal impulse (not the death drives/instinct emphasized by Freud), Freud is missing the relational needs, the needs for affective attachment, the need for confirmation of existence and value (Vetlesen, 1994, p. 264). According to Vetlesen, the problem is that the psychic mistreatment or suffering as a child will cause the inability to form affectionate relations with others (Vetlesen, 1994, p. 253). Vetlesen's main point is that such mistreatment leads to a self that is seriously damaged and as a consequence the faculty of empathy fails to develop. Vetlesen also strengthens Fromm's position that necrophilia involves a deficient sensitivity to the pain of others, which leads to stunted growth and psychial "crippledness" (Fromm, 1973, p. 483).

Biophilia is related to Gandhi's *ahimsa,* which strictly means non-killing. But for Gandhi, ahimsa means much more than non-killing. Ahimsa means that you should not violate any sentient being; you should not nourish an unmerciful thought, even with him or her that may regard you as an enemy. For a person who has cultivated ahimsa there is no room for any enemy. Ahimsa is a contrast to curses, spiteful words, unfair and biased language, and unfriendly thoughts. Above all ahimsa is compassion and feeling togetherness with all livings, a willingness to act to the best of mankind (Galtung and Næss, 1994).

Reading Fromm, we see the close relationship of the individual's social conditions and his attitudes and behaviour. The importance of such a social perspective is supported by Bauman (1989). One of the results from reading the literature related to evil and human destructiveness is nicely summed up by Vetlesen (1994), who argues that the ethics of care and the importance of moral sensitivity should obtain a much more central place among moral theories (see Gilligan, 1982).

Towards a holistic peace theory – A model

Peace studies, inspired by the work of the Norwegian founder of the Norwegian Peace Research Institute Galtung (2007, 2012a, 2012b), Fischer (2007), Santa-Barbara et al. (2009) have a dual focus. It focuses on the existence of peace as absence of violence as well as the existence of peace promoting structures. In this part, we will describe and discuss peace theory and connect it to the development of economy with regard to rebuilding a peaceful harmony between economy, nature, and society. We will address the relationship on both individual and structural levels. In addition, we suggest that peace theory is inextricably connected to cultural value systems. We argue that the connection between research in peace and economy, respectively, can contribute to fruitful developments in peace economics.

According to Galtung (2012a), peace could be described as a formula (see Figure 1):

The formula means that peace is holistic, and indicates that the more of the good (Equity × Harmony), and the less of the bad (Trauma × Conflict), the better.[2] The lower limit is zero, there is no upper limit. We define trauma in a wide sense, also involving mistreatment in childhood. Here Freud's metapsychological analyses and Fromm's be used to clarify and understand traumatic experiences. One way to reduce trauma is reconciliation, acknowledging wrongs, wishing them undone, and engaging in true dialogues about how to do it and about future solutions. Some trauma may be deep rooted and caused by the fact that the individual's basic needs as a child were not satisfied, in particular due to lack of love from a mother or father.

It is important to solve conflicts in a sustainable way and it is necessary to make incompatible, contradictory goals more compatible. To promote equity and harmony we have to develop fair background institutions and authenticity that leads to compassion and a sense of enjoying the joy of others and suffering the suffering of the other. Equity involves cooperation for mutual and equal benefits. It also presupposes working against structural violence, by which we mean the difference between the potential and the actual, typically a justice gap. A violent structure impedes the development of the group and the self through a structure that is generally invisible. For example, cultural violence is very hard to change since it is a deep rooted phenomenon that legitimizes different forms of inequality, quite often discrimination between sexes and class differences.

A fruitful concept in peace theory and peace economy is, therefore, a *joint project* in which the numerator of the peace formula is increased; for example, by putting people together joint projects invoke strong I-we cultures, extending structures based on weak individualist cooperation. The structures could be established through social pacts, contracts, or other more or less formal arrangements (Galtung, 2012a, p. 42). If the

$$\text{Peace} = \frac{\text{Equity x Harmony}}{\text{Trauma x Conflict}}$$

Figure 1. The peace formula.

denominator of the peace formula increases by extensive competition among individuals or economic actors, the result decreases the value of joint projects. According to Galtung, "The more pronounced I-culture at individual or collective levels, the less likely positive peace, the more likely direct and structural violence." (Galtung, 2012a, p. 42). Galtung's advice is to write simple rules for both negative peace, avoiding violence and positive peace, building peace with nature in self, couples, companies, states, nations, regions, and civilizations.

An economy based on this peace perspective has implications on both micro and macro levels, and the two levels are strongly interrelated. Concerning conflicts, we need sensitivity within persons. Otherwise frustration in one level may lead to aggression on the other level. In our opinion peace is more like a process than a product. An increase in the values in the numerator and a decrease of the values in the denominator of the peace formula leads to a peaceful structural change process. On the cultural level, according to Galtung, increasing values of the formula could be stimulated by introducing the arts; for example, music, theatre, poetry, painting, and sculpture in the public space. Art can lift individuals beyond the ordinary and unite them, and creating such unity may be peacebuilding (Galtung, 2012a, p.162).

Perspectives on mechanistic and organic worldviews

In this part we will dig deeper into the differences between mechanistic and organic worldviews as context for interpreting changes in micro and macro levels in the economy. The mechanistic perspective is characterized by the idea that pieces of matter are isolated individuals (atomism), related to each other only externally. In a mechanistic perspective, society (and the market) is nothing more than a mere mechanism based on the interplay between egocentric individuals seeking their own ends.

One of the most important consequences of the mechanistic worldviews is that the whole universe is completely causal and deterministic. Everything that occurs has a defined cause and gives rise to a defined effect, and the future of any part of the system may, in principle, be predicted with absolute certainty, if its state at any time is known in detail. There is no capacity for creativity, spontaneity, self-movement, or novelty in the mechanistic worldview. According to Whitehead an important precondition in the mechanistic world view is that coordination in nature is regulated through external rules of connection. This is called the doctrine of laws as imposed (Whitehead, 1967b, p. 113).

In much the same way, the norms regulating the interplay between individuals in an atomistic, materialistic society are based upon mechanical solidarity. Durkheim uses the term "mechanical" to illustrate that the social molecules "lack any movement of their own, as do the molecules in organic bodies" (Durkheim, 1991, p. 84). This does not mean that the term "mechanic" indicates that the solidarity represents an "analogy with the cohesion that links together elements of raw materials, in contrast to that which encompasses the unity of living organisms" (Durkheim, 1991, p. 84).

The mechanistic worldview does not leave much space for ethics or, for that matter, aesthetics. If nature is valueless, there is, on the other hand, no reason to feel deep respect and esteem for natural or artistic beauty. On the other hand, there is no reason to orient our practices around such values. In the field of economics, it is obvious that ethics is often reduced to purely instrumental values. In much the same way as for aesthetics, ethics is regarded as a competitive tool to increase the market value of the firm or the product (Porter and van der Linde, 1995).

Whitehead (1967a) argues that it is necessary to introduce an organic categorical scheme to establish a foundation for a holistic worldview based on ethical values. As a beginning we have to rethink the status of life in nature. The mechanistic paradigm presupposes a dualistic separation of "nature" and "life"; whereas, in the organic concept of nature, "life" and "mind" are interwoven with matter and motion. It is the essence of life that it exists for its own sake as an intrinsic value. The point is that we can understand neither physical nature nor life unless we fuse them together as essential factors in the composition of the whole universe.

Referring to Capra's *The Turning Point* (1982), an organic worldview refers to a reality composed of living entities having a high degree of "non linear" interconnectedness. The interconnectedness is non-linear in the sense that freedom is considered as the claim for self-assertion. Spontaneity and originality of decision are the supreme expressions of individuality. In a civilized society the general end is that the variously coordinated groups should contribute to the complex pattern of community life, presupposing that individual freedom within each group should be possible without the destruction of the ends of the whole society.

As we have mentioned earlier, stronger relations between the actors stimulate growth in the peace formula. This means that individuals and community make each other and require each other at the same time. Thurows develops this further when he argues that societies are not merely statistical aggregations of individuals engaged in voluntary exchange, but something much more subtle and complicated: "A group or community can not be understood if the unit of analysis is the individual taken by himself. A society is clearly something greater than the sum of its parts." (Daly and Cobb Jr, 1994, p. 7).

According to Whitehead a society is self-sustaining, having its own reason. A society is more than a set of entities to which the same class-name applies: "The self-identity of a society is founded upon the self-identity of its defining characteristics and upon the mutual immanence of its occasions (…) and the creative advance into the future." (Whitehead, 1967b, p. 204). Therefore, it is important to ask questions about patterns, organization, rhythm, and flow.

Within the organic perspective, it is reasonable to replace mechanistic solidarity with organic solidarity based on coordination and cooperation. According to Durkheim societies based on organic solidarity are based upon "a system of different organs each of which has a special role, and which are themselves formed of different parts" (Durkheim, 1973, p. 69). Organic thinking is based on the concept of culture as a collective phenomenon, not as the sum of individuals. Economy is like a living organism, which means that its order, structure, and function are not imposed by the environment, but are established by the system itself. Following this reasoning we can conclude that the doctrine of "laws as immanent" (Whitehead, 1967b, p. 112) represents an important condition in the holistic and teleological perspective of the organic worldview.

Shallow and deep authenticity

In this part we give reasons for claiming that developing authenticity could contribute to an increase in the value of Galtung's peace formula defined on p. 405. Taylor (1991) criticizes the prevalent individualization in the modern society as the "malaise of modernity". Taylor claims that we have ignored seeing an individual in a deeper sense, as a "self that has ties to communities" as well as a "horizon of meaning". For Taylor, authenticity is the courage to be true to oneself as a *moral ideal* within a horizon of important values. By a moral ideal Taylor means "a picture of what a better or higher mode of life would be, where 'better' and 'higher' are defined not in terms of what we happen to desire or need, but offer a standard of what we ought to desire" (1991, p. 16). The core aspect of moral ideal involves a contrast with the concept of narcissism as well as hedonism. In the first part, Hitler's personality was described as a narcissistic character. Taylor's definition eliminates shallow, self-centered narcissistic forms of behaviour like Hitler's with the above mentioned definition. Taylor strongly argues against "soft relativism" favoring "some forms of life are indeed *higher* than others" (p. 17). We believe that many lives have been self-absorbed, flattened, and narrowed in the modern individualistic culture. Even if many argue they are living in harmony with their own values and principles, we think it is important to pose the question: What does it mean to live an authentic life?

We believe (Ims and Jakobsen, 2011) in accordance with Taylor that traditional concepts of individuals are one-dimensional and instrumentalist and do not emphasize the importance of social and dialogical relations with others. Authenticity in our sense is a kind of self-fulfillment that is inseparable from a horizon of important values. According to Taylor a self has an identity defined in terms of certain essential strong evaluations that provide the horizon. And to be human is to have the capacity for second-order desires, that is strong evaluations of desires, which means the capacity for second order reflections on one's first order preferences. By posing the question, "Are my preferences worthwhile?" I may start to discriminate right or wrong, better or worse, higher or lower. This discrimination is dependent on a standard by which they can be judged. However, the deepest evaluations are the most hidden and most subject to illusions. Thus, the struggle for self-interpretation is not trivial. For most people it is important to live as authentically as possible; however, in our globalized, complex world it is not easy to achieve this goal. In the western world, the advertising pressure oriented towards individuals is very strong, with its main purpose to delude the individual into the belief that self-realization is dependent upon their consumption habits.

Taylor uses the term "deep reflection" as a kind of radical evaluation, as "a reflection which engages the self most wholly and deeply" (Taylor, 1985, p. 42). The fallacy is that individuals are not generally open to the horizon of significance. One consequence is loss of meaning. We suggest a conception of authenticity that interprets humans as a part of a web of life. The concept of "authenticity" has been described and discussed within philosophical existentialism for centuries. Kierkegaard (1978) maintains that we are able to choose ourselves, and we have the ability to live in different existential stages. We can move from living a purely aesthetical life, in the sense of being an ego, to a stage of being, which is called ethical. The aesthetic man lives in dread and despair because he cannot sense that he is meant for something higher. The ethical man in contrast truly chooses himself. Individuals who do not live authentically often lose the meaning of life and can suffer from chronic anxiety, boredom, or despair. We believe that in our modern societies many people avoid the responsibility of living authentically and, in the worst case, they end up anesthetizing themselves with drugs. For most people it is important to live as authentically as possible.

From an existentialistic point of view, authentic persons must see themselves as fundamentally free, thereby acknowledging responsibility for their actions and lives. The only way to experience freedom is to regard and treat other people as free. The acceptance of other people's authenticity is a condition of oneself being authentic. Some existentialists argue that authenticity is based upon intimate personal relations – love and friendship. We have elsewhere elaborated the concept of "self-understanding" within an organic worldview by distinguishing between "inner" and "outer" experience and "me" and "other" beings, emphasizing the need to be in a continuous dialogue with other people (see Ims and Jakobsen, 2011, pp. 218–221).

We have, like Taylor, problematized the concept of authenticity. We will proceed on this line of thought by drawing a demarcation line between "shallow" and "deep". This distinction indicates our inspiration from the deep ecology movement (Drengson and Inoue, 1995; Næss and Rothenberg, 1989; Witoszek and Brennan, 1999). The difference between shallow and deep ecology concerns the willingness to go to the root of problems and ask critical questions about oneself as well as about the society with its prevailing institutions. The second concept we will problematize is the concept of self. One danger is to confuse the "self" with the "egocentric self". Deep authenticity depends on our ability to switch to an organic worldview where awareness of an "eco-centric self" becomes possible. An eco-centric self may be understood as a self with a capital Self – including all the sentient beings in the world. This radical and necessary change in mindset is visualized in Table 1.

What follows from the previous arguments is that an organic worldview, which presupposes an eco-centric Self, is more suitable than the mechanistic worldview for interpreting authenticity. When we take into consideration the highly advanced technologies prevalent in our modern times, these technologies typically have many dangerous hidden and unintended consequences. One result is that what we do today will have a multitude of consequences for our selves, for others, for nature and for future generations. Once we discover that we as individual persons are in a sense interwoven – always relating to other human beings and society and nature, we have to clarify our positions in the web of life. A consequence of eco-centered awareness is that we become much more responsible in

Table 1. Shallow and deep authenticity

	Mechanic	Organic
Ego-centric	1. Shallow authenticity	3.
Eco-centric	2.	4. Deep Authenticity

the social and ecological worlds. To become more responsive, openness and dialogue are vital. We need courage to change our interpretative map to discover the interconnectedness in the social and environmental realities. The concept of deep authenticity is essential for grasping these new perspectives. In the context of deep authenticity we are free, and we are aware of the burden of responsibility that is laid upon us. Freedom is a burden because we are called upon to act in response to the highly interwoven world. Each single individual has to face this oneness of reality and, if necessary, break with the ruling norms that hinder taking the deeper reality into account. To handle the individual, social, and environmental challenges we are facing today, we have to make a transition away from the idea of "economic man" to that of "ecological man".

To be inauthentic is to choose the easiest way and thereby follow the avenue to self-deception – to circumvent and avoid facing up to the real challenges in the world. To be inauthentic is even to ignore inconvenient truths, crucial facts about the state of the world, in the hope that the problems, in the end, will be resolved by someone else. Seemingly, this ignorance is strongly present in the lack of willingness to counteract the massive emissions of CO_2 that lead to global warming.

In the perspective of marketing and management literature, assumptions about an organic worldview lead to new perspectives for all stakeholders and the relations between them. Responsibility towards all living beings illuminates the inherent value in the web of life. One obvious consequence is a necessary shift from maximizing shareholder values to increasing the self-realization and flourishing of life for all stakeholders in the context of a long-term perspective.

Competition and cooperation

In this part, we will explain competition and cooperation in the context of contrasting worldviews. Textbooks on business strategy often describe a continuum between hostile competitive behaviour and collaborative relationships. At one end of the continuum, the competitive game consists of atomistic players involved in a zero-sum game, and cooperation, if used, is reduced to an instrumental and tactical way of operation. When collaboration takes place, it is typically a conspiracy of the strong to restrain competition and eliminate competitors. At the opposite end of the continuum, the embedded perspective, partnerships leading to a win-win situation are developed. Based on this, strategists have two alternatives, and must choose between competition and cooperation.

Competition between autonomous actors in the market has a fundamental standing in mainstream liberalistic economics. Competition is based on the idea of conflicting interests between the actors in the market. According to Adam Smith, the logic of the competitive market implies that private vices will be turned to public virtues by means of the market's "invisible hand".

> Every individual necessarily labours to render the annual revenue of society as great as he can. He generally, indeed, neither intends to promote the public interest, nor knows how much he is promoting it … he intends only his own gain, and he is in this, as in many other cases, led by an invisible hand to promote an end which was no part of his intention By pursuing his own interest he frequently promotes that of society more effectually than when he really intends to promote it. (Smith, 1976, pp. 477–478)

Today, Smith's "invisible hand" has global influence. According to the OECD all member countries fundamentally rely on competition to organize their economic activities. The reason is that they believe that competition stimulates innovation and efficiency in the use of resources, and thereby leads to greater product diversification and lower prices: "Therefore, competitive product-markets are in the interest of all consumers." (<http://www.oecd.org/about/>) The rationalism of competition is linked to economic efficiency without any rationality claims linked to the ends of economic activity. The reduction of the human being to a consumer exemplifies the argument that our western concept of rationality is linked to economic ends (Allinson, 2004).

Cooperation means that different participants or partners try to find solutions in which all relevant information and values are considered. In an economic context, the participants are typically defined as stakeholders. Freeman defines the organization's stakeholders as "any group or individual who can affect or is affected by the achievement of the organization's objectives" (Freeman, 1984, p. 46). In practice, different stakeholders have their own values and aims, and the organization has to interact with the stakeholders for mutual benefit. Values are not reduced to preferences (weak evaluations) since the stakeholders' fundamental values (strong evaluations) are also taken into consideration (Taylor, 1985).

In cooperative decisions, the ideal is to establish a platform of consensus, that is to find solutions all stakeholders can agree upon. This way of thinking is different from voting in which the simple majority wins all power. In cooperation all the stakeholders have a common starting attitude saying, "You may be right and I may be wrong." (Habermas, 1990; Popper, 1983). Table 2 summarizes some of our main arguments so far, including the consequences of the two different coordinating principles in the market.

It can be reasonably argued that the principle of competition is insufficient to establish solutions based upon a long-term social and environmental perspective. Welford emphasizes that "productive cooperation always (will) be superior to blind competition and recognizing cooperative opportunities are part of recognizing interconnectedness" (Welford, 2000, p. 141). Hence, Welford's argumentation is based on the presupposition stating that the market cannot be defined as an aggregate of autonomous

Table 2. Some of our major concepts dichotomized for the sake of analysis

Mechanistic worldview	*Organic worldview*
The market consists of isolated, egocentric actors (atomism)	Individuals and society are interwoven (holism)
The market is regulated through instrumental, externally defined norms	Internalized values and norms are essential
Shallow authenticity	*Deep authenticity*
Ego-centric (economic man)	Eco-centric (ecological man)
Instrumental rationality	Free and responsible
Competition	*Co-operation*
Competition anchored in productivity and efficiency From private vices to public virtues	Communicative action to reach common goal

actors; instead, the market must be considered as an integrated whole. Korhonen lends support to this reasoning by stressing that "Competition is … a barrier to the efforts of increasing stakeholder cooperation and cooperation between the firms and its suppliers or the local community actors" (Korhonen, 2002, p. 70).

This indicates that the focus shifts from merely means to include ends. Through cooperative processes the members might agree on the priority of different ends as well as of the use of available means. Cooperation based on dynamic dialogue allows more integrated solutions than the mechanisms of an atomistic and competitive economy. Equality and mutuality among the involved actors are necessary conditions for constructive cooperation. When competition is replaced by cooperation as the main principle for interaction in the market the development of solutions based upon the common good will gradually take place. Cooperation presupposes that the partners disclose relevant and valid information without strategic action (Habermas, 1982, pp. 263–271).

Discussion

In order to develop a framework for a peace economy, we will construct a four dimensional matrix consisting of shallow and deep authenticity on the one hand and competition and cooperation on the other hand, as shown in Table 3. We assume that shallow authenticity combined with a competitive market implies a situation in which there exist a strong tendency to a war-like relation between economy and the environmental conditions. The reason is that both nature and individuals in society are reduced to instruments for economic growth (see cell 1 in Table 3). Mainstream economic literature presupposes that this constellation will lead to the most efficient use of natural and human resources. Both Adam Smith and Milton Friedman, representatives from two different epochs in western history, presupposed a mechanistic worldview, claiming that competition between actors in the market will lead to resource efficiency.

Table 3. Peace economy

	Shallow authenticity	Deep authenticity
Competition	1. Decreasing the numerator and increasing the denominator *War*	2. Reducing the denominator *Negative peace*
Cooperation	3. Reducing the denominator *Negative peace*	4. Increasing the numerator and decreasing the denominator *Positive peace*

In this perspective, competitive power is based upon the actors' preferences, or weak evaluations (Taylor, 1985).

If economic actors with shallow authenticity cooperate, they can come up with measures that could reduce the negative impact of egocentric behaviour. In other words, it is possible to reduce the negative side effects of a violent economy based on shallow authenticity by imposing rules and norms from outside, even if the conflict still exists on a micro level. This is a situation that could be characterized as *negative peace.* The value of the peace formula increases when the denominator becomes smaller (see cell 3 in Table 3).

Looking at cell 2 in Table 3, we can see that the constellation between economic actors characterized by deep authenticity and a competitive market does not eliminate the fundamental reason for the conflict between economy and the environmental framework. Hence, the situation is almost identical to the one in cell 3. The value of the peace formula increases when the denominator becomes smaller (cell 2 in Table 3). It is possible to reduce the negative side-effects of competitive market transactions, even if conflicts still exist on a macro level. This is a situation that could be characterized as *negative peace,* in the same way as we found in cell 3. We argue that even if the actors have an organic worldview as indicated with the concept of deep authenticity, on the economic arena the actors will be confronted with competitive macroeconomic and political forces, as in Webel's (2007) sense. Furthermore, there will not be any interpersonal peace because we assume that competitive behaviour in itself will increase the number and intensity of conflicts.

In cell 4 in Table 3 we assume an organic worldview in which the actors (micro level) are characterized by deep authenticity and the market structure (macro level) is based on cooperation. In those constellations the partners are perceived as integrated, through dialogical processes, and they share common long-term values and interests. The organic paradigm presupposes cooperation that leads to "fair trade", which is a core phenomenon within the "Equity" variable in the peace formula. The organic worldview involves collaborative projects as well as open and transparent communication processes between the partners in the market. Here the value of the peace formula increases because the numerator of the fraction increases simultaneously as the denominator becomes smaller (cell 4 in Table 3). When structures that eliminate the reasons for war are introduced, the situation is close to what Galtung calls *positive peace*. One central conclusion is that we have argued that the concept of cooperation has to be qualified. Cooperation within an organic worldview is quite different from cooperation within a mechanistic worldview.

Conclusion

With reference to nature, in which different organisms cooperate in the nutrition cycle, we advocate that the peace economy must be based on similar principles. In addition to synergic co-existence, different life forms living in the same niche in the ecosystem compete for the same food. The natural principle of "the survival of the fittest" decides which individuals and species win the competition for food. Interpreted in an economic context, this means that cooperation is the fundamental principle for coordination of activities, while competition has a subordinate function. When competition dominates as the principle for coordination (cells 1 and 2 in Table 3), the different players in the market enter into fierce competition, the level and numbers of conflicts increase, and the value of the peace formula is reduced.

We claim that the organic worldview provides a better and often more accurate description of the interplay in the market, and that market behaviour based on competition will often lead to disintegration and egocentric behaviour.

Based on the above argumentation, we reach the conclusion that the discussion concerning shallow and deep authenticity (on the micro level) and cooperation or competition (on the macro level) should not be isolated from a discussion concerning the contextual worldview. And we see the importance of qualifying the concept of cooperation. Cooperation within a mechanistic context is quite different from cooperation within an organic context.

Awareness of the ontological and ethical preconditions can help us to see the limitations of the mechanistic worldview. To grasp the whole human being both as an individual as well as a member of the society, it is necessary to expand the perspective. The organic worldview is more fundamental as the cosmos is seen as interrelated and connected manifestations of one inseparable reality – always in motion, alive, organic, spiritual, and material at the same time (see cell 4 in Table 3).

Webel's (2007) distinction between the three "dialectical, dynamic 'spheres' of greater or lesser peace" indicates the importance of the interaction between the "inner", the "outer", and the "interpersonal" spheres of peace. We believe that the ontological and ethical assumptions that we have dealt with might add new fruitful dimensions to Webel's framework.

To change the worldview requires a change in mindset. Such a change needs new thinking about the importance of peace building as part of education. Peace work, in which peace business and peace economics are central topics, should be integrated in the curriculum at different levels of education. In an education for peace building there should be a focus on what constitutes true personal development in the sense of obtaining more "inner" peace and more peaceful interpersonal interaction. On the "outer" spheres, the need for equity and joint projects is fundamental. In this "outer" arena economists can increase the understanding of what constitutes good business and fair trade. As we have argued, a good business, a business for peace, should not be valued only by a one-dimensional profit rate, but rather by how much it promotes deep authenticity and facilitates a society characterized by equality and fairness.

The concept of structural violence should be an important part of the peace work curriculum. It entails that the concept of peace business should expand from a focus on the role of business in violent conflicts to the structural violence against both people and nature, perpetrated by businesses as a normal part of their operations. The equity concept also emphasizes mutuality and equality in order to increase the importance of relations (not the importance of attributes) and co-creative responsibility. To reduce structural violence, it is important to establish equal relations, or relations based on symmetry, in which no one dominates the other. In this context competition means conflict, because victory for one excludes victory for the other, just as we find it in war or debates, but not in dialogues. Business for peace should play an important part in a society to satisfy basic human needs like survival, wellness, identity, and freedom.

Notes

1. When this exchange of letters was published in 1933, Hitler was already in power and the exchange of letters was only being reprinted in 2,000 exemplars in German and 2,000 in English.
2. Galtung writes that the equation should not be interpreted as a theory but may, rather, be seen as a kind of statement summing up some core concepts of peace research (Galtung, 2012a).

Bibliography

Allinson, R. (2004). The birth of spiritual economics. In L. Zsolnai (ed.), *Spirituality and ethics in managemenr – Issues in business ethics.* Dordrecht: Kluwer Academic Publisher.

Bauman, Z. (1989). *Modernity and the Holocaust.* Cambridge: Polity Press.

Capra, F. (1982). *The Turning point – Science, society and the rising culiure.* London: Flamingo.

Daly, H., & Cobb, J. B., Jr (1994). *For the common good – Redirecting the economy toward community, the environment, and a susiainablefiiture.* Boston, MA: Beacon Press.

Drengson, A., & Inoue, Y. (1995). *The deep ecology movement: An introductory anthology.* Berkeley, CA: North Atlantic Books.

Durkheim, E. (1973). *On morality and society.* Chicago, IL: The University of Chicago Press.

Durkheim, E. (1991). *The division of labour in society.* London: MacMillan Press Ltd.

Einstein, A., & Freud, S. (1933). *Warum Krieg? £in Briefivechel,* Norwegian edition 2003. Kolbotn: Bazar.

Fischer, D. (2007). Peace as a self-regulating process. In C. Webel & J. Galtung (eds), *Handbook of peace and conflict studies.* London: Routledge.

Freeman, E. (1984). *Strategic management – Stakeholder approach.* Boston, MA: Pittman.

Fromm, E. (1955). *The sane society.* New York, NY: Henry Holt and Company.

Fromm, E. (1956). *The art of loving.* New York, NY: Harper & Brothers Publishers. Fromm,

Fromm, E. (1960). *Escape from freedom.* London: Routledge & Kegan Paul.

Fromm, E. (1973). *The anatomy of human destructiveness.* Greenwich, CT: A Fawcett Crest Book.

Galtung, J. (2007). Introduction: Peace by peaceful conflict transformation – The transcend approach. In C. Webel & J. Galtung (eds), *Handbook of peace and conflict studies.* London: Routledge.

Galtung, J. (2012a). *A theory of peace: Building direct structural cultural peace.* Transcend University Press. Retrieved from <www.transcend.org/tup>

Galtung, J. (2012b). *Peace economics. From a killing to a living economy.* Transcend University Press. Retrieved from <www.transcend.org/tup>

Galtung, J., & Næss, A. (1994). *Gandhis politiske etikk.* Oslo: Pax forlag.

Gilligan, C. (1982). *In a different voice.* Cambridge, MA: Harvard University Press.

Habermas, J. (1982). A reply to my critics. In J. Thompson & D. Held (eds), *Jurgen Habermas – Critical debates.* Cambridge, MA: Macmillan Press.

Habermas, J. (1990). *Moral consciousness and communicative action.* Cambridge, MA: The MIT Press.

Ims, K., & Jakobsen, O. (2011). Deep authenticity – An essential phenomenon in the web of life. In A. Tencati & F. Ferrini (eds), *Business ethics and corporate sustainability* (pp. 213–223). Cheltenham, UK: Edward Elgar Publishing.

Ims, K., & Zsolnai, L. (2010). Self-realization in business: Ibsens's *Peer Gynt*. In R. Ghesquiere & K. J. Ims (eds), *Heroes and anti-heroes. European literature and the ethics of leadership*. Antwerp/Apeldoorn: Garant.
Kierkegaard, S. (1978). Enten – Eller. Samlede Verker. Bind 2, København: Gyldendal.
Korhonen, J. (2002). The dominant economics paradigm and corporate social responsibility. *Corporate Social Responsibility and Environmental Management*, *9*, 67–80.
Næss, A., & Rothenberg, D. (1989). *Ecology, community and lifestyle*. Cambridge: Cambridge University Press.
Popper, K. (1983). *The open society and its enemies*. London: Routledge Classics.
Porter, M., & van der Linde, C. (1995). Green and competitive – Ending the stalemate. *Harvard Business Review*, 73 (5), (September–October), 120–134.
Santa-Barbara, J., Dubee, F., & Galtung, J. (2009). *Peace business. Humans and nature above markets and capital*. Transcend University Press. Retrieved from <www.transcend.org/tup>
Smith, A. (1976). In E. Cannan (ed.), *An inquiry inlo the nature and causes of the wealth of nations*. Two volumes in one, first edition 1776, the citations are from the 5th edition. Chicago, IL: The University of Chicago Press.
Taylor, C. (1985). *Human agency and language, philosophical papers* (Vol. I). Cambridge: Cambridge University Press.
Taylor, C. (1991). *The ethics of authenticity*. Cambridge, MA: Harvard University Press.
Vetlesen, A. J. (1994). *Perception, empathy, and judgment. An inquiry into the preconditions of moral performance*. University Park, PA: The Pennsylvania State University Press.
Webel, C. (2007). Introduction: Toward a philosophy and metapsychology of peace. In C. Webel & J. Galtung (eds), *Handbook of peace and conflict studies*. London: Routledge.
Webel, C., & Galtung, J. (2007). *Handbook of peace and conflict studies*. London: Routledge.
Welford, R. (2000). *Corporate environmental management – Towards sustainable management*. London: Earthscan.
Whitehead, A. N. (1967a). *Science and the modern world*. New York, NY: The Free Press.
Whitehead, A. N. (1967b). *Adventures of ideas*. New York, NY: Macmillan.
Witoszek, N., & Brennan, A. (1999). *Philosophical dialogues – Arne Næss and the progress of ecophilosophy*. Oxford: Rowman and Littlefield Publisher Inc.

KNUT J. IMS

Happiness and Human Flourishing

In modern economics and business the crucial question is: What does the modern, affluent consumer want? Does he or she want what she should want in order to promote her or his happiness? However, happiness and the good life has to be understood in the perspective of a culture. Then what is a good life in the western culture? The Gross Domestic Product of a nation does not tell the whole story because GDP is based upon a welfarist approach. GDP gives a one-dimensional materialistic, economic view of human well-being, and is very destructive of the natural environment, that is pollution. Even when we have a number of "just background institutions" in the western world, we have to determine the role of government and the role of "market mechanisms" in order to promote individual well-being and secure the common good.

In the western world, the market expands continually to new domains in a culture where the individual consumer is assumed to be the "king" through his or her consumer sovereignty. We may well ask which values does the consumer reveal in the market? This question is essential when we, like Etzioni, assume that we are all embedded in a community; which is nicely put in the formulation "I and We". Etzioni writes: "The individual and the community make each other and require each other. Society is not a 'constraint' not even an 'opportunity' it is us" (Etzioni, 1988: 9).

Buscher (2002) gives a number of examples of how the ethics of the market have to be understood in relationship to society's "frame of reference". Since the 1980s, when economic policy has given the market a dominant influence, an ethical vacuum has emerged. As a consequence, we have to analyze the normative elements of the market to reveal the ideological elements of economic reasoning. Which values do we in fact support if we assume that the economy to a large extent is disembedded from society, and when the myth of an "ethics-free economic rationality" is not valid? (Etzioni, 1988). To explore this question, we have to study a number of assumptions that are taken for more or less granted in our western culture where the ideal of the "free enterprise" is hiding in the background.

We will take a critical look on questions such as: Is it ethical to satisfy all kinds of wants? What if the wants are social and environmentally destructive? What if what I want is "miswanting"? (Kahneman, 2011).

The chapter is divided in two parts. First, some issues on a societal level are described to problematize that markets lead to the greatest good for the greatest number of people. In the second part, we develop the view that citizens need a polity in order to act as responsible consumers.

Freedom, market, and governance

The economic problem of production and distribution are intertwined. However, economists have traditionally emphasized the problem of production and let the politicians and other social scientists take care of the problem of distribution. Today the gap between rich and poor citizens has grown to an extreme extent in countries in the West. Only a few western economists have given much attention to the inequality of income and wealth in society (Stieglitz, 2013; Piketty, 2014).

Another problem of distribution is the choice between the share of the production that should be invested or saved for future generations, and how much should be consumed. The more we consume today, the less will be left for the future. Using huge quantities of finite natural resources today will leave nothing for future generations, and there are also the non-intended consequences, often expressed as waste, polluted environments, and a high degree of carbon in the atmosphere, which worsens climate change.

There are strong opponents of an active government. Milton Friedman strongly argues for "the power of the market" and against the "tyranny of controls". In their book *Free to Choose* Friedman and Friedman (1980) write that ideas of human freedom and economic freedom working together came to their greatest fruition in the United States, and they warn about the "overgoverned society". For the Friedmans, consumer sovereignty is the sacred principle in the society.

The international trend is a strong belief in the advantage of the "free market". It is demonstrated in the politics of privatization, deregulation, and liberalization from the mid-1980s. Privatization involves the transfer to the private sector functions or services previously performed by the state. Deregulation entails the reduction of government regulations and rules that are applied to the private sector, and liberalization means partial opening up of the economy in response to competitive pressures. Accordingly, the role of the state has diminished and the private sector has expanded (Sales and Beschorner 2006). The consequences are that society is more and more market oriented, and to some extent may be characterized as a "market society" (Sandel, 2012).

The paramount value for Friedman is the individual's right to choose what she or he wants. For Amartya Sen this is only one dimension of freedom. For Sen individual freedom is is of little value to individuals that starve or have no capacity to participate in decisions that concern them. Participation in social life has intrinsic value. Before one can enjoy the freedom of choice, one needs sufficient food and housing, etc. (Sen, 1993).

The market as the solution to the problem of production and distribution is closely connected to private property rights. It involves that private property rights must be easily transferable. The market works when "rights are both privately held and easily transferable" (O'Toole, cited in Desjardins, 2006: 59). Friedman's position of freedom fits well into the position of market utilitarianism, where "every person should be free to do whatever he wishes" where his actions do not interfere with the interests of other human beings, and the faith that "market preserves individual freedom because it depends on voluntary exchange between market actors".

The image of Homo Oeconomicus, which depicts economic agents as rational, self-interest maximizing beings are misleading (Zsolnai, 2013). One implication is that it leads to a belief in massive use of incentives to control or "guide" human behaviour. Altruism, acting for the common good or for ethical commitments does not fit into this model of man. Those assumptions also have a number of bad effects with regard to management practice (Ghosal, 2005).

Economists declare that their goal is to help decision makers towards their goal of efficiency. Seemingly, economic efficiency fits the utilitarian goal of providing the greatest good for the greatest number. But will "an efficient allocation of resources" maximize the economic well-being of all members of society? And if it does, why should we maximize social utility? Efficiency implies optimal satisfaction of consumer preferences given their willingness to pay for different goods and services. However, as a society, and with regard to the environment, should satisfying all the individual preferences be the overriding goal? As Desjardins wrote: "Why should this be the goal of public policy when we recognize the obvious and acknowledge that many individual preferences are silly, foolish, vulgar, dangerous, immoral, criminal, and the like? Why should we think that satisfying the preferences of a racist, criminal, fool, or sadist is a good thing … What is so good about satisfying preferences?" (Desjardins, 2006: 65–66).

When preferences are strange, it is difficult to argue that satisfying preferences is good in itself. Why should we maximize the satisfaction of preferences regardless of their moral worth?

Sandel (2012: 88–90) writes that when market reasoning travels beyond the domain of material goods, it must "traffic in morality". When market reasoning is concerned with material goods, such as cars, toasters, and flat-screen televisions, the objections to satisfaction may not be too strong, but when market reasoning is applied to sex, procreation, child rearing, education, health, criminal punishment, immigration, police, and environmental protection, it is questionable if everyone's preferences are equally worthwhile. In morally charged arenas, some ways of valuing goods may be higher and more appropriate than others.

Economists uses terms such as welfare, utility, or happiness to explain the goal of satisfying preferences. However, it is wrong to believe that satisfaction of preferences always leads to happiness. As Desjardins (2006: 66) writes: "Whether what I want is a good thing depends on what it is that I want." Modern happiness research and positive psychology gives us an insight of what makes people happy.

Kahneman (2011) demonstrates with the "focusing illusion" that people exaggerate the effect of major purchases or "changed circumstances" on their future well-being. For example, when buying a nice new car and comparing the car with joining a group of people that meets every week, people believe that the car will increase their well-being most. At the start, both experiences will be exciting, but after a while they will pay less and less attention to the car. The group, however, that they attend every week will continue to be exciting. By the fallacy of "What You See Is All That Is" people put too much emphasis on the long-term benefits of the new car. This focusing illusion "creates a bias in favor of goods and experiences that are initially exciting, even if they will eventually lose their appeal" (Kahneman, 2011: 406).

The Aristotelian-Thomistic tradition and positive psychology

Where do we find frameworks that promote good characters that live good lives in a good society, and where individuals are employed through meaningful work in the context of a fair economy and fair background institutions?

One approach to address these issues is the modern happiness research (Seligman, 2002). We find a lot of scientific evidences that very rich people are not happier than people with a moderate income. This phenomenon leads to the so-called "happiness trap" since it contradicts common sense thinking, according to which "more is always better".

But what is "happiness"? The word is ambiguous and its use is manifold, as pointed to by researchers (see for example Gilbert, 2006). Happiness may mean expressing beliefs about the merits of things, about their position, and not of their feelings. When saying we are happy about, we note that something is a potential source of pleasurable feeling, but we do not feel it at that moment. We could reserve the word "happiness" to a class of subjective emotional experiences that do not indicate the morality of the actions that motivated them. However, we might wonder whether the happiness we get from helping a child is better or deeper than the happiness we get from eating an ice cream? In a new study of human emotion, Ketner (2009) finds that happiness is found in the rich landscape of positive emotions that until recently remained mysterious to science. Central to Ketner's work stand the emotions of compassion, sympathy, and awe, which are programmed into our minds and bodies. As he says, "Emotions are signs of our deepest commitments … (They are) intuitive guides to our most important ethical judgments. Our pursuit of the meaningful life requires an engagement with emotions" (Ketner, 2009: 34).

The hedonic view of happiness originally developed by Jeremy Bentham is used as the metric for economists to understand happiness as a felicity calculus, where pleasure minus pain is the gold standard for a good society. However, such a calculus represents a reductionist view of happiness and it is amoral because it does not take into account the human capacity and willingness to live a meaningful life.

The OECD (2013) has developed its *Guidelines on Measuring Subjective Well-being*, arguing that data of subjective well-being can provide an important complement to other objective indicators of well-being. The OECD framework consists of three concepts of well-being: life evaluation, affect, and "eudaimonia" (or flourishing) (OECD, 2013: 28–35). This conceptual model gives a richer and more accurate model of human functioning.

Aristotle viewed human life as consisting in the pursuit of ends. The object of Aristotle was to discover the nature of the end towards which man ought to aim, and the ultimate end was described as "eudaimonia", a word usually translated to English as "happiness". However, such a translation is not satisfactory since happiness is typically regarded as a state of feeling while the Aristotelian eudaimonia involves a certain kind of activity. Eudaimonia is carrying feeling, but it is not in itself a kind of feeling (Aristotle, 1925/1980: vi).

Aristotle explores eudaimonia in his "Nicomachean Ethics" (Aristotle, 1935). First, he criticizes three popular views about the nature of life: (i) a life of pleasure, (ii) a political life which aims at honor, and (iii) the life of money making. Then he mentions the contemplative life as a fourth way. This requires two conditions. First, the end of human life must be desirable in itself, chosen for its own sake. The other condition is that those activities of the soul had to be in accordance with the rational element in human beings – virtues, either moral or intellectual ones.

Eudaimonia is based upon a judgment of a whole life, while individual actions and projects may be judged as virtuous *(arete)* or not. One statement used by Aristotle is that a swallow does not give a summer, but a summer gives many swallows, meaning that a single good action is not enough to evaluate one's goodness. So, Aristotle does not reduce eudaimonia to happiness understood as a particular state or a positive mood. Even if pleasure is taken by a good man in virtuous activity, pleasure cannot be the purpose of man according to an eudaimonic view.

For Aristotle the good man is virtuous, and excels in virtuous actions. Virtues are not external to the good life, but an integral part of it. Among the intellectual virtues, wisdom or *phronesis* is essential, and among the moral virtues, courage, temperance, and justice are cardinal. While the intellectual virtues to a large extend are taught, moral virtues are the product of habit (*ethos*).

A virtue may be destroyed either by excess or deficiency. Health may be destroyed by too little food or too much food (Aristotle, 1935/1951: 183). It means that too much of a good thing may change the quality of the thing. As Coplestone writes; "all valuational elements, taken in isolation, have in them a point beyond which they are dangerous, that they are tyrannical, and that for the true fulfillment of their meaning (...) there is always a counterweight" (Coplestone, 1950: 337–338).

Aristotle also discusses happiness in relation to contemplation. He values contemplation as the noblest of activities because reason is "the noblest part of us, and the objects of reason are the best of all knowable things" (Aristotle, 1935/1951: 265). In particular, to contemplate truth (*theorein*) will give a man the possibility to do it in a continuously way. Aristotle argues that happiness ought to contain a blend of pleasure, and activity in accordance with wisdom (*sophia*) is the most pleasant of all virtuous activities. The "pursuit of wisdom" (*philosophia*) gives durable and pure pleasures. Contemplation is characterized by self-sufficiency (*autarkeia*) and is seen as the only act that is loved for its own sake.

Aquinas developed the Aristotelian virtue ethical approach further. He introduced three new virtues, the triad of theological virtues; faith, hope, and charity.[1] Aquinas might have problems with applying the Aristotelian doctrine of the golden mean for the theological virtues. Why could it be a problem to love God too much? How can St Francis of Assisi be taken as an example of "the mean"? (Coplestone, 1955: 217). Aquinas was not an admirer of fanaticism. His interpretations of St Francis were that he did not act out of superstition or of a desire to make a name for oneself. His actions "were performed in response to an invitation of Christ". Aquinas suggests that in one respect there cannot be an excess, however there can be defect. On the other hand, if we consider man's condition and the way he should manifest the love of God, we can talk about a mean. Aquinas had a "respect for the rule of right reason and for the proportion and measure demanded by right reason"(1955: 218). Coplestone concludes that Aquinas application of the mean was due to his dislike of one-sided exaggerations and his respect for the rule of reason.

Based upon Aristotle, Aquinas made a distinction between act and potentiality that is found in every finite thing. Potentiality is unrealized capacity and indicates a lack of possible perfection. Any finite being is a mixture of potentiality and act. Development is a movement towards the full realization of form (Coplestone, 1955: 99). Habits dispose a man to act readily and easily. Good habits are virtues and bad operative habits are vices.

Aquinas distinguishes between moral virtues and intellectual virtues. Moral virtue "inclines a man's sensitive appetite to act in accordance with right reason" (Coplestone, 1955: 215). Intellectual virtues perfect a man's rational powers. It is not possible to have moral virtues without the intellectual virtue of prudence, which inclines us to choose the right means to the attainment of the objective good. According to Aquinas, we need virtuous habits to act uniformly, readily, and pleasurably in accordance with the right reason (Coplestone, 1955: 216).

Baumeister et al. (2013: 505) write that "some positive psychologists have revived a conceptual distinction dating to Aristotle between feeling happy and living a good life". The latter is known as eudaimonia. Based upon empirical research Baumeister and his colleagues emphasize that meaningfulness is more central to eudaimonia than feeling good.

Happiness, according to positive psychology, has been conceptualized and measured in quite different ways: (i) as life satisfaction, which goes beyond momentary feelings and includes an integrative, evaluative assessment of one's life as a whole, and (ii) as affect balance, which indicates more pleasant than unpleasant emotional states, that is, on how one feels at different moments.

According to Steger (2009), meaning may be defined as purpose, significance, or some multifaceted construct. Frankl (1963) emphasized that each person has a unique purpose or overarching aim for life, understood in light of his or her values, and enacted in reflection of his or her community. However, meaning may also be defined from an informational significance point of view. In accordance with this view, people experience meaning in life "when their lives make sense". One related view of meaning is a kind of binding the self to the external world. Baumeister and Vohs (2002: 608) argue that "the essence of meaning is connection". While the first two approaches are described as motivational purpose-centered and cognitive significance-centered definitions, there are definitions of meaning in life that combine these two dimensions with an affective dimension, referring to peoples' fulfillment in their lives (Steger, 2009: 681).

Baumeister et al. (2013) held that meaning is not personal but cultural, using the analogy of a large map that is gradually filled in by countless generations. An individual may see himself fitting in a section of this map. Meaning can integrate events across time, and one aspect is that meaning allows people to think about the past, future, spatially distant realties, and even the possibilities of these realities (Baumeister et al., 2013: 506). Purpose implies that present events draw meaning from future ones. It implies that one may be motivated to work toward future goals that are desirable, even though the present actions may be highly unpleasant (Frankl, 1963). Even if people cannot articulate what it means, they may feel that their life is meaningful if they find it is rewarding in some way. The focus on meaningfulness and the meaning of life implies that it is possible to have a "highly meaningful life that is not necessarily a happy one (for example as religious missionary, political activist, or terrorist)" (Baumeister et al., 2013: 506).

A number of studies found that hedonic pursuits produced short-term benefits, while long-term pursuits depended more on eudaimonic activities. Seligman (2002) writes about his own experience in one of his classes. He asked his students to engage in one pleasurable activity and one philanthropic activity and write about both. The students found that "The afterglow of the 'pleasurable' activity (hanging out with friends, or watching a movie, or eating a hot fudge sundae) paled in comparison with the effects of the philanthropic and kind action." The students said that when "our philanthropic acts were spontaneous and called upon personal strengths, the whole day went better" (Seligman, 2002: 9).

Seligman concludes that the exercise of kindness is a gratification that calls on your strengths in contrast to a pleasure, which he defines as delights that have distinct sensory and strong emotional components like "raw feelings": that is, ecstasy, thrills, orgasm, delight, mirth, exuberance, and comfort. The gratifications are activities we very much like doing, but they are not necessarily accompanied by any raw feelings. They engage us fully, and we become immersed in them. The gratifications last longer than the pleasures as they involve much thinking and interpretation, and they are encompassed by our strengths and virtues (Seligman, 2002: 102).

In his recent work on flourishing, Seligman (2011) argues that a mood view of happiness will only concern 50 percent of the world's population who are "low-positive affective". Those people may have more engagement and meaning in life than merry people.

In Seligman's view, there is a shift from happiness to well-being, understood as five elements; Positive emotion, Engagement, positive Relationships, Meaning, and Accomplishment (PERMA). All the elements contribute to well-being, but no single element defines it. A short look on these five elements follows:

1. Positive emotions about the past are satisfaction, contentment, pride, and serenity. Optimism, hope, trust, faith, and confidence are future-oriented emotions. In comparison with his "authentic happiness theory" Seligman now makes two changes. Happiness and life satisfaction as subjective measures are no longer the goal of the entire theory.
2. Engagement is also measured subjectively, and since flow is a typical activity in this category the subjective state for pleasure is retrospective. Thought and feeling are typically absent during the flow state.

3. A relationship is about a man's connection to other people. Positive relationships are accompanied either by positive emotion or engagement, or by meaning or accomplishment.
4. Meaning emphasizes belonging to and serving something bigger than the self. It is not only a subjective state. Abraham Lincoln is used as an example since Lincoln was known to be a profound melancholic, but most people may judge his life filled with meaning.
5. Accomplishment pursued for its own sake. This is the drive to exert mastery over the environment. For example, bridge or chess players try to improve in order to learn and to be in flow.

Together with other researchers, Seligman developed a classification scheme and a way of measuring character in 1999. His research group studied Aristotle, Plato, Aquinas, Augustine, the Old Testament, ilie Talmud, Confucius, Buddha, Lao-Tze, Bushido (the samurai code), the Koran, Benjamin Franklin, and the Upanishads – some two hundred virtue catalogues in all. "To our surprise, almost every single one of these traditions flung across three thousand years and the entire face of the earth endorsed six virtues" (Seligman, 2002: 132). Based upon the writings of the major religious and philosophical traditions, the research group created the following list of virtues: Wisdom and knowledge, Courage, Love and humanity, Justice, Temperance, Spirituality and transcendence.

The contents of the different concepts varied somewhat in different traditions. Humanity in Confucius is not the same as *caritas* in Aquinas. The research group also found that "The commonality is real and, to those of us raised as ethical relativists, pretty remarkable" (Seligman, 2002: 133).

Seligman regards these six virtues as core characteristics. Viewed together they capture the notion of good character. These virtues may be operationalized in order to be measured, and there are several ways or routes to achieve each virtue.[2] With the notion of signature strengths, strengths of character that a person self-consciously owns, celebrates, and exercises every day in work, love, play, and parenting (Seligman, 2002: 160–161), Seligman presents a way each individual may identify, build, and choose to use the virtues in the main realms of his or her life (Seligman, 2002: 134). Seligman stresses that strengths are not the same thing as talents, since strengths are moral traits, while talents are non-moral, possessed, and cannot be chosen in the same way as strengths. A strength involves a choice about when to use it, and whether to build it, or not. The strengths can be acquired by effort and determination by ordinary persons. A strength is valued in its own right, and a culture may support strengths by providing institutions, rituals, role models, parables, maxims, and children's stories.

In order to have a good life, a person should use his or her individual signature strengths in daily life. A meaningful life needs an additional component – to use these strengths in the service of something greater than oneself. A full life consists in experiencing positive emotions, which may be defined as the pleasant life, plus gratifying activities using your signature strengths, and in addition using your strengths to contribute to something beyond yourself.

Daniel Kahneman (2011) suggests a distinction between experienced well-being and remembered well-being, referring to two different parts of the self. The experiencing self is the one that answers the question: "Does it hurt now?" The remembering self is the one that answers the question: "How was it, on the whole?" Kahneman's point is that memories are the only thing we can use when we think about what we have done, and our life is seen through the lenses of the remembering self.

Kahneman and his colleagues developed the so-called "Day Reconstruction Method" (DRM) expecting it would approximate the results of experience sampling (Kahneman, 2011: 392). They found many variants of positive feelings in the experience of a moment or an episode like love, joy, engagement, hope, amusements, etc. In a similar way negative emotions come in many varieties, and it is possible to classify most moments of life as ultimately positive or negative. Kahneman's suggestion is that "experienced well-being" can inform social policy in multiple ways. For the time being, measures of experienced well-being are routinely used in large-scale national surveys in USA, Canada, and Europe, and the Gallup World Poll has extended these measurements to millions of respondents in more than 150 countries.

Discussion

The market-based solution to individual, social, and environmental problems has major limitations. Willingness to pay is not the only adequate criterion for the allocation of resources. One problem is that market-based solutions involve a tendency to systematically ignore the well-being of poor and disadvantaged people, future generations, and the natural environment as a whole.

Aquinas appeared as an innovative theologian who applied and developed the Aristotelian framework to Christianity. He used Aristotle as an intellectual instrument, but was not a slave of the Aristotelian framework. Aquinas considered perfect happiness attainable only in the afterlife. For Aquinas the Aristotelian framework was incomplete because it lacked the transcendent. What Aristotle defines as happiness, Aquinas calls temporal happiness or imperfect happiness.

Both Aristotle and Aquinas treat the moral and intellectual virtues as habits by which a man lives rightly. Moral virtues consist in a mean in order to conform to the rule of reason and avoid extremes of excess and defects. They both promoted virtuous behaviour, emphasizing the development of man's character. Both insist on the importance of activities that perfect the highest faculty in man, activities directed to the highest and noblest objects.

Aristotle and Aquinas agree that material wealth is needed and desirable, but only to a certain extent. Material means should never take the role of the goal in a person's life. In particular, Aquinas argued that the means will be easily transformed into goals themselves. He asserts that man has dominion over exterior things as to their use. But they are to be used for a purpose, serving as *instrumenta* or *adminicula* in man's search for virtue and beatitude (Langholm, 1992: 208).

Aquinas feared that exterior things would impede the love and fear of God. Riches have a distinctive feature, "if one seeks more than what is sufficient to sustain life, one is easily entrapped by avarice, which tends to infinity. The desire for external things as a means to a higher end is subject to a natural limit ...", "... the highest good is such that it is loved the more highly the better one comes to know it, but when the appetite for one material good is quenched, the deluded materialist is already searching for some other thing. Such appetite for inordinate wealth is vicious also because it makes man a slave of wealth rather than its master. By placing his end in an infinite quest for worldly riches he makes them his God (for one's God is that in which one places one's end), ..." (Langholm, 1992: 209).

American psychologist Tim Kasser summarizes the problem of materialistic value orientation in this way: "research shows that the more that people focus on materialistic goals, the less they tend to care about spiritual goals. Further, while most spiritual traditions aim to reduce personal suffering and to encourage compassionate behaviour, numerous studies document that the more people prioritize materialistic goals, the lower their personal well-being and the more likely they are to engage in manipulative, competitive, and ecologically degrading behaviour" (Kasser, 2011: 204).

A number of studies report that religion and faith lead to higher feelings of well-being. Frey and Stutzer (2002) support the view that believing in God and happiness are positively related. They give the following explanations why religion raises happiness: (i) church attendance is an important source of social support, (ii) religion offers an interpretative framework that can instill life with meaning and purpose, (iii) religious people are better able to cope with adverse circumstances, and (iv) church members are on average of better health (Frey and Stutzer, 2002: 59). However, they argue that the causality may be reversed: Happier people are more religious, as in societies in which religion is regarded as important and where participation in church activities is a normal part of social life.

Within the positive psychology literature there are glimpses of interest in understanding awe and the importance of awe in people's lives. Haidt (2006) introduces a third dimension to illustrate what he calls "Divinity with or without God". Usually we look at a social world as two-dimensional; a flat land where the X-axis represents closeness and where the Y-axis represents hierarchy. However, when you see something extraordinary you might have an overwhelming experience of natural beauty and you feel lifted up. Haidt's claim is that the human mind perceives a third dimension, a specifically moral dimension, which he calls "divinity" (Haidt, 2006: 183).

Haidt cites Ralph Waldo Emerson saying: "He who does a good deed is instantly ennobled. He who does a mean deed is by the action itself contracted. He who puts off impurity thereby puts on purity. If a man is at heart just, then in so far is he God" (Haidt, 2006: 191).

Abraham Maslow studied peak experiences and suggested that all religions are based on the insights of somebody's peak experiences. Maslow finds that peak experiences make people nobler. Maslow also attacks science as becoming as sterile as organized religion. In the late sixteenth century European scientists began to look down on wonder, regarding wonder as the mark of a childish mind. And mature science separates fact rigidly from values and emotions. Maslow concludes that science had helped to desacralize the world, and left us with what *is*, rather than what is good or what is beautiful (See Haidt, 2006: 206).

Amartya Sen (1993) warns us of being fetishistic about freedom of choice. We should ask ourselves whether it nourishes or deprives us, whether it enhances self-respect or diminishes it, whether it enables us to participate in our communities or prevents us from doing so. According to Schwartz (2004) the reason is that an overabundance of choice may contribute nothing to the kind of freedom that counts. It may impair freedom by the energy and time we use in selecting between different brands. We could have used this time and energy differently.

Based upon a number of experiments, Kahneman (2011) finds that in many cases there are patterns of situations where "more is less". Barry Schwartz (2004) gives a number of explanations why this is the case. The problem arises in a market society with an overabundance of choices, for example in a supermarket where you have to choose between one hundred types of cheeses. As the number of available choices increases, the autonomy, control, and liberation is in one sense positive. The negative aspects "escalate until we become overloaded". Then "choice no longer liberates, but debilitates. It might even be said to tyrannize" (Schwartz, 2004: 2).

Money has been the primary goal of many people in western societies. However, can money buy happiness? Kahneman's answer is that "being poor makes one miserable, and being rich may enhance one's life satisfaction, but does not (on average) improve experienced well-being" (Kahneman, 2011: 396). Illness for the poor is much worse than for those who have enough money. The average increase of experienced well-being beyond household income of about $75,000 (US) per year was zero. This is a remarkable finding, all things considered. Earning such an income might give occasion to purchase many pleasures. Kahneman speculates whether there may be circumstances where a high income might reduce happiness, and finds that higher income goes together with a reduced ability to enjoy the small pleasures of life, like eating a small chocolate or enjoying an ice cream a warm summer day (Kahneman, 2011: 397).

In light of increasing inequality in the USA and Europe, where a small minority are super rich, the middle class have a reduced income, and the poor are worst off (see Stiglitz, 2013), Kahneman's insights present interesting perspectives. Even if there is a complex relationship between happiness and income, more fairness and more equality would be major positive contributions to increased average happiness in society. We should not forget that a 10 percent increase in income above the "natural limit" does not count as much as a 10 percent decrease in income. This follows the general principle that bad things have stronger effects than good ones (Baumeister et al., 2013; Kahneman, 2011).

One of Kahneman's main conclusions is that one cannot measure experienced well-being without taking goals into consideration. Goals that people set for themselves are essential to what they do and how they feel about themselves. Kahneman argues that we need a combined strategy in which the well-being of both the remembering self and the experiencing self is considered (Kahneman, 2011: 402). The problem is that the remembering self is sometimes wrong and, accordingly, we try to maximize our future memories, which are not always the same as our future experiences.

Concluding remarks

We have problematized hedonic preferences and indicated that Aristotelian-Thomistic vocabulary implies a self-realization view that is radically different. Kahneman found that well-being research was based upon the collection of Gallup data, and heavily relied only on the remembering self. After years of investigation, Kahneman developed a complementary way to measure how people are experiencing life from moment to moment. This approach to the good life was not ignored by Aristotle. Aristotle discussed pleasure as part of eudaimonia, and argued that pleasure as a kind of feeling is an important aspect of life. However, when it comes to selecting either pleasure or the ultimate end of life, pleasure has to be subordinated to meaning.

It is interesting to note that new research on happiness also supports Aristotle's reasoning. Baumeister et al. (2013) concludes that meaningfulness appears to be central to the difference between feeling happy and eudaimonia. Happiness is not all that people seek; "meaningful but unhappy life is in some ways more admirable than the happy but meaningless one" (Baumeister et al., 2013: 516). The quest for meaning is the essential part of what makes us human. And meaning requires a certain frame of reference that the market cannot supply. Construction of meaning needs to be part of something bigger than oneself – bigger than one's vector of goods and services that can be bought on the market.

Michael Sandel asks: "What is the moral importance of the attitudes and norms that money may erode or crowd out? Would the loss of non-market norms and expectations change the character of the activity in ways we would (or at least should) regret? If so, should we avoid introducing financial incentives into the activity, even thought they might do some good?" (Sandel, 2012: 90–91).

Economics should be seen as a moral science, where we have to decide the fundamental tradeoff between a sufficient satisfaction of wants, a just distribution of the economic goods and a sustainable production in relation to the ecosystem. It is all about political-ethical decisions that have major impact on the environment and for the future. One of the problems of today is that a small group of the world's population are behaving unecologically because they are rich, while the biggest group behave unecologically because they are poor.

Joseph Stiglitz (2013: 228–232) writes that we need to adjust GDP to measure other influences on the well-being of a nation rather than the turnover of goods and money. In particular, we should include national happiness and environmental conservation. GDP is calculating the fuel we consume in a given day, but does not account for the "externalities", the extra carbon emissions in the atmosphere, the waste produced, or the poison that leaks into the water, or destroys the ground. GDP accounts what is easily quantifiable by assigning monetary weights. In this chapter, we have pointed out some of the qualitative factors that are crucial to understand the ecological, social, and psychological dimensions of economic activity.

One inspirational source can be found in a small Kingdom in the Himalayas, Bhutan, where the Government is committed to implement a policy that promotes "Gross National Happiness". The Bhutanese government has developed a holistic view of happiness, which is measured in a multi-dimensional way so as to ensure that the inhabitants have sufficient material goods to live a meaningful life. The Government claims that if economic development is not matched with spirituality there is no development (Alkire, 2015). The Bhutanese government also inquiries into the degree of the social connections and the vitality of the local communities of the (about 700,000) inhabitants in the (Buddhist) culture.

Finally, to what extent does the market logic enhance human well-being? The market represents a kind of freedom and autonomy for the single individual where the individual can pursue his or her own self-interests. The market gives opportunities of freedom from interference, a kind of negative freedom. However, markets represent a kind of "Freedom *without and from* responsibility" (Smart, 2006). The market is impersonal and needs trust to function in an efficient way, but that trust must be developed in a civil society. Trust cannot be created in an impersonal market. We need virtuous consumers that have the capability to make wise decisions and institutions that can educate the consumers (the citizenry) to negate the inclination to "miswanting". We need communities where citizens can excel their virtues in serving something beyond theirselves. However, in western society we have, in fact, a surplus of individualism and a deficit of companionship. A number of empirical studies conclude that "humans seem to be genetically wired for community". That one is happier in a community is not just some sentimental affection. McKibben writes "economic efficiency improves if we shift people 'from places where they are less productive to ones where they are more productive', but your actual physical health suffers. If you buy everything at Wal-Mart, it's cheaper, …

But if you go to the farmers' market, you'll have ten times as many conversations" (McKibben, 2007: 110).

Notes

1. Macintyre (1966: 177) stresses that Aquinas cannot mean by the Latin names of the cardinal virtues entirely what Aristotle meant by their Greek equivalents. For example, Aquinas explicitly acknowledges the biblical virtue "humility". " … In Aristotle's account of the virtues where anything resembling humility is mentioned, it is a vice."
2. To give one example of an operationalization of virtues, consider the virtue of wisdom and knowledge. According to Seligman (2002) there are six routes to this special virtue: (1) curiosity/interest in the world; (2) love of learning; (3) judgment/critical thinking/open mindedness; (4) ingenuity/originality/practical Intelligence/ street Smarts; (5) social Intelligence/personal intelligence/emotional intelligence; and (6) perspective. For every route, the researchers developed different statements and the respondents were expected to rate the different items and add the score.

Bibliography

Alkire, S. 2015. *Well-being, Happiness, and Public Policy.* Thimphu: The Centre for Bhutan Studies & GNH Research.

Aristotle. 1925. 80 *The Nicomachean Ethics,* Translated with an introduction by David Ross. Oxford: Oxford University Press.

Aristotle. 1935/1951. (Containing Selections from the Seven of the Most Important Books of Aristotle, selected and translated from the original Greek by P. Wheelwright). New York: The Odyssey Press.

Baumeister, R. F., and K. D. Vohs. 2002. The Pursuit of Meaningfulness in Life. In *Handbook of Positive Psychology,* eds, S. J. Lopez and C. R. Snyder. Oxford: Oxford University Press.

Baumeister, R. F., Vohs, K. D., Aker, J. L., and E. M. Garbinsky 2013. Some Key Differences Between a Happy Life and a Meaningful Life. *The Journal of*

Positive Psychology: Dedicated to furthering research and promoting good practice 8(6): 505–516.

Buscher, M. 2002. Ethics of the Market. In *Ethics in the Economy. Handbook of Business Ethics*, ed. L. Zsolnai. Oxford: Peter Lang.

Coplestone, F. C. 1950. *A History of Philosophy. Volume II, Medieval Philosophy.* New York: Dobleday.

Coplestone, F. C. 1955. *Aquinas.* Harmondsworth: Penguin Books.

DesJardins, J. R. 2006. *Environmental Ethics. An Introduction to Environmental Philosophy.* New York: Thomson Wadsworth. Division of Macmillan, Inc. Washington Square Press.

Etzioni, A. 1988. *The Moral Dimension. Toward A New Economics.* London: The Free Press.

Frankl, V. E. 1963. *Man's Search for Meaning: An Introduction to Logotheapy.* New York: Pocket Books.

Frey, B. S., and A. Stutzer. 2002. *Happiness & Economics. How the Economy and Institutions Affect Human Well Being.* Princeton/Oxford: Princeton University Press.

Friedman, M., and R. Friedman. 1980. *Free to Choose. A Personal Statement.* New York: Harcourt Brace Jovanovich.

Ghosal, S. 2005. Bad Management Theories are Destroying Good Management Practices. *Academy of Management Learning & Education* 4(1): 75–91.

Gilbert, D. 2006. *Stumbling on Happiness.* New York: Vintage Books.

Haidt, J. 2006. *The Happiness Hypothesis. Putting Ancient Wisdom and Philosophy to the Test of Modern Science.* London: Arrow books.

Kahneman, D. 2011. *Thinking. Fast and Slow.* New York: Penguin Books.

Kasser, T. 2011. Materialistic Value Orientation. In *The Palgrave Handbook of Spirituality and Business,* eds, L. Bouckaert and L. Zsolnai. Houndmills/ New York: Palgrave Macmillan.

Keltner, D. E. 2009. *Born to be Good. The science of a Meaningful Life.* New York: W. W. Norton & Company.

Langholm, O. 1992. *Economics in the Medieval Schools. Wealth, Exchange, Value, Money and Usury According to the Paris Theological Tradition 1200–1350.* Leiden: Brill.

MacIntyre, A. 1966. *A Short History of Ethics. A History of Moral Philosophy from the Homeric Age to the Twentieth Century.* New York: Macmillan.

McKibben, B. 2007. *Deep Economy. Economics as if the World Mattered.* Oxford: A Oneworld Book.

OECD. 2013. *OECD Guidelines on Measuring Subjective Well-being.* OECD Publishing. <https://www.oecd-ilibrary.org/economics/oecd-guidelines-on-measuring-subjective-well-being_9789264191655-en>.

Piketty, T. 2014. *Capital in the 21st Century.* Cambridge, MA: Harvard University Press.

Sales, A., and T. Beschorner. 2006. Societal Transformation and Business Ethics: The Expansion of the Private Sector and its Consequences. In *The Moralization of the Markets,* eds, N. Stehr, C. Henning, and B. Weiler, 227–254. London: Transactions Publishers.

Sandel, M. J. 2012. *What Money Can't Buy. The Moral Limits of Markets.* London: Penguin Books.

Schwartz, B. 2004. *The Paradox of Choice. Why More Is Less.* New York: HarperCollins Publishers.

Seligman, M. E. P. 2002. *Authentic Happiness. Using the New Positive Psychology to Realize Your Potential for Lasting Fulfillment.* New York: Free Press.

Seligman, M. E. P. 2011. *Flourish. A Visionary New Understanding of Happiness and Well-being.* New York: Atria.

Sen, A. 1993. Capability and Well-Being. In *The Quality of Life,* eds, M. Nussbaum and A. Sen. Oxford: Clarendon Press.

Smart, B. 2006. Freedom without Responsibility: The Culture of the Market and the Demoralization of Social Life. In *The Moralization of the Markets,* eds, N. Stehr, C. Henning, and B. Weiler. New Brunswick: Transaction Publishers.

Steger, M. F. 2009. Meaning in Life. In *The Oxford Handbook of Positive Psychology,* eds, S. J. Lopez and C. R. Snyder, 2nd edn. Oxford: Oxford University Press.

Stiglitz, J. 2013. *The Price of Inequality.* London: Penguin.

Zsolnai, L. 2013. The Moral Economic Man. In *Handbook of Business Ethics. Ethics in the New Economy*, ed. L. Zsolnai. Oxford: Peter Lang.

KNUT J. IMS AND OVE D. JAKOBSEN

Quality of Life

In this chapter we elaborate on the concept "quality of life" by connecting it to different frames of reference. Quality of life is a relevant concept in philosophy, psychology, and economics. It is also relevant both in western and eastern contexts. We start by stating the obvious; quality of life is more than the absence of physical and psychical pain. Distinguished scholars, though, in different disciplines and cultures have argued that quality of life is a question of being in harmony with something bigger than one's self. It could be nature, culture, the planet, or the whole universe.

One consequence of this line of argument is that any quality of life is impossible if the Earth around us is sick and its human communities are in distress. But in addition to harmonizing our relationship with the outer world, it is necessary to develop a harmonious contact with our inner world and break out of "ego" and become part of "eco".

To establish stimulating conditions to develop quality of life, we have to break out of the atomized, materialistic worldview of modern society and try to link our inner and outer worlds. Even if the individual is starving in isolation, we argue that individualism is still important. The problem is egocentrism not individualism. To dig deeper into these complex issues, we start with a review of some of the challenges connected to well-being and quality of life from an eastern perspective inspired by Satish Kumar (2002, 2007, 2013). In the second part of the chapter, we discuss how these challenges could be handled from a western point of view, based on a transdisciplinary perspective. In the third part, we discuss how the different perspectives integrate and contribute to a deeper understanding of the concept of "quality of life".

An eastern perspective

Well-being or quality of life depends on three dimensions: inner peace, social peace, and ecological peace. Even if there are no easy solutions we can still do something because "what is created by humans can be changed by humans" (Kumar, 2013, p. 169). According to Kumar, we have to move slowly but surely and develop step by step, never losing sight of the three qualities integrated in everything: the "sattvic" (natural and simple), the "rajasic" (extravagant and excessive), and the "tamasic" (dark and depressing).

Sattvic relates to creativity, truth, that which is natural, nourishing, buoyant, and shining, has a purity of means and kindness and seeks fulfillment. The word is used to remind people to follow the way that feels true to them at the deepest level something which is natural and real. People are considered to be a part of nature – not apart from nature – and there is no separation and no dualism between people and nature. The individual understands and celebrates the intrinsic goodness of the natural world and in everyday life he or she does not require a lot of money and resources. The individuals are in dialogue with other people, nature, and themselves. Dialogue happens when one meets others with open minds and open hearts. In dialogue change comes about. From within emerges a shared understanding, instead of change being imposed from outside. The sattvic life leads to inner and outer freedom and the sattvic lifestyle is both sustaining and sustainable.

Rajasic relates to energy, change, achievement, excitement, stimulation, movement, and seeks success, and is a word used to warn people of the pitfalls of the glittering and the glamorous. The raw and rugged realities of the natural world can be reformed, changed, even engineered to make it better suited for human purposes. People are searching, seeking, and longing for nirvana or enlightenment. It is a way of life for the elite. They like big projects, such as massive dams, towering buildings, mighty power stations, and spectacular shows, and they concentrate on achievement, outcome, and success. They often pay lip service to such things as fairness and justice, but only as instruments to serve their own interests. They claim to favor freedom, but with practice they impose their rule through domination and control; dependent on excessive use of natural resources, they value nature only in terms of her usefulness to humans. People living in accordance with rajasic ideas believe in scientific progress, technological development, and economic growth. By presenting their views in a certain way, people can conceal a fixed and closed position and their own self interest. Outwardly, they may show patience, politeness, and peaceable intentions but, at the same time, they are trying to find a way to convince people, to win them over through argument and bargams, and may even engage in bribery. Finally, a rajacic life involves avoiding confrontation and communications.

Tamasic relates to inertia, materialism, and dominating behaviour. The end justifies the means and people seek control. The word is used as an alarm bell to alert people to danger ahead. People dominated by tamasic ideas see nature as an enemy and try to move away from nature as far as they can. People are dictatorial, cunning, fearful, and secretive. Tamasic ideas produce depression, dullness, apathy, and inertia, and are often associated with casinos, the underworld, black markets, crime, prisons and torture, factory farming, huge slaughterhouses, genetic engineering, and large-scale mining. One has blind faith in being right – I'm right, you are wrong. Monologue is the preferred form of communications.

Sattva, according to Kumar, grows into rajas and then degenerates into tamas. When the economy grows, but still operates within the parameters of social responsibility, we may speak of a rajasic economy. When humans and the Earth are used to serve the economy, it turns into a tamasic economy. Most multinational corporations operate in a tamasic way – "the pathological pursuit of profit and power". Paradoxically, most of us seem to desire sattvic outcomes such as beautiful surroundings, good health, and a comfortable and convenient existence, but we are also attracted by the rajasic way of living and we are split by our desire to achieve sattvic by using rajasic methods.

We want peace, but pursue the path of war. We want happiness, but follow the rajasic path of materialism, power, and money. We want to achieve community cohesion, but follow the rajasic temptation of self interest, and so we sink deeper and deeper into personal frustration and political misery.

There are no easy solutions. Sattvic methods are humble. They move slowly, but surely, and develop step by step. The genuine sattvic way of living can only emerge from within by respecting the nature of reality and following it with humility. In the next section, we discuss how these ideas are connected to western philosophy and psychology.

Connecting east and west

From the times of the Greeks right up to today people have wanted to live a good life. To live it, one should have a notion of what it is. A good life is often connected with happiness and meaning. But what is happiness? And what is the meaning of life? The first question is asked by philosophers and psychologists. One tradition describes happiness by referring to feelings and emotions linked to the absence of suffering and pain. Ethical egoism and utilitarianism are examples of an egocentric definition of happiness. Aristotle argued that happiness is related to what we do and how we live together with other people; "The good life in the good society." Virtue ethics focus on our ability to function as a good member of a good society. The challenge is how to practice the art of living. Philosophers in the Aristotelian tradition, such as Taylor and MacIntyre, argue that we have to reflect on life to be able to master it. Happiness on a deeper level is connected to the meaning of life, which cannot be separated from a context of interpretation. This means that the reflection must include not only the specific topic, but the frame of reference too.

If we have a fragmented frame of reference, for example, the atomistic competitive market economy (cf. rajasic), we may be misled into focusing one dimensionally on things such as money, a new car, or a big house, or we constantly search for new experiences to get a kick. Stress and hyperactivity are closely connected to happiness in a fragmented world. Leisure time without action is considered boring and problematic. Meaning and happiness, on a deeper level, are anchored in cooperative co-existence (cf. sattvic). When we do not have to compare ourselves constantly against others in the pursuit of status, it is possible to discover a more relaxed position in which we become integral parts of the environment instead of actors forever fighting for material successes. This distinction between happyology and a far deeper happiness – where the latter involves second-order reflection on our wishes and desires – characterizes mainstream economics and ecological economics, respectively.

A western perspective

We may reformulate the question of meaning of life into two subquestions: (1) what is the purpose of life? and (2) what is the purpose within life? If we follow the first subquestion we may ask: what is the purpose for which human beings were placed on Earth? One category of answers to this question would be some idea of a spirit, or a God-created world. Another category of answers is that we are all thrown into a material world in which matter and energy interact according to the principles of Darwinian evolution (Haidt, 2006, p. 217). If we follow the second subquestion we may ask: how ought I to live? How should I live to have a meaningful and good life? Aristotle asked this question and his answer was "arete", which means excellence or virtue. He claimed that people are like archers who need a target at which to aim. Without a purpose one will not be able to live a good life. Viktor Frankl, the Austrian psychiatrist, expressed the same message in a number of books.

Jonathan Haidt separates the two questions: the first question asks about life from the outside, it looks at people, the Earth (…) as objects; "why do they all exist?" (Haidt, 2006, p. 218). The second question looks at life on the inside from a subjective point of view; "how can I find a sense of meaning and purpose?" This question is really an empirical question that we can approach with scientific means. "Why do some people live lives full of zest, commitment, and meaning, but others feel that their lives are empty and pointless?" (Haidt, 2006, p. 219).

Life satisfaction

Kahneman (2011) stresses the importance of obtaining one's goals for life satisfaction. A person with a moderate income goal at the age of 18, and who obtains that goal as adult, may be more satisfied than another adult with a much higher income, but who does not consider that his/her more ambitious goal was obtained. "Teenagers' goals influence what happens to them, where they end up, and how satisfied they are" (Kahneman, 2011, p. 402). It means that the goals that people have are essential to how they feel, and this has implications for experienced well-being as well as how they think about their life and their satisfaction with it.

Shaw Achor (2010) uses concrete examples when he writes about the effects of positive and negative goals in a work setting. Chip Conley, a CEO of an innovative hotel group, tries to engage with his employees in different ways. One strategy he uses is to tell his employees: "forget about your current job title. What would our customers call your job title if they were to describe it by the impact you have on their lives?" (Achor, 2010, p. 80).

When you make these larger connections your mundane tasks not only become more palatable, but you perform them with far greater dedication.

Another strategy is to end his executive meetings by allowing one participant to talk for one minute about someone in the company who deserves recognition. Then another executive will call, email, or visit that employee to tell her or him what a "great job" he or she is doing. "The benefits here are far reaching. The recognized employee … feels great, as do both the executive who made the recommendation and the executive who delivered the praise. Everyone else gets a mood boost as well" (Achor, 2010, pp. 58–59).

Achor also emphasizes how easy it is to destroy a positive mood when reducing the meaning of work to only receiving a pay-check; once upon a time a senior level executive introduced Achor to a meeting by telling the audience that "you are all here at work to make money (…) To be honest, it has to be about money: we're not saving the dolphins here." With this short introduction, the executive managed to disengage the employees by telling them that their work was only meaningful because it was financially rewarded (Achor, 2010, pp. 80–81).

Later research found that unemployment shows the black side of meaninglessness. To be unemployed is far worse than merely losing income because unemployment is also related to a decline in self-respect and the social relationships normally created by work. Unemployment also reduces happiness by creating the feeling of not been able to contribute to the wider society. There are several studies into the factors that influence the extent of happiness people enjoy. The British economist, Richard Layard (2005), sums up the "Big Seven" factors which have the greatest influence on happiness; family relationships, financial situation, work, community and friends, health, personal freedom, and personal values. All the different studies find that family relationships are more important than any other single factor. The next important factors are work and health. Layard, who is an economist, argues that economists have widely underestimated the negative impact of being unemployed, and a German survey by Winkelmann and Winkelmann (1998) supports his argument.

The significance of meaning

The significance of meaning is eminently described by Victor Frankl in several books and articles (see among others, 1962, 1969). Frankl, a Jew, was deported and imprisoned during the Second World War. In spite of extreme suffering he survived several concentration camps and was able to write about his extreme experiences. Frankl observed that prisoners who experienced or constructed some kind of meaning were able to stand the extreme suffering and pain in the camps. Prisoners who were unable to "find" any meaning failed to survive.

Backed by positive psychology (see Seligman, 1990, 2002, 2011), a new strand of scientific research and a number of studies have elaborated on the concept of happiness. One distinction made by Seligman is the difference between pleasure and meaning. The empirical findings show that meaning is more important than positive emotions. Seligman defines meaning as "belonging to and serving something that you believe is bigger than the self" (Seligman, 2011, p. 17). Meaning has both a subjective and an objective component, and the latter can contradict the subjective judgment.

The example par excellence is Abraham Lincoln, who was a melancholic. He may have assessed his life to be meaningless, while we, in the judgment of history, may very well contend that he lived a most meaningful life. Haidt writes that meaning is constructed in a number of relationships or connections to things beyond you, something larger than your self. "If you are able to get the relationships right, a sense of purpose and meaning will emerge" (Haidt, 2006, pp. 238–239).

The third dimension

According to Haidt (2006), the social world has two dimensions: a horizontal dimension of closeness, and a vertical one of hierarchy or status. The closeness dimension makes distinctions between close versus distant relatives and between friends and strangers. The hierarchical dimension makes a distinction between leaders and followers, between high status and low status individuals, and between rich and poor. These distinctions are seen clearly in many languages in the way people are addressed. In French "vous" is used for superiors and strangers, but "tu" for subordinates as well as friends. In English, people who are superior are addressed by using their titles and last names, but first names are used for friends and relatives. However, to account for an extraordinary or overwhelming experience of natural beauty, where you feel "lifted up", we may need a third dimension – a kind of vertical dimension. Haidt claims that we need the third vertical dimension because the "human simply does perceive divinity and sacredness …" (Haidt, 2006, pp. 183–184).

This third dimension, which may be called "divinity" is, in fact, a moral dimension. It may be independent of the existence of a God. The Chinese philosopher, Meng Tzu (also known as Mencius 372–289 BC) calls this dimension noble versus ignoble and from it it follows that human beings are "impoverished (…) when we lose sight of this dimension" and reduce our world into two dimensions (Haidt, 2006, p. 184). Secular thinkers have ignored this dimension. However, this dimension is fundamental if we are to understand religions such as Hinduism and Buddhism. Purity and polluter practices can be seen as a means to an end where the purpose is to move up to the third dimension to reach moral advancement. Purity is not just about washing the body, it is related to virtue and morality. It is about the soul, it is an ethic of divinity (Haidt, 2006, pp. 189–190).

The beauty of nature

In his *Critique of Practical Reason* Immanuel Kant connected morality and nature when he wrote that the two causes of genuine awe are "the starry sky above and the moral law within". According to Haidt (2006), awe is the emotion of self-transcendence. In our modern desacralized world awe has been reduced to surprise plus approval, but in history it is something greater than the self. In the Bhagavad Gita, a prototype of awe is found. William James writes that the deepest truth was that "we experience life as a divided self, torn by conflicting desires" (cited in Haidt, 2006, p. 204). Religious experiences are common and the experiences often make people whole and at peace. In a rapid type of conversion, the experience of the old self "full of petty concerns, doubts, and grasping attachments" is washed away in an instant of profound awe. People feel reborn; fear, and sorrow are diminished, and the world seems "clean, new, and bright" (Haidt, 2006, p. 204).

These findings are in accordance with Maslow's peak-experiences that make people nobler. In his *Religions, Values, and Peak-Experiences*, Maslow (1964) attacks science for becoming as sterile as organized religion. Maslow stated that modern science has helped to desacralize the world by documenting only what is, rather than what is good or what is beautiful. Maslow called for the reformation of education and society. "Education must be seen as at least partially as effort to produce the good human being, to foster the good life and the good society" (Maslow, 1964, p. 58).

Keltner (2009) describes awe as a deep emotion and connects it in particular to transformative experiences. Like the Irish philosopher, Edmund Burke, Keltner liberates awe from religion and gives an account of the awe John Muir – one of the pioneers of the American environmentalist movements – experienced when he started his naturalist's pilgrimage in the United States. In 1969, when Muir was 20 years old, he was herding sheep through the Sierra Nevada mountains. Muir kept a dairy and on June 6 he wrote:

> We are now in the mountains and they are in us, kindling enthusiasm, making every nerve quiver, filling every pore and cell of us. Our flesh-and-bone tabernacle seems transparent as glass to the beauty about us, as if truly an inseparable part of it, thrilling with the air and trees, streams, and rock, in the waves of the sun – a part of all nature, neither old nor young, sick nor well, but immortal. (Cited in Keltner, 2009, p. 251)

Discussion

Sattvic, rajasic, and tamasic represent three different qualities inherent in all kinds of phenomena. The more sattvic, the more authentic quality of life, or true happiness. The road to sattvic lifestyle goes through a reflection on life experiences. This is in accordance with Taylor and Maclntyre's focus on reflection of our lives. Kumar and Haidt agree that awe has been reduced to no more than surprise plus approval in the modern society. Taylor argues that individualism is transformed to egoism, and intrinsic values are transformed to instrumental values. They all agree that the problem is closely connected to the loss of an extended self that is greater than an individualistic self; a self-oriented award, one's ego and self-realization means little more than an ego trip. The loss of an extended self is replaced with a maximal ego (Seligman, 1990). This new self, with its absorbing concern for its gratifications and losses, is much more concerned with feelings than with duty. This new and modern self is full of inflated expectations rooted in the expansion of choice and, in the western world, is characterized by an overabundance of choice. Barry Schwarts (2004) writes about the paradox of choice: more choices lead us into a situation where we never become satisfied with what we have because there are always glittering alternatives that we could have chosen.

Seligman reminds us of a painting by Giotto in which everyone but Jesus looks like everyone else. From the Middle Ages right up until the late Renaissance the self was minimal. However, toward the end of the

Renaissance, the self expanded as a concept and, in Rembrandt's and El Greco's paintings, bystanders no longer looked like members of a chorus (Seligman, 1990, p. 283). Along with new freedoms, this new maximal self involves dangers. One of the consequences is massive depression. According to Seligman "the waxing of the self (...) coincided with a diminished sense of community and loss of higher purpose. Lack of community and lack of a meaning proved to be rich soil for depression to grow in" (Seligman, 1990, p. 284). The fullest form of meaning is to be found in a united self. In the eastern tradition it is obtained when the brain, mind, and culture are in harmony.

Csikszentmihalyi observed that people do many activities without any external rewards. The activities were autotelic and characterized by enjoyment and intrinsic motivation. The concept Csikszentmihalyi coined was flow. One outstanding composer tells how he feels when he is writing music:

> You are in an ecstatic state to such a point that you feel as though you almost don't exist. My hand seems devoid of myself, and I have nothing to do with what is happening. I just sit there watching it in a state of awe and wonderment. And it just flows out by itself. And as chess players say: "time passes a hundred times faster". A dancer expresses well this paradoxical feeling of simultaneously being in control and being merged with the environment. If I have enough space, I am in control (...) I become one with the atmosphere. (...) A strong relaxation and calmness comes over me. (...) What a powerful and warm feeling it is! I want to expand, hug the world. I feel enormous power to effect something of grace and beauty. (Csikszentmihalyi, 1975, p. 44)

Research has found that flow activities lead to a stronger and more united self. One of the implications of flow activities on the mind and the body is being healed, recovered, and antidepressed.

Arne Næss (1989) also writes about the importance of experiences being connected to other sentient beings and nature. According to Næss, processes of identification are central. One may start with one's nearest, and then continue with circles of friends, local communities, tribes, races, humanity, life and, ultimately, unity with the supreme whole, the world in a broad and deep sense. This will be a kind of self-realization with ever wider wholes, up to the level of the cosmos as a whole. This self-realization is grounded in recognition of strong interconnectedness. We are all intensely connected to each other in a metaphysical sense. One critic of Næss, Val Plumwood, argues that Næss has not been able to step out of the anthropocentric framework since value is based on human values (cited in Vetlesen, 2015, p. 119). What we need is an alternative basis for ecological action, one which recognizes the other's incommensurability. What we need, according to Plumwood, is solidarity, and that means standing with the others in a supportive relationship in the political sense (see Vetlesen, 2015, p. 119).

According to Peter Pruzan, Buddhist and Hindu teaching offers a nondual perspective on reality. It means that we simply are at one with an underlying reality. Hinduism refers to this as Brahman, and Buddhism refers to Nirvana (Pruzan, 2015, p. 134). It implies that humans are encouraged to seek and see unity in the diversity that is perceived in the outer world. According to Pruzan, there is a Sanskrit saying that refers to this aspect as "Tat tvam asi", which is often translated as "Thou art that" or "That thou art". This may be interpreted as the Self is one with ultimate reality – that there is identity between "Tat", the Absolute ("That"), and "twam" ("thou"), also referring to the Self.

> According to the wisdom of the East, the greatest obstacle to being able to realize this identity, (…) is the ego mind that continually experiences a distinction between self and all other aspects of reality and thereby inhibits the realization that "thou" and "That" are one. Only when the ego is transcended can unity or union with (…) the Divine Ground, The Universal Consciousness, be achieved. (Pruzan, 2015, p. 134)

On one hand, Kumar argues that we are tempted by the economic system to move in the direction of more focus on materialistic and egoistic values. The consequence is the satisfaction of superficial consumption needs. Too much focus on spiritual values, on the other hand, could lead to apathy toward the physical reality. Therefore, a balanced lifestyle leads to a harmonious wellness. We find the same way of understanding happiness in humanistic and positive psychology, put forward by Maslow and Seligman, as they claim that quality of life depends on both materialistic and spiritual values. It is interesting to see that Seligman in his book, *Flourish* (Seligman, 2011), strongly argues that he was wrong in his book *Authentic Happiness* (Seligman, 2002). He had put too much emphasis on positive emotions, such as a cheerful mood as a prerequisite for happiness. Even if meaning and engagement were stated as elements of *Authentic Happiness*, the theory was confusing because neither engagement nor meaning refers to how humans feel. Furthermore, "life satisfaction" has a far too privileged position as a metric for measuring happiness. In his later book, Seligman (2011) also included positive relationships and accomplishments and calls it the "Well-Being" theory.

Quality of life is connected to meaning within life in both eastern and western traditions. They also agree that meaning depends on a wider context than the ego. The idea of the extended self, or self-transcendence, unites both perspectives and gives a deeper understanding of the value of work. Work has three different functions. First, it makes it possible to develop individual abilities. Second, working together with others reduces selfish tendencies. Third, work contributes by producing commodities and services that meet the needs of other people. These ideas, central to Buddhist economics, are close to what Achor described in his hotel example. Research also finds that people can approach their work either as a job – that means working only for the money – or see their work as a career where you work for larger goals of promotion and status – or third, regard work as a calling. A calling means to find work intrinsically fulfilling because it contributes to the greater good. In particular, if you see your work as a calling, work is a very meaningful activity.

There are two possible ways to counteract the tendency to move in the wrong direction from sattvic toward tamasic values. We can either change the system or we can develop human consciousness. Even if in this chapter we have focused on individual change, we remain convinced that deep changes at the system level are also necessary. Our idea is that individual change is a prerequisite for change in the economic system. By reversing the tendency to develop a narrower ego, as in egocentrism, we will remind the individual to follow a path that leads to what is true to them and feels natural and real. When individuals consider themselves as part of nature, and not apart from nature, the context of quality of life is reestablished. The separation between people and nature is one of the most important obstacles to happiness and finding a meaning of life.

Bibliography

Achor, S. (2010). *The happiness advantage.* New York, NY: Crown Business.

Csikszentmihalyi, M. (1975). *Beyond boredom and anxiety.* San Francisco, CA: Jossey-Bass Publishers.

Frankl, V. (1962). *Kjempende livstro.* Oslo: Gyldendals Norsk Forlag.

Frankl, V. (1969). *The will to meaning.* Oslo: Aventura pocket.

Haidt, J. (2006). *The happiness hypothesis. Putting ancient wisdom and philosophy to the test of modern science.* London: Arrow Books.

Kahneman, D. (2011). *Thinking fast and slow.* London: Penguin Books.

Keltner, D. (2009). *Born to be good. The science of a meaningful life.* New York, NY: W. W. Norton & Company.

Kumar, S. (2002). *You are therefore I am – A declaration of dependence.* Cambridge: Green Books.

Kumar, S. (2007). *Spiritual compass – The three qualities of life.* Cambridge: Green Books.

Kumar, S. (2013). *Soil – soul – society: A new trinity for our time.* London: Leaping Hare Press.

Layard, R. (2005). *Happiness: Lessons from a new science.* London: Penguin.

Maslow, A. (1964). *Religions, values, and peak-experiences.* Columbus, OH: Ohio State University Press.

Næss, A. (1989). *Ecology, community and life-style. Outline of an Ecosophy.* (Trans. and ed., David Rothenberg). Cambridge: Cambridge University Press.

Pruzan, P. (2015). The source of ethical competency: Eastern perspectives provided by a westerner. In K. J. Ims & L. J. Pedersen (eds), *Business and the greater good. Rethinking business ethics in an age of crisis.* Cheltenham, UK: Edward Elgar.

Schwarts, B. (2004). *The paradox of choice. Why more is less. How the culture of abundance robs us of satisfaction.* New York, NY: Harper Perennial.

Seligman, M. (1990). *Learned optimism. How to change your mind and your life.* New York, NY: Vintage Books.

Seligman, M. (2002). *Authentic happiness. Using the new positive psychology to realize your potential for lasting fulfillment.* New York, NY: Free Press.

Seligman, M. (2011). *Flourish. A visionary new understanding of happiness and well-being.* New York, NY: Atria Paperback.

Vetlesen, A. J. (2015). *The denial of nature. Environmental philosophy in the era of global capitalism.* London: Routledge.

Winkelmann, L., & Winkelmann, R. (1998). Why are the unemployed so unhappy? Evidence from panel data. *Economica, 6j,* 1–15.

KNUT J. IMS

Quality of Life in a Deep Ecological Perspective: The Need for a Transformation of the Western Mindset?[1]

Introduction

The Norwegian philosopher Arne Næss made a distinction between the shallow and the deep ecological movements (Næss, 1973). While the shallow movement was regarded as a technocratic approach that tried to minimize the symptoms of ecological destruction of the planet, the deep ecological movement sought to address the deep questions in order to find sustainable solutions to the alarming environmental problems. With a motto "live a rich life by simple means", in 1984 Næss and George Sessions formulated eight platform principles of the deep ecology movement (Næss and Sessions, 1995; Næss, 1995). The platform were articulated as a core to be accepted by persons coming from different philosophical and religious positions. Deep ecological reasoning might be seen as a forerunner of the degrowth movements heavily inspired by Georgescu-Roegen's (1971, 1975) *The Entropy Law and the Economic Process*, which criticizes a number of the taken for granted assumptions in mainstream economics. Georgescu-Roegen shows that the mechanical laws used in mainstream economics are far from the correct interpretation of facts, and that a biological model is relevant to understand economic processes. The most dramatic fallacy is that economic processes are often represented by a closed circular system, with a continuous inflow of low entropy available energy from the environment. The economic processes are in fact "unidirectional"; they are a continuous transformation of low entropy into high entropy. It means a transformation from the inputs of valuable resources (low entropy) into the final outputs of valueless waste (high entropy). The law of entropy entails that we cannot use the same energy or matter over and over again. Its quality will be reduced and, finally, it will end up in an ecological catastrophe of irrevocable waste, which might be characterized as pollution (Georgescu-Roegen, 1971, p. 281). Pollution is typically a surface phenomenon that is easy to spot, but pollution is inseparable from resource depletion which does not get so much attention. According to the law of entropy, the "bigger and better" washing machines, automobiles, and superjets must lead to more pollution. Georgescu-Roegen's conclusion is that "any use of the natural resources for the satisfaction of nonvital needs means a smaller quantity of life in the future" (1971, p. 21). The best use of iron is to produce plows not Rolls Royces.

Standard economic theory builds upon a number of myths (Georgescu-Roegen, 1975). One of the myths assumes that without economic growth western civilization will collapse. The theory of economic development is based upon the goal of maximal economic growth, anchored in exponential growth models. Even when distinguishing between growth and development, defining development as involving innovations, economic growth is a dynamic state. Development will induce growth, understood as "economic growth", and measured by Gross Domestic Product (GDP). The deep ecology and the degrowth movements regard the economic growth mania as a reflection of man's "Master" image that there are no natural limits. However, a tragic side effect of economic growth is an increase in the rate of resource depletion. Exponential growth as a vision is logical as well as practical impossible, but immensely destructive in terms of depletion of resources. According to deep ecology, economic growth will not lead to a higher degree of an overall self – realization and happiness. The degrowth movement is a newer political, economic, and social movement based on ecological economics and anti-consumerist ideas, and may be regarded as an "essential and urgent" economic strategy in "overdeveloped" western countries (See Assadourian, 2012). The deep ecology and degrowth movements have fundamental similarities. They both warn that a crucial error of the standard economic theory is that a growth and zero-growth state cannot exist in a finite environment. It is urgent that overconsumption in the western world be reduced. Deep ecology advocates that a change in lifestyle towards a more simple life might be richer as well as more fulfilling.

In this paper I will focus on deep ecology for several reasons. First and foremost, deep ecology is compatible with Næss's own framework, his "ecosopy T", which was Næss's own personal view of the unity of man and nature. Næss, who lived from 1912 to 2009, walked his talk, and lived as often as he could in his simple cabin high up in the mountains of Norway. The total estimated time Næss lived in his cabin is ten years. At the same time Næss played a decisive role as the first professor in philosophy in Norway, with a thirty year career as a full time university professor, but he retired before he was 60 years old to completely dedicate his time to protect the vulnerable natural environment. It meant that Næss lived unusually close to nature; he insisted that he did not "go out in nature, but into nature". Nature was his home. Næss did not look upon himself as separated from nature and emphasized gestalt phenomenology to explain his concept of nature which is a key to understand the unity between man and nature. Næss claimed that as long as one's attention is not deliberately focused upon perceptual gestalts, all experience consists of units of "apperceptive gestalts", which is "not sensory elements, not intellectual elements" (Næss, 1974, p. 202; Næss, 1989, p. 60). Næss also constructed a system of norms, where self-realization of every sentient being is the top norm. Næss's "ecosophy T" and its sub-category deep ecological reasoning is heavily inspired, named and framed from ideas from Hinduism, Confucius, and Buddha on the one hand and Aristotle, Heidegger, and Spinoza on the other hand (Næss, 1989).

Based upon deep ecological reasoning, the western world's extensive use of GDP is a delusion and a destructive metric of human and other sentient beings attempts to obtain self-realization and flourishing. According to Næss, what we need is to make a profound distinction between big and great. GDP is a measure of the size of market transactions in society, but neither tells about the degree of ecological sustainability nor the extent of self-realization among people or other sentient beings (Næss, 1989; Sen, 1999; Ims, 2011). Gamlund (2012) criticizes deep ecology and argues that the framework is not appropriate to build environmental ethics, which is urgently needed today. However, Gamlund acknowledges that deep ecology's use of the gestalt theory is an interesting perspective. This paper will argue that in spite of relevant critics of deep ecology (Gamlund, 2012; Vetlesen, 2015), deep ecological reasoning is an important contribution and position in contemporary western society that deserves to be discussed as a source of knowledge, wisdom, and inspiration because we need a transformation of our western ideology. Næss's use of gestalt phenomenology as an explanation of the appreciation of unity between man and nature is an interesting aspect to be explored. Deep ecology should be seen as a narrative and an outline that still needs to be refined, developed, and filled out with concrete examples. With such a non-finished view on deep ecology an exploration might also be valuable for the newer anti-industrialists' movements, like Degrowth (Assadourian, 2012). It might give supporters of Degrowth insight into an interesting ecosophical framework, and deep ecological enthusiast's better knowledge and understanding of how to proceed in the future to obtain a more harmonious relationship between nature and culture.

There are cases that negate that deep ecology is an ecotopia. By studying and learning from them, we may collect and analyze knowledge which fills out deep ecology. One case of strong validity is a Buddhist approach to quality of life proposed by the King of Bhutan in 1979. Two practical cases from Thailand, also a Buddhist country, will also be highlighted.

The main purpose of this paper is twofold: (i) to dig into some of the problematic aspects of our western mindset that more and bigger are better than "less is more" and a frugal lifestyle, and (ii) advocate a transformative shift in ideology in our western world. By elaborating some parts of deep ecological reasoning supported by Næss's ecosophy T, and by presenting some real life cases, we will analyze actual, real-life alternative cases to mainstream views. Ecosophy T challenges the one-sided material lifestyle as the best way to obtain a good life. One main purpose is to find different kinds of evidences that can expand our understanding with new concepts and frameworks. Hopefully, this knowledge may contribute to the urgent need for a transformation of western lifestyles into a more nonviolent path, to live in a mindful and fruitful harmony with nature (Buergelt et al., 2017).

Ideological change from standard of living to life quality

Why is "increasingly higher standard of living" to be avoided? According to deep ecology, increasingly higher standards of living will mean reduced life quality to most people, while at the same time it will destroy the Earth. A higher standard of living is a very violent path for most sentient beings, and will even contribute to the dangerous climate change (Næss, 1989). The main reason is that the rate of energy consumption and the level of consumption of material objects have been regarded and measured as societal "progress" (Næss, 1989; Georgescu-Roegen, 1971). Standard of living is part of a deep-seated production and consumption ideology in industrial states, which has been the surrogate measure of a good life. Within the standard of living framework, it is an important goal of business to create wants and entice customers to increase their material consumption. The underlying assumption is that if the growth of consumption stopped, economic crises and large-scale unemployment would result. For the consumers it has led to "satisfraction".[2] A materialistic oriented shopper, a "shopaholic" will never be fully satisfied (Kasser, 2011). However, the consumer will be inclined to work more in order to consume more in order to have a higher material standard of living than one's neighbor. There will always be a reference group to challenge the individual consumer to obtain a higher standard of living. However, the material affluence in the western world cannot be replicated in Africa or Asia because we have only one planet. The overshot in the western world is far more than the carrying capacity of our only planet (Assadourian, 2012).

Is increasing energy demand a basic need in the western world? Or should we look upon the demand as a kind of "miswanting"? Material acquisition above a certain level does not lead to greater well-being but might lead to less, since everyday small enjoyment loses its value when income levels are relatively high (Kahneman, 2011).

In 1979, the fourth King of Bhutan pronounced that happiness and well-being was the purpose of the state. He had observed that many nations had modernized successfully, but "often overlooked a universal desire that people had – happiness and peace in their life" (Ura et al., 2015, p. 4). The King initiated a project "to expand well-being and true happiness based on a holistic view on happiness" using the metric of Gross National Happiness (GNH). The Centre for Bhutan Studies has developed a sophisticated approach to measure human well-being, along nine different domains and a number of variables (Ura et al., 2012a, 2012b; Alkire, 2015). Bhutan's GNH's approach has been an important inspiration and a field of study for economists from different parts of the world.

In an earlier paper[3] I compared Bhutan's GNH with deep ecology in order to explore whether these two approaches to sustainability and deep happiness can inspire and learn from each other. The conclusion was that the two approaches complement and enrich each other in several ways. It means that deep ecology and a Buddhist mindset have common denominators, which increases the validity of each positions, and opens up for an explorative study of both positions in more detail.

One exciting difference between deep ecology as a way of reasoning and GNH is that the latter is operationalized, applied, and followed up for policy purposes in Bhutan. GNH is used to adjust how the Bhutanesian government acts. Thus supporters of deep ecology can learn from the implementation and experiences of GNH's application in many ways. While some authors have discussed whether deep ecology is an utopia (Anker and Witoszek, 1998), GNH is not. GNH is real and applied within a small Buddhist country.

The economic system in the western world has generated a number of crises. The financial and environmental crisis are only the tip of the iceberg (Ims and Pedersen, 2015). There is also a crisis of identity which may be related to the new malaise "affluenza" – the obsessive quest for material gains in an endless effort to "keep up with the Joneses" (de Graaf et al., 2005). Finally, we may speak of a moral and spiritual crisis wherein a materialistic value orientation becomes ever more dominant. Some of the consequences of this material orientation, Kasser (2011, p. 204) argues, may be lower personal well-being and more "manipulative, competitive, and ecologically degrading behaviours". The consequences of such crises include unstable social environments, which erode trust and create a loss of respect. But in any crisis there is hope for a new beginning.

GDP is the extensive measure of economic growth and relates to standard of living. However, it is a shallow and misleading way of measuring human happiness and ecological well-being and, as Næss has written, GDP might be translated as Gross Domestic Pollution and is not related to self-realization, which is the top norm in deep ecology. GDP represents an anthropocentric view, while Næss advocates an eco-centric view. To transform the western mindset into a more eco-oriented and sustainable way of living, in order to live more in harmony with nature, requires a radical shift in mindset The prevalent western mindset was formed during the Enlightenment (Vetlesen, 2015). A study of the Enlightenment might give us some keys to see the roots of the problem. Vetlesen (2015) argues that westerners suffer from a "mental loss" due to the marriage between positive science and technology. This marriage has led to a surprising economic growth in terms of material goods measured by GDP. The materialistic turn has been combined with a rationalistic way of thinking about and perceiving the world. By swallowing the worldview created during the Enlightenment, we have lost many enriching aspects of how to live a good life through tight connections with human beings, as well as animals and nature.

The platform level and appreciation of life quality and nature

In the late 1970s Næss and Sessions suggested that the platform level of the deep ecology movement consists of eight points (Næss, 1989, pp 28–32):

1. The well-being and flourishing of human and nonhuman life on Earth have value in themselves – independent of the usefulness for human purposes.
2. Richness and diversity of life forms are also values in themselves.

3. Humans have no right to reduce this richness and diversity except to satisfy vital human needs.
4. We need a substantial decrease in the human population.
5. The present interference with the nonhuman world is excessive
6. Policies which affect basic economic, technological, and ideological structures must be changed
7. *The ideological change is mainly that of appreciating life quality (dwelling in situations of inherent value) rather than adhering to an increasingly higher standard of living. There will be a profound awareness of the difference between big and great. (Written in italics for the purpose of dissecting it.)*
8. Those who support these points have an obligation to directly or indirectly try to implement the necessary changes in a non-violating way.

These eight points are interconnected, and it may be regarded as artificial to select only one of the points for a deeper analysis. However, by elaborating on point 7 we will gain a more profound insight into the needed transformation of the mindset (Næss, 1989, p. 29). To see the relevance of deep ecological reasoning, the sufficiency concept in deep ecology will be supported with insights and arguments from the Bhutanesian GNH perspective and Buddhist inspired economic models based upon sufficiency and simplicity in Thailand. The concept of mindfulness (Langer, 2009) and a mindful approach to the economy will be presented. Relevant insight from the operationalizing of the GNH Index, which aims to balance material wants with spiritual needs, will be given to illustrate alternatives to western ideological thinking.

Instead of ideological change we might like Buergelt et al. (2017) use the notion of *shift*. Change means more of the same, shift means a fundamental transformation in consciousness or fundamental beliefs. An ideological shift from standard of living to life quality is the most crucial aspect. By ideology I mean a system of ideas and beliefs which forms the basis of cultural, economic and political thinking, and policy. In this context my assumption is that there is a political hegemony of technocratic values in which economic growth, globalization, and profit is elevated to normative standards. It means that technical-economic values are dominant, and that private as well as governmental organizations are measured and prized whether they make profit or not. It also means that people perceive their relationship with the environment largely in anthropocentric terms (Paton et al., 2015).

The main outcome of a deeper understanding of point 7 would be to open up to ideas and thought patterns that should be part of an ecological mindset, including a more developed ecological consciousness. Obviously, this approach will challenge the mainstream view of economists who criticize the concept "quality of life" because it is vague. What economists consider vague is the non-quantitative nature of the term "quality of life". According to Næss and Sessions, "one cannot quantify adequately what is important for the quality of life … and there is no need to do so" (1995, p. 53).

Næss (1973) writes that an ideological change is necessary to shift from a framework based upon "increasingly higher standard of living" to a new framework based upon "quality of life".

But what is life quality in the perspective of deep ecology? A preliminary answer is given by Næss (1989, p. 81) in the following formula:

$$W = \frac{G^2}{P\,m + Pb}$$

W = Well-being
G = Glow (passion, fervour)
Pm = Mental pains
Pb = Bodily pains.

The formula means that with sufficient glow any amount of pain can be overcome. The usefulness of this expedition inspired "well-being formula" would be to make people explore "what they deeply and eagerly want and thus to make them risk some pain and discomfort in its pursuit" (Næss, 1989, p. 81). The essence of the formula is that what people might find meaningful is beyond feelings of pleasure and shallow happiness.

By negating hedonism and extreme *joy ala hilaritas* (in the words of Spinoza), Næss (1989) ends up with norms of perfection as the ultimate good. In particular, Næss states that an increase in perfection and joy into one totality, a gestalt, is his own personal ecosophical view. Perfection is the unfolding of that which lies deepest in human nature. Næss adds that those who advocate perfection argue that well-being in general stems from an active life, even if it entails much pain (Næss, 1989, p. 84). Næss ends up with his own ecosophy T and the top-norm "self-realization" for all beings to indicate a kind of perfection. It is not ego-realization as a selfish project, but includes personal realization in the meaning of the universal self, the absolute or the *atman*. Næss is heavily inspired by Spinoza concept of "self-perseveration", which requires sharing joys and sorrows with others. It means a deep identification with all life forms (1989, p. 85).

I interpret the glow aspect in the Næssian "well-being formula" to entail elements of flow, in the meaning of the psychology of optimal experiences pioneered by Csikszentmihalyi (1990). Flow has developed into an influential research topic over the last decades (see Nakamura and Csikszentmihalyi, 2002). Flow is a mental state "of which people are so involved in an activity that nothing else seems to matter, the experience itself is so enjoyable that people will do it even at great cost, for the sheer sake of doing it" (Csikszentmihalyi, 1990, p. 4). An often ignored aspect is that flow involves healing and may function as an antidote to depression. Flows imply "order in consciousness, and it may entail growth of the Self. In flow we are in control of our psychic energy, and everything we do adds order to consciousness" (op. cit. 1990, p. 40.). In addition, flow activities help to integrate the self since, in the state of deep concentration, consciousness is well ordered because all the senses, thoughts, intentions, and feelings are concentrated on the same goal (op. cit. p. 41). We might also relate the phenomenon and state of flow to an increase in quality of life, which may be costless in an economic materialistic sense. Optimal experience is based upon autotelic motivation, which demands that the doer is an active and creative being. Thus, flow activities do not need extrinsic rewards that involves cost to nature and society in terms of the use of scarce resources. On a systemic level, extrinsic rewards have hidden costs in terms of human economy since they drive out inner motivation (Frey, 1997). A special type of extrinsic motivation concerns power, fame, and esteem, which are based upon comparisons between persons. However, status and conspicuous consumption are following a zero-sum pattern, and there are always winners and losers, and the latter pay with decreased self-esteem (Frank, 1999). In contrast, autotelic actions, which may involve flow activities, puts less pressure on ecosystems, and leads to increased self-esteem.

How to change from an ever-increasing standard of living to a quality of life paradigm? The standard of living concept fits perfect into the prevalent western materialistic, anthropocentric framework. According to Næss (1999) it is necessary with an ideological shift in mindset and lifestyle. The necessary change is to move from a technological to an ecosophical mentality. To understand Næss's concept ecosophy, Næss elaborates on the concept of ecophilosophy, which is the utilization of basic concepts from the science of ecology – such as complexity, diversity, and symbiosis, to clarify the place of human beings within nature through the process of working out a total view (Næss, 1989, p. 3). An ecophilosophy can lead in two directions, either to: (1) develop a deep ecological philosophy or (2) give support to an international deep ecology movement, which includes all groups of people, scientists, activists, artists, and all who actively may work towards a change in anti-ecological political and social structures. Ecosophy is the reasoning process behind the philosophical side of ecophilosophy that investigates the particular methods of viewing the world in the perspective of the deep ecology platform. This shift in mentality is necessary, but not sufficient. What is needed is practical action that is directed towards the realistic possibilities of starting a new development based upon a total view.

One outcome of a transformation towards an ecosophical mentality is to find beauty in less is more. Instead of a materialistic outlook on the world, Næss suggest "an economic ideal of simplicity of means and richness of ends" (1989, p. 32). Newer research supports the view that people in general overestimate the value of material goods when they already live in an affluent context, and underestimate the value of investing in social relations (see Kahneman, 2011; Westacott, 2016). Another finding is that rich people, or people who are primed with money, will not appreciate small everyday "treats", such as a single scoop of ice cream, to the same extent that people with lower income. The conclusion is that to be rich in a materialistic sense implies a cost in terms of lower appreciation.

A different approach to "less is more" is taken by Schwartz (2004), who argues that the more the better in terms of choices or opportunities can be a grand mistake. His starting point is that westerners regard choice as an advantage. It represents a kind of freedom, a positive liberty to be the author of one's life, and to make life "meaningful and significant" (Schwartz, 2004, p. 3). When we can choose between two alternatives, it may be a good thing, but if we expand the alternatives, such as what we find in a modern shopping center, the sheer number of choices may be experienced as tyrannizing. Schwartz claims that the overabundance of choices in the western society is not a positive liberty, and gains support from the Nobel laureate in economics, Amartya Sen, who examines the conditions of freedom and autonomy (Sen, 1999). The issue, according to Sen, is whether freedom nourishes or deprives us, whether it enhances self-respect or diminishes it, and whether it enables us to participate or not in a community. Not all choices enhance freedom. Increased choice among goods and services may impair freedom by taking time and energy which we could devote to other activities (p. 4). More choice can even make us suffer when increased opportunities may lead to regret and feelings of inadequacy in comparison with others. Ultimately, Schwartz suggest that increased opportunities for choice may actually contribute to the recent epidemic of clinical depression affecting people in the West (Schwartz, 2004, pp. 99–217).

One theme discussed by Næss (1999) is whether more equality between different species is vital. He concludes that there should not be an absolute distinction between human beings and other sentient beings. Næss supports his argument with Spinoza, who writes about the awareness of the human role in the biosphere in a more egalitarian way (Næss, 1999, p. 113), while Descartes argues for an absolute distinction.

What might Næss mean by appreciating life quality? The word appreciating gives an important hint that it is not only the ability to experience quality of life, but also to acknowledge it, be aware of it, and enjoy it (Næss, 1989, p. 23). This follows from our intellectual capacity, with which we can perceive and care for the diversity of other forms of life. Næss states that with our biological heritage we are able to delight in the intricate, living diversity in our immediate surroundings, and we can perfect and facilitate our creative interactions with these surroundings. We can delight in the "value-creative aspects of equilibrium rather than glorifying the value-neutral growth which implies exploitation and killing of nature and other sentient beings. This awareness can be contrasted to consumerism and our industrial mega technology" (op. cit. p. 23). A distinction between growth and development may be relevant. While growth is connected to increases in material resources, development is connected to inner peace, imagination, and enjoyment of life in a wide sense, including being creative and developing empathy through voluntary activities, like going to the theater, concerts, and art exhibitions,

Buergelt et al. (2017) claim that living in close interaction with nature increase life quality. One example given is the presence of common natural spaces in urban settings, which will facilitate "community members to interact, develop a sense of community", and even lead to "individual and collective efficacy" (Buergelt et al., 2017, p. 5).

In point 7 there is a parenthesis which has rich interpretations, namely dwelling in situations of inherent value. One way to interpret it is that the practice of mindfulness may contribute to our understanding of "appreciating quality of life". Langer (2009) defines mindfulness in this way:

> It is a flexible state of mind – openness to novelty, a process of actively drawing novel distinctions. When we are mindful, we become sensitive to context and perspective; we are situated in the present. When we are mindless, we are trapped in rigid mindsets, oblivious to context or perspective. When we are mindless, our behaviour is governed by rule and routine. In contrast, when we are mindful, our behaviour may be guided rather than governed by rules and routines .... Mindlessness is not habit, although habit is mindless. (2009, p. 279)

Based upon research, Langer writes that "an increase in mindfulness results in greater competence, health and longevity, positive affect, creativity, and charisma and reduced burnout ..." (Langer, 2009, p. 280).

For Næss spontaneous experience is vital, which involves the depth, the intensity, and the richness of the content of experience in itself. The GDP and standard of living approaches try to replace deep human experiences with external indicators, such as measuring market transactions, levels of material consumption, the variety of goods, status, or successes in competition. Focusing upon those "goods" may not sensitize but, rather, desensitize the human capacity for spontaneous experiences. Næss believes in intuition, and defends our spontaneous, rich, seemingly contradictory experience of nature as more than subjective impressions. These experiences make up the concrete contents of our world and may lead to our ability to identify with other species.

How does Næss define gestalt? An apperceptive gestalt "synthesizes at least two elements into a higher (more comprehensive) unit, and at least one element is from a sensory area and at least one from a normative and/or assertive area" (Næss, 1989, p. 60; Næss, 1974, pp. 202–205). "All experience is apperceptive" when one does not deliberately focuses upon perceptual gestalts. Such gestalts are neither sensory nor intellectual elements, and they blur the usual distinction between facts and value.

Gestalt theory is central to explaining how our spontaneous experiences of reality occur (Næss, 1989, pp. 57–63). It is a negation of the inclination to look at objects in the world as separate and stable, and ignore the interconnectedness, the complexity, and changes that actually take place in the world. Then, how are the objects or entities related to each other? What we experience is more or less comprehensive and complex, and we have the capacity to think about relations and relationships between parts and wholes. The popular slogan of gestalt psychology, "the whole is more than the sum of its parts", neglects the hologrammatic part, even if it rightly negates mechanistic thinking and mechanical models. For example, a piece of music can only be understood if we include a hierarchy of higher-order and sub-order gestalts. A music fragment only gets its meaning from the whole piece of music (op. sit. 1989, p. 61).

Gestalt thinking present some challenges. First, it is difficult to learn to form gestalts "in an ordinary fashion" and, second, application of "scientific observation habits" will lower gestalt abilities. One way to destroy gestalt formation is by attempting to analyze fragments of gestalts consciously. It implies that "glorification of conventional scientific thought" is destroying gestalts. According to Næss, conventional scientific methods will "tears gestalts asunder" (1989, p. 61). Næss follows up with examples from the Sherpa people whose great mountain is called "Tseringma". The word is a geographical concept, but it is also the name of a wonderful white princess, and even "mother of long life". In a scientific discourse, this way of thinking may be said to be an ambiguity, self-contradictory,and "alogical". However, such unity between different entities can be said to be "mythic". A stone found by a European anthropologist was "simply a hard stone". However, in a mythic culture the stone was also a Spirit, "something not hard" (op. cit. p. 61). The important implications of this is that modern science identify fragments and not gestalts as contents of reality. Thus, mythic knowledge is largely unavailable in our culture. A has symbolic value B when A stands for B in conceptual experience. A has symbolic function in relation to B only if a gestalt is created that includes both A and B. In a symbolization of B through A, will A and B be bound together in an internal, not merely external, relationship.

There is a kind of participatory understanding in Næss's view, stating that: "I'm not much interested in ethics or morals. I'm interested in how we experience the world. If deep ecology is deep, it must relate to our fundamental beliefs, not just to ethics." (1989, p. 20).

Næss's gestalt view has a variety of important implications. One implication is that if we only have an instrumental relationship to nature, we might easily become alienated. And Næss advocates nature-therapy. The reason is that it is a very reductionist view to only "look at" nature. The genuine experiencing of nature as environment occurs by "doing something in it, by living in it, mediating and acting" (1989, p. 63). Næss stresses the necessity of interaction between elements of which we are a part. The reason is that "*Gestalts bind the I and the not-I together in a whole. Joy becomes, not my joy, but something joyful* ... I and something else are interdependent" (1989, pp. 60–61). Gestalt formation does not separate thinking and emotion.

The debasement of unreflective experiences

A profound critic of the western frame of reference is given by the Norwegian philosopher Arne Johan Vetlesen (2015), who argues for the importance of "unreflective experiences" and of the need for an intrinsic value of nature. Vetlesen discusses the "nature deficit" in critical theory (op. cit. pp. 52–95).[4] The proclaimed goal of the Enlightenment was the disenchantment, demythologization, and secularization of mythical, magical, and religious worldviews. It meant that the belief in scientific knowledge – the truth – can be obtained by appropriate methods. What can be measured, counted, and calculated should be the sole objects of science; the general is given primacy over the particular, the repeatable over the spatio-temporally situated, the abstract over the concrete first hand experiences, the quantitative over the qualitative (Vetlesen op. cit.).

Enlightenment meant that man's emancipation was to be free from spontaneity and natural instincts. "The subjugation of nature was man's purpose of life." Vetlesen (2015) writes that it is a "dialectic at work … between subjection of outer and inner nature:.." (p. 58). The deeper motif was self-preservation, and the price was denial and protest of every aspect of life that has us being alive and thriving …. "that threatens to defy the requirements of order over chaos, predictability over spontaneity, reason over affect" (p. 59). Autonomy, a central aspect of modern life requires self control, which presupposes repressing instincts, desires, and feelings. It means to restrain oneself from immediate needs and wants and let the inner and outer nature be mastered by rational calculation, to deny that man is nature, and treat our inner nature as an object of total domination. In the end, the outcome is the annihilation of life itself. So what is civilization? It is the denial of everything natural, spontaneous urge, or novel experiences – all aspects of life that have us being alive (Vetlesen, 2015). The ideology of the Enlightenment is that knowledge is inseparable from power and the essence of power is technology. We should learn from nature in order to dominate nature and other men.

The main argument from Vetlesen (2015) is that the Enlightenment has entailed looking upon and treating reality as an object rather than as a direct unreflected experience. Our direct lived experience of the world has been devalued in favor of abstract thinking about the world. Vetlesen also is critical of Næss's emphasis on identification with other sentient beings in larger and larger circles, finally ending up with an identification with all living on Earth. Vetlesen argues that identification is based on "sameness" (2015, p. 119) and, in a deeper sense, based upon an anthropocentric framework. It implies that the human is put at the center of value despite the biocentric commitments of deep ecology. In contrast, Vetlesen (2015, p. 119) writes that the basic concept of deep ecology should be solidarity, standing *with* the other in a supportive relationship. It requires sensitivity to the difference between seeing oneself *with* the other or *as* the other.

Economic models and real life cases aligned with deep ecological reasoning

There are empirical examples of alternative economic models and life styles in Asia that can be used as sources of inspiration for a transformation from a standard of living approach to an ecosophical mindset. Magnuson (2011) elaborates about pathways to a "mindful economy", based upon 2,600-year-old tradition of Buddhist practice. He criticizes "business as usual" in the US and argues for a more sustainable economy based upon local communities, networks, democracy, social justice, equity, and ecological sustainability. A mindful economy needs institutional change and a culture that evolves away from the existing growth oriented, profit driven capitalist machine. The concept of culture is essential since it consists of both material elements and non-material elements, of which institutions play an essential role. According to Magnuson we need to evolve the institutions through mindful practices into structures that will empower people and protect nature.

Sufficiency

There are interesting Buddhist economic models that practice an economics based on sufficiency. One is the royal Thai sufficiency economy model (SE) (see Essen, 2011). The SE model operates on the principles of moderation, reasonableness, self-immunity, wisdom, and integrity. The King of Thailand advanced the idea that introduced a theory which is more or less a negation of the growth theory suggesting, instead, a self-supporting economy interpreted as "have enough to survive". It did not mean a reversal to traditional subsistence living, but a model based on frugality. This represents well the Buddhist notion of the middle way, of not having too much and not too little. The model is not based on rationality, but reasonableness which implies compassion in analyzing reasons and potential actions and taking into account the immediate and distant consequences of those action (Essen, 2011, p. 66). In the Buddhist view this approach involves a moderate level of consumption and aspiration for wealth. It involves concerns for employees and for the environment, that is, genuine care for all externalities and for the long-term health of the business.

Another approach is adopted by the Santi Asoke Buddhist Reform Movement of Thailand (AM) (see Essen, 2011). The AM movement promotes the Buddhist ideals of releasing attachment to the material world and to attain spiritual freedom. AM communities are organized around the principles of meritism, specified in the slogan, "Consume Little, Work Hard, and Give the Rest to Society" (Essen, 2011, p. 68). Members value "to be content with little" and "to be satisfied with what one has", taking into account the Buddhist wisdom that desires cause suffering. Giving the rest to society is a training in selflessness or non-self, the pillar of Buddhism. From an environmental point of view, moderation is essential in order to change the path from destruction of the biodiversity and the biosphere to an ecofriendly way. Western wasteful lifestyle has much to learn from such a Buddhist movement.

Spirituality in Bhutan

Spirituality in Bhutan "can encompass belief in spiritual values like compassion, peace, and a sense of purpose and connectedness" and include "acts of compassion, altruism and selflessness …" (Ura et al., 2012a, p. 131). The concept of spirituality is measured in the GNH Index, which is based on balancing material wants with spiritual needs. The GNH research group argues that Bhutan is a "spiritual nation and the influence of spirituality is highly visible in the everyday lives of the population, in spiritual gatherings, and in the numerous spiritual landmarks such as sacred temples and monasteries, prayer flags, and prayer wheels. These provide a platform for people to develop spiritual maturity" (Ura et al., 2012a, p. 132).

For the purpose of GNH, spirituality is "intrinsic to development", which means that no meaningful development can occur without "inner spiritual growth along with a peaceful environment that allows spiritual nourishment. If material growth undermines the spiritual framework of society and its values of compassion and integrity, then development has not occurred." (Ura et al., 2012a, p. 132).

The spirituality indicator is based on four questions. One finding was that meditation practice had a very low loading. It may be mentioned that the government in Bhutan has recently initiated a school-based meditation curriculum, because of its ability to "provide balance, positive emotions and mental clarity"[5] (Ura et al., 2012a, p. 133).

Mindfulness is an important practice that, in the GNH Index context, measures psychological well-being as three components: (i) spirituality – meditation or mindfulness practices, and the consideration of the consequences of one's actions; (ii) emotional balance, which is the outcome of emotional intelligence, and the cultivation of positive emotions such as generosity, empathy, and compassion, and (iii) evaluative satisfaction with respect to different domains of GNH (Alkire, 2015, p. 93). One example mentioned by Alkire (2015) are the results of meditation in prison, in India (Tihar) and the United States. A program of meditation in prisons demonstrates that well-being is improved (Alkire op. cit. p. 97), and violence and racism are reduced. The Oxford Mindfulness Centre applies mindfulness techniques to patients with mental and physical problems. "There is great potential that widespread availability (of mindfulness tools) will have a beneficial effect on the general population, not just those who are diagnosed unwell" (Alkire, 2015, p. 98).

Concerning ideology, Alkire (2015) writes, "a key pillar of the new paradigm is sufficiency". This is in opposition to "many policies both public and corporate (that) seek to maximize wealth and profit, regardless of its opportunity costs on other domains of well-being or on well-being in future years" (p. 74). This pillar also emerges in the GNH index that measures the domain called "living standard". The living standard domain refers to the material well-being of the Bhutanese people, and ensures the fulfilment of basic material needs for comfortable living. In 2007, 23.2 percent of Bhutanese "still live in income poverty; some lack assets such as land or adequate housing" (Ura et al., 2012a, p. 168). They use household per capita income, assets, and housing conditions. Assets include livestock, land, and appliances, while housing conditions are measured by room ratio, roofing, and sanitation.

Sufficiency means that the GNH index does not use the poverty line, because the sufficiency threshold refers to higher conditions of well-being than poverty lines (Ura et al., p. 169). Furthermore "an absolute sufficiency threshold was chosen, since the GNH values encourage people to achieve happiness through their accomplishments and discourage a relative approach in which one is satisfied only if one has relatively more income … than one's peers" (Ura et al., 2012a, p. 169).

In principle living standards concern (i) meaningful and decent work and livelihoods and (ii) housing that sufficiently shields from the elements – cold and heat, rain, snow, and sun – and (iii) some form of currency – money, assets, or other tradeables. In accordance with Alkire (2015), all those aspects have an intrinsic value (p. 72). That money has an intrinsic value is contrary to Aristotle's view. For Aristotle, money should be valued as a means because wealth is only useful for something else. As Alkire (2015, p. 73) notes, whether money has an intrinsic value depends to some extent upon context. As a general-purpose resource up to a certain threshold, it gives freedom related to security, diversity, generosity, and sufficiency. Alkire argues that we should re-evaluate money. The essence is to find a balance between spirituality and acquisitiveness. One ideological element is to explicitly acknowledge and respect unpaid work (Alkire, 2015, p. 169). We might also add that sufficiency is in harmony with one of the four Hellenistic cardinal virtues. The virtue of moderation or temperance is seen as a golden means to avoid the extremes of too much and too little, to avoid greediness as well as stinginess.

Final remarks

We have made an exploration into some interpretations of two statements: (i) "appreciating life quality rather than adhering to an increasingly higher standard of living", and (ii) there is a "profound awareness of the difference between big and great". These statements refer to point 7 in the deep ecological platform, which requires a transformation of the western mindset towards a radically different ideology and lifestyle. We criticized the use of GDP as a measure of a standard of living tightly connected to the ideology of economic growth based upon materialistic values. As an alternative we have explored some aspects of the quality of life concept where often "less is more" and "more is less", and where self-realization of all beings is the top norm. There are many hindrances and challenges in order to obtain a transformation of western ideology towards an ecosophical mentality and lifestyle. We have limited our study to explore some concepts drawing on gestalt psychology, and ideas of sufficiency, simplicity, and mindfulness, supported by concrete cases. In particular, gestalt thinking is rewarding when we relate to nature. The appreciation of nature can best be understood via gestalt thinking. Via gestalt psychology the concept of nature takes on a new meaning, and the human-nature interaction gives benefits in terms of higher life quality. The importance that Næss gives to gestalt thinking cannot be overemphasized. Næss argues that the increasing meaninglessness of wealthy people is partly due to the indifference related to the symbols. We have to return to a living environment, which has value in itself, to avoid the problem that wealthy "self-made" men have been inclined to look at everything as merely means. In Næss's view the concept of nature is not something one uses for something else. It is independent of us but requires our attention unconditionally. The essence is experiencing, and experiencing nature and activity in nature is inseparable. To only look at nature is a very peculiar behaviour. The very experience of an environment happens when one does something in it, when one lives in it. Nature is much more that aestheticism, which is a passive hedonic enjoyment (Næss, 1974, pp. 204–205).

The sufficiency concept challenges the maximization view that is ingrained into the ideology of economic growth, and arrests the thinking that more and bigger are always better. The myth of limitlessness is a denial of the finite resources that are available on our planet, but fits well into the existing standard of living framework. The illusion of no limits is adequately described by Buddhist reasoning as ignorance, and it is emphasized in newer movements like Degrowth, inspired by Georgescu-Roegen's oeuvre (1971, 1975) of the laws of thermodynamics, which states that the cost of high material welfare is too high. We are overusing and exhausting the finite stock of mineral resources and, as Georgescu-Roegen writes, the greater usage today, will "cause a speedier decumulation of its dowry" … "every baby born now means one human life less in the future. But also every Cadillac produced at any time means fewer lives in the future" (Georgescu-Roegen, 1971, p. 304).

Vetlesen's (2015) critic of the western pattern of thinking and the marriage between science and technology (ST) gives a deeper understanding of how difficult it is for well educated people to understand what appreciation of life quality might mean. The ST marriage creates a peculiar western outlook. In western society there is a general high level of formal education based upon the Enlightenment's mechanistic framework, which fits well with a standard of living thinking. Economics as a field emerged out of this framework, with the consequence that it is insulated from nature. Vetlesen (2015) frames this as a dedevelopment and emphasizes that the natural world in our civilization is geared to technical control based upon a detached view and separation of nature from man. In such a view the intellect is highly appreciated, at the expense of feelings and inclinations of sensual-affective receptivity (Vetlesen, 2015, p. 92). Such a conclusion supports and corroborates the reasoning behind the need for a deep ecological reasoning and the need for a paradigm shift. Vetlesen's strong emphasize on "unreflective experiences" complement Næss's view of spontaneous experiences. However, while Næss emphasizes identification with other species, Vetlesen advocates the concept of solidarity. Vetlesen as well as Næss argue that other species should be treated with necessary respect because they have value in themselves.

Deep ecology is not a coherent framework, it is a narrative that needs to be filled out and complemented. Ideas, frameworks, and cases of alternative worlds might strengthen its legitimacy, but chasing happiness can be counterproductive. Empirical research gives weight to living a richer life based upon simple means. Recent empirical works in the tradition of human happiness and positive psychology also contribute to understanding why appreciation cannot be bought by a materialistic culture, by more of the same, but rather of sharing smaller treats with each other (Dunn and Norton, 2013). The main thesis is that deep ecology represents a significant frame of reference that inspires and challenges the western mindset. Deep ecological reasoning may also inspire degrowth, and GNH researchers to continue to raise deep questions on how to reduce consumption and foster an economy that is in harmony with the Earth's limits.

Deep ecology is not an ecotopia. The case of Bhutan has high validity. The two empirical cases from Thailand are also interesting empirical cases. A different stream of thoughts that might be researched is on the indigenous people of Australia (Buergelt et al., 2017; Paton et al., 2015). Drawing upon this research, we might learn from different worldviews that may benefit and challenge existing western ideology. Buergelt et al. (2017) write about indigenous people who have developed knowledge, sensitivities, and practices for living in harmony with each other and with nature over thousands of years. These "people hold a unified and egalitarian ontological worldview that has its core that all creatures are one and equal" (2017, p. 8). These groups of people probably live meaningful lives and can only be understood in relationship to their cosmology, ontology, and epistemology.

There are many sub-groups in the western world that live by consuming less and devoting more time to art, music, family, culture, community, and nature. Groups rooted in Quaker traditions have a lifestyle that is much more sustainable than the average. Could we learn about simplicity from the Quaker tradition? Michaelis (2008) investigates "Quaker simplicity" in history as well as in modern time. The Quaker ethic of frugality is partly a rejection of the status seeking that Veblen framed as "conspicuous consumption". There are empirical examples of several projects started by a network of Quakers. Michaelis (2008) presents evidence "that lifestyle change is possible, and that it can be a fascinating and joyful experience" (Michaelis, 2008, p. 115).

Bhutan's GNH measure means a more harmonic approach between material and spiritual values. The Bhutanese research on GNH presents a number of insights that have relevance for exploring alternative ways of life and different lifestyles compared to the prevailing western view. We should reflect deeply on these insights in order to be inspired and learn from them in order to pursue a more flourishing and sustainable life on Earth. The exploration of GNH represents an interesting alternative framework. The ideology behind GNH shows that deep ecological reasoning is realistic, even if it requires a radical transformation of western ideology. The basic ideological shift we need is "a change of revolutionary depth and size by means of many smaller steps in a radically new direction" (Næss, 1989, p. 156).

Notes

1. I am grateful for the substantive comments during my presentation of the paper on Contemplative Science & Management, International Transdisciplinary Conference, May 19–21, 2017, in Budapest. Two anonymous reviewers of my paper detailed very constructive comments from Petra Buergelt.
2. Holbæk Hansen, the first Norwegian professor in marketing, framed this notion, which may be interpreted as even if we buy and consume more and more materialistic objects, we are not fulfilled, but fragmented as human beings. We will continue to demand products, but will never be satisfied (from lectures at NHH).
3. Ims, 2016, *Deep Ecology and the relevance for Gross National Happiness and Bhutan*, presented on Bhutan's GNH conference in November 2015. The paper was published in hard copy in April 2018, and has been available electronically in the official CBS.
4. This section is drawing upon Vetlesen (2015).
5. Mainly the monks and nuns practice meditation in Bhutan. The GNH survey included twenty-five monks and nuns. This is not representative of monks and nuns who make up about 3 percent of the population of Bhutan. They live largely institutionalized lives in monasteries and "nunneries" and are not easy to interview.

Bibliography

Alkire, S. (2015): "*Well-being, Happiness, and Public Policy*". The Centre for Bhutan Studies & GNH Research.

Anker, P. & Witoszek, N. (1998): The Dream of the Biocentric Community and the Structure of Utopias. *Worldviews: Environment, Culture, Religion* 2: 239–256.

Assadourian, E. (2012): *The Path to Degrowth in Overdeveloped Countries, State of the World 2012, Moving toward Sustainable Prosperity* (pp. 22–37). London: Island Press. See also <http://www.worldwatch.org/system/files/SOW12%20Summary%20%28Chapter%202%29.pdf>

Buergelt, P. T., Paton, D., Sithole, B., Sangha, K., Prasadarao, P. S. D. V., Campion, L., & Campion, J. (2017): Living in Harmony with Our Environment: A Paradigm Shift. In Paton, D. & Johnston, D. M. (eds): *Disaster Resilience. An Integrated Approach* (2nd edn, pp. 289–307). Springfield, IL: Charles C. Thomas.

Csikszentmihalyi, M. (1990): *Flow: The Psychology of Optimal Experience*. New York: Harper and Row.

de Graaf, J., Wann, D., & Naylor, T. H. (2005): *Affluenza: The All-consuming Epidemic*. San Francisco, CA: Berrett-Koehler Publishers.

Dunn, E., & Norton, M. (2013): *Happy Money. The Science of Happier Spending*. New York, NY: Simon & Schuster Paperbacks.

Essen, J. (2011): Economic Sufficiency and Santi Asoke. In Zsolnai, L. (ed.): *Ethical Principles and Economic Transformation – A Buddhist Approach*. New York: Springer.

Frank, R. (1999): *Luxury Fever: Why Money Fails to Satisfy in an Era of Excess*. New York: Free Press.

Frey, B. S. (1997): *Not just for the Money: An Economic Theory of Personal Motivation*. Cheltenham: Edward Elgar Publishing.

Gamlund, E. (2012): Hva er galt med dypøkologien? Noen kommentarer til Arne Næss's Økosofi T. (What is wrong with the Deep Ecology? Some comments to Arne Næss's ecosophy). *Norsk filosofisk tidsskrift* 47(4): 229–244.

Georgescu-Roegen, N. (1971): *The Entropy Law and the Economic Process*. Cambridge, MA: Harvard University Press.

Georgescu-Roegen, N. (1975): Energy and Economic Myths. *Southern Economic Journal* January 41(3): 347–381.

Ims, K. J. (2011): Deep Ecology. In Bouckaert, L. & Zsolnai, L. (eds): *The Palgrave Handbook of Spirituality and Business*. New York: Palgrave, Macmillan.

Ims, K. J., & Pedersen, L. J. (2015): Rethinking Business Ethics in an Age of Crisis. In Ims, K. J., & Pedersen, L. J. (eds): *Business and the Greater Good. Rethinking Business Ethics in an Age of Crisis*. Cheltenham, UK: Edward Elgar.

Kahneman, D. (2011): *Thinking Fast and Slow*. London: Penguin Books.

Kasser, T. (2011): Materialistic Value-Orientation. In Bouckaert, L., & Zsolnai, L. (eds): *The Palgrave Handbook of Spirituality and Business*. New York: Palgrave.

Langer, E. (2009): Mindfulness Versus Positive Evaluation. In: Lopez, S. J., & Snyder, C. R. (eds): *Oxford Handbook of Positive Psychology* (2nd edn, pp. 279–293). Oxford Library of Psychology, Oxford: Oxford University Press.

Magnuson, J. C. (2011): Pathways to a Mindful Economy. In Zsolnai, L. (ed.): *Ethical Principles and Economic Transformation – A Buddhist Approach*. New York: Springer.

Michaelis, L. (2008): Quaker Simplicity. In Bouckaert, L., Opdebeeck, H., & Zsolnai, L. (eds): *Frugality. Rebalancing Material and Spiritual Values in Economics Life*. Oxford: Peter Lang.

Næss, A. (1973): The Shallow and the Deep, Ecological Movements. A List of Contrasts. *Inquire* (Spring) 16, pp. 95–100.

Næss, A. (1974): *Økologi, samfunn og livsstil. (Ecology, community and lifestyle):* Oslo: Universitetsforlaget,

Næss, A. (1989): *Ecology, community and lifestyle. Outline of an ecosophy*, Arne Næss. (Trans. and ed., David Rothenberg). Cambridge: Cambridge University Press.

Næss, A. (1995): The Shallow and the Deep, Long-Range Ecology Movement: A Summary. In: Drengson, A., & Inoue, Y. (eds): *The Deep Ecology Movement. An Introductory Anthology*. Berkeley, CA: North Atlantic Books.

Næss, A. (1999): *Økologi, samfunn og livsstil. Utkast til en økosofi. (Ecology, community and lifestyle. Draft to an Ecosophy)*, Fifth reworked edition, Oslo: Bokklubben Dagens Bøker.

Næss, A., & Sessions, G. (1995): Platform Principles of the Deep Ecology Movement. In: Drengson, A. & Inoue, Y. (eds): *The Deep Ecology Movement. An Introductory Anthology* (pp. 49–53). Berkeley, CA: North Atlantic Books.

Nakamura, J. & Csikszentmihalyi, M. (2002): The Concept of Flow. In: Snyder, C. R., & Lopez, S. J. (eds): *Handbook of Positive Psychology*. Oxford: Oxford University Press.

Paton, D. P., Buergelt, T., & Cambell, A. (2015): Learning to Co-exist with Environmental Hazards: Community and Societal Perspectives and Strategies. In *Advances in Environmental Research*, 43. New York, NY: Nova Publishers.

Schwartz, B. (2004): *The Paradox of Choice. Why More Is Less*. New York, NY: Harper Perennial.

Sen, A (1999): *Development as Freedom*. New York: Knopf.

Ura, K., Alkire, S., Zangmo, T., & Wangdi, K. (2012a): *An Extensive Analysis of GNH Index*. Thimpu, Bhutan: Centre for Bhutan Studies.
Ura, K., Alkire, S., Zangmo, T., & Wangdi, K. (2012b): *A Short Guide to Gross National Happiness Index*. Thimpu, Bhutan: The Centre for Bhutan Studies.
Ura, K., Alkire, S., Zangmo, T., & Wangdi, K. (2015): *Provisional Findings of 2015 Gross National Happiness Survey*. Thimpu, Bhutan: Centre for Bhutan Studies & GNH Research.
Vetlesen, A. J. (2015): *The Denial of Nature. Environmental Philosophy in the Era of Global Capitalism*. London: Routledge.
Westacott, E. (2016): *The Wisdom of Frugality. Why Less Is More – More or Less*. Princeton: Princeton University Press.

KNUT J. IMS

Nature, Economics and *Scream*

Introduction

I had the privilege to travel with Laszlo to a number of European cities. The main purpose was to teach, together with the CEMS (Community of European Management Schools – The global alliance in Management Education). Typically a visit to a museum was also on the agenda to watch one or a few masterpieces. The rule "less is more" was practiced in order to avoid being an unreflective consumer. Laszlo is a very devoted and selective art observer who, with great passion, pays high respect to the art and the artist, having himself the poetic insight of artists.

We have seen the *Last Supper* in Milan, watched great artwork in Florence, and visited the Munch museum in Oslo. Let us stop for a "spirited" moment in front of *Scream* (See Figure 1 – Scream by Edvard Munch).

Munch's painting *Scream* is one of the most famous paintings in the world.[1] It is an icon in modern art. Why is the painting special? A man stands on a bridge with two men in the background. The heaven is painted with dramatic colors, and the man in the foreground is screaming with his hands up to his face – or to the ears, possibly to mitigate the scream. Is it the man who cries in anguish? Or does the cry come from nature? Thus, we have at least two different interpretations of the painting. When we read the text that Munch wrote, which connects to the painting, we find that the second interpretation is in accordance with Munch's text in the explanation of the situation he painted.

Figure 1. *Scream* by Edvard Munch (with permission from The Munch Museum in Oslo, cf. Endnote 1).

> I was walking along the road with two friends – the sun set. The sky turned blood-red. And I felt a wave of sadness – I paused tired to death – above the blue-black fjord and city blood and flaming tongues hovered.
>
> My friends walked on but I stood still, trembling with fear – I felt an immense, infinite scream passing through nature. (The author's translation.)

However, there may not be any contradiction between the two interpretations. The text and the painting can complement each other and express together that man and nature penetrate each other. Toyner (2000) asks, with Munch's own words: Is the world around the human on the painting colored by the man's inner life, or is the man's inner life colored by a tone that comes from the outside? Neither the text nor the painting shows whether it is an internal or an external pressure.

After Munch's collapse in 1909, he was living like a monk, alone and frugal. He had already bought a small spartanly equipped summerhouse in 1898 for 900 NOK in Asgardsstrand. The house had a beach and Munch felt at home in this small house. Munch's "enemies" wrote that because he exhibited very strong emotional pictures with sickness and death as lead motifs, Munch had to be a madman. Several of Munch's exhibitions involved scandals. In spite of the public disdain, not least in his own country, Norway, Munch continued with his strong emotional painting style. In 1916 Munch was able to buy a property consisting of several buildings on 45,000 square meters of ground in Oslo, where he lived until his death in 1944.

No-one who has studied Much's paintings can avoid seeing that most paintings have nature as an important element. We may well ask whether Munch used nature as background and means, or whether nature was the foreground in his painting? Munch developed his style – first from natural to impressionist, and then symbolist where he cultivated his very strong but simple forms. In 2014, the Munch Museum in Oslo arranged a special exhibition called "Nature and Munch".

Unhappy with the crude and simple language of painting, Munch tried to transcend its limitations. *Scream* is the master example of this attempt. Munch was frequently practicing ekphrastic poetry, the old conversation between art and poetry, which goes back to antiquity. Poetry is the source to many of Munch's most famous paintings. When we look at *Scream* in this way, we find that *Scream* had a long history. In a poem Munch wrote some years before his first version of the *Scream* (in 1893), he described art as the "soul's diary". The poem stems directly from an entry in Munch's diary. It is interesting to note that Munch refined the poetry when he wrote it into the frame of one of the versions of *Scream* in pastel-on-cardboard in 1895. This refinement gives a hint of how Munch tried to crystallize his emotional experience through his expressionist and symbolist painting style.

Tøyner (2000) gives an insightful analysis of the reasons why Munch provoked scandals among the audience. Tøyner writes that it was not that Munch painted the unconscious soul-life, with all the hidden demons, but that Munch "made visible the life of the soul" (p. 12).

Munch's paintings draw the soul's abyss up to the surface, and lets it be there as an uncompromising brutal presence (op. cit. p. 13). For sure, Munch was not the first to paint dreams, death, and anguish, but the way he did it was regarded as repugnant. One of Munch's means was to use frontal figures, so that the observer looks directly into the gestalt's face. In the long preparation for *Scream,* we observe that Munch first draws figures that we can observe from the side, before he finally found his strongest expression – the front of the face. This is the essence of the Munchian painting style – it is not about something – it is not observing a gestalt from a distance – it is what it shows, directly, unmediated reality, a "depth in the surface".

Has *Scream* any implications for modern economics?

Munch's paintings are extremes, and they catch the onlooker so strongly that it is impossible to create a distance. The paintings do touch us in a deep emotional sense. Munch's mode of visualizing the world is in a contrast with the abstract way of envisioning the world we see in modern mainstream economics. Mainstream economics is an academic discipline and, as a scientific discipline, it has some limitations due to how it organizes knowledge. While Adam Smith had the privilege to see the economy as part of the whole human activity, and investigated it historically and empirically, modern economics has identified its subject matter in separation from the rest of reality. It has also adapted the methodology that defines it as a discipline. The methodology could have been historical case based, which had been more in line with humanistic academic disciplines, but in fact the end-result was that the natural sciences – to a large extent physics – were imitated, which implied a relatively abstract view of the world.

Abstract models dressed as mathematical formulae were given much prestige. In this way economists create their own professional discourse, which is very difficult for outsiders to understand and to be an engaged part of. In order to work on an abstract mathematical level, one needs to confine and narrow the world through a number of assumptions so that the world fits into the mathematical models. The clue is that with the right assumptions for the right time, the mathematization of economics could be fruitful and "innocent". However, with wrong or inadequate assumptions, it may give erroneous answers when the results from the models are applied to the real world. This phenomenon is explained as the "fallacy of misplaced concreteness" in economics, by Alfred North Whitehead. One is confusing the map with the real world. It means that even if it is possible to draw some conclusions from the mathematical equations, the applications of the result on the complex and dynamic world is a very risky maneuver (Griffin, 2007).

The use of advanced mathematical models may enforce the tendency to distance oneself from nature. Furthermore, nature is very often neglected in modern economics. In this perspective, land in the academic discipline of economics is typically reduced to rent. As Daly and Cobb argue (1994) in their *For the Common Good*, the development of the economy into an academic discipline implied ignorance of the Earth, focusing on the productive factors "work and capital", which have been extensively analyzed. The "Earth", as economists use the word, implies the forces in nature to the extent that they have economic impact. One problematic aspect is that the Earth is treated as a property relation – to a large part similar to other property relations in the market economy. The Earth is the passive, and human beings are the active factor. The Earth is typically treated as not being productive in itself. The need for caring about the physical reality and nature has hardly been dealt with by economists, even if the exploitation of nature via economic reasoning is in many cases unusually violent and creates much suffering. Efficiency has been the central means without discussing the brutality of means, the goals, and the institutional frame of modern economics (Daly and Cobb, 1994).

When modern economists in general have not been interested in land, neither as soil nor genuinely in the animals depending upon the soil, but reduced them to commodities, it follows that soil, mammals, birds, and fish can be exchanged like any other commodity. This commodification of nature shadows the violations of nature that are often a result of huge scale factory farms. Exploitation of soil and animals are submerged in the calculus of income minus cost. Vetlesen suggest that economists may have an autistic view of nature (2015).

When we read Singer's (2009) concrete descriptions of how animals are mistreated in research, in laboratories, and in farm factories, we might conclude that modern economists have not been able to "hear", see, or experience the suffering of animals. If they had, they would have introduced animal welfare into economics. Inspired by Munch, I will try to verbalize some of the pain and suffering of animals due to their commodification. First, an account of veal "production" based upon Singer (2009) will be presented.

Veal "production"

The economist's problem is to "optimize veal production". Central concepts in the discourse universe are "to produce a calf of the greatest weight in the shortest possible time and secondly, to keep its meat as light colored as possible to fulfill the consumer's requirement" at a profit that reflects the risk and investment involved (The *Stall Street Journal*, July 1972, cited in Singer 2009, p. 130).

What do these normative recommendations imply for the raising of veal? Veal is the flesh of a young calf, and the raising of young calves has turned into a modern industry. Veal has a flesh that is paler and more tender than that of a calf that has begun to eat grass. Traditionally a day or two after being born, young calves were pulled to the market where they were sold for delivery to the slaughterhouse. In the 1950s, veal "producers" in

Holland found a method to keep the calf alive longer without the flesh becoming less tender (Singer, 2009). The strategy was to keep the calf in highly unnatural conditions, and to prevent the calf from "romp[ing] around the fields, developing muscles that would toughen their flesh and burn up calories that the producer must replace with costly feed" (Singer, 2009, pp. 129–130).

To solve the "problem", the calves are taken to a confinement unit, each 1 foot 10 inches wide and 4 feet 6 inches long. The unit has a "slatten wooden floor". "Tethered by a chain around the neck" the calves are prevented from "turning in their stalls" (ibid., p. 130). The stall has no straw or other bedding since the calves might eat it, which will destroy the paleness of their flesh. The calves are fed a liquid diet, based on nonfat milk powder with vitamins, minerals, and growth-promoting drugs. The calves live for the next four months in this unit and, finally, the calves leave their unit only to be taken to slaughter.

"The beauty of the system" is that at this age the veal calf may weigh as much as four hundred pounds, instead of ninety pounds as a newborn calf, "and since veal fetches a premium price, rearing veal calves in this manner is a profitable occupation" (Singer, 2009, p. 130). This method of raising veal was imitated by Provimi Inc., a feed manufacturer in the United States in 1962. The company is by far the largest in the business and it tries to obtain "optimum veal production" (*The Stall Street Journal* July 1972, which is Provimi's news sheet, cited in Singer 2009, p. 130).

A research group from the School of Veterinary Science, University of Bristol, England, wrote that "Veal calves in crates 750 mm wide cannot, of course, lie flat with their legs extended. Calves may lie like this when they feel warm and wish to lose heat. Well-grown veal calves at air temperatures above 20 degrees C may be uncomfortably hot. Denying them the opportunity to adopt a position designed to maximize heat loss only makes things worse." Veal calves raised like this were unable to adopt a normal sleeping position, which is "a significant insult to welfare" (Singer, 2009, p. 131).

The veal crates are too narrow to permit the calf to turn around, which certainly is frustrating for the calf. Such a stall is also "too narrow to groom comfortably in, and calves have an innate desire to twist their heads around and groom themselves with their tongues".

Veal calves in crates cannot reach much of their body. The research group concludes that "denying the veal calf the opportunity to groom itself thoroughly is an unacceptable insult to welfare, whether this is achieved by constraining its freedom of movement or, worse, by the use of a muzzle"[2] (Singer, 2009, p. 132).

Calves are standing or lying on a slatted wooden floor, which is hard and rough on the calves' knees when they arise. A slatted floor is a type of cattle grid, which cattle typically avoid. The idea is that most of the manure falls through. It means that young calves "for some days (are) insecure and reluctant to change position" (p. 132).

After three or four days young calves are taken from their mothers. The calves sorely miss their mother and something to suck on. Everybody that has given a little calf a finger know that a calf has a strong urge to suck like a human baby. However, the calves have no teat to suck on, nor any substitute. Later the calves develop the need to ruminate and chew the cud, but the business idea is to keep the calves anemic. The color of veal is one of the primary factors involved in obtaining "top-dollar" returns from the fancy veal markets. "Light color" veal is a premium item much in demand at the better clubs, hotels, and restaurants. "Light color" or pink veal is partly associated with the amount of iron in the muscle of the calves (*Stall Street Journal* November 1973, cited in Singer, 2009, p. 133). It is necessary to hinder the calves from everything that contains iron, because iron will darken the flesh. Consequently, "digestive disorders, including stomach ulcers, are common in veal calves. So is chronic diarrhea. The calves are deprived of dry feed. This completely distorts the normal development of the rumen which may also lead to chronic indigestion" (Singer, 2009, pp. 132–133). A normal calf would get iron from grass, but "pale pink flesh is in fact anemic flesh". So, iron has to be avoided, "Neither the taste, nor the nourishment for humans are changed by the color, but the special demand for flesh of this color has obvious snob appeal".

The producers' formula for maximizing profit is to find a "trade off" between the quantity of iron that is necessary for survival and pale flesh. The use of hardwood stalls are used to hinder the calves from licking any iron in their stalls. Provimi writes that: "metal may affect the light veal color … Keep all iron out of reach of your calves" (Singer, p. 133).

Anemic calves have an insatiable craving for iron. Since urine does contain some iron, the calves' desire for iron is so strong that they would overcome the natural repugnance to go near their own urine or manure. To control veal calves drinking more iron than absolutely necessary, it is important to prevent the calf from turning around in the stall.

The color of the meat is only "one parameter". Another parameter is to maximize the growth by the cheapest means within a given time period. To maximize the food per calf, one means is not to give the calves any water. Then their only source of liquids is their food, which consists of powdered milk and added fat. A third parameter is to keep the buildings warm, because cold calves burn calories just to keep warm. Due to lack of water, the calves will overeat and sweat, and "in sweating, the calf loses moisture, which makes him thirsty, so that he overeats again next time" (ibid., p. 134). Such treatment of the calves is evidently unhealthy, but since the goal is to produce the heaviest calf in the shortest possible time, the health of the calves in the long-term is not relevant as long as a certain number of calves survive long enough to be delivered to the market. So Provimi can advise that "sweating is a sign that the calf is healthy and growing at capacity" (Singer, p. 134). A fourth parameter is to reduce the movements of the calves. By leaving the calves in the dark for 22 to 24 four hours a day, the visual stimulation and absence of contact with other calves leads to relative inaction. Such calves have a mortality rate of between 10 percent and 15 percent, but such losses can be considered a calculated "trade off" between the premium price for white tender veal at gourmet restaurants and the calculated loss of "raw material" during the production processes (This section is built upon. Singer, 2009).

Animals used in research, including farmed fish in aquaculture systems

Norway has 5.1 million inhabitants and used about 1,230,000 animals in research in 2015, of which 90,000 were not fish (Torp, 2017). In 2013,

Norway used 5,500,000 animals in research. The reason for the reduction in 2015 is that fewer fish were used in research. The great numbers of fish is partly due to vaccination tests and experiments with different medicines for the fish. The reason for the use of medicines for fish is that it is better with prophylactic measures than treatment of sickness. When we compare Norway with Sweden, where Sweden has almost ten million inhabitants, Sweden used "only" 260,000 testing animals in 2015, and two-thirds of these were mice. A global estimate is that in 2005, 115 million test animals were used. In Norway there are sixty-five different locations where they do research on animals (Torp, 2017). In order to use animals in research in Norway, permission from the Norwegian Food Agency (Mattilsynet) is necessary.

Pigs are often used by heart surgeons in Norway in order to train for heart surgery on humans. A new way to use pigs is in the training of trauma-teams deployed in war zones in Afghanistan and Syria. When a trauma team has to work together on a person with multiple and life threating harms, the team need training. Consequently, the teams train on pigs that have been shot in different places of the body (Torp, 2017). Another much used mammal in research is mice. Small, black mice have been "perfected as research objects" between 1950 and 1960. They are free from bacteria. Not one bacterium is to be found in their bowels or in their skin. The food and the water they drink are sterilized before use, and even the sawdust used in the cage is sterile. The mice do not smell anything, live in cages, and live longer than conventional mice. One example of the instrumental use of mice is to investigate the relationship between microorganisms and host organisms (Torp, 2017, pp. 4–7).

OncoMouse is a registered trademark of experimental mice that are "produced and marketed" as any other commodity. The mice was genetically modified to carry an oncogene. The gene made the mouse highly receptive to cancer, and thus an efficient means for scientists to study the development of cancer and appropriate treatments. The genetically modified mouse was regarded as an invention and the patent was granted to Harvard College. The patent rights were bought and marketed by the transnational company DuPont. In Europe the patentability of animals were first excluded, but finally the patent was granted in 1992. The patent expired in the US in 2005, and the genemodified Oncomouse is now free to be used by other interested "stakeholders", but "OncoMouse" remains a registered trademark.

Fish farming or "aquaculture" as big business

The global seafood industry (capture fisheries and aquaculture) is a vital source of food, income, livelihoods, and culture. Aquaculture supplied 42 percent of the world's fish in 2012 and is expected to pass beyond the capture fisheries production and supply 60 percent of the fish by 2020 (Lam, 2016, p. 37). Fish farming or "aquaculture" is a big business globally, and in particular in Norway. Global aquaculture produced 90.4 million tonnes in 2012, valued at $144.4 billion (US), including 66.6 million tonnes of fish food. Chinese aquaculture contributes more than 60 percent of the global aquaculture volume. The value from aquaculture in Norway is twice the value from marine fisheries. Norway is the world's second largest exporter of fish by value. More than 93 percent of the fish from aquaculture are Atlantic salmon *(Salmo salar)* farmed in marine cages. Norwegian fish farming/aquacultures produced 1.14 million tonnes fish in 2011. According to Lam (2016), "Fisheries and aquaculture governance are evolving to meet ethical standards; however, the compliance to the CCRF (The Food and Agricultural Organization of the UN Code of Conduct for Responsible Fisheries) is poor, with the majority of fishing nations, comprising 96% of the world's catch, failing to achieve even a passing score" (Lam, 2016, p. 37).

Farmed fish are different from wild fish. While wild fish are captured in the sea, farmed fish live all their lives in marine cages. Human beings have full control of the life span of the farmed fish, from embryo to slaughter. The human is the host with full responsibility for the welfare of the fish. Consequently, empathy and deep knowledge with the "guest" in all the meetings where the fish are going to live, is needed (Børresen, 2000, p. 31). Atlantic salmon was originally known as the king of the river, an *anadrom* specie that spent the first part of their lives in rivers and lakes, and when they grew to a length of about 10 to 15 cm they left the rivers in the spring and traveled to the salt ocean to grow. As a mature fish, after one to three years and a swimming trip of several thousand kilometers, they return to the river from which they left in order to spawn (Kristiansen, 2013).

Marine fish are typically farmed near shore and influence habitats and ecosystems. Aquaculture systems are not charged for discharges of waste into water or other adverse environmental "pollutions". Atlantic salmon is a high-value, high trophic-level specie that is carnivorous, which means that it eats other living organisms. According to Lam (2016) the ratio of wild fish input to total farmed fish output (fish in–fish out, FIFO) is as high as 4:1, that is fish farming is a net consumer of fish that might otherwise be consumed by the human population or provide essential ecosystem services. Significant progress has been made in aquaculture feed efficiency in recent decades, lowering the FIFO ratio to between 1.5 and 3.6, depending upon the species of fish (Lam, 2016, p. 50). As a by-product of confining a large number of carnivorous fish cages in coastal waters, aquaculture facilities can cause "unmitigated environmental damage"; that is, sea lice and other diseases can threaten wild fish and may destroy the indigenous fish population (Lam, 2016). It is stated that feed production is the largest factor of climate impact for carnivore fish species. Boats are needed to capture wild fish in the ocean (to be used as fish feed), and the bigger the boats the more CO_2 is produced. Fish feed is also based on at least 50 percent agricultural products, such as wheat, maize, and soybean, while industrialized farming is also based upon the burning of fossil fuel.

For the slaughter of farmed fish, the anthropocentric view is fatal. In Norway a regulation that "farmed fish shall be anaesthetized with CO_2 before slaughter" was implemented.

According to Børresen (2000), the fact is that fish are equipped with a chemical apparatus on their surface, which is far more developed than the human apparatus. When CO_2 increases in the environment (that is in the water in the cage), the fish tries to escape. The increasing content of carbon acid in the water may be experienced as similar to swimming and breathing in hydrochloric acids. Since the fish is confined in the cage, it will try to jump over the cage. That is why the water boils with panicking fish in the 5–6 minutes it takes before the fish lose consciousness and stop moving. Børresen (2000, p. 36) suggests that if the authorities had acted ichtiocentrically they would have known that fish are almost weightless life in the water. The connective tissue is, therefore, more vulnerable to blunt force than animals that live on land. The alternative "final solution" might be to give every fish one single blow to its head. Then, probably, the fish will be inflicted serious brain damage to the extent that it will not sense the succeeding slaughter.

Discussion

The details and concreteness of Singer's case of veal "production" is striking. However, other researchers and investigations support and corroborate the empirical validity of similar cases.[3] The farming and "production" of broilers might be even more screaming than veal "production" (Martinsen, 2010). Seemingly, the cruel "production" of veal reveals a complete lack of compassion. We may wonder how it is possible to be insensible vis-à-vis a small, newborn, and vulnerable calf? Calves have expressive eyes, relatively similar, but bigger eyes than human eyes, and calves have a searching glance that attempt to meet the human eye. Sartre writes that a human "glance" may kill, but a calf's glance is expected to create a warm, kind, and caring attitude from people physically close to him or her.

Levinas emphasizes the importance of "the face" (1990). In *Difficult Freedom,* Levinas states:

> "The face is not the mere assemblage of a nose, a forehead, eyes etc.; it is all that, of course, but takes on the meaning of a face through the new dimension it opens up in the perception of a being. Through the face, the being is not only enclosed in its form and offered to the hand, it is also open, establishing itself in depth and, in this opening, presenting itself somehow in a personal way" (p 8). The belief is that in front of another person's face, one cannot violate him or her.

Even if Levinas' concept of the face might originally be related to people that were close to them, that is in their proximity, there are a lot of examples where humans relate to the face of their pets and other animals and meet them with a caring attitude. Then, we may wonder why human beings can be so unsensitized vis-à-vis animals? Is it human greed – the addiction for money – that makes the producer blind for the calves' eyes and his welfare? In a series of experiments done by Vohs et al. (2006), they found that money alienates us.

Priming people with money entails certain behavioural consequences for the persons primed. Thinking about money creates a greater distance between people and makes them less helpful and less altruistic to other people. The research group suggests that when societies began to use money the need to rely on family and friends diminished and people were able to become more self-sufficient. The research group concludes that, "money enhanced individualism but diminished communal motivations, an effect that is still apparent in people's responses today" (Vohs et al., 2006).

Could it be that the language of profit and moneymaking desensitize the "producers" of veal? Using the word "it" and differentiating the animal with only numbers may contribute to the switching off of empathic attitudes vis-à-vis animals. To put a name on an animal would be one step towards a more *caring* attitude. Even if Bauman's (1989) focus on human behaviour towards other human beings concerns the the Holocaust, some of the same reasoning might be fruitful in different contexts. One statement held by Bauman (1989) is that most human beings have an Eichmann within themselves. To a certain context, this secondary ugly face might be revealed. Bauman explains why so many "ordinary" people were able to participate in the destructive project of "Endlosung". One fundamental reason is the mediation of action, that is, when one's action is performed by someone else, by an intermediate person. In such a situation, it is impossible for me to experience the consequences of my act directly. Could it be that the distance between they who manage the companies (and in fish farming we find big corporations as well as small companies) and what is managed imply that the managers become morally blind? That the managers are not concerned about the individual fish, but look at fish only as a commodity that should be operated according to a profit maximization calculus, like what we saw in the "the production of veal"?

Two recent empirical studies made in Norway, in large and small aquaculture companies, indicate that the managers and the administrative staff in the big companies are ignorant about the practical situation in the fish cages and of the slaughter of the fish (Bundli and Liltvedt, 2012; Bolstad and Gloppen, 2015). Following Bauman, we may interpret this result as the physical and/or psychic distance between their acts and their consequences, achieving "more than the suspension of moral inhibition". It extinguishes the moral significance of the act and thereby "pre-empts all conflict" between personal moral decency and the immorality of the consequences. The moral dilemmas are absent from sight due to long chains of causalities.

According to Bauman (1989), the technical-administrative success of the Holocaust was due to skillful utilization of moral blindness, which is supported by modern bureaucracy and technology. The specialization of tasks, organizational routines, the long distance to the site of the "raw material", loyalty to the chain of commands, and moral discipline substituted moral responsibility (Bauman, 1989, pp. 18–27). Bolstad and Gloppen (2015) found that the fish keepers who supported the cages daily, and fed and surveyed the farmed salmon, were more concerned about the well-being of the fish.

Gowdy (2006) emphasizes that at the core of traditional economic theory is a concept of human nature that is devoid of social context, mechanical rationality, and driven by an insatiable appetite for material possessions. Mahatma Gandhi stated that there is enough in the world for everyone's need, but not enough for anyone's greed. According to Schumacher (1999), the drive for greed is due to the rejection of wisdom. The cleverness used in "producing" veal to obtain maximum profit can only happen because cleverness has displaced wisdom, and without wisdom "man is driven to build up a monster economy, which destroys the world".

Schumacher asks: How could we "begin to disarm greed and envy"? One way is to scrutinize "our needs to see if they cannot be simplified and reduced", and we could support those "who work for non-violence: as conservationists, ecologists, protectors of wildlife, promoters of organic agriculture …, … An ounce of practice is generally worth more than a ton of theory" (Schumacher, 1999, pp. 24–25).

Bandura et al. (2002) describe several mechanisms of moral disengagement that enable "eminent members of the business community" who are not "dangerous, criminally oriented mavericks" to commit unethical acts without experiencing moral guilt. One strategy is euphemistic labeling. A Finnish study on consumer attitudes to farm animals, including fish, concludes that for some citizens fish are "semi-animals" and "that there is no urgent consumer pressure to improve the conditions of farmed fish in Finland". It means that animal welfare does not automatically include "semi-animals" like fish. Another strategy is the displacement of, or diffusion of, responsibility. One example from Bundli and Liltvedt's (2012) study is the outsourcing of the slaughtering of fish by big companies, and thereby placing the responsibility on another company. We see a number of examples of the strategies of disregarding or distorting the harmful consequences of the many accidents in the open sea cages used in aquaculture. One frequent phenomenon is when farmed fish escape from the cages into the sea. Another phenomenon is the dumping of waste from the fish farm at the bottom of the sea or the fiord. The main reason is that the security and the robustness of the sea cages is not important because the business interests in the short run have first priority.

The Norwegian veterinarian Børresen (2000, 2007) writes about "the hunters' insensibility", that is a kind of attitude that characterizes a person when taking on the role of a hunter. It means that humans have an innate emotional master switch that can create an impenetrable mental barrier against other species, unaffected by facts. Such insensibility gives the hunter a certain kind of blindness for the suffering he or she will inflict upon the animal that is the hunted object. It gives the hunter an ability to use logic and objectivity in the OFF position of the switch, but a possibility to show physical empathy and sensitivity in the ON position.

Børresen states that "A human is probably the only animal that has the ability to decide to create empathy across species boundaries by means of knowledge and thought" (2007, p. 56). It is obvious that the self-interest of "the producer" leads to the switching off of the emotional master when he/she strives to maximize profit using "animals as means". For Singer it is this conflict of interest that is the decisive barrier to take the interests of animal welfare seriously. According to Singer, animals' "interests are allowed to count only when they do not clash with human interests. If there is a clash – even a clash between a lifetime of suffering for a non-human animal and the gastronomic preference for a human being – the interests of the nonhuman are disregarded" (Singer, 2009, p. 212).

While courage is an important virtue in the western tradition, exemplified by a heroic hunter (Harari, 2011), compassion is one of the most important virtues in the Buddhist tradition. The "veal production" may be seen as the antithesis to a Buddhist economic approach. An essential concept within Buddhist economics is minimizing suffering (Zsolnai, 2011, 2014). To minimize suffering includes not only human beings, but extends to animals, that is, to all sentient beings. We saw that the *Stall Street Journal* recommended to "optimize the production" in order to maximize profit. The cultural frame of the west tries to maximize profit and desire. Through massive marketing and advertising budgets in global corporations, we see systematic attempts to pump up the desires of consumers. A rational Buddhist strategy is to practice the "middle way", which often involves minimizing desire. A Buddhist strategy would also be a negation of the western dogma that more is more, and bigger is better, and recommend living based upon the motto of "less is more". This is not an empty hypothesis. The Nobel laureate Daniel Kahneman gives several examples of less is more in his major book that presents his insight after forty years work with research in judgment and decision making, including what makes people happy or unhappy (2011).

However, with the best intentions, sometimes the treatment of animals has been a result of the fallacy of anthropomorphism vis-à-vis animals – using human norms and wrong assumptions about animals. A Buddhist concept for such an error might be delusion. The challenge is to look at the animal from the perspective of the animal, which is to use a therocentric perspective. The wolves are born canocentric, the lions are leocentric, and the fish are ichtyocentric. Our treatment of animals requires a multicentric view based upon correct knowledge and attitudes of respect for the interests of the animals.

Newer research by Daniel Weary and colleagues at the University of British Columbia, (see also Wagner et al., 2015), demonstrates that dairy cows possess surprising intelligence and emotional sensitivity. One implication is that calves are affected by the emotional pain of separation from their mother and the physical pain of dehorning. In dairy farming, calves may be separated from their mothers within hours after birth in order to harvest the cow's milk for human consumption, based upon the belief that early isolation is safer for the calves. But if the calves are allowed to be with their mothers after birth, a strong bond develops, and it may take about eight months for the calf to be independent of milk. The early separation of the calf is "distressing for both calf and cow", according to Weary. "The calves will engage in repetitive crying and become more active", he says, "and sometimes you'll see a decline in their willingness to eat solid food." Both separation and types of dehorning can result in a negative cognitive bias similar to pessimism. <https://www.wired.com/2014/06/the-emotional-lives-of-dairy-cows/>, <https://www.sciencedaily.com/releases/2015/04/150428081801.htm>, Even if animals have been domesticated through our modern civilization, the domesticated animals have the same "innate behaviour patterns and needs even if the animal has never known natural conditions" (Singer, 2009, p. 142). Therefore five principles of basic freedoms should be fundamental to mammals: (1) sufficient freedom of movement to be able to turn around without difficulty; (2) the ability to groom itself; (3) to get up; (4) to lie down; and (5) to stretch its limbs (op. cit. p. 142). If implemented, it will mean much less suffering in "veal production".

Even when fish are categorized as animals, and therefore protected by Norwegian animal welfare law, due to ignorance and anthropocentrism the ichtyocentric view has for decades been neglected in aquaculture. It has had fatal consequences, since the large companies have operated in the farming and slaughtering of the fish based on a number of dogmas about fish. These dogmas are strongly criticized from an ichtyocentric perspective (Børresen, 2000).

Some of the dogma include; "Fish have a low body temperature, almost no brain, and thus are without consciousness and thinking ability"; "The fish is deaf and has no sound language"; and "The fish has no sense of pain". Based upon such dogma a regulation that farmed fish shall be anaesthetized with CO_2 before slaughter was prescribed. Børresen (2000, pp. 31–36) provides ample evidence that these dogmas have no validity. Better knowledge about fish should lead to treatment that will reduce suffering. By acknowledging that fish are animals capable of pain, animals should be within and not without the circle of morality (Røcklingsberg, 2015). Even Jeremy Bentham, the father of the utilitarian approach, acknowledged that animals are moral beings because they can suffer. This contrasted the view of one of the greatest ethicist's in the eighteenth century, Emmanuel Kant, who claimed that only humans could be moral objects because only human beings can reason. However, modern science argues that fish have neuro-architecture analogous to mammals that can process and integrate complex information and organize behaviour as a response to stimulation from the environment. Fish also have cognitive and emotional capacities and behavioural patterns that vary between and within fish species (Braithwaite et al., 2013). There are also studies that individual fish can exhibit distinct "personalities" and stress coping styles (Kalueff et al., 2012). To include fish in the moral circle means that fish deserve to receive moral considerations in their own right and allow them some basic welfare principles in the form of five freedoms: freedom from hunger and thirst; freedom from discomfort; freedom from pain, injury or disease; freedom from fear and distress; and freedom to express normal behaviour (see Lam, 2016 for an overview).

In the last thirty years we have seen a slow but positive development in the protection of animals. In the early 1970s, there were almost no public voices against the violation of animals. There were no animal rights – or animal liberation organizations. Animal welfare and well-being were almost exclusively for pets such as cats and dogs. In the 1980s, there were debates in Europe, which were formalized as an EU directive in 1986 (European Commission 1986) on the protection of animals used for experimental and other scientific purposes. The focus was to improve the control and use of laboratory animals, setting minimum standards for housing and training of such animals. One of the goals was to reduce the numbers of animals used for experiments by requiring that an animal experiment should not be performed when an alternative method exists. In 2006 the Council of Europe developed guidelines for the accommodation and care of laboratory animals, and a new EU directive was adopted in 2010, with effect from 2013. The ultimate goal is to replace the use of animals, but nevertheless it is acknowledged that animals are still needed as experimental animals today.

However, the revision of the directive clearly acknowledges the intrinsic value of animals, which means that animals should be distinguished from "goods and other possessions". The new directive spells out the principle of the 3Rs: replacement, reduction, and refinement in Article 1. Refinement is not limited to scientific procedures; it extends to the care, accommodation, and breeding of animals. Refinement also means making small changes to reduce the risk of the animal dying, and reducing the number of animals used.

Animals may only be harmed if there is a significant and probable utility for animals, humans, or the environment. Cancer research and cancer treatment support the argument that utility for humans weighs more than the cost (stress, pain and suffering) inflicted on animals through the research. The OncoMouse is the "classic" example that human utility trumps any pain inflicted upon an animal, in this case an infected, vulnerable mouse. However, the oncomouse might also be seen as a symbol of the interconnection of animals and human beings. It reveals the little difference in genetic mapping between humans and other animals. It is not that other animals looks like human beings, but that human beings look like our much older siblings. The Atlantic salmon may have been swimming in the sea and in the rivers at least in 20 million years, while the *Homo sapiens*, the modern human beings have only been walking on the Earth the last 200,000 years. Thus, many of the profound neurological and biochemical processes that regulates the functions of human and fish bodies have basic similarities. For example, the stress physiology is very similar between fish and human beings. Signal molecules and reward systems that are connected to emotions and states of the mind, such as dopamine and serotonin, are also common between fish and human beings. Even if the fish brain and the human brain is very different in size and construction, it is reasonable to assume that fish also have qualitative experiences of the world (Børresen, 2007; Kristiansen, 2013).

In Norway there is an ongoing debate on reducing the pain and suffering of animals used in research. New ethical guidelines will be negotiated. Norway is not a member of the EU, but has developed ethical codes which accept that animals are moral objects that deserve respect.

Consequently, more people criticize the use of animals in research. One interesting sign is that people employed as researchers who use animals in their research regard the subject matter so sensitive that they do not like to talk about their work outside of their own organization (Torp, 2017, pp. 4–7).

The intention with the EU directive in 1986 was not to protect animals in general, but only animals used for scientific purposes. In Norway, the first law about animal welfare (Dyrevelferd) was enacted in 1935 (Frøslie, 2000). Before this law, punishment was restricted to torture of animals, and it was restricted to cows and horses. Only severe or evil mistreatment resulted in punishment. However, in 1920, the words "severe and evil" were deleted from the common law so that any mistreatment of animals was punishable. The Norwegian law on animal welfare was a pioneering law internationally. However, technological development within farming – larger scale, bigger markets, and more competition created new challenges. The first revision of the law was proposed in the 1960s, but only enacted in 1977. An important addition to the law was to take due consideration of the animals' instincts and natural needs. The central paragraph is Paragraph 2: "One shall care for the animals and take due consideration to the animals' instincts and natural needs in order that they should not suffer unnecessarily." The spirit of the law is that the animal should be protected and defended (see Frøslie, 2000).

But many cases show that the reality in Norway is not in harmony with this law, which is not effective in protecting and defending animals from mistreatment. Examples of outright mistreatment are not punished with imprisonment,[4] and many cases related to fur farms reveal many transgressions of the spirit of the law. Nevertheless, it is still "business as usual" in the fur industry in Norway. An example can illustrate the quality of the public debate. The Minister of Petroleum and Energy, Ole Borten Moe, (in 2013) stated that the arguments against the fur industry are based on feelings, and that people, to an increasing extent, have a "Walt Disney like view on animals" (Johansen, 2014, p. 60). At the same time, he owned a farm with 30,000 chickens. Obviously, he was making fun of those who do not share his anthropocentric worldview. For the Minister the animal right activists are victims of their emotions and should not be taken seriously.

Finally, he expresses his concern for what will happen if the animal rights activists are able to close down the fur farms; he is afraid that the chicken farm he owns will be the next to be closed down. In such a debate, even with a Minister, there is "no room for fact-based arguments" about the ethical issues of "keeping non-human animals locked in small cages, and exploiting their lives to make luxury products" (Johansen, 2014, p. 62). The public discourse in Norway is also dominated by taboos when it concerns "heroic traditions" connected to the "Norwegian identity", like seal hunting and whaling.

Experiments with zebrafish have been performed in the primary schools in Norway, but teachers responsible for the experiments, have not been given proper training in treating the fish as having intrinsic value (Kjetland et al., 2016). Furthermore, the equipment in schools is not satisfactory for paying the zebrafish appropriate respect for its life and fulfillment. The zebrafish *(Dania rerio)* is explicitly categorized as an animal in EU directive 2010/63/EU. If we had taken Hans Jonas's (1984) *The Imperative of Responsibility* seriously, "in dubio pro malo" the risk of inflicting suffering on the zebrafish would have been reduced (Kjetland et al., 2016).

However, the experiment with zebrafish in the primary school had some unpredictable results since it awoke the teachers' consciousness – an invaluable input for learning and teaching. During the experiment with zebrafish what happened might happen in any experiment. The small group of zebrafishes died, even if the procedures given by the EU directive were followed (Kjetland et al., 2016). However, in a surprising way the encounter between the dying zebrafish and the teacher lead to a higher consciousness and a deeper empathy with the fish. In fact, I observed that "encounter trumps knowledge" in having a decisive impact on the teacher as researcher.[5] This observation has obvious importance for the kind of data that should have validity, and we need to re-evaluate our attitude towards the value of "objective scientific knowledge" versus giving credence to encounter, which means that unreflective, unmediated reality deserves high validity. It means that we should question modern science and its methods. In order to re-evaluate our received view of what counts as valid knowledge, we have to understand how and under which circumstances the modern sciences were developed.

The western worldview was developed during the Enlightenment. This was a time of human disengagement from the world, in which the world became an "object" and participation in nature lost its scientific significance. Modern science, in a close alliance with technology, constructed and tested universal laws and theories about the world to know about the world, with the one-sided instrumental goal to control the world. Unmediated, unreflected, first-hand, "unbefangen" experiences of different qualities of the world were deemphasized (Vetlesen, 2015). Our relation to the world changed, not by nature, but by discursive patterns of our way of thinking and talking about the world.

That encounter should transcend knowledge is in accordance with Vetlesen (2015), who writes about his encounter with a wild animal, a bird. This encounter was for him a kind of experience different from all other experiences. The encounter "provided perspective on my being in the world". Vetlesen was looked at by an eagle, and "That look fixed me in a manner as elusive as it was commanding. Its intelligence captivating, irrefutable, making mockery of Cartesian doubt. It forced me to ask myself what I looked like from that point of view, being the object of a look from a bird who, in looking at me, clearly was not an object but fully a subject." This did more to repudiate Descartes' view that "animals have no souls, no thoughts or experiences than do dozens of books of philosophy." (Vetlesen, 2015, p. 204).

Such a one-to-one experience with an animal is well documented as a "mind shift" for several persons. Aldo Leopold's (1966) meeting with a dying wolf is one classical example. Leopold writes that: "My own conviction dates from the day I saw a wolf die … In those days we had never heard of passing up a chance to kill a wolf. In a second we were pumping lead into the pack, but with more excitement than accuracy … We reached the old wolf in time to watch a fierce green fire dying in her eyes. I realized then, and have known ever since, that there was something new to me in those eyes – something known only to her and to the mountain. I was young then, and full of trigger-itch; I thought that because fewer wolves meant more deer, that no wolves would mean hunters' paradise. But after seeing the green fire die, I sensed that neither the wolf nor the mountain agreed with such a view" (Leopold, 1966, pp. 129–130).

The Canadian activist, Paul Watson, tells about a similar strong encounter with a huge sperm whale. Watson recalled, "I looked up … into a massive eye the size of my fist – an eye that reflected back intelligence, an eye that spoke wordlessly of compassion, an eye that communicated that this whale could discriminate and understand what we had tried to do … From that day, I knew emotionally and spiritually that my allegiance lay with the whales first and foremost, over the interests of those humans who would kill them" (Watson quoted in Vetlesen, 2015, p. 205).

Those one-to-one encounters illustrates a communicative exchange between humans and non-human animals that had existential implications on people's lives, which resulted in stronger commitments and even a change in "life projects". Leopold, who eagerly shot the wolves became a holistic environmental thinker after his encounter (Leopold, 1966). Similar encounters sensitize humans and make them more open to the complexity and mystique in nature. Encounters stand in sharp contrast to share numbers and statistics. For example, we know that the loss of biodiversity was 52 percent from 1970 to 2010 (WWF, 2014). As a naked number, 52 percent is an abstract entity, which apparently is not enough to change the mindset and actions of human beings in general. However, encounters can change! Without animals the human lifeworld will be reduced. We need nonhuman nature to fully develop human beings. The unreflective experience is open ended and implies vulnerability, while knowledge provides closure, control, and security of the future (Vetlesen, 2015).

Can ethical theories teach us something about how we should relate to animals? Albert Schweitzer (1875–1965) may be called one of the first ecological ethicists. His primordial ethical principle was "reverence for life", which means that all living things have an intrinsic value that commands our awe and reverence (Schweitzer, 1990). Schweitzer's diagnosis was that modern industrial society had moved away from a worldview that connected the goodness of life with the goodness of nature, and it had destroyed what Schweitzer called the "world-and-life-affirmation". In the dominant worldview, nature was a value-free and mechanical force – a machine, governed by physical and mechanical laws. Schweitzer's cure was to re-establish the bond between nature and ethics. Reverence for life or *Ehrfurcht vor dem Leben* which was Schweitzer's original term, implies an attitude of awe and wonder, honor, and fear. According to Schweitzer, ethics begins when we become fully aware of the fact that "I am life which wills to live, in the midst of life which wills to live" (Schweitzer, 1990, p. 130). It is good to promote and preserve life, and let life rise to its highest value, its fulfillment. Evil is what destroys life, what injures and represses life, which is capable of development. This is the absolute, fundamental principle of the "reverence for life" (Schweitzer, 1990, p. 131; see also DesJardins for an overview, 2006, pp. 131–134).

Concluding remarks

Modern economics is built upon a mechanical worldview, imitating "natural sciences" as a paradigmatic example of what a real science should look like by constructing and using abstract theories and models of the world. We may ask whether emphasizing abstract models dressed in mathematical formulas might contribute to a general desensibilization vis-à-vis nature? Our cases of violence against animals mirror the need to sensitize economists to act with higher empathy and emotion towards other animals. The dream is to (re)educate economists in a way that they may respect and feel awe and wonder towards nature, of which they are a part of, not outside of. A view of and appreciation of nature as wonderful, and as a gift for human fulfillment and future generations, transcends immensely the one-sided view of nature as property and rent. The OncoMouse is the icon of the pure commoditization of nature, but also of a general hybridization where the border between the natural and the artificial is blurred. We may well ask whether farmed Atlantic salmon is identical with the original king of the river who has evolved over twenty million years. Industrialization, domestication, and "production" of calves and salmon implies a narrow view of these animals, which enables the infliction of massive violence upon them. Cases where animals are used as pure commodities in large scale farms, aqua production, and experiments in the name of research were described and presented. A number of similar cases of violations were presented and described; for example, chicken factories, egg production, the conditions of "production" cows, etc. (see Larsen, 2013 for a precise description of the "lifeworld" of a calf when she follows a calf daily, from its birth to slaughter four months later).

A case that deserves much more attention is the farming of Atlantic salmon from embryos to slaughter. The two empirical studies (Bundli and Liltvedt, 2012; Bolstad and Gloppen, 2015) shows a screaming ignorance about the welfare of fish. To kill fish with CO_2 has caused an enormous amount of pain for millions of individual fish.[6] If we had taken Hans Jonas' *The Imperative of Responsibility* seriously, "in dubio pro malo", we should have invested in much more scientific knowledge about fish from an ictyocentric view. The problematic anthropocentric worldview and the human inclination to anthropomorphize other species creates unnecessary suffering. Aquaculture is now the fastest growing food sector, projected to supply 60 percent of fish by 2020, while production from capture fisheries has declined since the 1990s (Lam, 2016, p. 37). Therefore, the importance to take the ethics of farmed fish seriously is of immensely practical importance. One fruitful approach is using methodologies like the ethical matrix (see Kaiser, 2005, 2012; Kaiser et al., 2007; Lam, 2016). With the emergent knowledge of science and ethics of fish, individual fish – not only groups of fish – is added in the ethical matrix.

What could be the cure to stop the mistreatment of animals? We have to replace some of the basic tenets of modern mainstream economics. The assumptions about efficiency as the primary goal should be challenged. Efficiency often conflicts with a caring attitude towards the "objects" to be economized. Throughput is a different conception of measuring economic performance, which means that frugality should be more prominent (see Daly, 2008; Bouckaert et al., 2008). This means that there will be less use of animals in the "production process", and it follows that instead of looking at animals as "resources", as pure means, animals should be paid respect because they have intrinsic value. Such a change in attitude might inspire a closer connection between man and other animals, and thus less violence against animals. The 3R formula regarding using animals in research is also a step in the right direction. Replacement, reduction, and refinement of animals to be used for scientific purposes are important goals. For research on mice and zebrafish, we might ask about the validity of the research for human.

Even if zebrafish has 80 percent of the same genetic material as human animals, what about the 20 percent gap? When we believe that medicines tested on animals might have the same effect on human animals, we take a very high risk. The reason to inflict suffering on research animals such as mice or zebrafish should be more profound than "nice to know". Concerning knowledge, we need scientific knowledge, and much more of a different kind than Humean empiricism. The question should not be "what can we experience", but "What have we experienced?" (see Griffin, 2007, pp. 32–35). Knowledge in the old sense of direct, unreflective experiences should be given appropriate status.

In this essay, I have partly been drawing upon Bauman's (1989) insights from his reasoning on the Holocaust and from Bandura et al. (2002) on moral disengagement strategies to explain the lack of empathy and mistreatments of animals. Their reasoning might be seen as an alternative, or perhaps complementary, to the prevalent view of a "hunters insensibility" and an inborn ability to switch on or off sensitivity towards animals (Børresen, 2000, 2007). Perhaps Bauman and Bandura's explanations will be more important when animal rights movements become stronger, granting animals inviolable rights? The conflict between human material self interest and the instrumental use of animals is almost nonexistent in our western culture, The hope is that we are open to learning from other cultures, and economic reasoning like in Buddhism (Zsolnai, 2011, 2014), to learn to live in a synergistic interaction with animals that will enrich our lifeworlds in a substantive way, and reduce the suffering of animals immensely.

Vetlesen's (2015) description of modern economists as autistic should be taken seriously, since we have moved into a human "created" world called Anthropocene (Crutzen et al., 2011). Human agency has had an immense impact on the climate and the environment all over the world, which urge for a renewal of business ethics. Singer's (2009) work on animal liberation has given a strong push to the animal movement. Singer uses the term speciesism, which implies that human animals regard themselves over and above other animals. Singer's strong claims for treating animals with respect and care are in accordance with Leopold's (1966) vision to look at animals as members of the community, where humans have not a prominent place, only a prominent responsibility. A central goal in the education of economists would be to transform the worldview from mechanic to organic. A parallel change would be to define economics in more substantive ways. The old definition of economics as the study of how humans use scarce means to satisfy insatiable and unlimited wants to obtain material abundance (Gowdy, 2006), should be replaced by concepts like "well-being", happiness, and enjoyment of life. Deep ecology (Næss, 1989), biology, and caring relationships with respect to nature and anthropology should be compulsory parts of the curriculum.

We have focused on husbandry and fish welfare. However, from a systemic perspective other important problems related to aquaculture should also gain public attention. Farmed salmon threaten the wild Atlantic salmon in a number of ways: via escape; via use of cobber to clean the cages; via tons of chemical and antibiotic used as medicines; and via the feed which to a large extent consists of other fishes lower down the food chain, which leads to overfishing and injustice for poor people living close to areas where mega ships catch fish in order to supply aquaculture systems with fish only affordable to "affluent consumers". Aquaculture is certainly one of the many contributors to micro plastics[7] that slowly kill many animals, and is the size of an island in the Pacific Ocean as large as half of Europe. One screaming example was a dead Cuvier's beaked whale *(Ziphius cavirostris)* found by the shores close to Bergen in January 2017. The whale's stomach was completely filled with large amounts of microplactics and thirty plastic bags. The whale was alive when they found it at the beach, but clearly very sick. According to one of the scientists who dismembered the remains, "there is no doubt that the whale has suffered immensely". Finally, the plastic must have destroyed the metabolism of the whale and led to a slow and painful death.

The real life case of the "plastic whale" was a shock for many people around the world. The movie of the plastic whale gave the whale a "face". The concreteness of the whale transcended descriptions and concepts such as complexity and diversity and moved us downwards and closer to the immediate and experientially sensually rich experience of nature, even if only a few people had direct unmediated access to the whale in situ and were able to see, hear, touch, smell, and feel the whale. The whale should remind us that nature is a gift for which we feel awe, a gift that can surprise us, but cannot be possessed or controlled – not dictated by our desire, but something to respect as a subject in itself (Vetlesen, 2015, p. 179). We might realize that we are indebted to our older siblings – the animals from our own evolutionary existence.

Art is also one of the means to confront people with real, concrete phenomena in the world. Art has the power to influence our consciousness and function as a wakeup call and make us more compassionate with the existential conditions of man and other animals. Such a wake-up call may help to see some of the implications of limiting, often narrow, assumptions.

Economics as a discipline is in fact an existential enterprise where the fates of all the animals (human and non-human) in the world, including future generations, are at stake. This view has for a long time been one of Laszlo's strongly held theses. Confronted with great artwork, the spectator/participant will be more sensitized and open to hearing the screams of pain and degradation inflicted on animals and nature.

Edward Munch's *Scream* was used as the introduction of this essay. Let us close the essay with another text of Munch, where Munch describes human brutality vis-à-vis calves:

> Can you imagine a calf, a common red sided calf? Have you ever looked into the eyes? Have you seen the depth of its eyes and the innermost goodness in them? And this being with these eyes we stun one day brutally in the forehead, stick a knife in the trout on it and eat blood food and meat of it … We are beast, are we (…).

Notes

1. There are four versions of *Scream,* not counting the lithographs. One version is a pastel-on-card board from 1885 with a stooping figure in the background and a self -authored poem which Munch painted onto the frame, describing the experience

that prompted *Scream*. The *Scream* used in this article is the painted version in The Munchmuseet, Oslo, probably painted in 1910. There has been a long discussion about the date of this painting. It is "Tempera and oil on unprimed cardboard", 83.5 × 66 cm, Munchmuseet, Oslo, (Photo): Munchmuseet.

2. A muzzle is a tool that binds the mouth, which prevents an animal from biting or eating.
3. <http://meatless.no/dyrene/kua-og-kalven/> <https://www.wired.com/2014/06/the-emotionaJ-Jivesof-dairy-cows/>, <https://www.sciencedaily.com/releases/2ol5/04/15042808l8ol.htm>
4. A case from 2017 illustrates that severe violence against a small (dog) puppy, who died after five months of mistreatment, did not lead to imprisonment, but only a ban from keeping a dog for five years. <http://www.nettavisen.no/nyheter/innenriks/32-aring-mishandlet-valp--dmttil-fengsel/3423378l54.html>
5. I was invited to watch a part of the experiment, which took place in the local school in Bergen, 2016.
6. An estimate is that 200 million Atlantic salmon were slaughtered in Norway in 2012, (see Kristiansen, 2013).
7. Plastic waste is slowly breaking down in smaller fragments known as "micro plastics". There are many sources of these plastic waste islands, from car tires to personal hygiene products. One calculation is that one billion (defined as plastic pieces that are smaller than 5 millimeters) of plastic pieces are generated every hour, and most of them end up in the sea. One of the problems is that micro plastics spread hazardous pollutants that pose a threat to the environment and human health since other environmental toxins latch onto micro plastics, and they are found up to 5,000 meters below sea-level.

Bibliography

Bandura, A., G. V. Caprara, and L. Zsolnai. 2002. Corporate transgressions. In *Ethics in the economy. Handbook of business ethics*, ed. L. Zsolnai. Oxford: Peter Lang.

Bauman, Z. 1989. *Modernity and the Holocaust.* Cambridge: Polity Press.

Bolstad, L. T., and T. G. Gloppen. 2015. *Er det fokus på fisken* i *norsk lakseoppdrettsnæring? Et studie av fiskevelferd i tre norske oppdrettsbedrifter,:* ("Is it a focus on the fish in Norwegian aquaculture of Salmon? A study of the well-being of fish in three aquaculture companies in Norway"). Master thesis, Bergen: Norwegian School of Economics.

Børresen, B. 2000. Menneskets medfødte forutsetninger som vertskap for produksjonsdyr in, *dyreetikk* ed. A. Føllesdal. Bergen: Fagbokforlaget ("The humans' innate conditions as hostage for production animals"). pp. 16–39.

Børresen, B. 2007. Fiskens ukjente liv ("The unknown life of the fish"). Oslo: Transit A/S.

Bouckaert, L., H. Opdebeeck, and L. Zsolnai. 2008. *Frugality: Rebalancing material and spiritual values in economic life.* Oxford: Peter Lang Publishing.

Braithwaite, V. A., F. Huntingford, and R. van den Bos. 2013. Variation in emotion and cognition among fishes. *Journal of Agricultural and Environmental Ethics* 26: 7–23.

Bundli, V., and K. Liltvedt. 2012. *Fish farm ecology – a conceptual framework and empirical investigation of the CSR performance of Marine Harvest ASA and Cermaq ASA.* Master thesis. Bergen: Norwegian School of Economics.

Crutzen, P., J. Grindevald, J. McNeill, and W. Steffen. 2011. The Anthropocene: Conceptual and historical perspectives. *Philosophical Transactions of the Royal Society* 369/A: 842–867.

Daly, H. E. 2008. Frugality first. In *Frugality: Rebalancing material and spiritual values in eco nomic life*, ed. L. Bouckaert, H. Opdebeeck, and L. Zsolnai. Oxford: Peter Lang Publishing.

Daly, H. E., and J. B. Cobb Jr 1994. *"For the common good" redirecting the economy toward community, the environment, and a sustainable future.* 2nd edn. Boston: Beacon Press.

DesJardins, J. R. 2006. *Environmental ethics. An introduction to environmental philosophy.* 4th edn. Belmont: Thompson Wadsworth.

EU directive. 2010/63/EU of the European parliament and of the Council of September 22, 2010 on the protection of animals used for scientific purposes, *Official Journal of the European Union.*

European Commission. 1986. The European Council of Ministers adopted the directive 86/609/EEC.

Frøslie, A. 2000. At dyr ikke ska! lide i "utrengsmål" om dyrevernlovens hovedparagraf. In *Dyreetikk* ed. A. Føllesdal Bergen: Fagbokforlaget. («That animals shall not suffer «unnecessary». About the animal protection main paragraph.) pp. 40–62.

Global Footprint Network. 2015. <http://www.footprintnetwork.org/en/index.php/gfn/page/footprint_basics>.

Gowdy, J. 2006. Business ethics and the death of "homo oeconomicus". In *Business within limits. Deep ecology and Buddhist economics*, ed. L. Zsolnai and K. J. Ims. Oxford: Peter Lang.

Griffin, D. R. 2007. *Whitehead's radically different postmodern philosophy. An argument for its contemporary relevance.* New York: State University of New York Press.

Harari, Y. N. 2011. *Sapiens. A brief history of humankind.* New York: Harper.

Johansen, H. 2014. *Ecological ethics in the debate about Norwegian Whaling.* Thesis submitted for the Degree of Master of Philosophy in Culture, Environment and Sustainability, Centre for Development and the Environment. Oslo: University of Oslo.

Jonas, H. 1984. *The imperative of responsibility: In search of an ethics for the technological age.* Chicago/London: University of Chicago Press.

Kahneman, D. 2011. *Thinking, fast and slow.* London: Penguin Books.

Kaiser, M. 2005. Assessing ethics and animal welfare in animal biotechnology for farm production. *Revue Scientifique Technique Office International Epiz.* 24 (1): 75–87.

Kaiser, M. 2012. The ethics and sustainability of aquaculture. In *The philosophy of food.* Berkeley: University of California Press.

Kaiser, M., K. Miller, E. Thorstensen, and S. Tomkings. 2007. Developing the ethical matrix as a desicion support framework: GM fish as a case study. *Journal of Agricultural and Environmental Ethics* 20: 65–80.

Kalueff, A. V., A. M. Stewart, E. J. Kyzar, J. Cachat, M. Gebhardt, S. Landsman, et al. 2012. Time to recognize Zebrafish "affective" behaviour. *Behaviour* 149: 1019–1036.

Kjetland, E., K. Myklebust, and R. Rajanlingam. 2016. *Oppgave om Mikrobiologi. Forskning på sebrafisk, naturfag 2, videreutdanning for lærere.* Bergen: University of Bergen.

Kristiansen, T. 2013. Historien om laksen – fra elvas konge til Norges viktigste husdyr i *Hvem er villest i landet her? Råskap mot dyr og natur i antropocen, menneskets tidsalder.* red. R. Sollund, M. T¢nnessen, og G. Larsen. Oslo: Scandinavian Academic Press.

Lam, M. E. 2016. The ethics and sustainability of capture fisheries and aquaculture. *Journal of Agricultural and Environmental Ethics* 29: 35–65.

Larsen, G. 2013. Kuas tildelte rolle som matobjekt, i *Hvem er villest i landet her? Råskap mot dyr og natur* i *antropocen, menneskets tidsalder,* red. R. Sollund, M. T¢nnessen, og G. Larsen, ("Who is the wildest in the country? Brutality against animals and nature in antrophocene, the human age") Oslo: Scandinavian Academic Press.

Leopold, A. 1966. *A Sand County almanac. With other essays on conservation from Round River.* New York: Oxford University Press.

Martinsen, S. 2010. Det lysere livet ("The lighter life") *NOAHs Ark,* nr. 2.

Næss, A. 1989. *Ecology, community and lifestyle: Outline of an ecosophy.* Cambridge: Cambridge University Press.

Røcklingsberg, H. 2015. Fish consumption: Choices in the intersection of public concern, fish welfare, food security, human health and climate change. *Journal of Agricultural and Environmental Ethics* 28: 533–551.

Schumacher, E. F. 1999. *Small is beautiful: Economics as if people mattered: 25 years later … with commentaries.* Point Roberts: Hartley & Marks Publishers.

Schweitzer, A. 1990. *Out of my life and thought* (A. B. Lemke, Trans.). New York: Holt.

Singer, P. 2009. *Animal liberation. The definitive classic of the animal movement.* New York: An Imprint of HarperCollins Publishers.

Torp, L. S. 2017. Dyrebar forskning. ("Valuable research") *Forskningsetikk* nr 2. 2017 (4–7).

Tøyner, P. E. 2000. *Munch. Med egne ord ("Munch. With his own words").* Forlaget Søren Fogtdal.

Vetlesen, A. J. 2015. *The denial of nature. Environmental philosophy in the era of global capitalism.* London: Routledge.

Vohs, K., N. Mead, and M. Goode. 2006. The psychological consequences of money. *Science* 314: 1154–1156.

Wagner, K., D. Seitner, K. Barth, R. Palme, A. Futschik, and S. Waiblinger. 2015. Effects of mother versus artificial rearing during the first 12 weeks of life on challenge responses of dairy cows. *Applied Animal Behaviour Science* 164: <https://doi.org/10.1016/j.applanim.2014.12.010>.

WWF (World Wide Fund for Nature). 2014. *Living planet report 2014: Species and spaces, people and places.* Gland: WWF International. <http://wwf.panda.org/about_our_earth/> all_publications/living_planet_report/.

Zsolnai, L. 2011. Why Buddhist economics? In *Ethical principles and economic transformation – A Buddhist approach*, ed. L. Zsolnai. London: Springer.

Zsolnai, L. 2014. Chapter 9: Buddhist economic strategy. In *Beyond self. Ethical and spiritual dimensions of economics.* Bern: Peter Lang.

KNUT J. IMS

Deep Ecology and Its Relevance to Gross National Happiness and Bhutan

Introduction – "Let the rivers live!"

When the Norwegian government started its modernization and rebuilding after the Second World War, in the 1960s, Norway needed energy. Due to a unique combination of much rain, high mountains, and powerful waterfalls, there were excellent opportunities to generate electric power. However, high mountains may also be seen as majestic homes and habitats for eagles and snow tigers, sentient beings with dignity, and high waterfalls might be seen as spectacular masses of water that collapses into wild, beautiful rivers which is the life-giver of Mother Earth. Generating power by electricity needs domestication of the rivers and the building of large-scale dams, which transforms the topography and threatens all life in the ecosystem. The Norwegian philosopher, Arne Næss, also enthusiastic mountain climber, together with a small group of environmentalists resisted the domestication of some of the most beautiful waterfalls and rivers, using the motto: "Let the rivers live!" In a non-violent way, the group held active demonstrations against the domestication of the waterfalls. This was the beginning of a new green movement and philosophy named "Deep Ecology".

In 2015 deep ecology is not a finished ready-made theory, but an outline open to be filled out. According to deep ecology, every person has a responsibility to work out his own ecosophy, a reasoned process of ecocultural harmony. This kind of *sophia* or wisdom is openly normative, and contains both norms, value priorities, and hypotheses. Wisdom is policy wisdom, prescriptions and hypotheses are not merely scientific descriptions and predictions.

How should we proceed to develop our own ecosophy, the wisdom to see ourselves as small actors in an amazing world where we have to choose between different roads? The road that "we have long been traveling is deceptively easy, a smooth superhighway on which we progress with great speed, but at its end lies disaster. The other fork of the road – the one "less traveled by" offers our last, our only chance to reach a destination that assures the preservation of our earth" (Carson, 1962, p. 277).

One might start by criticizing the common and very popular instrument to measure and compare human welfare, the Gross National Product (GNP), which is a one-dimensional measure of economic growth that reflects the size of the commercial market. Næss aptly writes that GNP is equivalent to Gross Domestic Pollution, emphasizing that GNP does not imply any progress along the course of self-realization, community vitality, or environmental health.

Another positive road is to look to Bhutan: a country that proclaims to measure the well-being of its population in a direct and holistic way. The intriguing question then is, how does Bhutan's government conceptualize human happiness and how do they ultimately measure it? Since 1972, Bhutan's government has attempted to expand the well-being and true happiness of its people and, accordingly, articulated the goal of Gross National Happiness (GNH). The Constitution of Bhutan (2008) directs the state "to promote those conditions that will enable the pursuit of Gross National Happiness", (Ura et al., 2012a, p. 6). Bhutan's concept of GNH merits sincere exploration since it is described as "holistic, balanced, collective, sustainable and equitable" (Ura et al., 2012, p. 7). Furthermore, they argue that it balances material and spiritual development in such a way that they complement and reinforce each other, making it very promising from a western point of view. Thus, it attempts to meet the strong critics of the prevailing measure for development and welfare, GDP.

Bhutan's measurement is timely since there is a growing interest in measuring the subjective well-being of people. The international governmental organization, the OECD (2013), has developed OECD Guidelines on Measuring Subjective Well-being, and states that it has a particular interest in encouraging international comparability of such data. The OECD report defines subjective well-being to encompass three elements: (i) life evaluation – a reflective assessment on a person's life; (ii) affect – a person's feelings or emotional states and (iii) eudemonia – a sense of meaning and purpose in life, or good psychological functioning.

It appears that both GNH and deep ecology are strong opponents of the use of GDP and the political and ideological context behind it. Deep ecology claims to be an alternative way to a sustainable society where the flourishing of all life on Earth is the final goal. Could deep ecology and Bhutan's GNH complement and enrich each other? Through their similarities and differences, I aim to explore whether these two approaches to sustainability and deep happiness can inspire and learn from each other.

First, a short overview of deep ecology, formulated by the Norwegian philosopher Arne Næss (1913–2009) will be presented. Thereafter a view on some of the inspirational sources of deep ecology, the "deep ecological tree", and the core level – the eight points of the platform will then be summed up (see Næss, 1989, 1995; Ims, 2011).

For each of the "eight points" some similarities and differences with the nine dimensions or domains used to measure happiness or well-being in Bhutan will be explored. GNH's nine dimensions are: (i) education, (ii) living standards, (iii) good health, (iv) environmental diversity and resilience, (v) good governance, (vi) time use, (vii) community vitality, (viii) cultural diversity and resilience, and (ix) psychological well-being. Since my knowledge about Bhutan and GNH is limited, I will draw heavily on the research *Well-being, Happiness, and Public Policy* by Sabina Alkire (2015), and the two reports by Ura et al. (2012a, 2012b) to gain insight into the nine dimensions of GNH. Alkire's research (op. cit.) emphasizes human flourishing, which is also a key element in deep ecology. With notions like sufficiency, mindfulness, and the multidimensional way of approaching happiness, Bhutan's way of thinking may bring new light and inspiration into the development and understanding of deep ecology.

One fundamental insight deep ecology and Bhutan's GNH approach have in common is to view the environment as fundamental to the survival of humanity (Alkire, 2015, pp. 78–79). Alkire writes that "like each of the other domains, the study of human happiness adds something new. For harmony with nature has intrinsic value … Also of intrinsic value are relationships with non-human life forms" (p. 79). Alkire stresses the need to change the underlying instrumental and materialistic mindset of humanity, from maximization of living standards towards sufficiency. This view is in accordance with "live a rich life with simple means" which is a central motto of deep ecology.

Deep ecology – roots and core concepts

When a "long-range deep" approach was what western societies really needed, environmental thinking in western countries was mainly concerned about short run measures. Arne Næss formulated this critique in the 1970s, making a distinction between the deep and the shallow approach to environmental problems. Shallow ecology represents a technocratic attitude to pollution and resource depletion, treating the symptoms through technological quick fixes, using brutal rules like making the polluter pay to reduce the ecological footprints. On the other hand, deep ecology assumes a relational, total field perspective that fits into a holistic, non-reductionistic, non anthropocentric worldview by focusing on the underlying causes, and represents a change in mindset. It means that to solve the environmental problems, the basic political, economic, and ideological structure have to be changed. Ultimately, it means to change ourselves.

There are many reasons why the deep ecology thinking has gained supporters in the last decades. The Living Planet report for 2008 gives sophisticated evidences that the exploitation of resources and the level of consumption in all the western countries has an overshoot of several hundred percent. The US is on the top of this dubious ranking of ecological footprints, using 800 percent beyond its sustainable state.

Deep ecology represents inspiring insights as an alternative to the western materialistic society. The essence of deep ecology is a fundamental respect to diversity, that all life on Earth should flourish, and that the very notion of self as a subject should be redefined. The new self should be an eco-self. The "unit of survival" is not organism alone, but "organism plus environment". In short, deep ecology is both a philosophical perspective and a campaigning platform.

The idea behind deep ecology is ancient. It is drawn from ideas from Hinduism, Confucius, and Buddha and, on the other hand. it draws on Aristotle, Heidegger, and Spinoza. It is inspired by Gandhi's metaphysics, which is based upon the concept of oneness and that everything is interrelated. Such a holistic worldview leads logically to non-violence. And, in accordance with Aristotle, every plant has a telos, a goal – and is expected to realize itself, that is to be in a state of flourishing. Spinoza is an ontologist and claims that we have the ability to identify with others and thereby come close to all kinds of life. Part of this claim is Spinoza's ideas about the circle of friendship that may continuously grow and finally unite everyone.

Deep ecology proceeds in two directions: it tries to create a change and it tries to develop an alternative philosophy called ecophilosophy. It is explained in four levels: the most basic level is the metaphysical level; the second is the platform level; and the third and fourth levels consist of the policy level and the level of political actions (see Figure 1).

Level two, the platform, is the core and unites all kinds of radical ecocentrists, such as ecofeminists, direct action groups, as well as religious groups. Activities range from "ecotage" (sabotage to liberate exploited ecology) to the support of politically oppressed people in countries under development.

Level I	Ultimate premises	Taoism, Buddhism, Hinduism, Christianity, Ecosophy T, etc.
Level II	Platform Principles Movements	Peace Movement, Deep Ecology Movement, Social Justice Movement, etc.
Level III	Policies	A, B, C, etc.
Level IV	Practical Actions	W, X, Y, etc.

Figure 1. Four levels of questioning and articulation.

The above chart is a simplification of Næss's Apron Diagram. (See Drengson and Inoue 1995, pp. 10–12.)

The logic behind the framework is that there should be a continuous dialogue between the levels to keep our philosophy and practice in harmony.

One core activity is to pose deep questions to explore ultimate premises and norms. When we have articulated our position on the first level, we may move toward the lower levels.

The framework admits a great diversity at the level of "Ultimate Premises" (philosophies). We do not have to subscribe to the same ultimate ecological philosophy to work cooperatively. According to Arne Næss "the front is very long" and each person may contribute on his own premises.

It is illustrative to give some hints on the particular position of Arne Næss. His view starts with only one norm, self-realization! This norm means "self-realization for all beings!" The self to be realized for humans is not the ego self (self with a small s), but the larger ecological Self (Self with capital S).

Arne Næss does not "difficultivate" the concepts of the self. His focus is on the human ability to identify with a larger sense of self. Humans naturally have this capacity. This can be observed cross-culturally.

The piece movement is also a part of Næss's philosophy, but he argues that social justice cannot be enough. We have to produce and consume less – tread lighter and wiser on the Earth. "Simple in means, and rich in ends" is his motto. Its implication is to put quality of life over and above standard of living and to celebrate the virtues of slowness and smallness, contrary to our western ideology of speed and scale.

To sum up, Næss's holistic worldview negates the dominant metaphysics which sees humans as essentially different from the rest of nature. Næss claims that humanity is inseparable from nature: if we injure nature, we injure ourselves.

Even if Gandhi was one of Næss's inspirational sources, there are certainly some differences between them. Metaphorically, Gandhi used to look upon everybody as drops of water and writes: "This ocean is composed of drops of water; each drop is an entity and yet it is a part of the whole; the one and the many. In this ocean we are little drops …" Arne Næss on the other hand, states that for him it is more natural to look on himself as a little tree in a large forest.

The core level is level two: *the platform level*, which is usually summed up in eight points:

1. The well-being and flourishing of human and nonhuman life on Earth have value in themselves – independent of the usefulness for human purposes.
2. Richness and diversity of life forms are also values in themselves.
3. Humans have no right to reduce this richness and diversity except to satisfy vital human needs.
4. We need a substantial decrease of human population.
5. The present interference with the nonhuman world is excessive.
6. Policies which affect basic economic, technological, and ideological structures, must be changed.
7. The ideological change is mainly that of appreciating life quality (dwelling in situations of inherent value) rather than adhering to an increasingly higher standard of living. There will be a profound awareness of the difference between big and great.
8. Those who support these points have an obligation to directly or indirectly try to implement the necessary changes in a non-violating way.

A comparison of deep ecological thinking and GNH

I will interpret the eight points of deep ecology (DE) into the multidimensional measures of the quality of life and well-being, as used in Bhutan, but not dig into the technicalities of the advanced and complex measurement that forms the basis of GNH index. My approach will be to explore similarities and differences mainly on the conceptual level. The findings from the GNH index will be used when it is appropriate to support my arguments.

The three first points of DE will be presented in the same section, called "The diversity and wildlife category", and point four to point eight of DE will be subsumed under the category "Human oriented points". The last point, point 8, will be treated separately. Point eight is about the responsibility to put one's beliefs s into political action.

The diversity and wildlife category

Point one in DE emphasizes the value of all life in general. This view may be a common denominator of DE and the Buddhist and Hindu traditions in Bhutan. However, there might be an important distinction in terms of an eco-centric perspective in DE and a moderate anthropocentric perspective that I interpret in the GNH measurement. Alkire (2015) writes "For harmony with nature has intrinsic value." She also adds that "… the natural processes of co-existence … a sense of harmony between people, the animals and the Earth; the deep respect for the land, reverence for a specific sacred grove, … a feeling of affiliation with nearby cliffs. Also of intrinsic value are relationships with non-human life forms, various animals we live with or alongside" (p. 79).

Point two in DE emphasizes that diversity has intrinsic values. Diversity, complexity, symbiosis, and unity contribute to resilience. In the GNH index, domain eight, "ecological diversity and resilience" is measured. The GNH index has four indicators for this domain. One of them is wildlife. The GNH index measures this as "damage to crops" (p. 166). It is a growing concern in Bhutan, since "Wildlife damage can have catastrophic economic consequences for farmers, especially for vulnerable households; it also disrupts sleep patterns and may create anxiety and insecurity" (p. 166). This is a farmer specific perspective, and it is based upon the fact that only 21 percent of farmers reported "no" wildlife damage in the past twelve months. The GNH index states that it is not easy to ascertain the true cost of the damage. On the other hand, Bhutan is undergoing rapid urbanization, and urban respondents were asked about "inadequate green spaces".

This way of measuring the wildlife as crop damage might be in line with deep ecological thinking, since *DE's point three* accepts that vital human needs may justify the right to reduce the richness and diversity of life and wildlife. However, we see deep conflict between the farmers in Norway who suffer crop damage, and whose sheep are killed by bears and wolves, and the Norwegian government which proclaims that bears and wolves should be part of the Norwegian fauna because they naturally belong to the Norwegian land. A test case has recently popped up, ever since a wild boar started to invade the southern areas of Eastern Norway. Due to the milder climate, the wild boar established itself in the southern part of neighbor country, Sweden, where there are several hundred thousand. This is a "new" species which challenges Norwegian hunters. Wild boar destroy crops; they come in large groups and are very effective at digging up crops during the night. The Norwegian government has declared that this new animal does not historically belong to Norwegian fauna and that we should, therefore, not allow this new species to enter Norway. At the same time, there is a new movement of "rewilding" the nature via accepting or even introducing species of wild animals that originally lived in wild nature. We see this trend in different countries in Western Europe.

Human oriented points (four to seven) of DE

What does DE require concerning necessary changes in mindset in terms of ideology, political participation, technology, and lifestyle?

Næss's own ecosophy T tries to balance the two main ecological concepts, unity and diversity. To find the appropriate balance of those two concepts is a great challenge.

Point four in DE states the need for a substantial decrease of human population. It is a logical step since most western countries have an overshoot of many hundreds percent of the carrying or bio capacity of Earth. If we extrapolate this consumption pattern it is too much for a sustainable Earth. We need to reduce our pattern of consumption, and the question of optimum population is relevant (1989, pp. 140–141).

Næss emphasizes that the view of nature has evolved from looking at the land in terms "empty" and "desolate" when there was no human settlement to using terms like "free nature" and "untouched" nature where individuals now see the "desolate" nature full of life. The evolution is due to the steadily shrinking areas of free nature, and that natural parks are slowly ruined through excess numbers of visitors. Næss writes that a negative reaction towards an increase in human population "is not to foster any animosity towards humans as such – on the contrary, human fulfilment seems to *demand* and *needs* free nature". Homocentricism" and "anthropocentrism" ... should be qualified by an adjective, "narrow homocentricism" (1989, p. 141). Human is part of nature, not beyond. The greatest challenge in the future will be to protect the planet for its own sake.

Point five of DE declares that present interference with the nonhuman world is excessive and has to stop. A new concept for this state of the world is the age of Anthropocene. The destructive pattern is clearly visible in many ways. One of the symptoms of a non-sustainable economy is human created garbage mountains found around many of the big cities today. We know about the level of pollution and the high degree of smog in many of the biggest cities in the world. Oslo and Bergen, the biggest cities in Norway, have been measured to be among the worst cities in Europe in terms of air quality during winter time. The main cause is dangerous emissions from cars, not least diesel engines, and using of oil to warm up houses and water. In the cities there is an urgent need of renewable clean energy. In Schumacher (1999) terms, we need intermediate technology that is appropriate to the context, simple and therefore understandable, suitable for maintenance and repair on the spot. This idea of intermediate technology opposes the enormous cost and complication of production methods for the sake of labor saving and job elimination, and favors small-scale establishments (Schumacher, pp. 148–158).

Point six of DE's platform states that policies that affect basic economic, technological, and ideological structures must be changed. Arne Næss gives concrete examples on what this might mean by deducing a set of fundamental and derived goals for economic policy. According to Næss, some of the basic norms in economic policy are (1989, p. 108):

- B1 Full employment!
- B2 High consumption now! (that is within the present electoral term)

B3 High consumption in the future!
B4 Much leisure time now!
B5 Much leisure time in the future!

Derived norms:

D7 High national product now!
D8 High national product in the future! (rapid economic growth = high growth rate in GNP)
D9 High investment!
D12 Hold prices stable!

Within this ideological frame, we find that economics concerns itself only with means, not goals. Næss emphasizes that "such a proposition is clearly untenable" (1989, p. 109) because it is necessary to work with goals. Næss ends up writing that this way of thinking is anti-ecosophical and reveals the gigantic illusion that modern industrial society guarantees leisure time (op. cit. p. 109). The cost of making economics a "science" (in a narrow sense) is, according to Næss, a certain "barrenness from the point of view of norms, barrenness from a point of view of humanity, and extreme danger from the point of view of ecosophy" (Næss, 1989, p. 110).

Næss criticizes the use of GNP and GNP growth, which is calculated by adding up the national accounts every year. GNP is the proxy of welfare and is used "as if it were a decisive ingredient of a successful economic policy" (1989, p. 111). GNP growth was relevant as a measure after the Second World War in Europe where the big project was to rebuild "Europe after five years with a destructive war", but after two decades it turns up that every activity, negative as well as positive, the number of prisons, the frequency of traffic jams, smog and traffic accidents, and everything to repair the undesirable sides of the society was included. GNP is in one sense "a value-neutral quantity, a measure of activity, *not* of *activity* of *any kind* of *value* … (p. 112). GNP is not related to meaningfulness of that which is created. It does not imply any growth in access to intrinsic values and progress along the course of Self-realization" (p. 112). The main conclusion is that GNP is not a measure of welfare and life quality. On the contrary, GNP growth favors hard and distant technologies, wants not needs; it discriminates against people working at home, supports irresponsible and unsolidaric resource consumption, and global pollution (Næss, 1989, pp. 113–114).

Bhutan's GNH index is one profound answer to this critic. Many of the domains behind the GNH measurement lead to a holistic perspective on happiness. Community vitality is one interesting domain. Through the related indicators we find answers that at face value give high credibility. For example, the probable general pattern of city life is confirmed; to live in cities leads to a high degree of autonomy, but at the same time to long for a community.

Another factor assessed in the GNH index is time use. To obtain a high score on this factor, a balance between work and leisure is necessary. On time use the notion of *sufficiency* is applied. The logic is to have "enough sleep" as well as "enough work" and "enough leisure". Alkire (2015) writes; "This embeddedness of sufficiency norms is interesting, because it also conveys with brilliant clarity the need for concepts of sufficiency to incorporate human diversity" (p. 85), since time use to different activities will depend upon an individual's age, family, and social and cultural patterns. For the destitute, time poverty is often endemic. For the materially rich, "good time balance is partly self-made". There may be many sociocultural pressures and needs to "accomplish" or seem "busy" for building self-esteem (Alkire, 2015, p. 86). Alkire is also stressing that we should balance the time in a way to be able to perform at our peak. Paternity leave for fathers in Sweden is mentioned. It involves greater freedom and emphasizes a stronger relationship between fathers and children. This has increased children's well-being, and even a drop in male mortality (see Alkire, p. 89).

Concerning ideology, Alkire (2015) writes; "a key pillar of the new paradigm is sufficiency". This is in opposition to "many policies both public and corporate (that) seek to maximize wealth and profit, regardless of its opportunity costs on other domains of well-being or on well-being in future years" (p. 74). This pillar also emerges in the GNH index that measures the domain called "living standard". The living standard domain refers to the material well-being of the Bhutanese people and "ensures the fulfilment of basic material needs for comfortable living". In 2007, 23.2 percent of Bhutanese "still live in income poverty; some lack assets such as land or adequate housing" (Ura et al., 2012a, p. 168). They use household per capita income, assets, and housing conditions. Assets include livestock, land, and appliances, while housing conditions are measured by room ratio, roofing, and sanitation.

Sufficiency means that the GNH index does not use the poverty line, because sufficiency threshold refers to higher conditions for well-being than poverty lines (Ura et al., p. 169). Furthermore "an absolute sufficiency threshold was chosen, since the GNH values encourage people to achieve happiness through their accomplishments, and discourage a relative approach in which one is satisfied only if one has relatively more income … than one's peers" (Ura et al., 2012a, p. 169.) In principle living standards concern (i) meaningful decent work and livelihoods; (ii) housing that sufficiently shields from the elements: cold and heat, rain snow and sun; and (iii) some form of currency – money, assets, or other tradeables. In accordance with Alkire, all those aspects have an intrinsic value (p. 72). That money has an intrinsic value is contrary to the Aristotle's view. For Aristotle, money should be valued as a means because wealth is only useful for something else. As Alkire notes, whether money has an intrinsic value depends to some extent upon context (p. 73). As a general-purpose resource up to a certain threshold it gives freedom related to security, diversity, generosity, and sufficiency. Alkire argues that we should re-evaluate money. The essence is to find a balance between spirituality and acquisitiveness. One ideological element is to explicitly acknowledge and respect unpaid work.

Another implication of point six of DE is strengthening local practices and local communities. The basic norm of self-realization for all beings implies a capacity for self-determination. This means that ecological policies will favor decentralization. Centralization will tend to lower self-determination. In terms of DE self-determination does not mean ego-trip, but others are essential to the realization of Self with a capital S. (Cf. Næss, 1989 pp. 141–142). It also means self-reliance. Næss acknowledges that international trade has had positive effects on material standards of living. However, "Lifestyle and entertainment import has led to a dependence upon international economic fluctuations, leading to uniformity, passivity, more consumption, less creativity" (1989, p. 143).

This trend to decrease cultural diversity on a global scale undermines the independence of different cultures and makes it difficult for them to be self-reliant. And the ecosophical position of DE wants "the possibility of maximum *self-activity*: creating, rather than consuming. Doing, not being done to. The basic ecosophical terms here would be activeness, inner and outer, in reaching goals" (1989, p. 143). Næss argues that "self-realization is not against cultural communication, but it favors intrinsic values, material and spiritual".

The realizations of local communities are important in DE. The German distinction between *Gemeinschaft* and *Gesellschaft* is indicative. Locality and togetherness are key terms in the DE movement. Nobody wishes to be absorbed into a society that is "big but not great". Næss puts forward several points to describe essential characteristics of a desirable local community (pp. 144–146). One implication is that differences in income and wealth should be small. Those at the "bottom" and those at the top should be "sufficiently near in ways of life so that they can go together and work together" (1989, p. 144).

Point seven of DE might be the most difficult point to understand from a western point of view. To see its relevance, it will be supported with insights and arguments from the Bhutan perspective. Literature on mindfulness (Langer, 2010) might be key, and special attention to the concepts of spirituality measured in the GNH Index, which is based on balancing material wants with spiritual needs, will be presented.

Ellen Langer (2009) defines what mindfulness is and is not: "It is a flexible state of mind – openness to novelty, a process of actively drawing novel distinctions. When we are mindful, we become sensitive to context and perspective; we are situated in the present. When we are mindless, we are trapped in rigid mindsets, oblivious to context or perspective. When we are mindless, our behaviour is governed by rule and routine. In contrast, when we are mindful, our behaviour may be guided rather than governed by rules and routines. Mindlessness is not habit, although habit is mindless" (2009, p. 279). Based upon research Langer writes that "an increase in mindfulness results in greater competence, health and longevity, positive affects, creativity, charisma, and reduced burnout" (Langer, 2009, p. 280). Mindfulness is a critical factor in determining individual performance and shaping learning experiences. Mindfulness appears to be crucial in helping us deal with uncertainties in our lives and environments.

Spirituality in Bhutan "can encompass belief in spiritual values like compassion, peace, and a sense of purpose and connectedness and include Acts of compassion, altruism and selflessness ..." (Ura et al., 2012a, p. 131). The GNH research group argues that Bhutan is a "spiritual nation and the influence of spirituality is highly visible in the everyday lives of the population, in spiritual gatherings, and in the numerous spiritual landmarks such as sacred temples and monasteries, prayer flags and prayer wheels. These provide a platform for people to develop spiritual maturity" (Ura et al., 2012a, p. 132).

For GNH, spirituality is "intrinsic to development", which means that no meaningful development can occur without "inner spiritual growth along with peaceful environment that allow spiritual nourishment. If material growth undermines the spiritual framework of society and its values of compassion and integrity, then development has not occurred" (Ura et al., 2012a, p. 132).

The spirituality indicator is based on four questions. One finding was that meditation practice had a very low loading. It may be mentioned that the government in Bhutan has recently initiated a school-based meditation curriculum because of its ability to "provide balance, positive emotions and mental clarity" (Ura et al., 2012a, p. 133). Mainly monks and nuns practice meditation in Bhutan. The GNH survey include twenty-five monks and nuns. This is not representative of the monks and nuns who make up about 3 percent of the population of Bhutan. They live largely institutionalized lives in monasteries and "nunneries" and are not easy to interview.

Mindfulness is an important practice that in the GNH Index context measures psychological well-being as three components: (i) spirituality – meditation or mindfulness practices, and the consideration of the consequences of one's actions; (ii) emotional balance, which is the outcome of emotional intelligence, and the cultivation of positive emotions such as generosity, empathy, and compassion; and (iii) evaluative satisfaction with respect to different domains of GNH (Alkire, 2015, p. 93). One example mentioned by Alkire (2015 is the results of meditation in prison in India

(Tihar) and the United States. A program of meditation in prisons demonstrates that well-being is improved (Alkire, op. cit. p. 97) while violence and racism are reduced. The Oxford Mindfulness Centre applies mindfulness techniques to patients with mental and physical problems. "There is great potential that widespread availability (of mindfulness tools) will have a beneficial effect on the general population, not just those who are diagnosed unwell." Alkire 2015 (p. 98).

Point eight; active participation in the value struggles in society

In Norway there is a representative democracy with several political parties and elections every fourth year. The Government implements laws and important decisions based on a simple majority in Parliament (Stortinget). The Norwegian political system is supported by a strong technocratic tradition with several experts in different fields, and does not always work in an appropriate way to take the interests of minorities into consideration. As a result, several civil disobedience actions have taken place to preserve the Norwegian wilderness. Arne Næss and other pro-ecologists were partly inspired by Gandhi's example and methods.

Good governance is an important domain in the GNH index, where four measures for good governance are used: (i) whether people knew their fundamental rights and felt they were protected; (ii) whether they trust public institutions; (iii) how people assess the performance of the governmental institutions, and; (iv) whether they vote in national elections and participate in local government meetings. These questions are supported by Sen's important contribution in his *Development as Freedom* (1999). According to Sen, participation in making decisions that affect people's life and the lives of others are fundamental to human well-being. Participation can also be regarded as having intrinsic value for the quality of life.

Deep ecology presupposes that every person takes a practical stand and acts on it. In this value battle science not enough, and ecology is not the ultimate science. Following Næss's example we should act in a non-violent way, and we should always argue for our position. Deep ecology draws heavily upon ecology as a science where the values of unity, symbiosis, and diversity are central. However, we must avoid ecologism, the view that ecology is the final authority. This would mean that we over-generalize and universalize ecological concepts. But ecology cannot be a substitute of philosophical analysis. We must "fight against depolitization". Ecological science, concerned only with facts and logic, cannot answer the essential ethical challenges we must face. How we should act as responsible deep ecologists and good citizens is partly a question about how we should live, and that is an ethical question.

Final remarks

Bhutan's GNH measuring is an attempt to develop a holistic measurement of human happiness. The Bhutanese research reports present several insights that have relevance for alternative ways of life and different lifestyles compared to the prevailing western view. We should reflect deeply on these insights to learn and to pursue a sustainable life on our Earth. Summing up the main findings from this exploration of deep ecology and GNH, both represent clear ethical positions, knowledge and wisdom, and can reinforce and enrich each other.

Deep ecology needs GNH because GNH's concepts of sufficiency, mindfulness, and spirituality contributes to an elaboration of DE. Deep ecology may inspire GNH researchers to continue to raise basic questions on how to contribute to human flourishing in Bhutan. However, the non-anthropocentric assumption in DE may challenge Bhutan's GNH measurement. Deep ecological thinking implies living a rich life with simple means and self-realization for all beings.

GNH is a milestone towards understanding and the elaboration of the deep ecological narrative, often regarded as a distant ecotopy. However, GNH is real; it is implemented in practice, represents a role model for the global community, and directly benefits the people of Bhutan. The GNH approach is an innovative as well as a humanistic way to pursue happiness in a deep sense and is a hopeful project towards a sustainable practice and a good life.

Bibliography

Alkire, S. (2015). *Well-being, happiness, and public policy.* Thimphu: The Centre for Bhutan Studies & GNH Research.

Carson, R. (1962). *Silent spring.* New York: First Mariner Books.

Drengson, A., and Y. Inoue, (eds) (1995). *The Deep Ecology Movement: An Introductory Anthology.* Berkeley: North Atlantic Publishers.

Ims, K. J. (2011). Deep Ecology, in L. Bouckaert and L. Zsolnai (eds) *The Palgrave handbook* of *spirituality and business.* New York: Palgrave Macmillan.

Langer, E. (2009). Mindfulness Versus Positive Evaluation, in S. J. Lopez and C. R. Snyder (eds) *Oxford handbook of positive psychology* (2nd edn). Oxford: Oxford University Press (pp. 279–293).

Næss, A. (1989). *Ecology, communiry and lifesryle. Outline of an ecosophy.* Arne Næss. (Trans. and ed., David Rothenberg). Cambridge: Cambridge University Press.

Næss, A. (1995). The Shallow and the Deep, Long-Range Ecology Movement: A Summary, in A. Drengson and Y. Inoue (eds) *The deep ecology movement: An introductory anthology.* Berkeley, CA: North Atlantic Books.

OECD (2013). *OECD Guidelines on measuring subjective well-being.* OECD Publishing. www.dx.doi.org/10.l787/9789264191655-en>

Schumacher, E. F. (1999) *Small is beautiful. Economics as if people mattered. 5 years later ... with commentaries.* Vancouver, BC: Harley & Marks.

Sen, A. (1999). *Development as freedom.* New York: Knopf.

Ura, K., Alkire, S., Zangmo, T., and Wangdi, K. (2012a). *An extensive analysis of GNH index.* Thimphu: Centre for Bhutan Studies.

Ura, K., Alkire, S., Zangmo, T., and Wangdi, K. (2012b). *A short guide to Gross National Happiness index.* Thimphu: Centre for Bhutan Studies.

PART III

KNUT J. IMS AND JUDITH A. WHITE

Learning Ethics in a Social Context: A Practical Guide Using Experiential Learning Theory[1]

Introduction

To the extent that our institutions and practices are extensions of ourselves they will change as we change, particularly as we change our ways of seeing, learning, and responding to one another. This paper introduces a new way of thinking about ethics. It uses experiential learning and a social context to underscore the need for a comprehensive and interpersonal approach to resolving ethical dilemmas, and to assist management students in their thinking and learning about ethics.

Traditionally, the instructor, as sole agent of transferring knowledge or skills, actively takes charge while students passively listen, observe, and absorb knowledge and information. Among the assumptions embedded within this traditional social practice is the belief that students have the same goals as the instructor, assuming they ultimately will be illuminated by what they find. Another assumption is that of the homogeneity of modes of learning, without regard for individual differences in how experiences, information, and ideas are absorbed and transformed for application.

The challenges facing management educators and trainers interested in teaching ethics raise several questions. Can ethics be learned and if so how? What are the problems with traditional classroom practices? Is there a particular ideological and methodological perspective in learning and teaching that can be used in other areas of management education? Can we legislate Judo-Christian values and business ethics and morality within a culturally diverse society and among multinational organizations?

Should values, which are ever present in all settings, be made more explicit in teaching ethics?

The experiential learning cycle for learning ethics

Kolb's (1975, 1984) theory of experiential learning provides a model to develop a solution or set of solutions with the requisite complexity to adequately address the uncertainty posed by many ethical problems. It does this by promoting the integration of all four stages in the learning cycle. Individual and group problem solving, communication styles, and personality characteristics emerge from these four ways of interacting with the environment. Experiential learning encompasses two primary dimensions of the teaming process: taking in ideas, information, and experience through either feeling (Concrete Experience) or thinking (Abstract Conceptualization) and processing or transforming the input through action (Active Experimentation) or reflection (Reflective Observation). See Figure 1, below. Understanding these four approaches to learning introduces the legitimacy of a variety of perspectives on ethical issues, thereby eliminating the need for a singular absolutist position while developing norms for generative and innovative learning (Senge, 1990; Botkin et al., 1979).

The experiential learning model facilitates the development of ethical reasoning and ethical comportment because it inherently contains the valuing of all aspects of experience. The theory proposes that, "learning, change and growth are best facilitated by an integrated process that begins with 1) here-and-now *experience* followed by 2) collection of data and *reflection and observation* about experience. The data are then 3) *analyzed and conceptualized* and the conclusion of the analysis is fed back to the actors in the experience for their use in the 4) modification of the *behaviour* and choices of experience. Learning is thus conceived as a four stage cycle as shown in Figure 1. Immediate *concrete experience* is the basis for observation and reflection. These observations are assimilated into a 'theory' from which new implications for action can be deduced" (Kolb, 1975, pp. 33–34).

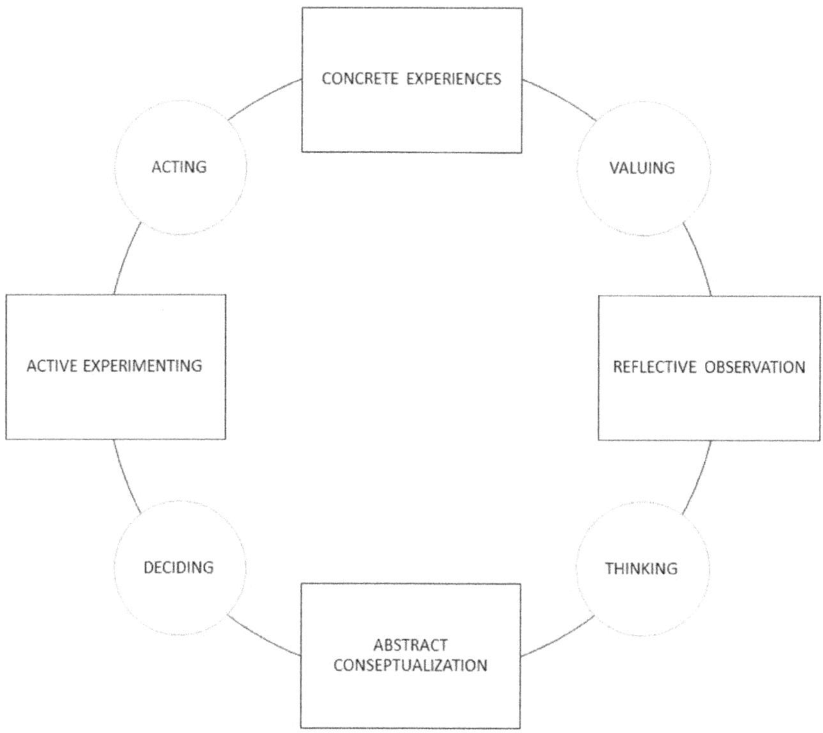

Figure 1. The experiential learning cycle inspired by Kolb, Rubin and Osland (1992).

The following discussion illustrates the different modes of the experiential learning cycle with their application to teaching and learning ethics.

Concrete experience

Concrete experience involves valuing and engaging in the here-and-now by being fully present in the moment with sense experiences, emotions, and interpersonal interaction. An understanding of self through the use of self-assessment instruments helps with self-understanding. Students take the learning styles inventory to assess their learning strengths: the Myers Briggs type indicator to assess their preference for thinking or feeling judgment and the ethical orientation questionnaire (White, 1992) to determine their orientation towards a consequentialist or communitarian ethic, and they write answers to questions concerning three real life ethical dilemmas to determine their use of an ethic of care or an ethic of justice (White, 1992).

The case method also provides students with concrete experiences to facilitate the learning process. Both the Ford Pinto (Ims, 1992) and the Bhopal case are particularly useful because in each case management actions led to disastrous consequences for many innocent people. They evoke strong feelings and raise intellectual, ethical, and practical dilemmas for students.

Learning ethics lends itself well to role-playing and the small learning groups provide safety for those individuals who would otherwise be shy or too embarrassed to participate. The role plays provide an opportunity to experience the issues surrounding an ethical dilemma by having students take on the perspective of one of the parties in the case. By enacting the role of one of the parties the student brings his or her thoughts and feelings into the situation, thus sensitizing him or herself to another person's perspective, assisting him or her to develop the managerial competencies of social objectivity and empathy (Boyatzis, 1994). Social objectivity, seeing a situation from the perspective of another person's point of view, and empathy, understanding the feelings of the other person at the moment of the interaction, are important skills in addressing ethical dilemmas and contribute to the development of affective and behavioural complexity (Fry, 1978).

Classroom discussions of ethics demonstrate perceptual and symbolic complexity, a relatively sophisticated understanding of the ideal and the real. Despite the availability of philosophical and behavioural models for ideal ethical comportment, behavioural and affective skills lag behind most students' perceptual and symbolic skills. In the interest of focusing on the action and the ideal rather than on reflection, understanding and acceptance of the negative emotions of greed, hatred, and anger have been neglected. Jung believes society has avoided the darker aspects of human nature, preferring to project these feelings and perceptions outward onto others in the form of mythology, war, and social and family conflict.

The instructor, through his or her own authentic behaviour, can simultaneously demonstrate ethical virtues and establish norms for ethical behaviour in the class. Ethical interpersonal behaviour, what we call ethical comportment, has similarities to the Buddhist virtues of right effort, right action, and right speech. Ethical comportment means enacting respect and openness to others, honest expression and truth-telling, and avoidance of harm to others. It requires the interpersonal competencies of active listening, including being authentic, empathetic, and sensitive to feelings within oneself and others, similar to those interpersonal skills Rogers (1961) describes as necessary for a helping relationship. A sincere interest in the other may involve meeting the other on his or her own ground, or at midpoint between the interests of the two parties, including an awareness of what is present in the moment and, if necessary, suspending one's own agenda to attend to the needs of the moment. The instructor may come to class with a planned agenda, but needs to spontaneously respond to the students' needs as they surface.

The instructor's role includes modeling the qualities of honesty, equality, reciprocity, care for the well-being of others, and putting oneself in the situation of others. Pride, dignity, self-respect, (Nordenstam, 1968) and autonomy (Hill, 1991) are important but, as Etzioni (1988) emphasizes, "we are also the community" and must see our responsibility in a broader societal context. How does the instructor act in relationship to students and what attitudes or values are explicitly or implicitly demonstrated? The essential issue is the degree of authenticity: will the instructor dare to be himself or herself, openly discuss conflicting issues, or question institutions, leaders, and even colleagues? By expressing feelings of vulnerability and courage the instructor explicitly legitimizes acceptable affective behaviour for the classroom.

Interpersonal skills for ethical comportment include building rapport, sensitive communication, respect for others, and embracing of differences. Sophisticated conceptualization and abstraction without the concomitant affective and behavioural competencies leave an obstructive gap in the interaction. For example, men and women sometimes use different forms of communication (Tannen, 1990) and while they may intellectually understand and agree with each other, these differences can interfere with good communication.

The role of emotions in learning ethics

Feeling, as an aspect of concrete experience, involves the being present in the moment, with an awareness of the experience of the body, the mind, and the emotions. Feelings can be transitory or long-lasting, and may be positive, negative, or neutral, and range in intensity from mild or slight to strong or intense. At the root of many negative feelings is fear from loss, pain, helplessness, and vulnerability. Ekman (1982) has categorized emotions into six basic areas: happiness, sadness, surprise, anger, fear, and disgust. Feelings have a biological basis: they are a function of sensation and emotion, and the function of emotion is located in the brain's right hemisphere (Gazzaniga, 1985).

In our highly rationalized culture emotional vocabulary is limited to a few stock feeling words. In teaching management students to give feedback we use the model of having them express their feelings about a particular situation, the reason for the feelings, and the consequences. With the students' limited forms of expression, anger encompasses frustration, resentment, disappointment, sadness, hurt, rage, irritation, etc. One way to have students expand their vocabularies is to have them make two lists of feelings, positive and negative, and then expand the lists for intensity.

The concrete experience of instructor role modeling, role plays and discussion of cases, expressions of authentic feelings, and working in small teams form the foundation for subsequent reflection, conceptualization, and enlightened action.

Reflective observation

Moving from concrete experience in the direction of abstraction, reflection is an intermediary step towards a more complex conceptualization of the formation of an ethical position. Reflection involves searching for understanding through observation, seeing a variety of perspectives and, with time, turning something over in one's mind for consideration. Reflection may occur through independent thinking, writing, reading, dialogue, and discussion. Reflection through dialogue includes staying with one's own center of ideas while seeking perspectives from others.

In the reflective observation stage of experientially learning ethics, learning groups can develop and respond to questions concerning the particular case at hand. In working with the Ford Pinto case (see Ims, 1992), the initial task is to recapture and review the case in detail so that the entire group starts with the same information. Some factual questions are: why was the gas tank so vulnerable so as to burst into flames so readily? Did Ford have any testing procedures? Was negligence of safety typical for that period of time, or was the Pinto an extremely bad manufacturer that only thought in terms of profit? Were any of the other similar models produced by competitors such as Toyota, General Motors, or Chrysler better? Ford used a cost-benefit analysis to define the trade-offs between the injuries and deaths to be expected by bad design and the costs of improved design. How did Ford arrive at the amount of $200,000 (US) as the value of a human life? What precedents have been set in terms of compensation for the loss of life in industrial accidents, particularly on this large scale? What has been going on at Ford since they produced the Pinto?

In the social context of the small learning group students can compare and contrast their viewpoints concerning what occurred in the Pinto case. Through role playing students experience the loss that resulted from some of the retrospectively unsatisfactory decisions. These role plays, presented to large classes and in small groups, provide students the opportunity to develop perceptual complexity as they see multiple perspectives and a more complex picture of the situation, giving them additional elements to consider when making decisions in the future.

Reflection and dialogue

Reflection involves examining experience as though the experience was "bending back on itself"; it is an inspection of one's experience, beginning with the inside and looking outward, and then looking from the outside inward (Hunt, 1987). Dialogue is indispensable for reflection because it is in the face of the other's reaction that the individual evaluates his or her utterances. As Maranhao (1991) claims, "reflection is deeply associated with the subject as a thinking person, or represented as internalized dialogue in which self and the other become encompassed by the activity of a selfs' mind".

To be reflective when dealing with ethical dilemmas requires a dialogue with self and others about what is considered ethical, what is appropriate for this situation, what feels right, what works, and what does not. The individual takes in ideas and perspectives from others, sorting through one's own thoughts and feelings and trying to make sense of the experiences, the feelings, and the actions. Dialogue is an explicit tool for reflection with oneself, with another and, in a small group, it allows each participant the time and support to explore meanings, attitudes, and values by raising questions. Hearing the viewpoints of others broadens one's world view and increases one's awareness of different values. We discuss dialogue in greater detail below.

Abstract conceptualization

Most of traditional higher education uses abstract conceptualization because the teaching in most classrooms is intellectual and cognitive, and more impersonal than interpersonal or affective. Through lectures instructors attempt to present a coherent, consistent, and logical argument, developing it in a linear manner while feeding information to students. While this is a valuable part of the learning process, for learning ethics it cannot stand alone.

In the Pinto case the instructor poses a series of conceptual questions to stimulate students' symbolic thinking and to help them make meaning of the case. Questions that facilitate learning included: What was the logic of management's decisions? What patterns emerged in the case? What were Ford's primary and secondary motives? What place, if any, did virtue have in the behaviour of Ford? Was the profit motive the main value guiding Ford's resolution of the situation?

Lee Iacocca set a goal of "The limits of two thousand", which meant that the Ford Pinto should not cost a nickel more than $2,000 (US) and that it should not weigh more than 2,000 pounds. The ingenious economists and engineers at Ford accomplished this goal, but at what cost? Did they trade safety for profit? Another of Ford's far-reaching goals was to build this new subcompact model more quickly than any other mass produced car thus far in the automobile industry. In this way Ford had to use a parallel manufacturing process, replacing a previously used sequential process. The former involved less testing and control of individual parts. From an economic and technical perspective, measuring success in terms of profits, the Pinto represented something new and became an absolute winner.

At this abstract conceptualization stage in the learning process, students have learned about their own learning, decision making and ethical orientations. They are ready for the introduction of theory. In addition to learning about the ethical theories of rights and justice, readings and lectures can include the ethics of care, communitarianism, consequentialism (White, 1992), feminist, gendered, and postmodern ethical perspectives.

Developing theory and building models are important elements associated with the abstract conceptualization stage of the learning cycle. The instructor can present the class with elements of a model so as to make ideas explicit and common. Model building is a creative and inventive process involving critical thinking, judgment, analysis, convergence, and closure. It is best to be first creative and then critical, otherwise critical thinking can stifle the creative process in building a model. The outline of the model presents students with a symbolic complexity that lends itself to repeated iterations through the learning cycle. An effective theory and model should result in practical ideas for action.

From the benefit of dialogue and exposure to ideas one can proceed to analyze and synthesize the perspectives and experiences in order to develop conceptual understanding and develop generalizations that could be used to test one's beliefs. In Flanagan's (1991) Principle of Minimal Psychological Realism (PMPR), "Make sure when constructing a moral theory or projecting a moral ideal that the character, decision processing, and behaviour prescribed are possible, or are perceived to be possible, for creatures like us." To talk about moral or ethical values that involve actions towards others, one must consider the realistic enactment of the belief or value and the possibility of implementing it. Generalizing about the nature of the value, generating ideas about the when, where, and how the belief or value could be enacted and placing it within a larger context, model, or theory, along with other ideas and values, are all skills of abstract conceptualization.

Active experimentation

This stage of learning ethics involves putting into practice what has been gained from the earlier three stages, the empathy, valuing, and feeling that was generated from the role play and discussion; the reflection, dialogue, and looking at multiple perspectives; and the development of a practical model of the issues. With understanding and meaning gained from the cases, the next step is "learning by doing". Students can take action on the basis of their understanding of the case by sharing with others their newly formed viewpoints or, perhaps, engaging in further dialogue or action within or outside of the classroom (Hunt, 1987). Commentary and role-playing, and writing personal application assignments (Kolb et al., 1991) are other kinds of active experimentation that allow students to heighten their awareness by enacting their deeply rooted and frequently tacit values. In order to elicit values, one has to act them out.

Following Kolb's (1992) model of the valuing process, testing of beliefs follows the development of generalizations. For example, does kindness towards others occur when one is physically threatened by another, or in times of war, or be authentic and sincere with one's negative feelings when hurt by the other? These situations require testing ethical values in order to act in accordance with one's ethical beliefs. Acting on tested beliefs is a stretch for persons more comfortable with reflection than action. Comfort may come about from an earlier reflection on the unintentional harm the other has caused oneself. Hearing similar experiences from others, one could say the harmful act was not intentionally directed at oneself and should not be taken personally. These various perspectives contribute to the development of behavioural and affective complexity, appropriate to match the previously developed perceptual and symbolic complexity.

Feedback follows action, information that can be incorporated into the next concrete sensory and affective encounter, beginning the learning cycle again.

Social context and dialogue

Openness to learning and development requires a minimal degree of security. A social context is implicit in a classroom situation, creating an opportunity for the development of a sense of community, security, and openness. The classroom as a social context for learning supports self-awareness and understanding, and the opportunity to experiment with new ways to frame experience. This is the concrete experience stage of the experiential learning cycle: direct, personal, expressive communication. The classroom as a learning organization concerns a relationship and sense of responsibility to others, seeing the self as part of a greater whole.

Closely related to the concept of community is dialogue. A dialogue is not a discussion nor a unilateral teaching effort. While discussion can be a ping-pong game where individuals hit the ball back and forth, dialogue eliminates the competitive element. Through dialogue individuals reflect on their own ideas and express themselves. As Weick (1979) says, "How can I know what I think until I see what I say." The heart of learning through dialogue (Ims and Singh, 1992) is free exploration that "brings to the surface the full depth of peoples' experience and thought, and yet can move beyond their individual views" (Senge, 1990, p. 241). A good dialogue implies that the participants may exchange roles in an evolutionary and dynamic process and the teacher may step down from the lectern and comfortably engage with students as a learner, permitting the students to teach.

The teacher as learner provides a role model for students, setting a concrete example of learning through dialogue. Participants share experiences and explore individual values, forming a basis for respecting voiced differences. *Logos* – the subject matter or content – is another aspect of dialogue. Content is framed by the way questions are asked and by the criteria used to judge the responses to the questions. Questions that invite mutual inquiry, exploration, and new perspectives, direct the dialogue in a different direction than questions that converge on the right (that is, the teacher's) answer.

The instructor has the responsibility to develop a learning community within the classroom through "barn raising" (McCormick and Kahn, 1982). As a collaborative group process barn raising is both form and norm for dialogue and discussions, of meeting and joining in on the other's idea or project, joining the other in seeing the project from his or her perspective, bringing one's skills and tools to help develop and shape another's idea or theory rather than drawing attention to one's own ideas. It can include finding a middle ground of common forms of language, and/or creating, through consensual means, new forms of meaning (Hazen, 1987; Ims and Singh, 1982).

Hierarchy or network

Dialogue forms a bridge for students and teachers to come together. This deviates from the usual hierarchical class organization in which students are expected to understand what the teacher unilaterally delivers. From any position in a hierarchy, persons have limited contact with others and can only see and talk directly with those immediately above, below, or next to them (Tannen, 1990). Their access to ideas, information, feelings, and activities of those considerably lower or higher on the hierarchy is limited (Belencky et al., 1986; Wilson-Schaef, 1988). In a flat network or web-like structure learning is more efficient because of the easy access to ideas, information, experiences, and feelings of others anywhere in the network. Communication is open and accessible to all, and ideas flow freely without having to cross barriers of status, position, esoteric vocabularies, or obtuse abstractions.

While a network can serve as a condition for a rich learning environment, it will not automatically lead to a sense of community. The small group structure is the most effective setting for a free flow and exploration of ideas and behaviour, and requires time for team building to develop a productive and comfortable learning situation.

A community of voices – The small learning group

The physical and social arrangement of the learning environment, in particular alternating between the large class and small groups, is vital and fruitful for learning. Small groups secure involvement and participation while maximizing expressiveness for the greatest number of people. By forming groups with a balance of students with each of the four learning styles from the learning style inventory individual members and the group draw on strengths from each style while developing weaker areas. Membership of the learning teams remain permanent throughout the class. In discussing cases, students share their experiences, ideas, attitudes, and prejudices, getting to know one another.

Just as the small groups become a laboratory and context for thinking about ethical issues, the larger class forms a community as it works on ethical dilemmas. For example, to the extent that individuals are concerned with one another, are honest or would consider cheating, work cooperatively, or take advantage of others, they experientially explore the ethics of care and justice, and consequentialism and communitarianism, directly in the classroom. Entire class meetings offer the opportunity to hear a wider range of experiences and ideas.

Conclusion

The management classroom, like most organizations, provides an ideal social context for learning ethics and values because it contains leadership, technical processes, communication, reward systems, structures, and interactions with the environment. The classroom's culture as a learning community has its own beliefs, values, norms, artifacts, and social and technical processes that allow for the learning and practice of ethics and values.

Ethics inherently concern self-concept as it addresses how individuals are able to live with themselves; that is, "everything that I can regard as mine", my body, my relatives, my ideas, my convictions, etc. (James,

1936). Because of the reflective ability of human beings all actions can be valued or judged against a standard. If actions are not judged as good or appropriate the self feels diminished and suffers from a gap between aspirations and achievements, a gap that struggles for resolution. Individuals can engage in the adjustment process, either pulling one's actions toward the standard, or lowering the standard towards what is achieved. Learning ethics through the experiential learning model and the process of valuing (Kolb, 1992) needs to be made explicit in order to be able to teach someone to abstain when he or she is tempted to do "only a small act of harm". Management students can heighten their awareness of their beliefs and values in order to avoid "Catch 22" situations, those unfortunate entrapments where before they are aware of what they have agreed upon, they are sitting deep in the mud. In retrospect, and in an attempt to reduce cognitive dissonance, individuals can acknowledge that they have been dragged into a situation where they gradually justified some degree of unethical behaviour.

Authenticity is an essential element in the teaching and learning of ethics. The teacher and the students can both model particular values through their explicit behaviour. The human mind is capable of developing ideal models for action, and all disciplines describe normative behaviour. Because of the complex and unpredictable interplay of emotions and cognitions, behaviour lags far behind the ideals for ethical comportment. It is easier to preach than to practice; often it is difficult to "walk the talk".

This paper has attempted to address some of these complex and challenging implications for teaching ethics in the context of professional development and management education. It is hoped this model will prove helpful and stimulate ideas for future research and teaching.

Note

1. This paper is a collaborative effort. The authors' names are listed alphabetically.

Bibliography

Belencky, M. F., B. M. Clinchy, N. R. Golberger and J. M. Tarule. 1986. *Women's ways of knowing, the development of self, voic, and mind.* New York: Basic Books.

Bernstein, R. J. 1986. *Philosophical profiles. Essays in pragmatic mode.* Philadelphia: University of Pennsylvania Press.

Botkin, J. W., Elmandjra, M., and Malitza, M. 1979. *No limits to learning.* New York: Pergamon.

Boyatzis, R. E. 1994. Stimulating self-directed change: A required MBA course called Managerial Assessment and Development. *Journal of Management Education.* 18(3): 304–323.

Ekman, P. 1982. *Emotion in the human face.* Cambridge: Cambridge University Press.

Etzioni, A. 1988. *The moral dimension. Toward a new economics.* New York: The Free Press.

Flanagan, O. 1991. *Varieties of moral personality. Ethics and psychological realism.* Cambridge, MA: Harvard University Press.

Fry, R. E. 1978. *Diagnosing professional learning environments: an observational framework for assessing situational complexity.* Unpublished PhD thesis, Massachusetts Institute of Technology.

Gazzaniga, M. S. 1985. *The social brain. Discovering the networks of the mind.* New York: Basic Books.

Goodman, P. 1971. Decentralization. In *Individualism. Man in modern society.* L. Original, A. Gross, and P. Osterman (eds). New York: Dell Publishing Co. Inc.

Gordon, T. 1974. T. E.T. *Teacher effectiveness training.* New York: Peter H. Wyden.

Hazen, M. 1987. Dialogue as a process of integration in human systems. *Consultation,* 6(2): 119–130.

Hill, T. E. 1991. *Autonomy and self-respect.* New York: Cambridge University Press.

Hunt, D. E. 1987. *Beginning with ourselves. Practice, theory, and human affairs.* Cambridge, MA: Brookline Books.

Ims, K. J. 1992. Teknostrukturens moral. Om spesialister som ikke stiller spørsmål utenfor egen profesjon. In S. Linge Solberg and A. Danielsen (eds) *«Teknologiledelse – Markedsorientert bruk av teknologi»* Oslo: TANO.

Ims, K. J. and P. P. Singh. 1992. *Learning through dialogue: an approach to some problems in management practice and management education.* Case Western Reserve University, Cleveland, OH.

James, W. 1936. *The varieties of religious experience.* New York: Modern Library.

Kolb, D. A. 1975. Toward an applied theory of experiential learning. In C. Cooper (ed.) *Studies of group processes.* New York: John Wiley & Sons.

Kolb, D. A. 1984. *Experiential learning*. Hoboken, NJ: Prentice Hall.

Kolb, D. A. 1992. The process of valuing: an experiential model. Personal communication.

Kolb, D. A., I. Rubin, and J. Osland. 1991. *Organizational behaviour. An experiential approach*. Englewood Cliffs, NJ: Prentice Hall.

Maranhao, J. 1991. Reflection, dialogue, and the subject. In F. Steier (ed.), *Research and reflexivity*. London: Sage.

McCormick, D. and M. Kahn. 1982. Barn raising: collaborative group process in seminars. *The Organizational Behaviour Teaching Journal*, 7: 16–20.

Nordenstam, T. 1968. *Sudanese ethics*. Uppsala: The Scandinavian Institute of African Studies.

Rogers, C. 1961. *On becoming a person. A Therapist's view of psychotherapy*. Boston: Houghton Mifflin.

Schumacher, E. F. 1973. *Small is beautiful. A study of economics as if people mattered*. London: ABACUS, Sphere Books Ltd.

Senge, P. M. 1990. *The fifth discipline. The art & practice of the learning organization*. New York: Doubleday.

Tannen, D. 1990. *You just don't understand, men and women in conversation*. New York: Ballantine Books.

Weick, K. 1979. *The social psychology of organizing*. Reading, MA: Addison-Wesley Publishing.

White, J. 1992. *The role of individual characteristics and structures of social knowledge in ethical reasoning using an experiential learning framework*. Unpublished dissertation. Case Western Reserve University, Cleveland, Ohio.

Wilson-Schaef, A. 1981. The female system and the white male system: new ways of looking at our culture. In A. Wilson-Schaef (ed.) *Women's reality*. San Francisco: HarperCollins.

KNUT J. IMS

The Consultant-Client Relationship: Personal Autonomy and Development Through Dialogue

Introduction

In western societies consultants are highly visible. They perform activities and rhetoric at different organizational levels and are either blamed or blessed for their advice. They are expected to give advice to clients. To be effective, the quality of the consultant-client relationship is assumed to be important. The consultant-client relationship, even though dealt with in prior research is, however, not without ambiguities and difficulties. Due to the assumed importance of the advice, a more thorough understanding of the dyadic relationship is needed. This chapter brings a new perspective to this literature. Three assumptions will be basic:

1. Knowledge cannot be treated as a material resource. Knowledge consists of mental facts or personal constructs. Knowledge transfer, therefore, cannot be separated from the psychological makeup of the persons involved in the information process.
2. The actor's self is of central importance for the knowledge construction process and a person's self is a relational phenomenon, that is the self is constituted and sustained in an interactive, social process with important others. It means that the substantive content of the knowledge exchange process cannot be understood in a vacuum. Parallel with the primary information process the actors signal secondary information elements that are interpreted as either disrespect – or self-respect enhancing – and, as Rawls (1971) holds, self-respect is "perhaps the most important primary good". So respect and self-respect need to be part of the analysis.

3. The most important part of the process of knowledge creation does not have a problem solving character, but does have a problem finding character. This means that the question to be asked is not given, but is in itself a core issue (Pounds, 1969; Schon, 1983).

With these three basic assumptions in mind, many issues need to be confronted in order to illuminate aspects of the consultant-client relationship in an ethical perspective. What does it mean to receive advice while at the same time acting as an autonomous person? How should the client avoid gliding unconsciously into the consultant's world – thinking, seeing and speaking on the premises of the assumed expert? And for the consultant, what does it mean to signal respect for the others and at the same time advise them? In the following the quality of the relationship between the consultant and client will be investigated. It is not only a question of substantive content, that is a question of what is to be exchanged or constructed, but most importantly how the process is going on between the persons involved. To protect and enhance each other's self-respect is one thing, but how should the participants behave in order to pose the right questions, that is how to "frame and name", a situation that initially is regarded as a mess?

Literature review

To consult (latin consulto) means to seek the opinion or advice of another person. By consultant (counsellor) I mean persons who make it their business to give advice. When using the verb "consult", I think of the client as asking for advice, and when using the noun "consultant", I think of the giving of advice. There are consultants within many professions – law, computers (system analysts, programmers), auditing, and accountancy, to mention a few.

In general, consultants have overlapping roles. In practice they often act as social scientists, (collecting, interpreting, and analyzing data), as psychiatrists (clarifying distorted perceptions) and educators (supporting and stimulating learning of new skills) and, to cite Benne (1969), "in the effort to integrate these discrepant but overlapping roles, ethical problems arise".

How is the dyadic relationship between a sender and receiver of advice depicted in the literature? Parts of the literature with explicit ethical focus describe different models, listing them as either paternalistic, contract, agency, friendship, or fiduciary (Bayles, 1988; Ellin, 1988). Ellin (1988) distinguishes between five models:

- Adversary model – a competitive relationship between the parties, where they bargain at arm's length and where it is necessary to compromise the conflicting interests.
- A cooperative model where there is partnership and often mutual trust.
- An agency model in which the professional (consultant) carries out the wishes of the client: The client determines the ends to be attained, and the professional acts to achieve these ends, being an instrument of the client's will.
- But if the professional is superior (p. 135), then he/she may use the client for his/her own ends, that is an exploitation model.
- The professional may work for the client's end, in which case there is a fiduciary model.

Ellin's surprising point is that in a strict sense trust is only necessary in the last model.

The self and related concepts

Benne (1961) makes an interesting distinction between ethical and technical problems. When the self is at stake, the issue is said to be ethical or, as Benne puts it, "What makes the problem of the consultant ethical is that it involves in its solution the future career as a consultant self", and "it is an ethical conflict since its resolution involves recommitment in the self and some new inter-accommodation of the associations significant to the self and of their normative demands upon members" (pp. 600–601). Benne poses the question of fit or misfit between consultants' image of their role as a good consultant and the client's image of what a helpful consultant should (or shouldn't) do; that is what proper help is. Furthermore, he discusses what kinds of influence are involved when the consultant belongs to various associations, is a citizen, and is a member of more "personal" associations such as families and churches. Therefore, role segregation is difficult in practice.

A related issue is the problem of loyalty. According to Tranøy (1974), the professional (consultant) must give due interest to the "new" stakeholders – clients and society. This widening of the horizon is important since in some professions there has traditionally been a one-sided and exclusive loyalty to members of the same profession. A changed focus of loyalty is also argued by Winslow (1988) in his analysis of the nursing profession.

In the management science tradition, there has been an ongoing concern for getting the results of operational research (developing computer based models) into the hands (minds) of managers (see for instance Schultz and Slevin, 1979). In this literature there are different views on successful "implementation" that might be held by different groups. A management science researcher (a consultant) may think of successful implementation in terms of technical validity; that is, if the model developed works. The user (client) on the other hand, may see the problem in terms of organizational validity, that is the extent to which the model is used in making decisions – how the model fits the organization. The top manager may see the problem in terms of the contribution to the organization's effectiveness. This means that the advice is good if it produces a behaviour change that is desired and is in accordance with the value system of those in top management. A classical contribution is made by Churchman and Scheinblatt (1965). They suggested a clarification of four relationships or "positions" between expert/researcher (consultant) and manager (client) called mutual understanding, communicator, persuasion and separate function. The ideal situation, called mutual understanding, exists when the manager (client) understands the researcher (consultant) and vice versa.

This means that the manager reacts to what the researcher tries to do in a way that improves the manager's possibility to reach relevant goals. In the communicator position it is important for the researcher to make the manager understand the researcher better. As a contrast to this position, the researcher in the persuasion position attempts to understand enough about the manager's position in order to persuade him/her to accept the researcher's results/models. This is a "selling of ideas" case, and implementation means to sell one's own recommendations to the manager. In the separate-function position the view is to look upon management and research as two essentially separate functions. The researcher's (consultant's) task is to work out a complete plan and then present it to the manager (client), whose task it is to either accept or reject the plan.

More subtle forms of manipulation are described by Boland (1979) and Bråten (1983a, 1986). Boland (1979) challenges the system analysts' (experts') value neutrality by describing how they create the users' (clients') world through the design of information systems. Thus he focuses upon the system developers' ethical responsibility. Through a variety of important examples, the Norwegian sociologist Stein Bråten (1983a) illustrates how model power mechanisms work. He describes how a model strong actor, even with the best intentions, influences a model weak actor through their symbolic interaction. By distinguishing between symbolic ("mental") resources and material resources, Bråten illustrates in a brilliant analysis how model power mechanisms may reinforce existing influence gaps between persons when they begin to interact. Bråten's analysis of model power mechanisms and his listing of five ways to transcend ("aufheben") the power of the model strong, are particularly relevant for the client-consultant relationship. In order to minimize the consultant's role, partly based upon Bråten's frame of reference, Ims (1987) used a version of the Kelly Role Rep Test (1955) methodology, permitting the client to select the elements for reflection as well as using the clients' (electricity directors) own language.

Interesting contributions focusing on the dyadic relationship between the consultant and client have been made by Rogers (1969), Håland (1986), and Løvlie (1982a, 1982b). These authors maintain that the relationship is more important than the methods. Rogers (1968) distinguishes between what he calls a "manufactured" relationship and a "helping relationship". A helping relationship involves growth for the client while the optimal helping relationship is created by a helper who is a psychologically mature person. A psychologically mature helper is "congruent". This means a unified or integrated person in the sense that whatever feeling or attitude he/she is experiencing is "matched by his/her awareness of that attitude" (p. 298).

Rogers states that a process of personal development is essential to a good helper, otherwise words may convey one message while, at the same time, the consultant's possible annoyance will be communicated in subtle ways. This may confuse the client. Rogers also comments on the problem of distance between the consultant and client, which is often typical of the professionalization of most fields. The consultant may, therefore, keep from experiencing the caring "which would exist if … (the consultant) recognized the relationship as one between two persons" (p. 299).

An important aspect of a helping relationship is also to keep it free from judgment and evaluation as it will "permit the other person to reach the point where he recognizes that the focus of evaluation, the centre of responsibility, lies within himself" (Rogers, 1968, p. 302).

Negative as well as positive evaluation may be threatening. Valuative procedures work in such a way that they transform a person into an object. Rogers' final point is in accordance with Martin Buber and focuses on the importance of "confirming the other". Buber says, "Confirming means … accepting the whole potentiality of the other … I can recognize in him, know in him, the person he has been … created to become … I confirm him in myself, and then in him, a relation to this potentiality that … can now be developed, can evolve" (Rogers, 1968, p. 302). The main point is to accept the other person (the client) as a process of becoming – not as something fixed, already classified, already shaped by the past. The importance of the self-concept as a relational and a dialectic construct is excellently developed and discussed by Løvlie (1982a, 1982b). (The self-concept will be developed in the frame of reference that follows.)

Openness, trust and accurate information flow between consultant and client are recommended by Argyris (1969, 1985) and Argyris et al. (1985) in their distinction between model I behaviour and model II behaviour. Part of the problem is taking feelings and emotion to be central to the resolution of organizational problems, while most managers (clients) regard feelings and emotions *as* "bad" or "immature". According to Argyris and his co-authors (1985), the main rule is to be open and face the problems. Due to the "defensive routines", "the brittleness syndrome" and other subtle defense mechanisms of clients, this is not obvious. Consultants are in a dilemma whenever their values are unacceptable to the client. If the consultant states this, he/she may become an unwanted person in the organization, but if acting in accordance with the client's values there will be no change. In short, this literature problematizes the client's willingness and ability to learn. A related problem is described by Maslow, who pinpoints the tension between the fear of knowledge and the search for knowledge. This fear of knowledge violates the overly mechanistic assumption of knowledge creation – that we always strive for more knowledge, and that more knowledge is always preferable to less. This is also observed in cultural analyses of organizations (Schein, 1992).

Argyris et al. (1985, p. 387) argue that a non-directive approach may involve more manipulation and dependence than the forthright approach due to the difficulties of placing responsibility on the client to come up with ideas about how to proceed.

Concerning values of the consultant or a consultant group, Argyris (p. 434) distinguishes between two types of competence, technical and intellectual competence versus interpersonal competence. Both types or levels of competence are important. However, Argyris hypothesizes that if either or both is low, if either is significantly higher than the other, difficulties will tend to arise (p. 435).

Reflections on the review of literature

My review of literature has been eclectic – broad, but relatively superficial and fragmentary. I have searched within quite different fields and my findings differ and are more indicative than authoritative. Hence my research cannot be called comprehensive within any field. My purpose was not to epitomize any field but, rather, to generate problems/statements and probe for possible cross fertilization. I am left with many questions and issues which need to be resolved.

When Ellin (1988) asserts that trust is only necessary in the fiduciary model, we may be led to think that trust is a clear and well-defined concept.

Butler (1991), however, found that trust was a multi-dimensional concept with sixty subdimensions.

In parts of the explicit ethical literature I have cited, the main theme is framed as a problem of choosing the appropriate model. I find this unsatisfactory. It may well be of a certain analytical help, but it is not enough. Every model needs to be adapted to the specific concrete reality. We need a methodology as well. There may also be a cross-fertilization by combining different theoretic schools; that is, organizational-oriented theory might be complemented with ethical concepts and reasoning. Important ethical concepts such as dignity, pride, and self are not discussed in mainstream organizational literature. In particular, some of the concepts found in the psychotherapy literature, especially the self-concept and the emphasis of relationship, seem to be lacking in the other fields. My contribution synthesizes the ethical, the organizational, and the information- and psychotherapy-oriented literature. Developing the first rudimentary elements of a methodology of consultant-client relationships with a highlight on the self-concept is the focus of the next section.

A conceptual framework: A sketch of a tentative theory of client-consultant relationship in an ethical perspective

This section will develop a frame of reference for the ethical issues of the dyadic relationship between client and consultant. Due to the preliminary character of the theory, the emphasis on relevant concepts will be stronger than the relationships between the concepts. One central concept will be the self. The self-concept is nicely introduced by the following Taoist story from ancient China:

> When Yen Ho was about to take up his duties as tutor to the heir of Ling, Duke of Wei, he went to Ch'u Po Yu for advice. "I have to deal," hf said, "with a man of depraved and murderous disposition … How is one to deal, with a man of this sort?" "I am glad," said Ch'u Po Yu, "that you asked this question …. The first thing you must do is not to improve him, but to improve yourself."

A challenge to the dyadic but asymmetric relationship between consultant and client

A consulting relationship assumes the existence of an asymmetric relationship in terms of knowledge between the consultant and the client. The traditional perspective states the problem as a transfer of knowledge from the consultant to the client. However, many problems are involved. One pitfall is the generalizing of expertise, when the "experts" talk about knowledge that they are not in a privileged position to talk about. Another problem is manipulation. Manipulation may be explicit as strategic behaviour (Habermas, 1982; Ims, 1987) or implicit through model power mechanisms (see Bråten, 1983a). Therefore, a high level of awareness for both consultant and client concerning their respective values is necessary. In the following I will focus on how different value systems may be elicited and made more explicit in order to reduce the risk of non-intentional influence (manipulation).

The analysis is limited to two parties (two persons). The quality of the relation ship will be the unit of analysis. The task of personal development, which may or may not be part of an organizational development project, is the subject matter and it is obvious that the two parties will be involved in a sensitive project.

Consultancy as intervention

An intervention is any kind of research into an organization, a group of people, or a person. To put into writing thoughts, values, and assumptions concerning an organization or a person is quite a new way of looking into a reality otherwise taken for granted. When written down and introduced to the people in the organization in the form of a report, this may represent an intervention. It is also important to argue our values, otherwise we fail to take responsibility for the ethical dimension in a relationship (Schein, 1985). Furthermore, it is important that the professional consultant has formulated ethical guidelines to regulate these interventions. These guidelines should be communicated to the clients for their well-being, for instance as a written description to the person who will be researched.

Practical judgment (phronesis) will be seen as a central component of decision making and action. Therefore, advice based exclusively on a technocratic understanding of the situation is seldom satisfactory (Nordenstam, 1983). One needs something additional that is part of the context. The more people believe in themselves (have adequate competence and a well-established self-concept), the more faith will they have in their own judgment. Faith in own judgment may strengthen the consultant's independence and autonomy, which in my opinion are ethical values.

Autonomy and integrity defined

Autonomy is not independence from other persons in an absolute sense. On the contrary a person has a continuous need to be seen and valued by other persons for affirmative mirroring and "anerkennen". Thus we should acknowledge every person's relational reality and our dependence on other people throughout our lives. With autonomy I mean self-determination as the ability to say no to wants and claims put forward in the environment. This presupposes a separated self that is able to represent one's own will (Gullestad, 1992).

Integrity is necessary, but not sufficient; for virtues of integrity allow for most any content. An evil tyrant could display integrity to a high degree, so it is important that an impartial spectator could say that the person's notion of the common good is in correspondence with well accepted principles; for example, ideas of the original position (Rawls, 1971). Virtues of integrity are truthfulness and sincerity, lucidity, commitment, and authenticity.

The self-concept

The concept of self captivates the person when experiencing the inner and outer world. This is a phenomenological view of the person and includes feelings of self, that is self-esteem. A person's self-respect and dignity are assumed to be central. In correspondence with Rawls (1971), self-respect consists of two aspects: (i) the person's sense of their own value, "his secure conviction that his conception of his good, his plan of life, is worth carrying out", and (ii) the confidence in one's ability to fulfill one's intentions. As Rawls holds, without self-respect "nothing may seem worth doing – and we may sink into apathy and cynicism" (Rawls, 1971, p. 440). Further I believe that one's self-respect is contagious and effects people with whom one interacts (Nordenstam, 1968). Bennis and Nanus (1985) wrote about the management of own and others' selves: "What we observed was that our ninety leaders induced (stemming from their own Self-regard) positive other-regard in their employees. And this turns out to be a pivotal factor in their capacity to lead" (p. 62).

As Løvlie writes, "The self is the totality – and cannot be localized here or there - it is inside my skin, but also in the relationship to persons and objects in the world with whom and which I interact ". Just where the self begins and ends is an open question; the arbitrariness of self "delineation", according to Gregory Bateson (1972, p. 30). The essence of self is relationship. In accordance with the one-world view, man-in-the world constitutes a unity and it is up to us to choose how we divide this unity. Løvlie writes, "We are relationship" (1982, p. 110). We usually think of two persons as separate bodies because they are separate in a physical sense. However, when two persons interact they transcend their physical separation. Bråten's theory of model power, illustrates negative consequences in terms of autonomy of one type of monological relationship. Through interaction we can also be acknowledged (in Hegel's sense "anerkannt") or we may experience alienation; that is, the feeling of not being acknowledged and not regaining one's self (Løvlie, 1979).

Does the consultant make the client feel inferior and incompetent, or give the client a feeling of pride and self-esteem, stimulating an enthusiastic exploration of the world and self? This question relates to the dialectic view of the self-concept. Anyone's self is defined as a relation to other selves. This involves consultants creating the client through their ongoing patterns of interaction, which is filtered through all their past patterns of interaction – and as he creates the other, he also creates himself.

Through the self-concept we can account for the dialectic and reflective nature of humans. The subjective and objective aspects of humans presuppose each other. If I can reflect, I can also be responsible; that is, I can answer for my experience – or I can choose not to respond to certain parts of my experiences. According to Løvlie (1982b, p. 36), one type of irresponsible communication is to send unclear messages which I can, therefore, disqualify, that is use constructs that are "loose" (Kelly, 1955).

The quality of relationship

Authenticity is a property of interpersonal relationship. This means that the person must behave genuinely, that is in accordance with personal values (Argyris, 1969). According to Pollak (1976), methodologies may be a barrier between two persons. This statement needs to be corroborated. A proper methodology, used in a flexible and sensitive way, may be helpful, depending upon the subject matter to be investigated and, in general, the context. Therefore, I maintain that the most decisive factor in the consultant-client relationship is that the first sequence (part) of the dialogue must be free of any kind of technology. This is the I-You part in which the relation ship is characterized by connectedness and presence (Buber, 1964). The second sequence may involve purposive-rational action, by which I understand either instrumental action or rational choice, or their conjunction. Instrumental action is based on technical rules using empirical knowledge and implies empirical predictions about observable phenomena (McCarthy, 1978).

Dignity and freedom to choose

Following Benne (1969), I will illustrate a methodology that tries to maximize the client's freedom to choose; that is, maximize client autonomy. It is important to be aware of model power mechanisms and the threat of cognitive imperialism. Dialogue and dialectics may be used to enhance awareness of one's own values through value clarification, (to protect and enhance the personal autonomy of the client).

I suggest the following steps concerning how to reach value clarification (Benne, 1969, pp. 590–591):

Consultants (practitioners) should be aware of the fact that they are controlling the client (Benne, 1969, p. 591). The professional consultant's methods and solutions are not based upon objective reality, but on a particular set of values that might fruitfully be discussed. The institutional settings and the methodology in which the consultant operates may favor certain kinds of values. Consultants introduce their own values into the relationship both in the definition of the situation and in the setting of standards. Consequently, it is useful for the consultant to "have certain values about what needs to be done in the situation itself", and to communicate these values to the client. By recognizing that they are engaged in a certain degree of control and that this is an ethically ambiguous act, the consultants provide some safeguards against the control. This will make it easier for the client to "talk back" to the consultant and to argue about the appropriateness of the values introduced into the situation.

Table 1. Steps to reach value clarification

Desirable steps	Role of consultant
1)	
Increasing awareness of manipulation	Labelling own values to self and clients; allowing client to "talk back"
2)	
Building protection or resistance against manipulation	Minimizing own values and maximizing client's values as dominant criteria for the change process
3)	
Setting enhancement of freedom of choice as a positive goal	Using professional skills and authentic relationship to increase client's range of choices and ability to choose

How to build in procedures that may protect against manipulation? One step is to encourage the client to explore personal values and relate what is learned to their own value system. As a matter of fact, the client should determine the direction of the process. The crucial point is that the client's own values should be at the centre of attention. These values should be as criteria against which any change can be measured. In order to stimulate this value clarification process the consultant should help the client in the value exploration process, often by eliciting and contrasting his/her own values.

The third step is to enhance the freedom-of-choice as one of the positive goals of the change effort. This means providing the client with data and information that widens the range of choices and creates new experiences for the client which increases the ability to choose (and thus maximize his/her own values).

New experiences may be made through dialectics of meaning-making and action. One is a presupposition for the other (Løvlie 1982a, 1982b). A person acts and creates meaning and self. The work will be perceived as meaningful if you understand, predict, or master it to a minimal degree.

Empathy versus inclusiveness

Inclusiveness "is the complete realization of … the 'partner' not by the fancy but by the actuality of the being" (Buber, 1965, p. 97). Empathy means to glide with one's feelings into the dynamic structure of an object, a tree, an animal, or a human; it means to "transpose" oneself over there and in there. "Thus it means the exclusion of one's own concreteness, the extinguishing of the actual situation of life …." On the other hand, inclusion is "the extension of one's own concreteness... … the complete presence of the reality in which one participates." Inclusion presupposes a relation between two persons, an event experienced in common, in which at least one of them participates actively, and one person who, at the same time, lives through the common event from the standpoint of the other. And Buber concludes, "A relation between persons that is characterized in more or less degree by the element of inclusion may be termed a *dialogical relation*" (My emphasis, p. 97).

Real dialogue

A real dialogue presupposes a symmetry between the partners that does not permit either one to dominate (Bråten, 1983). Løvlie (1984) distinguishes between persuasion, which presupposes a subject-object relationship, that is an I-It relationship, and convincing, which presupposes a subject relationship, that is an I-You relationship. Consultants should have a communicative attitude towards the client, which means that the consultant should be able and willing to reach a common understanding with the other part (Habermas, 1982; McCarthy, 1978).

I will distinguish between power, which means threats of punishment or punishment and manipulation, by which I mean hidden influence, for instance when the consultant, partly on the premises of the client, arranges the conditions in such a way that the client makes the consultant's (master's) words their own (Løvlie).

In terms of Bråten's model power theory I am particularly aware of how to transcend ("aufheben") the model power of the most powerful part.

Contract

A contract is any arrangement in which the parties talk and negotiate how to proceed during the interaction process between consultant and client. One important aspect of the contract is to create security. The critical issue is not whether the contract is formal or informal, written or verbal, but the way in which the negotiations are experienced. If the intention of the dialogue is to break down interpersonal barriers, this may create a fruitful platform for confidential, dialogical, and growth enhancing conversations (Håland, 1985). In this situation we may well be aware of Bråten's theory of model power. Even with the best of intentions, one of the participants may slide into a dominated relationship by "swallowing" the premises of the expert. In this way the consultant and the client create a monological state (see Figure 1).

In order to avoid this unconscious "mono-perspective", one may transform the conditions into a dialogical state (Bråten, 1986b). By crossing the border of the meaning horizon or changing the domain, one may establish

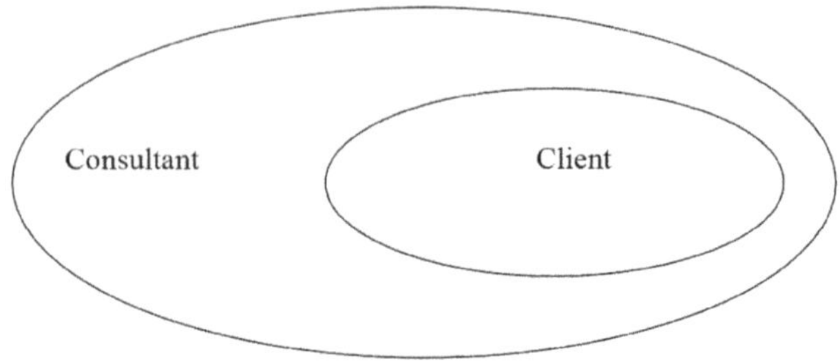

Figure 1. Consultant and client in a pseudo dialogical state.

a symmetric, cognitive conversation in which the complementing and crossing of perspectives are permitted. Crossing of perspectives is a presupposition for awareness and gives an opportunity for an extended creative horizon. In accordance with Bråten (1983, 1986) I postulate that every human being has abilities to enter into a self-reflective dialogue within their own lifeworld. In order to create a real self-reflective dialogue, both parts (or perspectives) in the dialogue must represent something in common as well as something unique (see Figure 2).

During the contract negotiations consultants must be aware of their responsibility. They have a professional duty to make the client aware of the consequences of entering into the relationship. This may be a process involving explorative research of the client's personality and culture (which may be analogous to a person's character), leading to organizational change. It is very important for the consultant to talk about the possible consequences in advance in order to give a free choice (informed consent) before the research begins. This also involves the client having the authority to stop the process at any time if it is felt that the organization's integrity is

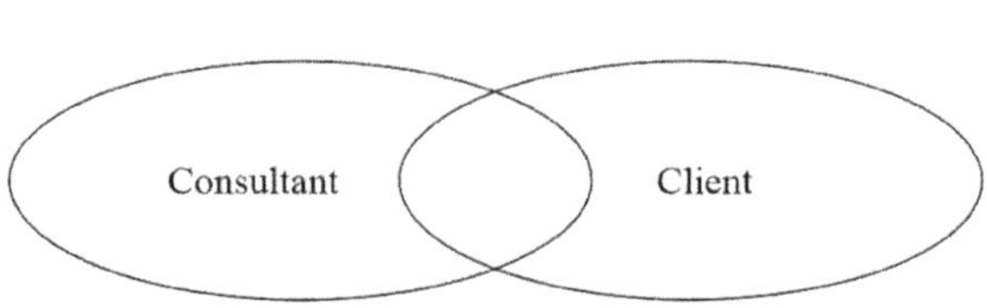

Figure 2. Consultant and client in a dialogical state.

threatened. This right to say no, to stop the project, should be explicitly agreed. To stop a development project that has gained momentum may be problematic. Therefore openness, sensitivity, courage, and honesty between the participants have to be critical factors in the relationship. It is essential that the development process should be anchored in the individual participant's dignity and his or her inviolable right to stop and drop the project at whatever stage it may have reached. To work through the contract as early as possible, stating these factors, is important.

The integrity-flexibility issue

How flexible should the consultant be when accommodating the client's claims? This is a problematic issue, particularly when accepting a task that may be on the edge of one's professional expertise, or builds upon quite different values. When discussing a possible contract, the consultant may experience great pressure to accommodate the wishes of the client. Furthermore, if the consultant is in great need of a contract, the degree of free choice may in fact be limited. Thus, during the contract negotiations the client may often turn out to be the model strong actor.

Toward self-organized learning

Up to now single elements have been presented as part of the conceptual model, but what about the processes? With the "two-community theory" as a point of departure, a self-organized learning approach will be encouraged.

The two-community theory of knowledge use (Caplan, 1979; Dunn, 1980) states the problem as a gap between the consultant's and the client's way of looking at the world (often in terms of different ontological and epistemological assumptions). The theory states implicitly that the main problem is a transfer of expert knowledge from the professional to the client. As I see it, personal development is the crucial issue. The purpose is

Knowledge ⟶ Consultant ⟶ Client

Sentence 1. The traditional relationship between a resource, consultant, and client.

to attain autonomy in the relationship with the expert. A pragmatic issue is involved as well: Will the expert's advice improve anything at all (the effectiveness/ efficiency question)? Due to uncertainties in every structure of the world, the problematic character of causality, and the importance of practical judgment, it will be difficult to rely on any expert's technocratic knowledge (Schon, 1983). Reflection-in-action is the alternative to the notion of traditional expertise.

The traditional relationship between a consultant, a client, and a resource like knowledge, may be conceived as involving three elements visualized in Sentence 1.

The prototype of this pattern occurs when the consultant interprets the public knowledge of the topic and "injects" this knowledge into the client, who efficiently absorbs the well-organized knowledge. I reject this approach which presupposes a hierarchical relationship. It may work as a short-term solution, but it creates dependence and has little to do with effectively taking control of one's own situation.

Following Thomas and Hani-Augstein (1985), a different perspective on relations that might be called conversational, self-organized learning may be put forward. The term "conversational" points to the fact that no person can know their self unaided: We all need help to exploit our infinite potential – and through conversations and dialogues with others, we can stimulate each other and pool our experiences. By learning, I mean the "construction, reconstruction, and exchange of personally significant, relevant, and viable meaning" (after Thomas and Harri-Augstein, 1985, p. XXIV). "Self-organized" refers to clients' right to define their own learning purpose and that learning may occur naturally and spontaneously from experience through creation of personal meaning. It is important that the meanings

Client$_p$ ⟷ Consultant$_p$

Sentence 2. The client and consultant as reflecting and exploring partners.

are personally significant for some part of the person's life. This perspective enables the client as well as the consultant to participate in a mentally and emotionally active way in order to cultivate and develop their common resources. This perspective involves the consultant helping the client to:

- reflect on self as a person using the client and the consultant's experience
- explore and experiment on some part of the social or physical environment.

The letter p indicates that both consultant and client are concerned with the client as a person and "not merely as a receptacle of knowledge" (Thomas and Harri-Augstein, 1985, p. 320) (see Sentence 2).

Dialogue as a language creating process

A communicative dialogue is not governed by egocentric calculations oriented towards one's own success. On the contrary, the interaction is directed at realizing an agreement or a consent as a platform in order to obtain personal goals (Habermas, 1982). As Bråten (1983a) emphasizes, such an interaction may have a dialogical flavor, but is not necessarily a dialogue. It may well be a pseudo dialogue; that is, a monologue on the premises of the model-strong person. So what is the critical factor?

The crucial point for the consultant is to give the client an opportunity to be expressive. To talk about complicated phenomena that have not yet been put into concepts or categories (at least for one or both of the participants) is seemingly impossible. First, the participants must be aware of the dialectic between phenomena and concepts. Concepts are not identical with the phenomena, they are dialectic. Phenomena exist and concepts are constructed in an effort to understand and master the world. We cannot divide the phenomena, but we can divide concepts (Løvlie, 1982a, p. 109). (The notion of the interrelation of everything means that it is up to us – the participants – to choose how we divide this unity. Any division is arbitrary.) In this process the client/participant must have initial help to engage in an ongoing language-creating process.

When the client expresses a word/concept (or even better, a construct), consultants should not think of themselves as clairvoyant experts who immediately know what the client means. The biography of the client will probably be different from that of the consultant. Therefore, the two parts may well:

- put different meaning on the same expression (Case 1)
- put the same meaning on different expressions (Case 2)

Case 1 is an example of a rich or a loose construct. As a consequence of arbitrariness in the articulation of constructs, one may well have many examples of case 1. In case 2 we can speak about pseudo-agreement or -disagreement in the subject matter (Næss, 1961). A pseudo-disagreement can be resolved when the participants during the conversation define the meaning of the expressions.

In order to reach each other's lifeworld, one must define concepts and communicate each other's interpretations. This necessitates learning and openness from both parts through a conversation which must be a mixture of I – You and I – It. Consultants should not regard themselves as the know-all before entering into this process. Their image should be rather that of a student prepared to learn. No doubt this is a demanding and a partly painful process for a well-educated, technically oriented professional.

Kelly's (I 955) concepts of loosening and tightening may be helpful in this phase of the dialogue. "Loosening" is one strategy that makes conceptualizations so elastic they cover everything. As a consequence, loosened constructs hinder validity tests of the applicability to reality. "Tightening" or constricting one's construct system makes a conceptualization so tight and limited that most phenomena fall outside of a person's (the client's) mind (construct system). It will be important to change between loosening up a construct in the beginning and tightening it up later in the dialogue. When grasping the other part's meaning with a concept, one can tighten the concept in order to increase the level of precision (See also Ims, 1987).

I – you, then I – it

What kind of consultant/client relationship is preferable? Following Bråten (1983b), I would suggest that it is essential to distinguish between two different types of interaction. First, we have an external interaction in which the other part is reduced to an "it", an object, a thing in the sphere of "it" (also see Habermas, 1982). The opposite is an I-You dialogue. This is an interaction characterized by here and now experiences, including sincerity and inclusion (Buber, 1964; Bråten, 1983b). This is a type of inspiring, deep dialogue, and is usually not associated with matter of fact (objective) conversations. On such a basis I would postulate that I-You should precede I-It. Therefore, one should first establish a covenant of mutuality and equality between the dialogue partners. Then one may start objective conversations and instrumental thinking and action. This means that achieving each other's lifeworlds is one premise for planning and designing somebody's future. Thus an external analysis of a person, a group or an organization has to be partly based upon a perspective from within; that is, from the client's world of meaning.

When shall the process be punctuated? When has a mutual understanding in a pragmatic sense been reached? Regardless of the interpreter's capabilities, an attempt to interpret other persons' meanings, based upon words, hints, and expressions is hazardous. At this point the dialectic and debate moment should be brought into play. Interpretations of another person's expressions should not be regarded as a final interpretation, but as a possible interpretation. Thus an interpretation should not be regarded the "one and only" and undiscussed, but as an appropriate point of departure for reflection. My point of view is that any elicitation of important concepts should be debated and be a means in an ongoing dialogue (Shaw, 1982, in particular p. 65). I will conclude by showing a structured methodology which I have applied and experienced as useful in stimulating an exploration dialogue in the actors' own premises.

The conceptual model in review

In the center of my frame of reference, I have pointed out processes and factors that are important in order to create a dialogical relationship from which both parties should benefit. The essence is that the constructs of both participants should be elicited and be an object for clarification and finally evaluation.

This might occur within a process in which gestures and words are exchanged between the parties. If both parties have an experimental attitude towards words and concepts, but a humble attitude towards the other person, a dialogue may be the way. The dialogue may be viewed as a mutual journey of exploration into the unknown, where goals and information requirements can be formed and developed. Instead of only looking upon "solutions" and information requirements as something that can be uncovered, we should regard it as a process of construction, articulation, and formulation. Usually one does not find something that is hidden in the expert's head or in the books, but through reflection and talking one creates important aspects of the world. Attitudes in terms of conceit, self-assertion, and mastering of technology may be a handicap, whereas openness and ability to communicate are central virtues. As Bennis (1989) strongly argues:

> Free self-expression is the essence of leadership: "Leaders are people who are able to express themselves fully … the key to full self-expression is understanding one's self and the world, and the key to understanding is learning from one's own life and experiences." (1989, p. 3)

Nobody has described the power of expression clearer than Ralph Waldo Emerson: "Man is only half himself, the other half is his expression" (cited in Bennis, 1989, p. 2). The central elements of the preferred relationship are depicted in Figure 3 as a description of attitudes and behaviour that are preferable in order to facilitate a dialogue as a platform for development and autonomy.

At this point it is important to be aware of the importance of the ladder of inference (Argyris, 1985). We all go through a ladder of inference to make

Consultant	open and communicative humble and experimenting (debating and reflecting) listening and exploring loosening and tightening	*Client*

Figure 3. Central attitudes and actions between the dialogue partners.

sense of our world and to act within it. Due to our skillful behaviour, our reasoning process is automatic and effortless so we don't pay attention to it. Nevertheless, we build our inferences on different levels of data which may be analogous with rungs on a ladder. The first rung consists of relatively directly observable data, the second rung consists of culturally understood meanings, the third rung of meanings imposed by us, and the fourth rung of the theories we use; that is, abstract entities difficult to subject to a test.

Due to skillful behaviour and a willingness to please, and not say what one thinks, one communicates at a generally high level. The more sensitive the evaluations and attributions of others, the stronger the tendency to discuss it only at the most abstract level (rung four). Therefore we typically skip the lowest level of the ladder, thus creating learning avoidance. We avoid testing our theories in use, and thus we reinforce our old behaviour instead of changing it.

In order to learn one has to be willing to test one's own theories and assumptions with valid and concrete data. Thus, the participants should be aware of the fallacy of discussing meanings, evaluations, and attributions, ignoring the building blocks of theory; that is, the empirical data.

Conclusion

A main issue in this chapter has been to increase the consultants and clients' awareness of the many traps of manipulation during a physical and symbolic process of interaction. Even with the best of intentions of both consultant and client, there is a danger of gliding into an expert's world view and solutions. Being aware of model power has been central. Establishing the self-concept as a platform to understanding a person and giving them due respect and dignity have been essential. Two methodologies have been presented along with a three-step model of value clarification, indicating the importance of eliciting the consultant and the client's values during the interaction process. This process might preferably take place within a broader context of dialogue in the spirit of the labels of Figure 3.

I have attempted to view the information exchange process as a dialogue in which the persons involved are partners in a mutual growth enhancing process, thus criticizing the traditional one-sided instrumental approach to knowledge transfer. Our present understanding of how professionals should interact with dignity and respect for the other in such a dialogue – and at the same time act in an autonomous manner – is at best rudimentary and, hopefully, it should be a topic for further theoretical and empirical research.

Bibliography

Argyris, C. 1969: Explorations in Consulting-Client Relationships, in: W. G. Bennis, K. D. Benne & R. Chin (eds) *The Planning of Change,* 2nd edn. London: Holt, Rinehart & Winston.

Argyris, C. 1985: *Strategy, Change and Defensive Routines.* Boston: Pitman.

Argyris, C., R. Putman & D. McLain Smith 1985: *Action Science.* San Francisco: Jossey Bass Publishers.

Bateson, G. 1972: *Steps to an Ecology of Mind.* New York: Ballantine Books.

Bayles, M. D. 1988: The Professional-Client Relationship, in: Joan C. Callahan (ed.) *Ethical Issues in Professional Life.* New York: Oxford University Press.

Benne, K. D. 1969: Some Ethical Problems in Group and Organizational Consultation, in: W. G. Bennis, K. D. Benne & R. Chin (eds) *The Planning of Change,* 2nd edn. London: Holt. Rinehart & Winston.

Bennis, W. 1989: *On Becoming a Leader.* Reading, MA: Addison-Wesley Publishing Co.

Bennis, W. & B. Nanus 1985: *Leaders. The Strategies for Taking Charge.* New York: Harper & Rows Publishers.

Boland, R. J. 1979: Control, Causality and Information Systems Requirements. *Accounting, Organization and Society*, Vol. 4, no. 4, pp. 259–272.

Bråten, S. 1973: Model Monopoly and Communication. *Acta Sociologica,* Vol. 16, pp. 98–107.

Bråten, S. 1983a: *Dialogens vilkår i datasamfunnet.* Oslo: Universitetsforlaget.

Bråten, S. 1983b: Modeller av dialogiske systemer: Tid og selvrefleksjon. In B. W. Hennestad & F. E. Wenstøp (eds) *Bedriftsøkonomi og vitenskapsteori.* Oslo: Universitetsforlaget.

Bråten, S. 1986: Paradigms of Autonomy: Dialogical or Monological? In G. Teubner (ed.) *Social and Legal Autopoiesis.* European University Institute Publication Series. Berlin/New York: De Gruytes.

Buber, M. 1964: *Jeg og Du.* Danmark: Munksgaard.

Buber, M. 1965: *Between Man and Man.* New York: Macmillan Publishing Company.

Butler Jr, J. K. 1991: Toward Understanding and Measuring Conditions of Trust: Evolution of a Conditions of Trust Inventory. *Journal of Management*, Vol. 17, no. 3, pp. 643–663.

Caplan, N. 1979: The Two-Communities Theory and Knowledge Utilization. *American Behavioural Scientist*, Vol. 22, no. 3, pp. 459–470.

Churchman, C. W., & A. H. Schainblatt 1965: The Researcher and the Manager: A Dialectic of Implementation. *Management Science*, Vol. 11, no. 4, February, pp. 69–87.

Dunn, W. N. 1980: The Two-Communities Metaphor and Models of Knowledge Use. An Explorative Case Survey. *Knowledge: Creation, Diffusion, Utilization,* Vol. I, no. 4, June, pp. 515–536.

Ellin, J. S. 1988: Special Professional Morality and the Duty of Veracity. In Joan C. Callahan (ed.) *Ethical Issues in Professional Life.* New York: Oxford University Press.

Gullestad, S. E. 1992: *A si fra. Autonomibegrepet* i *psykoanalysen.* Oslo: Universitetsforlaget, Det Blå Bibliotek.

Habermas, J. 1982: A Reply to My Critics. In J. Thompson & D. Held (eds) *Habermas, Critical Debates.* Cambridge, MA: Macmillan Press.

Håland, W. 1986: *Psykoterapi: Relasjon, utviklingsprosess og effekt.* Oslo: Universitetsforlaget.

Ims, K. J. 1987: *Leder i dialog. En studie av informasjonssøk med metoder for personlig utvikling.* Bergen: Universitetsforlaget.

Kelly, G. A. 1955: *The Psychology of Personal Constructs*, Vol. I and II. New York: Norton & Co. Inc.
Løvlie, A. L. 1982a: *The Self – Yours Mine or Ours. A Dialectic View.* Oslo: Universitetsforlaget.
Løvlie, A. L. 1982b: *The Self of the Psychotherapist: Movement and Stagnation in Psycho-therapy.* Oslo: Universitetsforlaget.
Løvlie, L. 1979: Dialektikk og pedagogikk. Arbeidsmanuskript. *Oppland Distriktshøgskoles Skrifter* 22.
Løvlie, L. 1984: *Det pedagogiske argument. Moral, autoritet og selvprøving* i *oppdragelsen.* Oslo: J.W. Cappelens Forlag A.S.
May, W. F. 1988: Contract or Covenant? In Joan C. Callahan (ed.) *Ethical Issues in Professional Life*. New York: Oxford University Press.
McCarthy, T. 1978: *The Critical Theory of Jurgen Habermas.* Cambridge, MA: The MIT Press.
Nordenstam, T. 1968: *Sudanese Ethics.* Uppsala: The Scandinavian Institute of African Studies.
Nordenstam, T. 1983: Datautvecklingens filosofi. Två oforenliga traditioner. In B. Goranzon (ed.) *Datautvecklingens Filosofi*. Stockholm: Carlsson & Jonsson Bokforlag AB.
Pollak, O. 1976: *Human Behaviour and the Helping Professions.* New York: Spectrum.
Pounds, W. F. 1969: The Process of Problem Finding. *Industrial Management Review*, Vol. II, no. I., pp. 1–19.
Rawls, J. 1971: *A Theory of Justice.* Cambridge, MA: Harvard University Press.
Rogers, C. R. 1969: *Freedom to learn*. Ohio: Charles E. Merill.
Schein, E. H. 1985, 1992: *Organizational Culture and Leadership.* San Francisco: Jossey Bass.
Schon, D. A. 1983: *The Reflective Practitioner.* New York: Basic Books Inc.
Schultz, R. L. & D. P. Slevin 1979: Introduction: The Implementation Problem. In R. Doktor, R. L. Schultz & D. P. Slevin (eds) *The Implementation of Management Science.* Tims Studies in the Management Sciences, Vol. 13. Amsterdam: North-Holland Publishing Company.
Shaw, M. L. G. 1982: The Extraction of Personal Meaning from a Repertory Grid. In E. Sheperd & J.-P. Watson (eds) *Personal Meanings*. Chichester: John Wiley.
Thomas, L. F. & E. S. Harri-Augstein 1985: *Self-organized Learning. Foundations of a Conversational Science for Psychology.* London: Routledge & Kegan Paul plc.
Tranøy, K. E. 1974: Yrkesetikk, personlig moral og offentlig ansvar. Næringsutbygging og samfunnsverdier. *Bedriftsøkonomen* nr. 7.
Winslow, G. R. 1988: From Loyalty to Advocacy: A New Metaphor for Nursing. In Joan C. Callahan (ed.) *Ethical Issues in Professional Life*. New York: Oxford University Press.

KNUT J. IMS AND MORTEN HUSE

Community in the Classroom: The Philosophy and Practice Behind a Course in Ethical Action

Introduction

Can ethics be taught? This is the title of a book that describes an ambitious effort to make the study of ethics and social responsibility essential to the mission at Harvard Business School. The authors' answer is based upon research and shows that students as well as faculty members are meaning makers who are influenced through a number of processes and structures (Piper et al., 1993). This meaning-making activity is dependent upon and stimulated by the conditions in the environment. In particular, students are dependent upon mentoring, guiding authorities who can help them to form a vision of their contributions to the business world. The most influential mentorship consists of a mentoring community – a "we" that share a vision as well as a practice.

Piper et al.'s (1993) contribution is significant. Nevertheless, they leave teachers in business ethics with the essential questions: Which ethical tradition should be used as a base for moral development? Among the great philosophers, who is to be selected? What is true anthropology? In the following we will show our ethical position.

We believe that morality in the last resort must be based upon *each individual's personal responsibility*. We have sympathy with existential thinkers like Pascal, Kierkegaard, and Sartre. Also we find the thinking of Zygmunt Bauman (1993) intriguing. According to Bauman morality has moral impulse as the ground on which it stands, and this is *an elusive foundation.* The moral self as moral impulses is not very popular among philosophers, nor the administrators of order. Moral impulses are frequently seen as "subjective, elusive, erratic". Thus, the guardians of order cannot tell with authority what will happen, and ethical theories typically seek to create order, dogmas, rules, systems, and consistencies. In correspondence with Bauman we held that *the ethical demand* is vague and confused, barely audible. The moral self has to be his/her own interpreter and leaves the moral subject with the questions; *Could I not do more? Is it right? Or should I have acted differently?* This way of thinking holds that it is moral anxiety that gives the only substance the moral self could ever have – the urge to solve the problem at hand.

This point of view is also taken by the Danish philosopher, Knud Løgstrup (1971), who sees the unspoken, radical, ethical demand as a foundation in all his work. He points to the danger of conventions and rules because they make life comfortable. In fact, they safeguard life lived in the pursuit of self-interest because conventions are used as a means for keeping aloof from one another and for insulating ourselves.

Martin Buber (1992) expresses that our self-concept depends on how we look upon and treat other human beings. He distinguishes between I-Thou and I-It. In the latter meaning, the other does not appear as a moral subject. He/she is reduced to an object, a thing that may well be manipulated. On the other hand, I-Thou is *a relationship,* it is – *a between* – a living relationship *with* the other. This I-Thou lives in the relationship which is expressed in dialogue.

We take the position of Bauman and Løgstrup and see *moral impulses as the critical element to be developed.* We believe that community building based upon dialogue may play an important role in moral development. Thus, in the next section primary importance is given to dialogue as means to foster community in the classroom. In the third section the example from an ethics course in a business school is presented and, finally, we present the reflection on the community building dialogue in this classroom setting.

Dialogue

> "To exist, humanly is to name the world, to change it"
> "Men are built in action-reflection"

Freire (1972) criticizes the dominating way of teaching, framing it as the banking concept of education. The teachers "bankier-role" is to regulate the way the world "enters into" the students. The teacher "cognizes a cognizable object" while he prepares his lessons. Thereafter he "expounds to his students on that object". The task of the students is to memorize the contents narrated by the teacher. This approach to education poses many problems from a learning perspective as well as a development perspective. Freire suggests a different approach to a more "liberating education", which he terms "problem-posing education". In this approach the teacher is no longer only the-one-who-teaches, but he is himself taught in dialogue with the students, who in their turn also teach. "They become jointly responsible for a process in which all grow" (1972, p. 53). Arguments based on formal authority are no longer valid. The essence is to create dialogical relations.

Martin Buber's *I-Thou* illustrates the substantive attitude of the dialogical character of an encounter. From the start there is a symmetry in attitudes between partners, which are expected to incessantly exchange roles, address each other, and respond to each other.

Bird (1991, p. 45) outlines the marks of what he calls *good conversations in order to ground ethics in the ordinary language* and "overcome the pervasive prejudice that links ethics far too much exclusively to formal discourse …". His purpose is to avoid making ordinary people *feel ethical illiterate* … Bird stresses the importance of working on a number of skills that help people in practical interaction in their daily life and performing their job. According to Bird good conversations are "*vocal, reciprocating, issues oriented, rational, imaginative, and honest*" (1991, p. 35).

Community and empathy

Community and empathy are related to moral sentiments and ethical action. Community building is a vital phenomenon because many forces and structures disrupt the social bonds between people. And when the social emotional bond of empathy is severely weakened, the phenomenon of numbing takes place. Then there are no limits to the harm against others that can be brought about. The Norwegian philosopher Arne Johan Vetlesen (1994) argues that there is a logical sequence prior to moral action described by the following concepts: *Empathy, Perception, Judgment, Action*. The logic is that *one has to first see, to be sensitive, to identify a situation that affects the weal and woe of others before one assesses and weighs how to act.* Moral judgment is therefore concerned with the interests of the parties that are affected by the situation and which norms and principles are at hand that may guide me, as moral subject, in the deliberation.

Empathy is intersubjective or interpersonal in a twofold sense. Firstly, empathy is other-regarding. In empathy "there is always a Thou, never only me" (Vetlesen, 1994, p. 8). Empathy sets up and sustains a *relation, a between* (zwischen), I and thou, and a connection between a moral subject and his or her addressee. Secondly empathy is a faculty that *depends upon being stimulated by others.* If as a child I had a cold relationship with other, I am unable as adult to have a warm relationship towards other. I tend to do to others what others have done to me. I have to have been the object of empathy in order to become its subject (Vetlesen, 1994, p. 260).

However, empathy as a concept is not so much corroborated in the mainstream moral-philosophy, which has in some degree a cognitivist one-sidedness following a tradition from Kant. Our position is more in line with David Hume's moral theory, which states that "reason is, and ought to be, the slave of the passions" (1980, p. 1) Reason does not provide a motive for action. Motivation requires passion (hope or fear, or grief or joy, or despair or security, sentiment or desire). However, reason and sentiments co-operate to determine what to do (cf. Adam Smith, 1976).

We think that it is through community that we can encourage the cultivation of social bonds between people (students). A stronger *feeling of connectedness means an solid start on the logical chain of empathy, perception, judgment, and action.* In accordance with Naylor et al. (1997) we define a community as "a partnership of free people committed to the care and nurturing of each other's mind, body, heart, and soul through participatory means" (p. 42). Community is "about cooperation, sharing, commitment, communication, trust, justice, empowerment, adaptability, and tension reduction". Empowerment is seen as the right of each member (employee or student) to share equally the power to influence and shape the direction of the organization. In order to become "a true community", the leader must be prepared to risk total loss of control.

How to build a community? The essence of community building, as we see it, is to be engaged in dialogue, to share a sense of "we-ness" and to make a commitment to connect and to stimulate local leadership, initiative, interdependence, and responsibility of the persons concerned. There are different kinds of communities; residential, monoethnic or integrated, and work-related communities. We define the schoolroom as a work-related community.

Schools may play an important role in community building. Perhaps more through personal example and institutional culture than through classroom lectures. We agree with Etzioni, who writes: "The whole school should be considered as *a set of experience generating situations* in which young people learn the values either of civility, sharing, and responsibility to the common good or of cheating, cutthroat competition, and total self-absorption" (1993, p. 259).

In the classroom it is important to stimulate courage, openness, trust, and a stronger feeling of we-ness. It is vital to encourage the cultivation of social bonds between the students – not so much because it fosters a good atmosphere and security in the learning situation, but because it is important in itself. More basically social bonds – the glues of community – *develops the moral sentiments which are seen as the main door into the moral domain.* For students in business schools the development of the moral sentiments is an attempt to bring a more balanced relationship between the "cold brain" and the "warm heart". We need both elements in ethical action. However, we assume that the moral sentiments typically are underdeveloped in our setting.

Praxis

The setting

The Norwegian School of Economics and Business Administration (NHH) is Norway's leading teaching and research institution in the field of economics and business administration. It is located by the west coast in Bergen. It is the oldest business school in Norway, but still no more than sixty years old. The NHH is a member of the prestigious Community of European Management Schools (CEMS). The program of studies offered at NHH developed from the siviløkonom degree course (the equivalent of a Master of Science in Business to a doctoral level course). NHH has 3,000 full time students. The course *Ethical Action* is an elective (thirty-six hours) in the fourth and last year of the siviløkonom program. All the students have already been through a required course, *Business Strategy and Ethics*, in their third year.

Before the Ethical Action course started it was uncertain whether the course was to be graded based upon a scale from 0 to 9 (0 and 1 fails). A slow, but steady change in the direction of more detailed grading had taken place over the last twenty years. This semester the school established a new regime in which all courses shall be graded from 0 to 9. Nevertheless, the teacher thought that this course would be allowed a pass–fail grading due to the peculiarities and intentions of the course.

Classroom practice

What was done in the classroom and how did the students react? The students selected coregroups (3–5 persons) who were expected to work together through the semester, and all the plenary meetings took place in a room that had similarities with a board room. The teacher and the students were all seated around the same table.

Following Freire (1972), the teacher expressed that it was through inquiry and praxis that all men and woman are becoming truly human.

Demonstrably, the teacher tried to practice the attitudes of respect, courage, openness, trust and we-ness. Respect was shown by inviting the students to enter into a dialogue about ethical issues and dilemmas, a dialogue in which all voices were expected to be listened to. Courage and trust were demonstrated by giving the students freedom and responsibility to choose the themes to work on. The teacher was himself quite outspoken concerning his own subjective viewpoints on issues in the global and local area. Thus, the teacher did not take a neutral position. In this way it was believed that the students were encouraged to hold their own position, especially minority positions that might enrich our view of the world.

It was made clear that in order to have a dialogue about ethical matters it was not necessary to use a vocabulary that was either specialized or technical. The use of an ordinary vocabulary without using philosophical terms was expected (cf. Bird, 1991) The essential fact was that we as participants in the dialogue were authentic, that we dared to voice concerns that were important for us as individual persons – here and now – in the role of students, and in the role of citizens outside the schools.

Concerning authenticity, the teacher cited Freire's position (1972, p. 60): "Action without thinking or thinking without action creates an unauthentic world. When a word is not meant to create action, the word is changed into verbalism." "There is no true word that is not at the same time a praxis. Thus, to speak a true word is to transform the world."

In Bird's (1991) terms the teacher used a personal "formative ethical discourse" which was intended to arouse the students' emotions, to stir their feelings. The teachers calling to the students involved the language of virtues and vices. The grand purpose was to inspire and motivate in order "to connect a person's own sense of worth with proposed and past moral actions" (Bird, 1991, p. 32).

The students were invited to *reflect on their own ethical dilemmas.* The teacher stressed that it was not typical to reach a definitive solution to a moral dilemma. In a postmodern world we are placed in *aporetic situations* in which contradictions cannot be overcome – conflicts are not resolvable (cf. Bauman, 1993). However, the essential point is to be able to continue to interact – to maintain on-going relationships, thus reinforcing a sense of reciprocity.

In general, the students in business schools feel ethically illiterate because they have only a minimal formal discourse training based on ethical theories. The time in the plenary sessions was used to practice a series of skills and attitudes that was expected to help the students to become more competent at voicing concerns, more attentively listening to others, and more clearly formulating their arguments and imaginatively considering alternatives. An important means to develop relevant skills and gain experiences was through role play.

Role playing

The students wrote their own role play. In all the plays a young aspiring economist was the main character and his overriding question was: Should I blow the whistle? Themes which were worked on were, among others:

- sexual harassment at the workplace
- a company that polluted the environment
- an au pair's ethical dilemma
- hiring processes (the students played applicants or managers)

Role play 1

In the following the first role play that was written and played will be presented. It was played with an extraordinarily vivid seriousness. The story deals with a company which pollutes the environment and is written by Ann Christin Dahl Johansen. The speech (lines) were not written out. They only hinted at the subject matter in order to make the play spontaneously.

Actors

Per, a fresh economist; Per's wife, Kari; Per's friend, Espen. The manager of the company (Asta Ltd).

The story

Per was out of work, newly married, and the young couple were awaiting their first child. Per has just been to the bank to ask for a loan in order to buy a house. However, the bank was not willing to lend him money because Per had no work. Per tells his friend, Espen, about his problem. Espen has a high position in the local company, Asta Ltd – a cornerstone company in the small society, and Espen promises to try to help Per. Per gets a job and is grateful to Espen. After three weeks Per discovers that the company pollutes the river which passes through the community. Per voices the issue to Espen, who tries to cool him down. Espen explains the wrongdoing in this way:

> ESPEN: "The leadership team is aware of the pollution but has decided that they can't afford to do anything just now. You know our business has recently experienced serious problems, so we have to be disciplined. We have to be patient. We can't afford to use too much money to solve this difficult problem just now."

Per feels uneasy and discusses the issue with his wife, who too becomes anxious. However, she does not really know how to support her husband, whether she should recommend him to tell about his uneasiness to the manager of the company or not.

Finally, Per asks for a meeting with the manager and is (to my surprise) very well received. The manager who is a woman understands Per's uneasiness completely, and assures Per about their goodwill:

> THE MANAGER: "We will certainly do something with the pollution as soon as we can afford it."

However, Per can't avoid mentioning the paradoxical situation that the company is in. It works hard to create a profile as a green company which, to a large extent, considers environmental issues.

PER: "I find the situation rather paradoxical. You saw the headline in the local newspaper the previous day in which you, as the manager, proudly said that the employees even had replaced the sandwich paper with food boxes in order to save paper as a means to save the rain forest".

THE MANAGER: "We are very pleased with your performance in the company so far, and we would like to give you a salary increase."

Per is surprised, offered his thanks and walked out of the manager's office amazed. He reflects on what has happened. The manager had promised to give him a nice salary increase, and he feels, absolutely, that the manager respected him and listened to him. Nevertheless, he is still uneasy and Per wonders: "How much can the environment suffer to the advantage of more profit? How long can I really wait before something is done? When is enough enough?"

Discussion and feelings

The actors come up with some very interesting comments after the play. Per said this: In the moment of whistleblowing I was fearful. As member of a group of three students, one year earlier he had nearly blown the whistle due to ignorance by a major oil company when they discovered that the oil company was maximizing profits without due consideration to the environment. Then he and the other students confronted the leadership in the oil company and warned that they were ready to go public with the story. After a while, however, they decided to wait and see, and concentrated on their more mundane roles as students. Per declared to me just after the play: "*My earlier experience of nearly blowing the whistle was so living in the role play situation that I was frightened*!"

Espen tells this about his feelings: "I have earlier never been an advocate of roleplays as a form of education, and only regarded it as pure entertainment. During the time of this roleplay I changed my point of view. I am sure that the word by Confucius is true: "I hear – and I forget, I look – and I remember, I do – and I understand."

"Before the roleplay began, I was skeptical about the false character I was to represent. I thought it would be easy to clearly distinguish the properties or "identities" between Espen (the rolefigure) and Rune (myself). During the play I became aware that this was not true. I could have taken the position of Espen in real life, and that scared me. In fact, I was into the play to such an extent that I added arguments to what was written in the text, and in this way I identified myself completely with the role I was given. In particular, this was true with the sequence of arguments that Per could not do anything, because we could only engage a new person in his position.

What scared me was how easy it was to play Espen, who was not willing to deal with the dilemma, but only would brush it under the carpet. I felt that, in a way, Espen wanted to enter into a dialogue with Per about the problems in order to help Per, but also because it was something closed or locked in Espen that would like to become aired: I felt that the pollution had niggled Espen to a certain degree, but he was too cowardly to challenge the issue. He was cowardly in the sense that he feared reactions from the environmental lobby, but also cowardly because he did not feel secure enough in himself as to how he might react if he became engaged in a moral dilemma. It caused pain to meet myself, realizing my lack of courage. This was supported by the fact that I continuously referred to the pollution some years ago as something "unavoidable". "At that that time I experienced that it was important not to open up for a discussion that would deal with what was already done."

The author of the roleplay played the manager of the company. She wrote that she was fascinated by the empathy and engagement of the other actors. The teacher was also captured by the dramatic play and immediately saw it as an analogy to Henrik Ibsen's *The Enemy of the People.* In Ibsen's play Dr Stockmann, a physician, discovers that the water in the public bath of his little village is polluted. Because it was too expensive to clean the water, the directors of the bath decided not to do anything about it in the short run – and the directors urgently requested Dr Stockman to remain silent. Every summer the tourists traveled to the town in order to take a healthy bath, and the town needed the money badly. Dr Stockmann's dilemma was: Shall he blow the whistle in spite of the will of the directors?

Presentation of the students own ethical dilemma

The students were asked to present their own ethical dilemma in the class. One of the dilemmas presented was whether to refuse military service. One of the students revealed that he actually was deliberating about refusing the military service and go for "civil service" as a conscientious objector. (In Norway males have to do a required military service that lasts twelve months. Meals and housing are covered.) The student wrote up one page with his considerations pro et contra, and this was distributed and read in the class. Then the teacher opened up for a dialogue about the dilemma. The result was striking. The theme evoked an intense debate that hardly could be categorized as a friendly, good dialogue. Much aggression was clearly expressed. Particularly, two male students who really had done the service and felt commitment to do it argued strongly that to refuse was not a legitimate action. They defended the argument that military service was a responsibility that all men in Norway should undertake.

The turn

After some weeks we got a sad message. Our good faith and our efforts to have the course ungraded was not accepted by the men in power at the school. A new turn had to be taken revolving around preparation for a written, individual, anonymous three-hour examination. The students were furnished with new texts: excerpts and written interpretations from Henrik Ibsen's works, (*Peer Gynt, An Enemy of the People*); Sophocles (*Antigone*); poems (in particular a poem by Hans Børli called *Nothing*). We discussed the material through different sessions and related it to norm-conflicts, self-interests, freedom, responsibility, and whistleblowing in particular. These texts were discussed in the class.

A student's feelings of the new turn

One student wrote this account based upon her reaction to the examination: "When I was told that we were to be graded I was at first *surprised*, however also *motivated* and *eager*. I was surprised because I believed with almost 100 percent certainty that the administration at NHH (the business school), would understand that ethics cannot be measured on the bases of some written pages on a scale from 0–9." "I was a bit *frustrated* because I felt that NHH did not understand what ethics really was about. I felt that we were not taken seriously – that such a course in ethics was nice to have in the curriculum. However, the school was not willing to satisfy the spirit of the course within the school system. In addition, I was a bit afraid of my ability to express/formulate myself correct in a written form concerning my ethical values and beliefs. On the other hand, I was also *motivated* and eager: I thought that now I really had to try to verbalize what I feel and think in different situations …"

"On Friday the May 15 (three days before the examination), when we had a meeting in the class, I felt myself relaxed and easy concerning the examination. It was good to spend the time together, as I got a kind of overview on what was to be expected … It was also good to discuss "ethical maturity". After the meeting I was *secure* that I would be able to perform well in the examination on Monday. In fact, I was delighted – I felt *satisfaction*. It would be good to use three complete hours exclusively on ethics. However, after dinner on Sunday and in the evening, I suddenly got a feeling of *uncertainty*. I felt a bit *resigned*. I was *afraid of* not being able to express myself well enough and that I should not be able to name and frame the feelings and thoughts I would have during the time of the examination. This was a disgusting feeling. However, after speaking with my mother on the telephone, I became calmer.

After the examination I felt *despair*, – I felt that I had not had time to answer all the tasks I would like to, and besides, I felt that I had not been able to "give enough". It was so much more than I could and would have “said”. The time had passed all too fast. Besides it was incredibly difficult expressing with written words, the thoughts and feelings I had in my heart and head! (The feeling I had in advance about the examination was confirmed.) I would rather like to have an oral examination, at least as a complement. Then I would have “been able to give much more” of my thoughts and feelings.

After the examination I felt extreme *exhaustion.* I felt myself empty of words. When shortly afterwards I was going to speak with somebody, I had difficulties in finding the right words. Some hours after the examination, I also felt pleasure. I was alleviated because I finally had finished a long and demanding period of examinations. That was a good feeling.”

Reflections on practice

Different kinds of reflections were built into our everyday classroom life. The following reflections became institutionalized:

- Teachers reflections. After each plenary meeting the teacher wrote up his reflections on what had been discussed and what had been left out. Thoughts and texts to be discussed in the next meeting were included in the memos, which were delivered to the students.
- Students reflections: Students wrote (most often one page) up cases and dilemmas to be discussed. These “onesiders” were distributed to the class. At the beginning of each meeting two reporters were appointed who were asked to write up (ca. three pages) a summary of the plenary dialogues. The reporters were also responsible for preparing a “starter” for the next meeting (one “3 minute speech”). The “starter” had as primary purpose to stimulate our ethical reflections.

Some results from the student’s evaluations

Were the students’ level of trust, openness, self-confidence, and self-respect really influenced? And did the teacher in effect act as a mentor? The evaluations given by the students indicate that these aims, to some extent, were obtained. All through the course the students were given written questions about different ethical dilemmas and how they evaluated the course along a number of dimensions. As a consequence, the students own written evaluations can be used as data. However, the final analysis is not yet completed. Our purpose with presenting our classroom practice is to give students and colleagues a point of departure for discussions and critics in order to continue the dialogue for improvement. Some of the student's evaluations of *Ethical Action* are presented below:

All students answered that they, in comparison with other courses, experienced this course as more vivid and engaging and stimulated more reflections. The students were asked about how the course affected them concerning their feelings. Most of the students answered that their dignity, self-esteem, and self-respect increased. They were also asked why? A typical answer concerning dignity was as follows: "My dignity increased because we have discussed values, and I have been obliged to make up my mind and listen to my 'heart' about what I feel and mean about what is right". A typical answer on the question of self-esteem was that an increase was due to the realization of good dialogues: "It has been good to discuss ethics in the plenary session as well as in the core groups. Everybody has shown interests in listening and taking my own and others meanings seriously. It has been good to receive feedback and confirmation on my feelings and meanings." A typical answer concerning an increase in self-respect is this: "I have been more skilful to express my meanings and to voice them when I think that something is wrong, for example if a person is not treated in a fair manner." One question was about their development in relationship to other courses. The typical answer was that they have developed in a different way. One answered that "during the intensive sessions in the weekends I have particularly experienced development". All students answered that they had more interaction with the teacher than in other courses, and also that they were taken more seriously as human beings than in other courses.

A transfusion of practice?

Could our practice be conveyed to other courses, other schools, and other settings? As far as the course *Ethical Action* is built upon elements that we have experienced in other courses, we think that in general a number of practices can be transferred into other courses. We believe that the realization of a dialogue in the spirit of Freire and Buber is the most important single element in our approach. In fact, the teacher has participated in similar courses of "*Personal development*" at PhD level at a prestigious business School in the USA. In those courses we discussed the fact that most graduate school programs fostered some measure of infantilization and often unwittingly stimulated the students to operate on a lower level, that prevented them from flourishing. The programs encourage passivity and foster a habit of mind and being that matters of ethical significance are taken care of by someone else – either in the organization or outside (cf. Piper et al., 1993). In order to realize real dialogues small classes are needed, preferably less than twenty students in the classroom. Otherwise dialogue is easily perverted into monologues. As a consequence of the failure to foster personal agency, we strongly believe that students should be given more extended responsibility for creating the curriculum.

Role play is a powerful means in all management courses, given that leadership is realized through communication. Core groups are also used in MBA programs with students from diverse cultural backgrounds (for example HEC, Paris). In the core groups the students learn to know each other, they discover each other's personal idiosyncrasies and *talents*!

Since the course Ethical Action had a special content – ethics – and a special purpose – to develop ethical maturity, it is uncommon. On the other hand, we believe that all courses at all levels should discuss ethical values to a certain degree. We hold that to learn and to teach are personal processes in which the teacher's personal values and attitudes are significant. In correspondence with Piper et al. (1993), we believe that we as teachers have important roles to play as exemplary models and mentors. This is not necessarily an unpleasant task to fulfil. Mentorship may mean community. We know that community strengthens social bonds and improves quality of life. We therefore recommend community building in the classroom with the teacher as one of the participants. We also had good experiences with a community of teachers who discussed values, attitudes, and classroom practices *qua* colleagues, accepting criticism but also given support.

Acknowledgement

The authors would like to thank Anita Skogen, Gunhild Wickstrøm and the participants of the session "Teaching Innovations That Leverage Links with the Larger Community" at the 1998 IABS Conference, for stimulating comments and dialogues.

Bibliography

Bauman, Z. (1993) *Postmodern Ethics*, Blackwell, Oxford.

Bird, F. B. (1991) Good Conversations: A Practical Role for Ethics in Business, in *The Role of "Good Conversation" in Business Ethics,* James A. Waters Colloquium on Ethics in Practice, Boston College, Boston.

Buber, M. (1992) *JEG OG DU,* Oslo: J. W. Cappelens Forlag A.S. (Qrig. *lch und Du*. lnsel-Verlag, Leipzig, 1923).

Etzioni, A. (1988) *The Moral Dimension: Toward a New Economics*. The Free Press, New York.

Etzioni, A. (1993) *The Spirit of Community. The Reinvention of American Society,* A Touchstone Book, Simon & Schuster, New York.

Freire, P. (1972) *Pedagogy of the Oppressed,* Penguin Books, New York.

Løgstrup, K. E. (1971) *The Ethical Demand*, Philadelphia: Fortress Press (orig. 1956).

Mackie, J. L. (1980) *Hume's Moral Theory*, Routledge, London.

Nylor, T., W. H. Willimon, & R. Osterberg (1997) The Search for Community in the Workplace. *Business and Society Review*, No. 97, pp. 42–47.

Piper, T. R., M. C. Gentile, & S. D. Parks (1993) *Can Ethics Be Taught? Perspectives, Challenges, and Approaches at Harvard Business School,* Harvard Business School, Boston, MA.

Smith, A. (1976) *The Theory of Moral Sentiments,* Liberty Classics, Indianapolis.

Vetlesen, A. J. (1994) *Perception, Empathy, and Judgment. An Inquiry into the Preconditions of Moral Performance*, The Pennsylvania State University, University Park.

KNUT J. IMS, LARS JACOB TYNES PEDERSEN, AND MARIA PRESTMO

Dialogue-Based Problem Solving: A Discussion Based on Experience from Teamwork among Ethics Students[1]

Diary: A positioning

Kaufmann and Kaufmann's book *Psychology in Organization and Leadership* (1996) was prescribed for the core course *Organizational Theory 100* at NHH[2] in 1999.

A few years later, two of the students attended the course *Ethical Action* at NHH. Teamwork was an integral and very important part of this course and the two students participated in a core group[3] of four persons. The teacher who was responsible for the course assumed the role of a coach rather than that of a traditional lecturer, which is common at the school. The co-operation between the members of the core group and between the core group and the teacher/coach turned out to be highly fruitful. Consequently, two of the members of the group and the teacher began to question why the core group had been so successful. We began to search for literature to address the question, and suddenly Kaufmann and Kaufmann came to mind. This was perhaps not so strange, since Kaufmann and Kaufmann taught the first course themselves, and were employed at the same institute as lecturers/coaches, who inspired colleagues for many years.

The Kaufmanns' book was a great help for us in our search for a vocabulary to describe our experiences with teamwork. Although the book gave us valuable insights, it also made us realize that vital aspects of the dynamics of the group were still unaccounted for in traditional theories on the subject.

For a long time the behavioural perspective was predominant in the field of psychological research. However, towards the end of the 1950s it became increasingly evident that the theoretical models grounded in behaviourism could not adequately explain essential aspects of intelligent human behaviour, for example the development and utilization of language. As a reaction to this, the so-called "cognitive revolution" of psychology took place at the beginning of the 1960s.

The cognitive perspective on learning has been criticized for its one-sidedness in focusing on the individual and the mental side of learning. In the last ten years, many of the cognitivists have put more emphasis on the social and cultural contexts. In the Kaufmanns' book (1996, pp. 197–200), social cognitive learning is mentioned as one of the most complex ways of learning. It uses the cognitive aspect in the social context and is especially concerned with learning through observing others.

This theoretical framework proved very helpful in our efforts to identify the various factors that had contributed to the success of our group. In spite of this, *we still felt that the social cognitive model of learning could not capture some vital aspects of our experiences with group work.* We then turned our attention to a newer perspective on learning, which is called the *socio-cultural perspective* (see for example Dysthe, 2001; Dysthe and Westrheim, 2003). This perspective, which embodies elements of both the dialogism of Mikhail Bakhtin and the theory of situated learning (Lave and Wenger, 1991), basically states that learning is a fundamentally *cooperative process* and places great emphasis on the vital importance of context and interaction.

Towards the end of Chapter 10, the Kaufmanns write:

> The use of the group as a tool of problem solving has become increasingly frequent. This is in part due to the increased emphasis on participation in the decision process and the growing complexity in the problems that organizations and companies face. Research also shows that the most productive organizations are characterized among other things by group decision procedures.
>
> Psychological research indicates that the group has an important potential for problem solving, but this potential is rarely fully realized. It is therefore a very important task to identify the factors that promote and inhibit problem solving in groups, in order to make possible the development of appropriate procedures for problem solving. (Kaufmann and Kaufmann, 1996, p. 293, published in Norwegian)

Thus, the Kaufmanns state that the potential of groups is seldom exploited. Therefore, it should be a great challenge to find different factors that promote as well as prevent problem solving in groups. Based on our general experiences from group work, we are of the opinion that this viewpoint is fundamentally correct. The course, *Ethical Action*, has teamwork as a fundamental requirement and an essential activity among the students.[4] The purpose of the course is to promote reflection about ethical dilemmas, sensitize the students to ethical values, and develop empathy. *Our experiences from these learning processes based upon teamwork might corroborate some of the results in Kaufmann and Kaufmann* (1996). Based upon self-reports and case studies with naturalistic methods, we will build upon richly textured information from a natural context. It is our aspiration that we might complement findings from laboratory experiments to the extent that they have high *ecological validity* (Sternberg and Ben-Zeev, 2001). However, we are well aware that the power of case studies is more appropriate for formulating hypotheses than testing them. Our position in general is confidence in the strength of a synthesis of different views or methods.

In the group work we practiced a dialogue-based approach to ethical dilemmas. We believe that different kinds of dialogues might enable the students to appreciate the nuances and subtleties not only of ethics and language, but also of human interaction in the broadest sense. In the following, we will primarily discuss the relevance and applicability of the model of dialogue-based problem solving in ethics. However, we believe our conclusions might also be of relevance to other fields of research and education.

Frame of reference

This study is grounded in the practical experience from one of the core groups in the Ethical Action class of 2002. After the course was finished, the lecturer and members of the group initiated a dialogue in order to understand and explain the success factors of the group. An understanding slowly emerged that the excellent performance by the group might partly be explained by using a combination of different theoretical perspectives. Some of the group's activities corresponded well with normative recommendations of optimal problem solving, with some interesting deviations, while other elements were better explained by a different theoretical perspective – the newer *socio-cultural perspective that draws heavily upon the dialogical perspectives of Bakhtin and Buber. We have extended this perspective by using the Aristotelian concept of friendship.* In this study, we will attempt to explore and partly extend these two perspectives in the light of the learning processes of the core group. Our main argument is that it might be fruitful to combine these different theoretical traditions. The first part of the paper therefore presents the above-mentioned approaches as a frame of reference.

Problem solving

Problem solving can be defined as a general category of activities organized to promote goal attainment (Kaufmann and Kaufmann, 1996). In addition to these activities, there is also an element of problem finding related to any problem (Kaufmann, 1980). This refers to identifying the problem as well as defining it, and is particularly relevant for unstructured problems. The difference between problem finding and problem solving is essentially whether the problem is approached within a predefined frame or not. Problem finding involves questioning what the boundaries are, for example by examining radically different strategies (Ims, 1987). However, problem finding is usually looked upon as an integral part of the problem solving process.

The term *problem* is often associated with something negative; however, a positive attitude and focus on opportunities and challenges rather than limitations can turn this around. These conceptions of a problem are related to problem horizon and problem valence, respectively. Problems are classified in a three-dimensional model using these two aspects, as well as problem structure (Kaufmann and Kaufmann, 1996).

Approaching problems as stated above is a common characteristic of creative problem solving. It involves using new perspectives and reformulating the problem, and it can increase the number of possible solutions. Creative problem solving is also characterized by novel thinking that can be transformed into practical innovations (Kaufmann and Kaufmann, 1996).

The Kaufmanns (1996) present four conditions that promote and four that restrain creative problem solving. The four conditions that promote creativity in problem solving are: (i) exploring the problem situation, which is related to problem finding; the main idea is to focus on the problem rather than the solution, in order to avoid concluding the search too early; (ii) distinguishing idea development from idea evaluation; evaluating ideas before they are properly developed can lead to anxiety, and can be a significant obstacle to creativity; therefore, evaluation must not take place before ideas are fully developed; (iii) conflict of ideas can be valuable, for example thinking in contradictions or opposites; and (iv) strong motivation and persistence. One example of this is called double problem solving, referring to individuals who renew their efforts to solve a problem after quitting.

The restraining conditions are respectively: (i) the Einstellung effect, which refers to a type of rigidity where the individual continues to use a problem solving method that has traditionally worked well, but is less suitable for a current problem; this prevents the individual from seeing new solutions; (ii) functional fixation, meaning that the individual cannot get past the original function of an item, and therefore does not see alternative ways of utilizing the same item; (iii) so-called hidden assumptions, which involve attempting to solve a problem while implicitly assuming how it should be solved, thereby narrowing the scope of possible solutions, and (iv) the tendency to seek and prefer verification rather than falsification of hypotheses. After falsification, the problem must be reworked, making room for creative thinking. However, this is often a lost opportunity.

Problem solving in groups

How should we define general principles for working in groups in order to find a formula for successful problem solving? This is a recurrent theme in the literature. A general checklist for what constitutes a successful group will probably never incorporate the variety of tacit and transient factors that lead to a good result. Nevertheless, it is possible to identify certain factors that ought to be present and principles that ought to be followed for problem solving in groups. There are three important conditions that make group work particularly appropriate: (i) many people are affected by the decision that will be made, (ii) there is no unambiguous solution to the problem, and (iii) the intrinsic value in forming a group, because the situation is such that the process is as important as solving the problem (Robson, 1995).

Kaufmann and Kaufmann (1996) state that there are also two structural dimensions that are important for the performance of the group. The first concerns the degree to which the group is heterogeneous or homogeneous. Research shows that complementary traits are important to make a group function. In spite of this, personal unanimity is also important for good cooperation, for example with regard to social mentality and attitudes to important topics. Hence, groups that are somewhat heterogeneous in skills and work styles but homogeneous in general attitudes and personality traits will often perform well.

The other important dimension is the degree of unity within the group, meaning how attached group members are to their group. Group unity is strengthened by close interaction, physical proximity, previous success in the group, as well as joint goals and similar personalities. According to research, group unity can be both positive and negative, and it depends on the focus of norms within the group. If there are norms aimed at high performance within the group, a high level of unity will have a positive effect on how the group works. If norms are aimed at low performance, the group will not work well.

Kaufmann and Kaufmann present additional promoting and restraining factors for problem solving in groups. A heterogeneous combination of individual traits is positive, since it allows variation in perspectives and strategies for problem solving. Another promoting factor is cognitive conflict, as long as the level of conflict is functional. This necessitates alternative solutions, and thereby promotes originality. A third positive factor is activation of skills and competence by other members of the group. Such stimulation is necessary to make group members perform at a high level. The final promoting factor is acceptance of decisions, something that can be arranged by involving all group members in the decision process.

With regard to restraining factors, social pressure and conformity are the first. Since behaviour in groups is highly affected by norms, there is often a movement towards conformity in any group. This is particularly true when there is high uncertainty within the group. Of course, such social pressure is not optimal with regard to creativity and dialogue. Asymmetric participation is another restraining factor. It refers to group members bringing different levels of input into problem solving; for example, domination of problem solving activities by one or more group members. This leads to a monological problem solving process where the competence of other group members is unused. A distinct case in this regard is groupthink, a phenomenon that is characterized by high unity and perceived unanimity, even though there might be covert dissension in the group. Groupthink can lead to, for example, insufficient information search, inadequate evaluation of alternatives, and maintenance of the *status quo* (Janis, 1982). A third restraining factor is valence, which implies that group members develop an emotional state of attraction or rejection to new ideas. This leads to the creation of a minimum value, or threshold, for alternative solutions, which often is quite low. Any alternative that exceeds this level is accepted but is not necessarily an optimal solution. A final restraining factor is related to group size. The essence of group size is that marginal productivity of group members diminishes as size increases, and that few groups can handle as many as 10–12 group members.

Dialogism

The socio-cultural perspective holds a dialogical view on learning and problem solving (Dysthe, 2001; Lave and Wenger, 1991; Ims and Jakobsen,

2003). This is manifested in its characterization of learning, in which social and communal processes are emphasized. Fløistad (2000) states that a community without dialogue is no community; thereby underlining that dialogue is inherent in communal processes. Dialogism is an alternative perspective on human cognition and communication, and it implies that communication in learning processes is inter-subjective, taking place *between* those involved. In this interaction, meaning is created (Dysthe 2001; Ims 1987). Dialogism is concerned not with the transmission of knowledge, but with the transformation of understanding (Dysthe et al., 2003); hence, more than one perspective is needed.

Dialogism is the opposite of monologism, which has been the prevailing paradigm in this respect. The monologue involves a one-eyed perspective, allowing room for one conception alone. Opposed to this, proper dialogue implies different conceptions being exchanged in order to create meaning. In such a dialogue, those involved take on the perspective of others (Ims, 1987). This underlines a main benefit of the dialogue – it provides an opening up rather than a closure. In proper dialogue, there is also a recognition of the *Other* as an equal, creating a relation in line with Martin Buber's (1992) notion of "I-Thou". Here the partners respond to each other, and their interaction is characterized by symmetry (Ims, 1987). There is a qualitative difference between the "I-Thou" and the relation noted as "I-It" (Buber, 1964). It involves an *objectification of the other, thereby depriving the relation of equality, symmetry, and dignity.*

This lack of recognition of the Other as an equal subject involves a monological relation, where there is room only for the perspective of the "I". The relation is merely instrumental and leads to alienation of the Other. However, in the "I Thou" relation the Other is regarded as a subject with unique, intrinsic value. The relation is thereby formative, in the sense that it involves critical reflection upon one's own perspective and that of the Other (Ims and Jakobsen, 2004). This characterizes dialogue-based problem solving, which involves partners who are continuously challenged and forced into testing and developing their arguments and attitudes.

The dialogue also has relevance with regard to *personal identity.* Charles Taylor (1992) points out the tendency of modern individuals to make their choices in life solely based on their own personal views on morality, without acknowledging common ethical boundaries. This is the result of the extreme focus on self-realization in modern societies. Taylor states that personal identity should not be developed in monologue, but through dialogue with others. In dialogue the individuals can reflect upon their own as well as communal values. This will have implications for the character and identity of the individual, and express commitment to his or her community.

The alternatives to dialogue-based problem solving

One of the alternatives to dialogue-based problem solving is very common and, in our opinion, suboptimal compared to the dialogue-based approach. Group work, particularly among students, often takes the form of what can be called *sequential monological problem solving*. This means that the problem is divided into different tasks that are solved independently by each group member. Such *partitioning* is done to save time from making the group work and solve the problem together, which may take more time than the aggregate time spent by each of the members in solitude. This problem solving is merely instrumental, in that it is oriented towards efficiency and the utilization of each group member's comparative advantage.[5] In sequential monological problem solving, person A solves task X, passes the work on to person B who expands it by solving task Y, etc.

When all parts of the problem are solved, the group tries to form a unified product from the single parts. Opposite to dialogue-based problem solving, this type of problem solving does not include opposite voices[6] for solving each task. Instead, each part is solved by utilizing the competence and conceptions of one group member only. There is no external dialogue between group members, only the *internal dialogue* conducted by each group member while solving his/her dedicated part of the task. Hence, the real synergistic advantages of group work are not realized.

The other alternative to dialogue-based problem solving is to some extent a cooperative process, but equally lacking in proper dialogue. This problem may be conceptualized as the opposite from partitioning, since all group members work closely together throughout the whole problem solving process. Thereby, *all dialogue is external.* Bråten (1986a) described in his theory on model power how participants that interact with each other may be *swallowed by one perspective,* where seemingly the process is dialogic, but the exchange of meanings is merely monologic. Every participant confirms the others point of view in a superficial manner. It easily ends up in group think-like processes dominated by only one perspective (Janis, 1982). These processes are often characterized by jumping too quickly over the problem finding phase, running more or less directly into the problem solving phase. This behaviour corresponds with Kaufmann and Kaufmann's notion of threshold for acceptable solutions instead of searching for more optimal solutions.

Interaction and dialogue

Habermas (1982) presents a traditional classification of types of (inter)actions and distinguishes between non-social and social types. Instrumental action is of the former type and its orientation is to success. Such instrumentalism is not well suited for interaction because of its objectification of those influenced. The social types of action are strategic action and communicative action, the former of which is oriented to success. It differs from instrumental action by including rational opponents seeking to influence decisions. Such strategic interaction can be either open or covert. The other type of social action is the communicative, which is oriented toward reaching understanding. This is a co-operative achievement of all participants and cannot be imposed, but is based on intersubject recognition. Consequently, communicative interaction can be of dialogic nature.

These three types of interaction are classified by orientation and imply that there is a goal to be obtained. Two other forms of dialogic interaction can be added. The first is the "I-Thou"-relation presented by Buber (1964), where the orientation is the dialogue itself and the interaction is characterized by sincerity.

The final type of interaction is based on Bråten (1986a, 1986b) and is in line with the concept of proper dialogue presented above. Such interaction has room for complementary perspectives and the participants engage in self-reflective dialogue (for an overview see, for example, Ims, 1987)

As a complement to this classification, it can be instructive to apply the Aristotelian distinction between types of friendship – *useful, pleasurable,* and *perfect* – onto the concept of interaction and the dialogue (Aristotle, 1987). We are of the opinion that the inclusion of pleasure as motivation for dialogue is of value when considering the concept of dialogue. To explain why this classification can be valuable, we will briefly present the three Aristotelian types.

Usefulness is not a permanent state, because it relates to obtaining constantly changing ends. Hence, useful interaction will take place because the participants pursue something that they can get from their partners. They interact *for themselves.* Pleasure is also detached from the interaction partners and is, likewise, a fickle value; because of the instability of what is perceived as pleasant, different from usefulness, participants engage in pleasurable interaction for what it is to *them.* The final type, perfection, is characterized by its connection to virtue. Partners in perfect interaction or dialogue are alike in virtue and pursue their relation because it is good in *itself.* Therefore, such relations will be highly stable.

We launch the idea of this distinction because it can give a good description of the work in student groups. Of course, the primary and initiating reason for forming student groups is that this is useful. It serves the function of solving problems as defined by the curriculum or the students themselves. This utilitarian aspect will be part of any group's existence. Furthermore, there is reason to believe that most groups will also have a parallel social development, where the element of pleasure will be central. In any group interaction there is room for pleasurable dialogue and action, and this is enriching vis-à-vis the one-dimensional utilitarian aspect.

Long-lasting relationships can possibly transcend into the realm of *perfection,* where the two former aspects are combined with a sincere commitment for the sake of the relationship itself. This will presuppose that there is a similarity in virtue between the individuals, and congruence between the values of the individuals and of the relationship. In a sense, such a relationship can involve an extension of the individual's self-concept. In a perfect relationship, group members form a unit, where each individual regards the others as a part of his own self. However, transcendence of this kind will take a long time and require the development of trust within the relationship.

We believe that this is a fruitful description of how interaction can develop. The element of pleasure has, up to now, been largely ignored in the classification of interaction and dialogue, even though it can be an important part of any given relationship. We will later show how this can be applied to our case.

Contextualization – The course *Ethical Action*

The context in which this paper is placed is the course Ethical Action; an elective on the Master in Business Administration program at NHH. The course has a duration of one semester and is partly intensive due to its content. Students work in groups throughout the semester with tasks such as creating a role-play, opening classes with short presentations, writing reflective papers on relevant topics, as well as involvement in plenary discussion.

These tasks are intended to make the students reflect upon their own view points on ethical questions, as well as the viewpoints of their co-students. The purpose is to practice ethical action and develop ethical sensibility and phronesis (prudence and practical wisdom, see Aristotle, 1985). Thereby, students should aspire to reach "moral *bildung*"[7] (Tønnesen, 1992). In order to achieve this, the course is based on dialogical interaction between participants.

Course participants are evaluated based on a written exam done at home (75 percent), as well as on reflective papers (25 percent). The exam is based on the topics students work on during the semester. This evaluation is formative, since the students get feedback on what they write during the course and follow their own development arid learning throughout the course. Formative evaluation also allows a positive type of assessment, where the goal is to identify strengths, opportunities, and needs of development (<http ://itslearning.nhh.no>).

The human being's experience of oneself as a moral subject and moral addressee takes place in the I-Thou-relation within a frame of proximity between the persons (see Baumann, 1993; Levinas, 1987; Løgstrup, 1992; Ims, 1995). Such proximity is relevant to the work in groups. The moral of Levinas is based upon the face of the Other. Access to the face is immediately ethical, and the face of the Other is the most vulnerable aspect of the Other. The ethics of proximity is the main philosophy behind the course.

The problem

The actual problem that the group had to solve was to develop a case containing one or more ethical dilemmas. The case had to be thoroughly elaborated in order for the group to create a role-play based on the case. This role-play was supposed to be performed in class, and to be the starting point for a discussion on the ethical dilemma at hand.

Of course, solving this problem involved searching for an appropriate case, since the ethical dilemma had to be sufficiently complex to serve as the basis for an extensive discussion and crossing of perspectives. The problem also had to be appropriate for role-play and the group had to ensure that its content and form would emotionally touch the audience – a prerequisite in the perspective of the ethics of proximity. An essential aspect of the problem is that it demands openness on the part of the participants and the audience in order to establish a genuine I-Thou-relation in a Buberian sense (Buber, 1964). In an open and real meeting the participants can be emotionally touched and develop empathy. Therefore, it is *necessary that the participants accept vulnerability and ambivalence.*

The group

The group had four members, two of each gender. The age difference was rather small, although members differed on personalities, interests, and skills. The members did, however, state that they had shared values. Furthermore, there was a high level of mutual respect between the members. The group members were not previously acquainted, except for the fact that one member knew all the others. However, since all four members had been students on the same program for some time, they were all aware of each other before entering the group.

All group members agreed that the work itself – the process – was much more important than "getting the job done" or performing particularly well on the exam. All members contributed verbally and in writing, but there was considerable flexibility with regard to how much each member contributed. The emphasis lay on the creation, not on who created.

The group wanted to create an environment that was supportive and rewarded creative ideas. Then the process could serve as an extrinsic motivation factor that increased the intrinsic motivation and so had positive effects on creative performance tasks. They therefore decided to meet at a cafe once a week, where sessions were opened with dinner and trivial talk. This put the group in a positive mood that facilitated idea generation, which was a very important part of the creative problem solving process. The positive mood facilitated idea generation in a way that made the group members very productive. Weisberg (1994) pointed out an important distinction between productivity and creativity. According to his studies, positive mood could produce increased productivity, in the sense of quantity of products, but it did not generalize to a higher quality of creativity.

The initial phase of unrestrained idea generation left the group with a huge body of different thoughts and ideas. In order to proceed with the problem solving process it was necessary to examine each idea in a more methodical manner. Traditionally, an unconditional positive causal link between positive mood and creativity had been postulated. Novel research, however, shows that the kind of nuanced, fine-tuned critical thinking that takes place in this stage would more likely be disrupted by positive mood (Kaufmann, 2003). Kaufmann further states that positive mood promotes broad and shallow processing, whereas negative mood leads to more constricted but deeper processing. Thus, positive mood may be an advantage in a task that requires the generation of new ideas, but may be detrimental when the task requires careful considerations and deeper processing. The group soon discovered that some of the ideas initially thought to be very good would be very hard to put into practice. In spite of this they were not immediately discarded, and were often found to be fruitful points of departure for further reasoning.

In this part of the process, the objective was to ascertain, through focused and more thorough analysis, if any of the different ideas generated could actually help to solve the problem at hand. The group alternated between working together and working separately, thereby ensuring that all members could reflect upon and revise each topic on their own. This ensured internal dialogue, and provided the group members with distance from the group process. This was done to avoid groupthink. In group processes, positive mood can often lead to a "no problem" attitude to the task at hand, but the internal dialogue also prevented the group members from overlooking important features. Hence, the group combined internal and external dialogue throughout the problem solving process. In so doing, the group tried to realize the potential of problem solving through group work, with regard to individual competencies and reflections, as well as the dialectics of group processes.

Kaufmann (2003) states that the detrimental consequences of positive mood can be moderated by frequent feedback from someone outside the group. It is in fact expected that positive mood subjects, when continuously informed of their success at the task, may be more persistent in pursuing the tasks, and may also be more flexible at shifting strategies rather than negative mood subjects. In our case, there was frequent contact between the group and the lecturer, so that any adjustments or improvements suggested by the lecturer could be integrated as the process developed. This prevented the positive mood in the group from adversely affecting the problem solving process.

Discussion

When comparing the problem solving literature (Kaufmann and Kaufmann, 1996) to how the group performed, there are many similarities. A main point that was highlighted in the discussion of the group's performance was the fact that there was heterogeneity with regard to skills, work styles, and personal traits, but homogeneity in general attitudes. Furthermore, there was a continuous drive toward finding new and creative solutions, in line with the notion of cognitive conflict. Efforts were also being made to activate the skills of the different group members.

The most *interesting point is, however, one that deviates from the prescriptions in the problem solving literature.* Asymmetric participation was described as a restraining factor for problem solving in groups since, as the theory goes, different levels of input from group members in the problem solving process can lead to dissatisfaction. In our case, the group stated that throughout their work, they differed in levels of participation. In fact, due to illness, one group member was absent for a long period of the problem solving process. It was evident that there was a significant difference in the total level of participation and input from each group member for the process as a whole. This asymmetry did not, however, have a negative influence on the group, since all members showed acceptance and ownership of the final product.

Since the group's reaction to the asymmetric participation differed from mainstream theories on problem solving, we looked for an explanation elsewhere. We found that dialogism was appropriate for explaining this "deviation". In the actual core group, the group as a whole decided upon how the participation of group members should vary during the process. This is not to be seen as a mere division of labor. It rather implies recognition of each group member as a unique individual with his or her own skills and character. Such recognition is closely related to the Buberian dialogical "I-Thou"-relation (Buber, 1964). The group members worked with the tasks that they were best suited for, so that they could realize the best of their competences and abilities. They developed excellence (arete) in an Aristotelian sense. This meant that the group was more focused on realizing the *best* possible competence of each individual rather than on ensuring that the level of input of each group member was equal.

This dialogical view on asymmetric participation is closely related to one of the promoting factors for problem solving in groups, namely activation of skills. The dialogical description of asymmetric participation has an activating effect on each individual in the group. It does, however, take the problem solving element one step further, since it also implies that there is something above merely the instrumentality and usefulness of helping each other in order to reach a common goal. In fact, *through the recognition of each other as persons with an infinite dignity* (in a Kantian sense, see Kant, 1959, p. 53), this interaction creates something over and above a technocratic viewpoint of the assessment of any input of the group members. The confirming aspect of the I-Thou dialogue implies that people are valuable in themselves and cannot be reduced to a quantitatively measurable input. The comparison of the input of one group member vis-à-vis another has its limitations. It is worth noting that a dialogical approach to asymmetric participation can have an explanatory power in problem solving.

The other point we will highlight is the *pleasure component* in *the dialogue.* As mentioned, this is not included in mainstream problem solving literature. However, a key value in the development of the relationships in our group was pleasurable socializing. This was manifested in how meetings took place over dinner, for example. Group members stated that this element was imperative for their satisfaction with the group work. This underlines how such interaction enriches the otherwise utilitarian process of problem solving in groups. It should be looked upon as a natural fact that pleasurable interaction is a positive element in the group. Seeing how its effect on mood had positive consequences, particularly in the idea generation phase, it should not be ignored in the literature. The positive impact of pleasurable interaction and positive mood should, nevertheless, be seen in relation to whether a person outside the group provides feedback.

We stated earlier that long-lasting groups can move into the realm of perfection (in friendship), something that we believe might have an even stronger positive impact on performance. When group members regard the other members as an integral part of their own self-concept they will naturally do whatever possible to help the other group members. This will enhance the quality of the group work further. Thereby, the distinction of types of interaction is relevant in this regard, since the different types will have important and different implications for the quality of the group work.

An important final point refers to the concepts of internal and external dialogue. We have described alternatives to dialogue-based problem solving that contained either internal or external dialogues. The alternation and interplay between these two types of dialogue is the essence of dialogue-based problem solving. Internal dialogue is stimulated by external dialogue (Ims and Jakobsen, 2004), and it *is* a certain *mental distance* from the external dialogue that allows each member of the group to reflect upon his/her own premises and develop their own meanings through internal dialogue. This interplay enriches the dialogical aspect of the problem solving processes, and is, we believe, the main element that distinguishes dialogue-based problem solving from *its* monological oriented alternatives.

A final comment

One swallow does not make a summer; neither does one study of one core group represent the whole truth. Nevertheless, some preliminary conclusions may be drawn from this study. The present study corroborates what is assumed in the mainstream problem solving literature. The findings suggest that the socio-cultural perspective (Dysthe, 2001; Dysthe and Westrheim, 2003) may be a complementary and a fruitful avenue to explore in addressing questions such as what are the critical success factors behind high performance problem solving? In particular, it might be profitable to draw attention to the literature on ethics and dialogues, not the least the notions of pleasure and perfect friendship, as well as confirming dialogues. The mutually stimulating effect of the interplay of internal and external dialogue in problem solving processes is also an important concept to address. Our findings, based upon our personal experiences, show that group members exhibit altruistic behaviour, which have implications for the design of incentive structures. The relationships between incentives and group performance are important, and our study indicates that group members should not be considered mere instrumental tools. The notion of groups as communities of practice (Lave and Wenger, 1991) in which the group members have dignity and are confirmed as ends in themselves (Kant, 1959) might be studied more extensively, while at the same time the notion of groups as management tools, as effective instrument in the managers toolbox, might be minimized.

Acknowledgements

We are indebted to the stimulating and constructive comments from Professor Tore Helstrup on an early draft of this paper.

Notes

1. The names of the authors are alphabetically listed.
2. NHH is the acronym for the Norwegian School of Economics and Business Administration in Bergen, Norway.
3. A core group does not exchange group members throughout the semester.
4. The requirement for the course is working in core groups with one project, which is expected to culminate with staging role-plays. It is expected that the challenge of making and acting in role-plays acts is a strong motivator, which makes the participants feel a common responsibility for a very visible product – the final role-play.
5. See Brooks 1975 for theorizing about under what circumstances it can be fruitful to partition different tasks among workers/It-personnel.
6. We have experienced that the term "crossing of perspectives" evokes unfortunate connotations in English. Therefore, we also use synonyms like "different perspectives" and "opposite voices" (from Bakhtin) because we find the term "intersection of perspectives" impotent.
7. We use the German term *Bildung* because we have not found an adequate English term. English synonyms might be moral formation or moral education, neither of which we find satisfactory.

Bibliography

Aristotle (1985), *Nichomachean Ethics,* translated by T. Irwin, Indianapolis: Hacklett Publishing Co.

Bauman, Z. (1993), *Postmodern Ethics,* Oxford: Blackwell Publishers.

Bråten, S. (1986a), Paradigms of Autonomy: Dialogical or Monological? In: *Social and Legal Autopoiesis,* (ed.) G. Teubner, New York: European University Institute Publ. Series.

Bråten, S. (1986b), The Third Position: Beyond Artificial and Autopoietic Reduction. In: *Sociocybernetic Paradoxes,* (ed.) F. Geyer and J. van der Zouwen, London: Sage.

Brooks, F. P. (1975), *The Mythical Man-Month. Essays on Software Engineering.* Reading, MA: Addison-Wesly.

Buber, M. (1964), *Jeg og Du,* Danmark: Munksgaard.

Dysthe, O. (2001), "Sosiokulturelle teoriperspektiv på kunnskap og læring". In: *Dialog, samspel og læring,* (ed.) O. Dysthe, Oslo: Abstrakt Forlag.

Dysthe, O. and K. Westrheim (2003), *The Power of the Group in Graduate Student Super-vision. An Empirical Study of Group-Based Supervision Combined with Student Groups and Individual Supervision,* Faculty of Psychology, University of Bergen.

Fløistad, G. (2000), *Utfordringer. Studier i langsomhetens filosofi. Bind 2,* Oslo: Ledelsefilosofi.

Habermas J. (1982), A Reply to My Critics. In: *Habermas: Critical Debates,* (eds) J. Thompson and D. Held, Cambridge, MA: Macmillan Press.

Ims, K. J. (1987), *Leder i dialog.* En *studie av informasjonssøk med metoder for personlig utvikling,* Bergen: Universitetsforlaget.

Ims, K. J. (1995), Mismeetings. An Essay about Money, Urban Life and the Ethics of Proximity. In *Essays in Marketing and Management. A Festschrift in Honour of Kjell Grønhaug,* (eds) T. B. Holmesland, K. J. Ims and A. Pedersen, Bergen: Fagbokforlaget.

Ims, K. J. and O. Jakobsen (2004), Moralsk dannelse eller etisk instrumentalisme? En kritisk vurdering av etikkurs innen profesjonsutdanning, *Nordisk Pedagogik,* Vol. 24, pp. 162–172.

Janis, I. L. (1982), *Groupthink. Psychological Studies of Policy Decisions and Fiascos,* 2nd edn, Boston: Houghton Mifflin Company.

Kant, I (1959), *Foundation of the Metaphysics of Morals.* Translated with an introduction by L. W. Beck, New York: Macmillan Publishing Co.

Kaufmann, G. (1980), *Imagery, Language and Cognition. Toward a Theory of Symbolic Activity in Human Problem-Solving,* Oslo: Universitetsforlaget.

Kaufmann, G. (2003). The Effect of Mood on Creativity in the Innovative Process. In: *International Handbook on Innovation*, (ed.) L. V. Shavina, Kidlington: Elsevier Science Ltd.

Kaufmann, G. and A. Kaufmann (1996), *Psykologi i Organisasjon og Ledelse,* Bergen: Fagbokforlaget.

Lave, J. and E. Wenger (1991), *Situated Learning. Legitimate Peripheral Participation,* Cambridge: Cambridge University Press.

Levinas, E. (1987), *Collected Philosophical Papers. Nijhoff, Dordrecht.*

Løgstrup, K. E. (1991), Den etiske fordring, 2. ed, Copenhagen: Gyldendal.

Robson, M. (1995), *Effektivt teamarbeid: Håndbok for problemløsning i grupper*, Oslo: Bokklubben bedre ledelse.

Sternberg, R. J. and T. Ben-Zeev (2001), *Complex Cognition. The Psychology of Human Thought,* New York: Oxford University Press.

Taylor, C. (1992), *The Malaise of Modernity*, Cambridge, MA: Harvard University Press.

Tønnesen, R. Th. (1992), *Demokratisk dannelse i tysk perspektiv. 20 års diskusjon om Herman Gieseckes syn på den politiske oppdragelsen,* Norges Allmennvitenskapelige Forskningsråd, Studier i jus og samfunnsvitenskap, nr.7.

Weisberg, R. W. (1994), Genius and Madness? A quasi-experimental test of the hypothesis that manic-depression increases creativity, *Psychological Science*, Vol. 5, pp. 361–367.

OVE D. JAKOBSEN, KNUT J. IMS, AND KJELL GRØNHAUG

Faculty Members' Attitudes Towards Ethics at Norwegian Business Schools: An Explorative Study

Introduction

Due to the pervasive influence of economic thinking in our societies, business schools play an important role in today's education. This tendency is reflected in the following quotation: "the schools where many of the leaders of organizations have been and are being educated, schools of business are strategically positioned to have a positive or detrimental impact on the values, beliefs and behaviours that executives to-be will bring to organizations" (Schwartz et al., 1991, p. 466). Bearing this in mind it is not surprising that business schools are often targets for critics when corporate scandals unfold in the media. Discussion about the content of business schools' curricula is of general concern, partly as a response to this critique and partly as a response to the need for ethical reflection in business, as well as in society as a whole.

This paper addresses opinions and attitudes held by faculty members of Norwegian business schools. We do so because the faculty plays an important role in any business school, and because faculty members attitudes influence their behaviour. More precisely, faculty members influence to a large extent what is thought and what is emphasized, and thus the ethical perspectives and values disseminated.

A brief look at recent research reveals that there is a growing interest worldwide for knowledge concerning the opinions and attitudes toward ethics among business school faculty members (Cowton and Cummins, 2003; Enderle, 1996, 1997; Hosmer, 1999; Matten and Moon, 2004). Attitudes can be defined as "learned predispositions to respond in a consistently favorable or unfavorable manner with respect to a given object" (Fishbein and Ajen, 1975, p. 6). Such learned predispositions are influenced by the subject's knowledge and experience. Knowledge about the opinions and attitudes toward ethics among faculty members of business schools is thus of great importance, because such information reveals what is emphasized in their teaching and reflects their ethical values. An investigation into attitudes will also indicate who are the "drivers" of the business ethics agenda in business schools, whether it is the leadership of the school or individual faculty members (Matten and Moon, 2004). In a case based approach in eight leading business schools in Europe the drivers of business ethics and the degree of involvement appear (Zsolnai, 1998). This study documents the diversity of the level of development of business ethics in Europe and emphasizes "some *critical distance* from the *mainstream American approach* to business ethics". An important finding is that European business ethics is "deeply rooted in culture and less influenced by abstract principles or ideas …" (Zsolnai, 1998, p. viii). However, Norway was not represented in the study. All things considered, due to the effects of education the ethical values of the faculty members are believed to influence future managers' values.

Today, ethics is a fast emerging and dynamic field of research, partly as a result of a growing awareness of the potential value conflicts between economics, environment, and culture. Because ethical issues are multiple, and several perspectives exist, it is reasonable to assume that faculty members may differ in their ethical views. To better understand the ethical attitudes and values held by faculty members, we introduce a frame of reference based on central perspectives in today's education in the field of ethics.

The remaining section of the article is organized as follows. First, we present and discuss a theory-driven frame of reference. We do so because such a reference influences both what will be emphasized in the empirical study that follows, as well as our interpretation of the findings. An explicit theoretical frame of reference also makes it easier for the readers to evaluate and criticize our study. Next, we report the research methodology underlying our empirical study, because the choice of research methodology and how the study is conducted impact the findings. Then we report and discuss our findings and finally draw our conclusions.

Four theoretical perspectives on ethics and economy

This study addresses opinions of and attitudes toward ethics among members of business school faculties. Attitudinal research shows that attitudes are relatively stable, that they are influenced by past experience and learning, influence actions and thinking, and that attitudes and behaviour covary. In this paper we are particularly interested in whatever, and how, the faculty members are influenced by various disciplinary perspectives and how these influences may impact their attitudes towards teaching business ethics.

Figure 1 categorizes four disciplinary perspectives: neoclassical economics, strategic management, corporate social responsibility, and socio-economics. In mapping these perspectives/theories we use two scales; that is, whether they reflect economics as a positive or a moral science (x-axis) and whether they addresses the micro, meso, or macro level (y-axis). This framework is partly similar to Enderle's (1996), "A basis of comparison for business ethics" in distinguishing between three qualitatively different levels of acting (individual, organizational, and systemic level). A look at Figure 1 shows that neoclassical economics reflects a positive science, as well as spanning over several levels. Strategic management reflects a positive science on the micro level. Corporate social responsibility reflects a moral science at the micro level. Socio-economics spans several levels, and reflects a moral science.

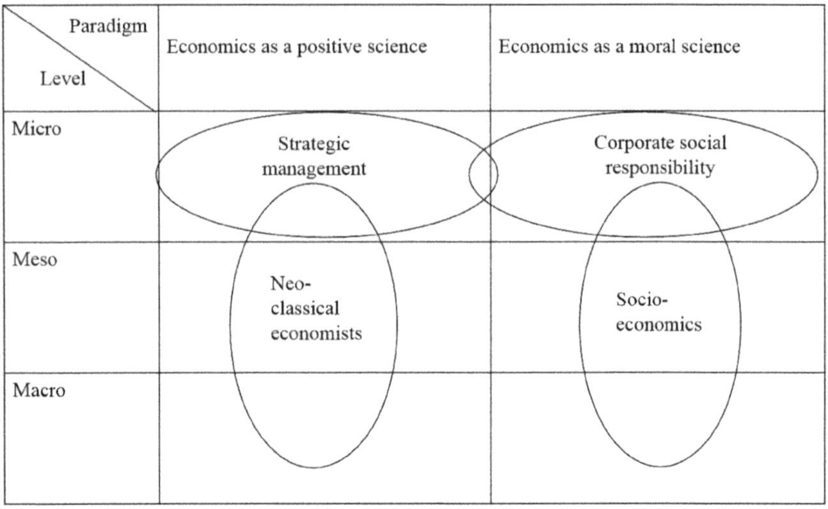

Figure 1. Faculty members attitudes – classifications.

(Neo)classical economics

Adam Smith, the founder of classical economics, claimed that economy functions best when the agent's freedom is greatest. He also postulated that general affluence will be the greatest when all industrialists and merchants seek to maximize their own profit, as reflected in the following quote:

> It is not from the benevolence of the butcher, the brewer, or the baker, that we expect our dinner, but from their regard to their own interest. (Smith, 1976b, p. 18)

Another well-known statement by Smith says that the logic of the market implies that private vices can be turned to public virtues by means of the market's " invisible hand":

> [E]very individual necessarily labours to render the annual revenue of society as great as he can. He generally, indeed, neither intends to promote the public interest, nor knows how much he is promoting it. He intends only his own gain, and he is in this, as in many other cases, led by an invisible hand to promote an end which was

> no part of his intention ... By pursuing his own interest he frequently promotes that of society more effectually than when he really intends to promote it. (Smith, 1976b, pp. 477–478)

Since the central idea of Smith's economic theory corresponds with classical liberalism it is not surprising that Smith meant that the market must be regulated by law and justice. Smith was also well aware of the importance of ethical values and norms as necessary conditions for market economy cofunction. For example, without trust between the actors in the market there would be no economic development. Smith (1976a) also took for granted that all human beings should act on the basis of sympathy and concern for other people. According to Smith, the bonds of sympathy between actors were a necessary condition for a well-functioning market economy.

Max Weber (1930) attempted to identify the specific conditions, during and after the Reformation, that prepared the way for the triumph of capitalism in the West. According to Weber, "the Reformation led to a radical change in the traditional ethical notions of duty and prepared the way for an ethics which legitimized a new rational way of life" (Skirbekk and Gilje, 2001, p. 405). From this we may conclude that Protestant ethics made up a necessary but not sufficient condition for the rise of capitalism. Weber also argued that the rise of capitalism was an unintended effect of ethical-religious ascetic attitudes developed among different Protestant sects.

According to the present authors most (neo) classical economists define economics as an "objective" science on the one hand. On the other hand, we do not have sufficient basis to claim that they overlook ethics, but there has been a serious distancing of ethics from economics. According to Sen (1987) this has weakened the base for both descriptive and predictive economics. The extremely narrow assumption of self-interested behaviour that modern economists use as a short cut to human behaviour may have often led economists astray. Modern economists have narrowed Adams Smith's view of human beings, emphasizing both sympathy and prudence (including self-discipline), with its roots going back to the Stoics (Sen, 1987, pp. 22–23). Therefore, we find it reasonable to conclude that most (neo) classical economists look upon ethics as a cultural phenomenon outside the sphere of economics.

It is reasonable to characterize (neo) classical economics as a positive science, because ethics is imposed from society and is not an integrated part of economics.

Strategic management

Friedman argues in much the same way as Smith. For example, he claims that in free enterprise a corporate executive is an employee of the owners of the business, and hence, his/her prime purpose is to take care of the owners' interests or, as stated in his own words, "That responsibility is to conduct the business in accordance with their desires, which generally will be to make as much money as possible while conforming to the basic rules of the society, both those embodied in law and those embodied in ethical custom" (Friedman, 1963, p. 87).

Today there is increasing pressure to generate higher short-term sales and profits. This is justified by economists like Friedman who argue that "few trends could so thoroughly undermine the very foundations of our free society as the acceptance by corporate officials of social responsibility other than to make as much money for their stockholders as possible" (Friedman, 1963, p. 133). Central scholars in strategic management often state that investments and innovations drive economic growth, and thereby contribute to satisfying the demands of the consumers. Economic growth is a key objective and a well-functioning market is the major contributing factor in the realization of growth. Strategic management reflects firms (and managers') efforts to consciously improve in order to prosper and survive. Such strategic management reflects economic activities at the micro level, which of course can have macro level effects (Schelling, 1978).

Due to changing preferences in emphasizing the ethical dimensions of firms' products and services many scholars in strategic management claim that taking ethics into account is important in order to become competitive. In the years to come, ethics is predicted to become one of the most important factors for market success. It is also interesting to note that in many businesses the traditional anthropocentric perspective regarding ethics has been extended into an eco-centric perspective, which implies that product development should also satisfy ecological requirements in the future. Michael Porter, one of the most influential contributors to the literature on business strategy, also claims that there is a positive relationship between ethics and profitability, based on arguments such as, "companies operate in the real world of dynamic competition, not in the static world of much economic theory" (Porter and van der Linde, 1995, p. 120).

This indicates that social and economic goals are deeply connected in the real world, not inherently conflicting as indicated by Friedman. Porter and Kramer argue that firms can improve their long-term business prospects by introducing ethics as a competitive tool, or as claimed by these authors, "corporations can use their charitable efforts to improve their competitive context – the quality of the business environment in the location or locations where they operate" (Porter and Kramer, 2002, p. 58). Their point of view corresponds to what Ulrich and Maak characterize as "functional business ethics", which is a purely economic theory of morals (moral economics), and contend that morality can be "worthwhile in business" and may be "a useful idea" (Ulrich and Maak, 1998, p. 66).

From this line of thought we can conclude that strategic management is a position in which economics might be defined as a positive science, because ethics is looked upon as either imposed from society, or as a competitive tool.

Corporate social responsibility

Wood argues that "business and society are interwoven rather than distinct entities" (Wood, 1991, p. 695). She emphasizes "the interdependence of social institutions rather than their functional specialization, supporting the idea that firms should be socially responsible because they exist and operate in a shared environment" (Wood, 1991, p. 696).

Warrick and Cochran define "corporate social behaviour" as "a business organization's configuration of principles of social responsibility, processes of social responsiveness, and policies, programs, and observable outcomes as they relate to the firm's societal relationships" (Wartick and Cochran, 1985). This definition indicates that business organizations must know and take into account important aspects of social reality. In this respect, social values and norms have special importance.

Based on the assumption that social systems are integrated parts of society we find it reasonable to broaden the notion of corporate responsibility. This is in accordance with what Matten and Moon (2004) do in their research, in which they use the term corporate social responsibility (CSR) as an umbrella label that includes "business ethics", corporate citizenship, business and governance, corporate environmental or ecological management, as well as business and society relations. We find Buchholz's five key elements appropriate for defining corporate social behaviour, that is:

1. Corporations have responsibilities that go beyond the production of goods and services at a profit.
2. These responsibilities involve helping to solve important social problems, especially those they have helped create.
3. Corporations have a broader constituency than stockholders alone.
4. Corporations have impacts that go beyond simple marketplace transactions.
5. Corporations serve a wider range of human values than can be captured by a sole focus on economic values (Buchholz, 1991, p. 19).

These five elements imply that corporations are integrated parts of a broader societal community characterized by a diversity of values. Frederick (1986) strengthens this argumentation by saying that the essence of corporate social responsibility is that corporations, in addition to making a profit, have an obligation to work for social betterment (philanthropic responsibility).

Welford and Gouldson argue that it is of great importance to find ways of meeting the needs of both current and future generations. To reach this goal, it is not sufficient to develop new technologies and more efficient methods of production. According to Welford and Gouldson (1993, p. 1): "There is a need for a change in attitudes towards both production and consumption" . There is a need to improve the companies' environmental performance, but so far we have not experienced any radical shifts in business practice contributing to a lasting reversal of trends towards environmental destruction. Welford also argues that "it would be unacceptable to engage in any business activity where there was no thought of the consequences of the activity with regard to impacts on the world in which we live" (2000, p. 139).

Corporate social responsibility is a position where economics is looked upon as a moral science since ethics has immanent standing. But it is also possible to argue that some spokesmen defending this position consider ethics to be a competitive tool. Hence economics could be defined as a positive science.

Socio-economics

In *The Moral Dimension* Etzioni opens the way for moral theorizing in the paradigm of neoclassical economics. He states that social scientists, as well as intellectuals draw on overriding sets of assumptions or paradigms in order to organize and understand the world. Etzioni warns against the ethical limitations of the prevailing neoclassical paradigms because, as he argues, the study of our world will suffer and the efforts to administer its ills will be deficient (Etzioni, 1988, p. 1). In his contribution he also contrasts the neoclassical paradigm (which is utilitarian-based) of radical individualism with a more collective paradigm, which presupposes that "the economy is a subsystem of a more encompassing society, polity and culture" (Etzioni, 1988, p. 5).

This more comprehensive way of thinking is also in line with Zsolnai's (2002b) description of what he calls *The Moral Economic Man*, and his critique of the "homo oeconomicus" model both on empirical and normative grounds. He also contends that economic behaviour requires a more complex and dynamic framework which takes into consideration that economic behaviour is both multi-faceted and context dependent. According to Zsolnai (2002a, 2002b) contextual factors are factors such as the power of collective belief in ethical norms and the pro-social orientation of economic behaviour. Zsolnai's emphasis on social norms is also consistent with Hofstede's (1993) view of the concept of culture as a collective phenomenon.

Costanza argues that, "Because all science must be based on a pre-analytic vision that is inherently subjective and judged for utility and quality against criteria that are inherently subjective, there is no such thing as scientific objectivity" (Costanza, 1991, p. 27). Daly follows up this argumentation by claiming that it is "important to be very clear on the paramount importance of the moral issue" in economics (Daly and Townsend, 1993, p. 155). To illustrate this, Daly draws attention to the famous assertion by Leibnitz, "If geometry conflicted with our passions and interests as much as do ethics, we would contest it and violate it as much as we do ethics now, inspite of all the demonstrations of Euclid and Archimedes, which would be labelled paralogisms and dreams" (Daly and Townsend, 1993, p. 355).

Using this metaphor, Daly demonstrates that ethics is an integral and inseparable part of all science, including economics. Looked at from this perspective, ethics is both a necessary and sufficient condition of economic theory and practice. According to Sen, economic theory would have been even more productive if economists had paid greater and more explicit attention to the close contact between ethics and economics. In fact, modern economics is largely an "offshoot of ethics" (Sen, 1987, p. 2).

We argue that it is reasonable to assume a position in which the economy of nature and the ecology of man are inseparable. To separate them could be more than misleading, it could be destructive.

Since socio-economics defines ethics as immanent in economic theory and practice, it is obvious that, in this perspective, economics is a moral science.

We have characterized four perspectives on ethics and economy above. It is believed that the perspective held by teachers is related to their attitudes towards ethics, and thus also what they emphasize in their teaching. Based on the above discussion we can hypothesize as summarized in Figure 2.

A look at Figure 2 reveals that faculty members with attitudes corresponding to the theses in "(Neo-) Classical economics" will pay no (or

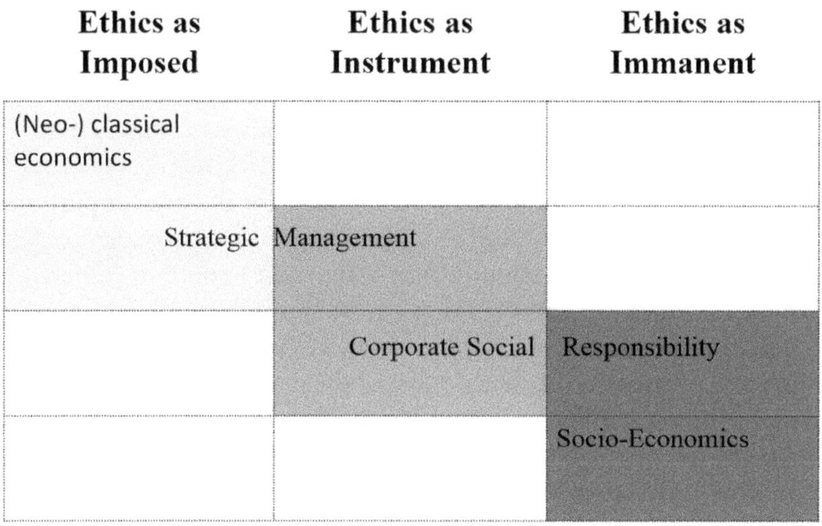

Figure 2. Economic paradigm and ethical attitudes.

modest) attention to ethics as a teaching subject in business schools. The main reason is that values and norms are looked upon as imposed from outside economics (Friedman, 1963; Smith, 1976a, b). This means that the economy is disconnected from the social world. The same arguments are believed to be important in "strategic management". In addition, for instrumental reasons (Porter and van der Linde, 1995; Porter and Kramer, 2002) the faculty members could take a positive view of compulsory courses in ethics. Faculty members with attitudes corresponding to the ideas in "Corporate social responsibility" are supposed to be positive to teaching ethics in business schools, either because economics is a moral science (ethics as an integrated part of economics), or because ethics is considered to be an important tool in the competitive market (instrumental reasons). Those, sympathizing with "socio-economics" are also supposed to have a positive attitude to teaching ethics as a compulsory part of the curriculum because ethics is an immanent part of economy. According to Whitehead (1967), the doctrine of immanence represents an important condition in the holistic world view.[1]

Research methodology

To empirically examine our assumptions or hypothesis as summarized in Figure 2 data is needed. Data based on a survey among teachers associated with business schools was considered adequate for the present purpose. A study was conducted during a conference, the FIBE (research conference in business administration); that is, the most important conference in business administration disciplines in Norway, which is held annually.

To capture various ethical dimensions, we developed a questionnaire, consisting of twenty statements. Most of the questions were designed to identify the main differences between the four theoretical perspectives. We asked questions to capture the respondents' attitudes towards economics as a positive or as a moral science. For example: (q2) "The market forces exempt the actors from ethical responsibilities"; (q17) "Questions regarding noms and values should not be included in economic science", and; (q5) "Market economy presupposes mutual trust between actors." Several questions were designed to distinguish between respondents with an instrumental view regarding ethics and respondents, defining ethics as a prerequisite for economic theory and practice. For example: (q11) "It is profitable to behave ethically"; (q4) "Ethical and environmental responsibility represent competitive advantage", and; (q9) "To behave ethically should be equally important for a manager as making money."

We ask questions to distinguish between the micro and macro level. For example: (q12) "All firms should contribute positively to the development of society (in addition to paying taxes)" and (q7) "It's the obligation of the government, not firms, to show ethical considerations." Finally, we also designed questions regarding the respondents' attitudes towards ethics as a part of the business schools' curriculum. For example: (q1) "Subjects like ethics should be included as part of the curriculum" and (q5) "Ethical issues should be discussed within all teaching areas."

For each statement the respondents were asked to mark the alternative "that best coincides with your opinion on a five-point Likert scale with end points "completely disagree" (=1) and "completely agree" (=5)".

The questionnaire was completed by 127 of the conference participants who had a research and teaching background in business administration and came from most of the undergraduate and graduate educational institutions in Norway. Even if the conference had open admission it did not attract individuals with a background in the humanities, such as history and language, nor law, or hard-core economics. The questionnaire was completed by the participants of the plenary session in which ethics (for the first time in the history of the conference) was one of the main topics. Nearly 60 percent of the respondents were aged between 35 and 55 years, and approximately 60 percent were men. An important question is whether the actual situation might have influenced the reported data. We would suggest that this is not the case. During the conference various views on the ethics and economy were presented. In addition, the completed questionnaires were anonymous. In the present case the empirical test relates to whether it is possible to identify patterns among the reported responses, coinciding with the perspectives on ethics emphasized above, and not in the traditional way by confronting the observed findings against null-hypotheses (that is, no effects on differences).

Findings

In this section we report the findings of the empirical study. The presentation of findings is organized in the following way. Firstly, we report the degree of expressed agreement with the various statements. Then we inspect covariations among statements. To examine whether patterns corresponding to the four perspectives could be identified we conducted exploratory factor analysis. Exploratory (and not confirmatory) factor analysis was chosen due to the explanatory nature of the research, in which the possibilities of more and/or mixed perspectives may have been held by some of the respondents. Finally, we examined attitudes towards ethics as part of the curriculum.

Opinions

Table 1 summarizes reported opinions concerning the various statements. Questions aimed at finding out attitudes towards ethics as part of the curriculum were only given to participants employed by educational institutions.

Table 1 is to be read as follows. Column 1 shows the percentage expressing agreement with the various statements. Agreement is interpreted as reporting either "4" or "5", also indicating that low reported percentage reflects the proportions disagreeing or neutral to the statements.

A look at column (1) in Table 1 reflects the following. First, ethical issues are perceived as important. For example, 79.5 percent agree (q1) "that subjects like ethics should be included as a part of the curriculum" and almost 52 percent claim that (q5) "ethical issues should be discussed within all teaching areas". In addition, 62 percent claim that (q10) "courses in ethics should be offered for teachers and researchers at business schools".

Table 1 also reflects the opinion that it may provide an advantage for firms to assume ethical responsibilities. For example, more than 54 percent claimed that (q4) "ethical and environmental responsibility represents competitive advantage". We also see that to behave ethically is perceived as important. Almost 65 percent also held the opinion that (q9) "to behave ethically should be equally important for a manager as making money".

The importance of ethics is also reflected in the finding that less than 4 percent claim that (q2) "market forces exempt the actors from ethical responsibility", and that less than 5 percent held the opinion that (q13) "to secure profitability and jobs, minor violations of the law can be accepted". In addition, only 7.1 percent (q19) "accept that a person shows less ethical considerations in her or his job than in private life". Support for this is also reflected in that only a few respondents, 11.3 percent, held the opinion that (q14) "when Norwegian firms do business abroad it is sufficient to act within local norms and laws". In addition, we see that only 10.2 percent held the opinion that (q16) "Norwegian firms cannot afford to take the lead on ethical responsibility". These findings also demonstrate that ethics may exceed laws and norms.

Table 1. Expressed agreement (percentage, mean, and standard deviation)

		Agree	Mean/S.D.
1.	Subjects like ethics should be included as part of the curriculum	79.5	4.17/1.03
2.	Market forces exempt the actors from ethical responsibilities	3.9	1.34/1.03
3.	The person who knows his teaching subject does not need education in ethics	9.5	1.76/1.03
4.	Ethical and environmental responsibility represent competitive advantage	54.3	3.55/0.99
5.	Ethical issues should be discussed within all teaching areas	51.9	3.4/1.26
6.	Persons obeying the law don't need education in ethics	3.9	1.5/0.89
7.	It's the obligation of the government, not firms, to show ethical considerations	5.5	1.29/0.64
8.	Violating ethical principles can be accepted as long as they are not in coflict with the law	14.9	1.61/0.86
9.	To behave ethically should be equally important for a manager as making money	64.6	3.8/1.23
10.	Courses in ethics should be offered for teachers and researchers at business schools	62.2	3.76/1.23
11.	It is profitable to behave ethically	45.6	3.43/1.05
12.	All firms should contribute positively to the development of society	65.4	3.80/1.14
13.	To secure profitability and jobs, minor violations of the law can be accepted	4.8	1.70/0.88
14.	When Norwegian firms do business abroad it is sufficient to act within local normas and laws	11.3	2.18/1.05
15.	Market economy assumes mutual trust between actors	63.0	3.68/1.24
16.	Norwegian firms cannot afford to take the lead on ethical responsibility	10.2	1.95/1.08

(*continued*)

Table 1. Continued

		Agree	Mean/S.D.
17.	Questions regarding norms and values should not be included in economic science	3.2	1.48/0.79
18.	Without mutual trust between actors, costs due to regulations and control will undermine	66.7	3.75/1.19
19.	I accept that a person shows less ethical consideration in her or his job than in private life	7.1	1.6/0.98

Our findings also demonstrate that mutual trust between actors is perceived as important. For example, 63 percent claim that (q15) "market economy assumes mutual trust between actors" and almost 67 percent agree that (q18) "without mutual trust between actors, cost due to regulations and control will undermine the market-based economy". An additional observation is that more than 65 percent claim that (q12) "all firms should contribute positively to the development of society (in addition to paying taxes)". This finding is surprising and reflects a norm about what a firm ought to do, as reflected in literature.

In column 2 we have also reported mean scores for the various statements, reflecting very much the same as the percentages in column 1. It should be noted, however, that great variations in opinions were observed among the respondents. This is reflected in the fact that, for every statement, all response alternatives were used. This is also reflected in the numerical values of the reported standard deviation, column (3) in Table 1.

Bivariate analysis

Below we focus on covariations between some of the statements. For this purpose, we apply correlation coefficients (r). As expected we find a positive covariation between emphasis on ethics and the opinion that courses in ethics should be offered ($r = 0.327$ $p < 0.001$). A positive correlation is also observed between the opinion that it is important to assume ethical responsibilities and that ethics as an educational subject is important ($r = 0.274, p < 0.002$), and that courses in ethics should be offered ($r = 0.327, p < 0.001$), and that it is profitable to behave ethically ($r = 0.327, p < 0.001$).

We also find that there is a positive correlation between the statement saying that market competition exempts the actors from ethical responsibilities and that ethical education is unnecessary ($r = 0.324, p < 0.001$) and agreement with the statement that, provided the law is obeyed, ethical education is unnecessary ($r = 0.329, p < 0.001$), and that it is the government and not the firms' obligation to be concerned about ethics ($r = 0.495, p < 0.001$), and that violating ethical principles can be accepted as long as they were legal ($r = 0.414, p < 0.001$). We also observe a positive correlation between the statements that it is sufficient to obey legal norms when operating abroad and that Norwegian firms cannot afford to take a lead regarding ethical issues ($r = 0.408, p < 0.001$).

These findings demonstrate that responses to the various statements are not necessarily independent. In other words, a respondent answer regarding one statement may influence her or his response to other statements. This also shows that various attitudes may be related, and that any one individual may hold a coherent set of attitudes.

Factor analysis

To examine whether the statements reflect a set of underlying dimensions, we conducted exploratory factor analysis. We did so due to the exploratory nature of this research; that is, we did not have a priori knowledge of how many perspectives might be applied among the respondents. This approach allowed us to identify dimensions not directly observable, and then determine which dimensions the various variables reflect. In our analysis we applied varimax rotation, that is the extracted factors are orthogonal. This was mainly done to ease the interpretation. (We also applied factor analysis with oblique rotation. This did not, however, add to our insight).

The factor analysis extracted six factors with eigenvalues equal to or greater than one (1). Table 2 reports a four factor solution. This solution was chosen because extracted factors with the lower eigenvalues did not yield specific meaning.

Table 2 is to be read as follows; h^2 are communalities; that is, the fraction of the variance extracted by the factor solution. For question 2, for example, we find that $h^2 = 0.58$; that is, 58 percent in this variable is extracted. When moving to column 2, we observe the number 0.76 for this variable. This is a factor loading; that is, the correlation between the variable and the underlying factors. On squaring the factor loading, we obtain $0.76^2 = 0.58$. Simple calculations show that this factor loading represents the total extracted variance for this variable $0.58/0.58 = 1.0$, that is, 100 percent.

In Table 2 we have only reported factor loading < 0.50 (with one exception at 0.42). This is done to ease interpretation. Moving to the second baseline we see that the total variance in the four extracted factors is 49 percent, of which the first factor counts for 24 percent. In the last line, extracted variance is reported in the various factors as a percentage of total extracted variance. Here we also see that the first factor represents almost half (49 percent) of the total variance extracted.

When reading down the column for the first factor, we observe that (q2) "market competition exempts actors from ethical responsibility" loads on this factor. Furthermore, we observe with regard to (q3) "persons who know their subject do not need ethical education", (q6) "persons obeying the law do not need education in ethics", (q8) "violating ethical principles can be accepted as long as they are not in conflict with the law", and (q16) "Norwegian firms cannot afford to take the lead on ethical responsibility", and that this factor reflects ethics being less important than other concerns. We label this factor "(Neo-) classical economics".

For the second factor we include the following questions: (q1) "subjects like ethics should be included as a part of the curriculum", (q5) "ethical issues should be discussed within all teaching areas", (q4) "ethical and environmental responsibility represent competitive advantage", (q9) "to behave ethically should be equally important for a manager as making money", (q10) "courses in ethics should be offered for teachers and researchers at business schools", and (q12) "all firms should contribute positively to the development of society (in addition to paying taxes)". This factor reflects that ethics is important, and is labelled "corporate social responsibility".

Table 2. Underlining dimensions

		H^2	**(Neo-) classical economics**	**Corporate social responsibility**	**Socio-economics**	**Strategic manage-ment**
1.	Subjects like ethics should be included as part of the curriculum	0.66		0.60		
2.	Market forces exempt the actors from ethical responsibilities	0.58	0.76			
3.	The person who knows his teaching sub-ject does not need education in ethics	0.44	0.51			
4.	Ethical and environmental responsibility represent competitive advantage	0.54		0.36		
5.	Ethical issues should be dis-cussed within all teaching areas	0.55		0.63		
6.	Persons obeying the law don't need edu-cation in ethics	0.62	0.62			

(*continued*)

Table 2. Continued

		H^2	(Neo-) classical economics	Corporate social responsibility	Socio-economics	Strategic manage-ment
7.	It's the obligation of the government, not firms, to show ethical considerations	0.62	0.76			
8.	Violating ethical principles can be accepted as long as they are not in coflict with the law	0.54	0.64			
9.	To behave ethically should be equally important for a manager as making money	0.42		0.48		
10.	Courses in ethics should be offered for teachers and researchers at business schools	0.67		0.79		
11.	It is profitable to behave ethically	0.64				0.56

Table 2. Continued

		H^2	(Neo-) classical economics	Corporate social responsibility	Socio-economics	Strategic manage-ment
12.	All firms should contribute positively to the development of society	0.46		0.58		
13.	To secure profitability and jobs, minor violations of the law can be accepted	0.42				0.52
14.	When Norwegian firms do business abroad it is sufficient to act within local normas and laws	0.53				
15.	Market economy assumes mutual trust between actors	0.64				
16.	Norwegian firms cannot afford to take the lead on ethical responsibility	0.59	0.5			

(continued)

Table 2. Continued

		H^2	(Neo-) classical economics	Corporate social responsibility	Socio-economics	Strategic manage-ment
17.	Questions regarding norms and values should not be included in economic science	0.49				0.57
18.	Without mutual trust between actors, costs due to regulations and control will undermine	0.57			0.74	
19.	I accept that a person shows less ethical consideration in her or his job than in private life	0.46				0.53
	Percentage common variance	49.0	24.0	9.8	7.7	7.4
	Part of common variance	100	49.0	19.8	15.8	15.2

For the third factor we use the following questions: (q15) "market economy assumes mutual trust between actors" and (q18) "without mutual trust between actors, costs due to regulations and control will undemine the market-based economy". We label this factor "socio-economics".

Furthermore, (q11) "it is profitable to behave ethically", (q13) "to secure profitability and jobs, minor violations of the law can be accepted", (q17) "questions regarding norms and values should not be included in economic science", and (q19) "I accept that a person shows less ethical considerations in her or his job than in private life", load on the fourth factor. This factor is somewhat confusing, perhaps due to interpretation difficulties. We label this factor "strategic management".

Discussion

The reported findings deserve some further comments. We start our discussion by an effort to contextualize our sample in relation to important contributions about differences in business ethics in North America and Europe. Enderle maintains that there does not exist a well-established paradigm of business ethics on either side of the Atlantic. However, he states that "German-speaking and Scandinavian countries are likely to be more concerned about taking seriously the contributions of the social sciences" (Enderle, 1996, p. 42). This means that the technical and scientifically-inspired natural science regime might be enriched by a "more humanity driven approach which includes philosophy, theology, literature, the classics, and sometimes history" (Enderle, 1996, p. 43). He also identifies significantly different approaches to business ethics on each side of the Atlantic.

One of the main differences is the stronger emphasis on collective actions in Europe, by using considerable ethical energy in order to develop laws and institutions to regulate the market. This de-emphasizing of the individual actor has previously been mentioned by van Luijk (1990). Even if there arc some deep resemblances across the Atlantic ocean in that European business ethics has a derived identity, profiting from what was developed fifteen years earlier by pioneers in the field of business ethics in the US (van Luijk, 1990), there are obvious differences. We agree with van Luijk that sometimes "differences between single European countries are as large as those between a specific European country and the United States" (van Luijk, 1990, p. 541). In particular, the *political tradition of experience with social democracy* and mixed capitalism with powerful trade unions have made the striving for consensus a major force in Norwegian society. We believe this experience has created its decisive "corporate social responsibility" stamp on the moral way of thinking about business in Norway.

In comparing our findings with those from Hosmer's (1999) study, we will more precisely contextualize our Norwegian sample in relationship to other countries, and particularly in relation to the USA. Then we will discuss our findings in relation to our theory-driven frame of reference.

Although our respondents expressed different attitudes towards ethics, we can conclude that most teachers and researchers in Norwegian business schools acknowledge that their responsibility is not limited to "preparing graduates to be astute managers, analytical financial executives, or creative manipulators of large amounts of money and data" (Sims, 2002, p. 395). Our generally positive findings stand in remarkable contrast to the negative findings in Hosmer's (1999) investigation of the position and respect of business ethics and/or social responsibility in business schools. Hosmer reports that varying emphasis was given to ethics across different types and sizes of institution. According to Hosmer, liberal arts colleges and private religious universities respect business ethics more than public and private non-sectarian universities.

In our sample, predominantly public universities were represented. Hosmer explains variations in respect of ethics referring to the *size of the institutions* as the most important factor. He found that large institutions (10,000–29,000 students) and, in particular, very large institutions (over 30,000 students) reported relatively lower respect for business ethics than smaller ones. In Norway most business schools are medium-sized or small. Moreover, Hosmer found higher respect for ethics among deans and administrative personnel than among faculty members in general. This finding fits in well with our own experience. Hosmer's study was initiated by the negative attitudes towards business ethics he had experienced himself.

Although the spirit of our study was unlike Hosmer's, it is easy to recognize some faculty members in business schools holding some of the same negative feelings towards business ethics. As the economist Robert Frank once said, scholars in ethics at business schools have to pay a premium by doing ethics and not business.[2]

Our surprisingly positive results compared to Hosmer's (1999) can, perhaps, be partly explained by an "after Enron effect"; that is, the collapse of big companies has exposed the instrumental value of ethical behaviour in large corporations and the market. The size effect that Hosmer identified is indeed very interesting from an ethical point of view, and in accordance with the expectations of Schumacher (1999), expressed in his book *Small is Beautiful.* Bauman (1989) also argues that large corporations tend to produce moral indifference because organizational size increases the physical and/or psychological distance between the act and its consequences, resulting in the suspension of moral inhibitions as well as obscure moral dilemmas. Smaller organizations offer more interaction between members, which we believe contributes to reducing stereotypical attitudes towards ethicists.

At the outset we introduced four perspectives related to ethics. In our discussion we also indicated that knowledge and beliefs are reflected in individuals' attitudes. This is also the case here as knowledge and beliefs held by faculty members influence their attitudes towards ethics. Going back to our findings, the following was found.

First, we observed a group (Figure 2) that did not think that ethics deserved a place in the business school's curriculum. Intuitively, this category has much in common with the theoretical perspectives "neoclassical economics" and "strategic management". These perspectives reflect the main concern of economists that business should limit itself to profit making and that the government should take care of the societal aspects of business. Friedman articulated this in his saying that "there is one and only one social responsibility of business – to use its resources and engage in activities designed to increase its profits ..." (Friedman, 1970). It is reasonable to assume that respondents subscribing in principle to his perspective would be negative about ethics as a part of the curriculum in business schools.

It is important to note that this does not mean that they are necessarily negative to ethics as an (imposed) precondition to business practice. Both Smith (1976a, b) and Friedman (1970) argue that values and norms anchored in society should be respected by the actors in the market. However the leaders, according to Friedman, should not focus on ethics because they lack the necessary ethical competence and the necessary information. Instead, the leaders should make every effort to do what is within the scope of their authority – to maximize profits for the shareholders. We conclude by referring to Friedman's saying that the only responsibility for business in addition to making profit is to obey the law. This reflects the attitude that profit and law are superior to ethical values and norms.

Second, the group of faculty members (Figure 2) characterized by attitudes corresponding to some of the main ideas in "strategic management" and "corporate social responsibility" indicate that business ethics is supposed to result in stronger competitive power. As a consequence they are positive to introducing ethics as a compulsory course in business schools. Supporters of this pragmatic perspective are positive towards ethics for instrumental reasons. They argue that ethics is an important issue because corporations will gain competitive advantage by including ethics in their competitive tool bag.

The key point for this group is that knowledge and competence in the field of ethics is regarded as an important instrument for profit making in the future. Porter, a prominent advocate of this position, argues that economic goals harmonize with ethical responsibility, exemplified. by firms making profit by acting ethically in the market (Porter and van der Linde, 1995; Porter and Kramer, 2002).

Third, our data also indicates that a group of faculty members (Figure 2) agree with the statement that ethics should be part of the business schools' curriculum for immanent reasons. They subscribe to one of the theoretical perspectives, "corporate social responsibility" or "socio-economics". In sum, our empirical findings correspond to the main points in these theoretical positions.

Faculty members in these categories hold a positive view to introducing ethics as a part of the curriculum for reasons independent of utility evaluations. Subjects subscribing to "socio-economics" also stress the importance of trust and mutual responsibility, in addition to ethical values and norms, as necessary conditions for the market economy. This also implies that they tend to conceive economics as a moral science (Costanza, 1991; Daly and Townsend, 1993).

In comparing our findings with those reported by Matten and Moon (2004), clear differences emerge. Our study reveals that 97 percent of the faculty members in our study want ethics as part of the curriculum. Matten and Moon (2004), in contrast, reported that less than 30 percent of European schools had in fact implemented ethics education (SCR) in compulsory modules. There can be multiple factors explaining this apparent gap. First, emphasis on ethics may, of course, vary across nations and institutions. Probably more important is what is really captured in the two studies. Our study addresses interest and opinions, while Matten and Moon's (2004) study addresses behaviour as implemented by educational institutions. It is well known that there is a long way to go from expressing opinions and interests to implementing programs. In addition, our study was conducted among a self-selected sample of faculty members in which economists were not included – due to the scope of the conference during which the data was gathered. It may well be the case that faculty members with an economic academic background may be less in favor of ethics as part of the curriculum.

In sum, we conclude that knowledge concerning the faculty members' disciplinary background is important in order to explain their attitudes toward ethics as a compulsory part of business schools' curriculum. Our study indicates that it is necessary to make a distinction between ethics as immanent, imposed, or as an instrument for reaching economic goals in order to understand the faculty members' underlying motives concerning teaching ethics.

It is not unreasonable to believe that the majority of faculty members are positive for instrumental reasons, and that only a small minority hold the opinion that ethics is of inherent value and consider ethics to be an immanent part of economics.

Further research in the area is needed. An important piece of future work is to investigate whether differences in motives among faculty members influence the way ethics are taught in business schools. We expect to find differences in the degree of involvement, as well as topics, depending on faculty members' attitudes towards the relationship between ethics and economics. Depending on the positions held by faculty members we think it is reasonable to suppose that different courses in ethics can be characterized as either moral "bildung" or ethical instrumentalism. In order to gain support for introducing ethics for inherent reasons, we presuppose that faculty members' attitudes towards ethics involve a paradigmatic change from looking at economics as a positive science to a moral science. Finally, we would like to offer a comment on Friedman's position. If business schools take their responsibility to educate executives through compulsory ethics courses more seriously, business leaders may obtain better ethical competence. In which case one of Friedman's arguments for leaders to distance ethics from economics and push ethics aside when making decisions loses its footing.

Acknowledgements

The authors gratefully acknowledge the very helpful comments and suggestions by the editor, Professor A.C. Michalos and the anonymous advisors. We also want to thank Geoff Moore for motivating us to write the first draft of this manuscript and Peter Glanfield for his friendly proofreading of the manuscript.

Notes

1. The concept of imposed and immanent are inspired by Whitehead's use (1967) where he emphasizes that the doctrine of imposition follows from the dualistic worldview. In our context this means that economy is disconnected from the social world.
2. Citation from Robert Frank's lecture on Business and Economics at the Center for Ethics and Economics at the Norwegian School of Economics and Business Administration autumn 2003.

Bibliography

Bauman, Z.: 1989, *Modernity and the Holocaust* (Polity Press, Cambridge).

Buchholz, R. A.: 1991, July-August, "Corporate Responsibility and the Good Society: From Economics to Ecology", *Business Horizons* 34(4), 19–31.

Costanza, R.: 1991, *Ecological Economics: The Science and Management of Sustainability* (Columbia University Press, New York).

Cowton, C. J. and J. Cummins: 2003, "Teaching Business Ethics in UK Higher Education: Progress and Prospects", *Teaching Business Ethics* 7, 37–54.

Daly, H. E. and K. N. Townsend: 1993, *Valuing the Earth, Economics, Ecology and Ethics* (The MIT Press, London).

Enderle, G.: 1996, "A Comparison of Business Ethics in North America and Continental Europe", *Business Ethics. A European Review* 5(1), 33–46 January.

Enderle, G.: 1997, "A Worldwide Survey of Business Ethics in the l990s", *Journal of Business Ethics* 16, 1475–1483.

Etzioni, A.: 1988, *The Moral Dimension. Toward a New Economics* (The Free Press, New York).

Fishbein, M. and I. Ajen: 1975, "Attitude and Prediction of Behavior", in M. Fishbein (ed.), *Readings in Attitude Theory and Measurement* (Wiley, New York), pp. 11–12.

Frederick, W. C.: 1986, *Theories of Corporate Social Performance: Much Done, More to Do* (University of Pittsburg Press, Pittsburg).

Friedman, M.: 1963, *Capitalism and Freedom.* Phoenix Books (University of Chicago Press, Chicago).

Friedman, M.: 1970, "The Social Responsibility of Business Is to Increase Its Profits", *New York Times Magazine,* September 13.

Hofstede, G.: 1993, *Kulturer og organisasjoner* (Bedrift, okonomcns Forlag, Oslo).

Hosmer, L. T.: 1999, "Somebody Out There Doesn't Like Us: A Study of the Position and Respect of Business Ethics at Schools of Business Administration", *Journal of Business Ethics* 22, 91–106.

Matten, D. and J. Moon: 2004, "Corporate Social Responsibility Education in Europe", *Journal of Business Ethics* 54, 323–337.

Porter, M. E. and C. van der Linde: 1995, "Green and Competitive: Ending The Stalemate", *Harvard Business Review,* 73 (5) September/October, pp. 120–134.

Porter, M. E. and M. R. Kramer: 2002, "The Competitive Advantage of Corporate Philanthropy", *Harvard Business Review,* 77 (6) December, pp. 2–16.

Schelling, T. C.: 1978, *Micromotives and Macrobehaviour* (Norton and Company, New York).

Schumacher, E. F.: 1999, *Small Is Beautiful. Economics as If People Mattered. 25 Years Later …* with Commentaries (Hartley & Marks, Point Roberts, Wach.).

Schwartz, R. H., S. Kassem and D. Ludwig: 1991, "The Role of Business Schools in Managing the Incongruence Between Doing What Is Right and Doing What It Takes to Get Ahead", *Journal of Business Ethics* 10, 465–469.

Sen, A.: 1987, *On Ethics and Economics* (Basil Blackwell, New York).

Sims, R. R.: 2002, "Business Ethics Teaching for Effective Learning", *Teaching Business Ethics* 6, 393–410.

Skirbekk, G. and N. Gilje: 2001, *A History of Western Thought* (Routledge, London).

Smith, A.: 1976a, "The Theory of Moral Sentiments", in D. D. Raphael and A. L. Macfie (eds), (Oxford University Press, Liberty Classics, Indianapolis).

Smith, A.: 1976b, *An Inquiry into the Nature and Causes of The Wealth of Nations*, in E. Cannan (ed.), *Two Volumes in One* (1st edn, 1776, the citations are from the 5th edn) (The University of Chicago Press, Chicago).

Ulrich, P. and T. Maak: 1998. "Integrative Business Ethics – A Critical Approach in St Gallen", in L. Zsolnai (ed.), *The European Difference. Business Ethics in the Community of European Management Schools* (Kluwer Academic Publishers, London).

Van Luijk, H. J. L.: 1990, "Recent Developments in European Business Ethics", *Journal of Business Ethics*, 537–544.

Warrick, S. L. and P. L. Cochran: 1985, "The Evolution of the Corporate Social Performance Model", *Academy of Management Review*, 10(4), 758–769.

Weber, M.: 1930, *The Protestant Ethic and the Spirit of Capitalism* (HarperCollins Academic, London).

Welford, R.: 2000, *Corporate Environmental Management – Towards Sustainable Development* (Earthscan, London).

Welford, R. and A. Gouldson: 1993, *Environmental Strategy and Business Strategy* (Pitman Publishing, London).

Whitehead, A. N.: 1967, *Adventures of Ideas* (Macmillan, New York).

Wood, D. J.: 1991, "Corporate Social Performance Revisited", *Academy of Management Review*, 16(4), 691–718.

Zsolnai, L.: 1998, *The European Difference. Business Ethics in The Community of European Management Schools*, L. Zsolnai (ed.), (Kluwer Academic Publishers, London).

Zsolnai, L.: 2002a, "New Agenda for Business Ethics", in L. Zsolnai (ed.), *Ethics in the Economy – Handbook of Business Ethics* (Peter Lang, Oxford).

Zsolnai, L.: 2002b, "The Moral Economic Man", in L. Zsolnai (ed.), *Ethics in the Economy – Handbook of Business Ethics* (Peter Lang, Oxford).

KNUT J. IMS

"Take it Personally"

They (people) do not realize that anonymity,
as the most absolute expression for the impersonal,
the irresponsible, the unrepentant, is a fundamental
source of the modern demoralization.

– Soren Kierkegaard[1]

Three women took serious professional and personal risks when exposing corrupt organizations in 2002. The giant organizations Enron, WorldCom and the FBI had exploited public trust and the women employed by these organizations. All decided to blow the whistle on their employers, even though they were well aware of the harmful consequences for their leaders, their colleagues, and themselves. In spite of expected personal costs, they "did the unthinkable" in the service of the common good.[2]

Since whistleblowing behaviour is generally known as supererogatory behaviour, it is associated with saints. However, these women did not look upon themselves as saints. Afterwards, they stated they only did what they had to do. The kind of burden the women took upon themselves is termed *deep personal responsibility.* This term emphasizes that *character* – and consequently attitudes, feelings, thoughts and actions – is decisively important.

Personal responsibility might be seen as the opposite of collective responsibility, in which the subjects involved prefer to look upon themselves as small replaceable cogs in a big machine-like system. As a cog, it is opportunistic to blame the system for its own unethical behaviour. To blame their own organization, their leaders, or the system for their own wrongdoing is *responsibility-taking in a shallow sense.* In this essay Adolf Eichmann is presented as the paradigmatic example of a person who tried to avoid personal responsibility by referring to obedience to German law during the Nazi era. Once upon a time obedience was seen as a moral virtue, and obedience may still have relevance for small children and close-minded fanatics. However, in the modern world obedience should be replaced by responsibility.[3] As Hannah Arendt put it: It is quite different to ask why someone supports than why someone obeys (2003, p. 47). To support an activity under the name of obedience is merely buck passing.

Following Arendt's reasoning on the banality of evil: Doing evil without acting out of strong feelings such as hate or anger, but by looking upon oneself as just a little cog, is due to *thoughtlessness, and lack of judgment.* To be able to judge in a way that takes due regard of other beings and a moral community is another key concept in personal responsibility. It is assumed that any moral community has some standard norms and rules that may be called "*common morality*". Common morality is assumed to be compatible with such fundamental moral norms as justice, honesty, and compassion, and to ultimately serve man, community, and nature, as well as future generations. In this sense, personal responsibility has parallels to deep ecological thinking. Taking personal responsibility means seeing oneself as an organic part of a community in which one is engaged in an ongoing moral dialogue (Niebuhr, 1999). This moral dialogue expresses that we as individuals are connected with one another in the same existential communion, so to speak. Such a deep responsibility may also be related to the *concept of oneness.*

A component of great practical importance is *role-mediated behaviour in professional responsibility.* An individual's professional role, be it as manager, economist, or employee in a company, gives the role-taker a collection of rights and duties. However, as a general rule *it does not give the occupant of the role any institutional excuse to trump common morality.* In order to reduce pain or injury, or to save life, a surgeon can use a knife in another's body, since the surgeon as a professional is legitimated and justified in so doing by the Hippocratic Oath. On the other hand, the practice of some lawyers to withhold truth in order to zealously advocate their clients' interests is problematic, because truth telling is a vital norm within any community. As one of the advisors of President Truman,

Albert Carr (1968), wrote that business behaviour might legitimately be compared with playing poker. Supposing that to be the case, the business community has a serious problem. What will be the consequences if businessmen are permitted to bluff and lie? According to Carr, bluffing and lying is necessary because if businessmen do not, they will lose the game (unethical behaviour legitimated from "any hand's" perspective) because they will not make a profit. But it follows from the stated principle above that professional behaviour that violates common morality is not justifiable. The essential principle is that common morality trumps professional morality when different norms conflict.

Analytically, we may put the three concepts of personal responsibility, professional responsibility as role-mediated behaviour, and common morality in a framework as exhibited in Figure 1.

As depicted in Figure 1, personal responsibility is the groundwall within all kinds of role-mediated behaviour. It has a constructive and critical role to play in the relationship with common morality, to secure a continuous and critical dialogue in the community regarding moral behaviour. Thereby, common morality is adjustable over time.[4] On the other hand, personal responsibility also means continuous reflection on the limits of role-mediating behaviour. The latter involves a critical reflection upon the limits of one's professional role in relationship to common morality, arresting all possible transgressions by the question: Are my actions legitimate from the point of view of common morality? Interactive relationships among all three concepts are assumed.

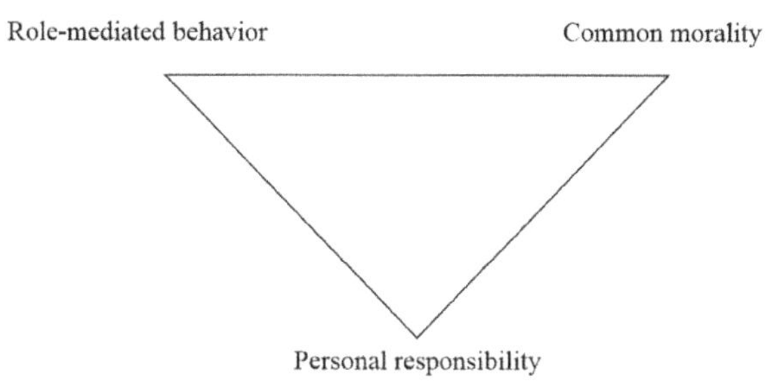

Figure 1. The responsibility triangle.

The crux of this conception of personal responsibility is that the person's character is unique and basic and, within limits, s/he will put her personal footprints on his/her professional roles as well as on his/her ongoing moral dialogues in the community.

The individual's *personality* is important for different reasons. First, much of the writing about responsibility focuses on the outcome and the product of behaviour. However, to be personally responsible is a process, not a fixed state.[5] Personal responsibility emphasizes *processes and subprocesses* as a precondition for responsible outcomes or products, as well as an inclination to go to the roots of important problems and *involving one's self-worth, identity, and dignity in the professional work.* It concerns the "Me", the authentic man or woman behind the social mask and acknowledges that I am not an isolated atom. It means having deep commitments in the processes, and being committed means using all the human senses. This implies not only using the cognitive competencies, but also feelings and experiences in order to find wise solutions. It also implies avoiding both simplistic theories and exclusionary objectivity in order to reach a judgment (Ghoshal, 2005).

To be committed also means that one typically acts out of high *personal ideals*[6] or *ground projects* that give one's life meaning. These "justified" values go beyond narrow self-interests and involve the well-being of an extended self or a community. Values inspire and root our lives. They give us something we hold precious as humans. We should not betray those values. If we tamper with our integrity, we risk destroying the virtue of self-respect. To be a committed professional also means to *advance excellence.* This is the craft motive in professional ethics (Martin, 2000). In our image of the good professional, the craft motive, together with intrinsic and extrinsic motives, are the motives that propel a man. These motives are often mixed. However, there is a serious risk with strong commitment. One may be too sensitive to the problems of others and behave paternalistically in relationships that should be symmetric. By being over-responsible, one violates the other's dignity by making the other under-responsible (Løgstrup, 1956). The other extreme consists of over-distancing, of being too detached – a behaviour typically seen when leaders operate as pure instruments for others ("hired guns") without empathy and personal involvement with those affected. Professional distancing signifies a golden mean between these two extremes and involves an *appropriate distance* – living constantly within the tension of detachment and sensitivity of affected persons (Martin, 2000).

Appropriate distance is also important when somebody without a legitimate reason blames one for something. In such a case, one should not take the criticism personally but interpret it as an expression of something underlying the blamer's situation as a whole or his character. A professional should be trained to know what legitimate blaming is and what is not.

True responsible action starts with the intra-personal processes rooted in an extended self, which encompass one's role in the community and in the organization. Thus, the solution to the most complicated issues and dilemmas are not to be found in professional roles or organizational structures, but typically in the *character and attitudes* of the subject.[7]

Summing up, *Why be personal?* Because it may counteract use of moral disengagement strategies, practices that form a slippery slope to corporate transgressions. There are a lot of mechanisms of moral disengagement that enable "eminent members of the business community" who are not "dangerous, criminally oriented mavericks" to commit unethical acts without experiencing moral guilt. Bandura et al., (2002) denote some of the strategies as euphemistic labeling; advantage comparison (comparing one's own behaviour with still more evil acts); displacement, or diffusion of responsibility;[8] disregarding or distorting the harmful consequences; dehumanization of other people, and attribution of blame onto others or on circumstances. Eichmann unquestionably used many of these strategies. The second reason for taking it personally concerns the life quality of the person and whatever social or professional role s/he is tied up with. It means to counteract the validity of one-sided belief in scientific, technical, and economic knowledge that has given rise to many social and psychological maladies (Taylor, 1991; Sheffler, 1992). It means to recognize the moral value of friendship and close personal relations, of belonging to a community, and religious and cultural traditions that may enrich "our lives in the way as earlier communities and traditions enriched the lives of earlier generations" (Scheffler, 1992, p. 15).

The approach taken is that we should not devalue the important human capabilities of sensitivity, perception, imagination, and the role of individual judgment. Rational decision making procedures can never duplicate or replace the capacity of individual personality. Human feelings and emotions can never be reduced to a skilled and mechanical use of moral theories (Bauman, 1993). And since traditional bearers of moral authority, such as churches and schools, have lost much of their power, a democratization of morality has resulted. Thus, the character of employees and managers becomes essential. It moves the heart of corporate citizenship to the individual manager.[9] In addition, modern organizations adapting to the changing conditions in the global market need a transformation from mechanistic to more organic forms. It requires involved workers with personal commitments to the organizational goals, and a close identification of the whole person with work (Burns and Stalker, 1991). Mechanistic organizations of the past preferred distanced workers that followed the rules, but the shift to organic forms demands much more of each individual. Organic organizations need persons who have the courage to initiate actions, and who use themselves as whole beings *building* stakeholder relationships (Lozano, 2005).

Historical and contemporary examples

Adolph Eichmann, a German Nazi, escaped from Germany after the war. However, fifteen years later he was kidnapped and *held accountable for* his actions during the war. He was charged with genocide of Jews, of being directly involved in the Holocaust. However, to every charge offered Eichmann replied: Innocent *(unschuldig)*. He said he had never killed a person. He had only obeyed the law – the German law made in the spirit of Adolph Hitler. Is this a case of *institutional excuse* and permissible violation of persons (Jews) due to dutiful performance of a bureaucratic role in a nihilistic organizational context? Does Eichmann's reference to "my station and its duties" [10] as well as the pervasiveness of the Nazi organization lessen each Nazi individual's personal responsibility?

The trial was held in Jerusalem in 1960, and Eichmann was sentenced to death.[11] The trial rested upon the Nuremberg trial in 1945–1946, which concluded that every adult has a personal responsibility not to harm anybody and cannot claim protection behind any law for wrongdoing.

What kind of *character* had the man who was held personally responsible for the death of six million Jews – a project the Nazis themselves euphemistically termed "Endlosung" (the Final Solution)? Was he a living incarnation of the Devil, an evil tyrant? Were his acts deliberately evil? Or was he, as he himself insisted, only a bureaucrat who performed "paperwork" – a cog in a huge machine like organization, and a useful nonentity? Could Eichmann have done otherwise? Did he, in practical terms, have a free choice? If he had no free choice, we cannot ascribe responsibility to him in moral terms. The trial was well-covered by journalists and experts from all over the world. What was the sanity quotient of this evil man? Several psychiatrists evaluated Eichmann and found him as moral as anybody else, and one of the psychiatrists added: "Anyway more normal than myself" (Arendt 2000, p .28).

Arendt (1980, 2000) concludes that the lesson to be learned from the Jerusalem trial is that actions with the most evil consequences (genocide) can be performed by persons who are not motivated by strong affections, but instead by such "banal" motives as doing a good job and being loyal to a given purpose. Eichmann was more "normal" than we like to believe. Most people could have done this, if circumstances had warranted it. Eichmann's evilness was not a break with the norms of society, but rather the opposite. The wickedness of Eichmann was a socially and legally demanded evil – an evil that reflected a community that was on a morally deranged track. It was not the result of a deviant individual.

In 2002, three American women were selected as "person" of the year by *Time* magazine.[12] They were all seen as courageous persons who exposed huge business scandals in the USA. What kind of persons were these women? Why did they blow the whistle and effectively put an end to their leaders' wrongdoing at the risk of losing their jobs and suffering other personal losses? It is a well-known fact that whistleblowing typically has the destructive side effect of retaliation. So these women expected retaliation in spite of doing what was right in a community perspective.

What are the main similarities and differences between Eichmann, a man who defended himself as "innocent" because he only did his job, and these women who violated a loyalty to colleagues and even opposed the norms of hierarchy?

In the court in Jerusalem, Eichmann insisted that all he had done was to obey the law. He only did his duty and followed the orders of his "Fuhrer" (leader), and by doing that he also obeyed the law. "My honor is to do my duty! It is wrong to accuse me for murder [homicide]. I have never had anything with murdering to do – I have never killed a Jew, and I have never killed a non-Jew." Later in the court, one of his reflections was that "it turned out that I by chance was not ordered to commit a murder a single time." But Eichmann did not leave any doubt that he would have killed his own father if he had been ordered to do so (Arendt, 2000, p. 24).

As a law-abiding citizen and professional bureaucrat, Eichmann performed the Fuhrer's (Hitler's) order as best he could. Eichmann stated that he would have had a guilty conscience if he had not done what he had been ordered to do. But he maintained that he did not have any negative ill will toward his victims. He did not hate the Jews – he had private reasons for not hating them. The private reasons were, among others, the existence of Jews in his own family (Arendt, 2000, p. 32).

Looking upon himself as a perfect bureaucrat, he admitted he had made two exceptions to the law, which made him uneasy because he regarded it as inconsistency. Eichmann was proud of having acted against his own human inclinations, in the manner of either sentimentality or self-interest, by always doing his duty.

Germany's capitulation on May 8, 1945 had a special significance for Eichmann. He felt that from that time on he had to live without a leader – without being member of a community (even if the Nazi community was a perverted one), without receiving directives, orders, commands, or guidelines. Suddenly he was confronted with a completely unknown life – the fear of living *as an individual*!

In 2000, Enron the energy giant was the seventh largest company in America. But in autumn 2001 the technology bubble was bursting and Enron's vice president, Sherron Watkins, warned the chairman that the accounting methods used were not justifiable. Watkins revealed that Enron had concealed billions of dollars of debt in dubious partnerships. Watkins' act may be described as whistleblowing, sounding an alarm from within the very organization in which she worked, aiming to spotlight neglect or abuse that threatened the public interest (Bok, 1988). Watkins was fully aware that her voicing was likely to create retaliation and that her career and her ability to support her family could be "unjustly impaired". As a result Watkins "cloaked herself in her family and church. Her faith was sharpened. She was disappointed with human nature and the power of greed and denial" (*Time* December 30, 2002/January 6, 2003).

In June 2002 at WorldCom, for a time the twenty-fifth biggest company in America, Cynthia Cooper, the head of WorldCom's internal auditing department, told the audit committee of the board that the company had rotten accounting practices. She knew very well what would happen, and within a week the famed chief financial officer (CFO) was fired. The largest example of accounting fraud in history became public knowledge. WorldCom had inflated its profit earnings by $3.8 billion (US). Shareholders lost $3 billion (US), and 17,000 employees lost their jobs.

What was the driving force behind Cooper's whistleblowing? After a hint from a colleague about creative accounting, Cooper went to the external auditor, the prestigious Arthur Andersen company, and was told: *It is not a problem.* Next, Cooper confronted World Com's CFO Scott Sullivan, who in 1998 enjoyed the fame of being one of the best CFO's in the US. At the age of 37 he earned almost $20 million (US) a year, enjoyed great prestige. and was considered the "gold standard"; he was an icon in himself. When Cooper confronted Sullivan with the example of creative accounting, Sullivan told Cooper that "*everything was fine and she should back off. He was furious at her …" (Time* December 30, 2002/January 6, 2003).

But Cooper did not back off. Known as a perfectionist who loved her job, she continued her investigation by night and directed her team members to re-do parts of Arthur Anderson's job. They worked late into the night to keep their investigation secret.

Coleen Rowley, employed by the FBI as a staff attorney, caused a sensation when she wrote a memo to FBI Director Robert Mueller about how the FBI ignored pleas from her office that one person be investigated. Later chis man was indicted as a September 11 co-conspirator. Rowley also took important initiatives to inform Congress of the FBI's dysfunctional internal organization in light of September 11. As a result, Rowley was looked upon as a leper inside the FBI. She did the "unthinkable", and she was condemned directly by many of her colleagues. They compared her with the newly convicted super-spy Robert Hanssen, who betrayed everything "we [the FBI] stand for". According to her FBI colleagues, Rowley's crime was the same, except that she went to Congress instead of the Russians. Her whistleblowing was against the common culture in the FBI, where the appeal to absolute loyalty was a top priority. In the words of an indicator on the wall of the FBI: "Stand by the institution you represent. *An ounce of loyalty is worth a pound of cleverness*" (*Time* 2003, p. 43, my emphasis).

What were the main similarities and differences between Eichmann and these three whistleblowers in terms of responsibility? Was Eichmann the responsible person (albeit with nihilistic values) he claimed to be? Eichmann was implementing his leaders' orders and used much consideration in selecting the best combination of railway wagons placed at different points to minimize the number of kilometers in bringing Jews to the concentration camps.

Did Eichmann, in a sense, act as an efficient manager by always economizing with scarce resources? There is decisive evidence that the economic interests were subordinated to the policy of extermination (Arendt, 1980). Eichmann's ground project was the "Final Solution", not efficiency. How was Eichmann able to live with and legitimate such a nihilistic project? Was the social and moral acceptance of this ground project among the Nazis so basic that it was unquestionable? In formal documents the Nazis consciously treated the Final Solution with as much secrecy as possible (Lang, 1990).[13]

Based upon my readings about Eichmann and the three women, they differ on several important dimensions, including the objects of responsibility, feelings of responsibility, ground projects, basic values, symbolic universal beliefs about human dignity, the social environment, versions of the categorical imperative, honor, and the self-concept (See Table 1). In the following, I will try to support these assumptions with some evidence.

Eichmann was only seemingly responsible to the Nazi organization, since there is strong evidence that Eichmann only felt unconditional responsibility to Hitler and not to the Nazi organization as a whole.

Table 1. Eichmann versus the whistleblowers

	Eichmann	**The whistleblowers**
Object of responsibility	One person	Many stakeholders
Feeling of responsibility	"Shallow" organizational	"Deep" personal
Ground project	"Endlosung"	To be a good citizen
Basic values	Nihilistic	Caring
Symbolic universe	Very limited, mainly professional	Rich, varied over different domains of life
Belief about human dignity	Certain people are dehumanized (beasts)	The other's dignity is confirmed
Social environment	Limited and ruled by "slavish discipline"	Several communities beyond work-organization (family, church)
Categorical imperative	Act as if Hitler is a spectator of your action	Do for others as you want others to do for you
Honor	Obeying the law	To live up to the expectations of the group/community
Self	Cliché-based individual self with a narrow, shallow, and perverted moral horizon	Dialogue-based extend Self

In contrast, the three female whistleblowers – Coleen, Cooper, and Rowley – were performing their organizational roles in huge modern corporations. But they placed serving the public interest (multiple stakeholders) above serving the interests of their organizational leaders and colleagues. They obeyed moral principles grounded in a large human community rather than individual persons.

"The Final Solution" as ground project

Was Eichmann a thoughtless, blind, uninspired bureaucrat who lacked ethical reflection, or was he conscious of his evildoing? Was his self-defense in 1961 based upon the argument of his "station and its duties", or only a shrewd attempt to appear as a bureaucratic idiot? New research complicates Arendt's concept of Eichmann's mindlessness, and thus the concept of the banality of evil.

Eichmann described his own obedience to the law as blind discipline *(Kadavergehorsam)*.[14] When Eichmann received the commission (task) to implement the "Final Solution" (Lang, 1990, p. 85), he stated that he could no longer live by Kant's categorical imperative, but consoled himself by thinking he was no longer "master of his own actions". He could do nothing to change events (Arendt, 2000, p. 142). During the Nazi regime, Kant's categorical imperative was rewritten by replacing the Fuhrer (Hitler) with the common law and reformulated as "Behave in such a way that der Fuhrer would have accepted your action if he had been a spectator." Eichmann called this the little man's version of Kant for home purposes.

Eichmann's organizational role was not always performed with loyalty. Near the end of the war, Eichmann ignored the call to stop the extermination of the Jews. At that moment, it became evident that *his deepest loyalty was to Hitler and their ground project of "Final Solution", and that he fanatically and personally wished that the Jews should be exterminated.* Thus, the relationship to the Nazi organization as such may be characterized as having elements of opportunistic behaviour. Setting this "pragmatic" order aside is a sign of fanaticism.

Supererogatory behaviour

In contrast, the three whistleblowers knew very well that acting against their organizational leaders put their careers in peril. But they did it nevertheless. It was more important to the three women than blind loyalty to their colleagues and leaders. We may wonder whether their professional values determined that a company is far more than the inside interests of its workers and managers. Stakeholders outside of the company were without doubt taken into consideration. The women probably took seriously the values that give legitimacy to every profession – to serve the interests of the larger community through special skills and competence. They were doing the right thing, even though the personal costs were immense. This is a hallmark of *persons with high integrity.*

Were they all acting from different ground projects that can be conceptualized as an idea or a vision about something deep and strongly wanted? It is a fact that there is a danger in acting on the belief of an idea, a faith, or a nation, which may create more evil than acting in one's own self-interest. Ideological evil appears not as evil, but as good; not as immoral, but as the highest form of morality (Vetlesen, 1997). In a sense, there are similarities among these four persons. They were all deeply committed to an ideal. Eichmann was dedicated to one person whom he admired deeply. The three women also had a deep commitment to their organizations in a professional sense.

Since she had been a child, Rowley had nourished a deep wish to become an agent for the FBI. But at that time women were not allowed to become agents. The times changed and she finally got her dream job – agent in the FBI. She was dedicated to the mission of the FBI, but not to the colleagues and persons inside the hierarchic organization, entwined in a culture where it was not proper to make a call to a person in a higher position. She was not loyal to her colleagues in terms of Hoover's old concept of loyalty, which protected the colleagues' vested interests in not changing the petrified hierarchical organizational culture (*Time* 2003).

Watkins was regarded as the most self-reflective of the three women, and she had a "booming" personality *(Time* 2003). As vice president she earned a good wage and thus had much to lose by voicing her discontent with the organization's accounting practices. We can speculate on what it might have meant for her to be part of a Christian congregation which gave her support and protection during the difficult times. She probably had a faith in a spiritual reality that goes beyond the values of human greed and power. She cited Martin Luther King, Jr.'s aphorism to explain her whistleblowing behaviour: "Our life begins to end the Day we become silent about things that matter" (*Time* 2003).

The third woman, Cooper, was very proud of her profession (internal auditing) and work, and she was met with the most active, direct, and visible resistance in her work-organization. She attributed her stubbornness to a saying from her mother: "*Never allow yourself to be intimidated, and always think about the consequences of your actions*" (*Time* 2003, p. 52). In retrospect, Cooper said that it was an unbelievably tough situation, and that she had read and reread the 23rd Psalm: "Yea, though I walk through the valley of the shadow of death, I will fear no evil: for thou art with me." Like Watkins, Cooper also had a Christian community that gave her support when the conflict level within her work-organization became too intense. Was it her religious faith that made it possible for her to say: "I only did my job. I am not a hero."

The above expression is similar to Eichmann's defense when he said: "[I am not a villain], I only did my job." The example of Eichmann gave the world a new understanding of human wickedness described by the terms *intellectual Eichmannism, petrified legalism, and the banality of evil.* But this trick was tried earlier by leading Nazis in the Nuremberg trials during 1945 and 1946 when the German Nazi leaders and German military commanders were held accountable for their actions before and during the Second World War. The message from the Nuremberg courts was that one could not do inhuman acts with reference to the state's positive law.

The lack of ethical reflection

According to Arendt, Eichmann was not the tyrant that most people wished to believe but, rather, an uninspired bureaucrat, an ordinary citizen, and a beloved husband and father.[15] How could this dichotomous behaviour – loving his family and at the same time exterminating millions of other children and women – be explained? One aspect of the understanding may be found in Eichmann's lack of ethical reflection.[16] He was an engine in the use of cliches. Using a variety of cliches he escaped thinking about himself, his actions, and his relationships to the people around him. He was thereby able to resist different points of view and apparently keep his self-esteem and self-image intact. The judge in Jerusalem finally said: "You have only been giving idle talk." And this was true. Eichmann always used the same phrases, the same words in speaking and writing. He only had a professional vocabulary, not a personal one.

He had always a phrase of hubris to fall back upon; for example, "I will jump laughing into my grave." It was impossible to communicate with him, not because he was lying, but because he was protected against the presence of others – the reality. He consoled himself with shallow phrases until his last breath.

In this sense, Eichmann expressed a *serious symbolic poverty.* He lacked the facility with symbols and metaphors to express his feelings, thoughts, and actions. This may partly explain his ability to engage in techniques of self-deception *(mauvais fois* in Sartre's terminology). This lack of symbolic ability to express personal relationships may also partly explain why he showed no guilt or compassion with the millions of Jews while efficiently transporting them to the concentration camps and gas chambers (see also Bauman, 1989). Arendt maintains that the evil of Eichmann was that he did not think of himself in terms of good and evil, only in terms of what he ought to do in order to make a career in the Nazi organization. Fifteen years after his crimes, he still recounted with explicit detail what he should have done in order to promote his career. Other parts of his memory were not so impressive. In short, Eichmann's crime was that through his clichés he suffocated his ability for ethical reflection and failed to develop his capacity for empathy with the millions of Jews that he fanatically deported, with the final purpose of extermination. His other concern was in being a clever manager in order to promote his career and thus obtain greater power and control. The Nazi organization also represented a kind of idealism, but it had even stronger nihilistic traits. The morality of work became supreme and higher than all other forms of morality.

So far, the lesson learned is that we *always carry a personal responsibility for our acts, regardless of society's positive law.* But what do we mean by personal responsibility, how far does it go, and how do we distinguish personal from organizational and professional responsibility?

Does gender partly explain the three women's whistleblowing? It is a fascinating question because we also had courageous female whistleblowers in Norway in 2002, as well as bold Norwegian women in the boardroom of Statoil in 2003. The female members of the board did not accept their leaders' corruption, while the men did. The small number of brave women does not give any decisive evidence in a statistical sense, but there are hints that there is a relatively strong inclination among women in male-dominated organizations to blow the whistle. Women are frequently outsiders in work organizations where the leaders are male and are typically kept separate from the inner circle of members, and thus they do not develop strong personal ties to their male colleagues and leaders. We may say that women have weak ties in their jobs but *strong ties outside their work,* where they participate actively in social *personal networks* as well as in their families. On the other hand, men may have a stronger focus on job-related networks that primarily have an instrumental character. An additional element may be that women value their work organizations relatively more in the perspective of the societal values, and women may be more likely to expose faults. Two of the three women said they hated the term whistleblower. And they persistently maintained they would never do it again, because they were "cornered: isolated and made irrelevant".

Whistleblowing in the ethical sense is a *supererogatory behaviour.* Individuals are generally unwilling to take high personal risks to warn the community about their company's wrongdoing. The costs for the whistleblower are usually high. Thus, many whistleblowers, such as the Swede Yngvar Bratt, choose to alert the public anonymously. While employed by the Swedish arms company Bofors, Bratt chose to disclose secret weapons sales to India by contacting the media directly. This alarm was effective and the government stopped the weapons sale. But some time after the incident Bratt chose to expose his own whistleblowing involvement. The retaliation from colleagues and leaders of the company was so strong that Bratt and his family had to move to a new village, and he had to start a new career. We might ask whether personal responsibilities had to be publicly exposed under such conditions. As a rule, what matters is to stop the wrongdoing, not demonstrate or announce personal responsibility for the sake of responsibility. To risk retaliation unnecessarily or to unduly engage in exhibitionistic behaviour, which whistleblowers are frequently accused by their colleagues of doing, is not virtuous behaviour.

The whistleblowing of the three women may be characterized as supererogatory behaviour. They were all married but served as the chief breadwinners in their families. Rowley and Cooper had husbands who were full-time, stay-at-home dads. For these women, the choice to confront their leaders meant jeopardizing a salary their families depended on (*Time* 2003).

Personal responsibility

Whistleblowing is the prototypical ethical dilemma, where the choice to blow the whistle or not is characterized as a personal struggle within the person's internal subjective world among different loyalties. This internal conflict is perhaps best described by Kierkegaard's fear and trembling, or in Sartre's terms as anguish, abandonment, and despair. These existentialist authors held that there are no genuine tests in morality; that is to say, there is no formula that can solve our moral problems.[17] Our moral standards can only be chosen. The authority, according to Kierkegaard, emanates from the fact that the person has *chosen* to utter his morality, and the greater the engagement of the deliberations and sincerity of the subsequent actions the better. "Subjectivity is truth" states Kierkegaard, pointing to the necessity of personal heartiness and engagement.

The conflict is personal because rational arguments can do nothing more than inform us and present us with different alternatives. In any chain of reasoning, an individual reaches a point where he has *to choose* his position. Kierkegaard contrasts different ways of living with his terms "Either/Or"; the ethical and responsible life or the aesthetic life which, ultimately, is very boring: Either you can satisfy your own immediate passions (the aesthetic mode of life) or enter into lifelong and inescapable duties. This latter way of life might be characterized as being committed to a *groundproject* in Bernard Williams's sense; viz. one or several projects that are closely related to one's existence and which to a significant degree give a meaning to one's life (Williams, 1981, p. 12). The idea of a person's ground projects provides "the motive force that propels him into the future and gives him a reason for living" (ibid., p. 13).

The internal struggle localizes responsibility in "Man the maker" (actor or agent), and not in the environment. The struggle may also be described as a dialectical relationship between moral authorship and social accountability. In the person's inner subjective lifeworld she is perceiving, interpreting, and judging what is going on in the objective world around her (see Niebuhr, 1999). Finally, she has to make a choice – which involves not simply drifting away like a piece of wood in the stream.

Harmon (1995) has further developed this idea of a personal struggle between our inner subjective world and the external objective world. In Table 2, following Harmon, I distinguish between some bipolar concepts in order to indicate the depth of the soul, but at the same time warn against dualism.[18] A key phenomenon of organic life is that there is typically an intimate connection between the world and the self. The key is to see the two sets of terms in Table 2 as a *dichotomy*; *that is, one concept with two poles, which may be complementary to each other like yin and yang.* These different sides of the same concept require each other and should be regarded as a state of creative tension (Harmon, 1995).

Localizing the responsibility in man (the maker), we presuppose that we as human beings have an internal world, a subjective lifeworld, within which ethical reflection may take place that influences our attitudes and actions in the external world. Basic questions are these: Who am I? What kind of person do I want to be? How should I act to preserve my integrity? Which identity should I strive for? Such questions may not always be explicitly asked. But the feelings related to identity and integrity may work as "inner voices" and give directions as to what I should do. An interesting case concerning integrity is cited by Harmon (1995) using Captain Hornblower's diary. This gives us some glimpses into a sea captain's inner struggles during the Napoleonic war. Should he as captain do his duty by hanging one of his crew, or should he admit him grace? The British captain felt a strong

Table 2. A dialectic concept of responsibility

Internal world (life-world)	versus Public, external world
Man the maker	versus Man the answerer
Man's feelings	versus Rationalistic thinking
Can do/the deliberate act	versus Must do
Self-reflection/authenticity	Versus Self-deception/inauthentic

loyalty to the navy and the Queen. On the other side, he also felt the meaninglessness of abstract duties when it means taking the life of a good person – a person he felt a special sympathy towards, a man who had only defended his dignity against a "bully" officer who "sadistically abused his authority" (Harmon, 1995, p. 89). Finally, it turns out that Captain Hornblower's feelings became his bottom line. In spite of the navy's strict, abstract rules of right and wrong behaviour during war, he permitted his friend to escape and saved his life. But after the event, Hornblower experienced strong and ambiguous feelings regarding his dissonance with the conventional beliefs. And he worried about the consequences to his self-concept. Had he, the heroic sea captain, now been transformed into a "softhearted sentimentalist"? He sat for hours in his cabin, "consumed in despair and self-loathing" (p. 90). "He had violated his sworn oath, and he had [done] so from mere personal reasons." He had committed a breach of duty and felt he had lost his integrity and his self-respect. As a matter of fact, Hornblower was ultimately listening to his inner voices of sympathy and friendship – not to the rational calculations of different duties.

The opposite of the captain's behaviour may be termed rationalistic thinking: calculating costs and benefits in order to arrive at actions that are acceptable from a public, external point of view. Such purely instrumental thinking may easily turn into emotional numbness because it involves the habit of suppressing one's inner feelings. One obvious consequence may be the development of cold relationships with people in general.[19]

In the realm of man the maker, freedom is basic. In this sense, we are the uncontested authors of our own lives and may write our own narratives through our chosen ground projects. As authors we always have the opportunity to, in Sartre's terms, step back, stop and reflect, and then choose. Thus, as human beings we are always free: "Man is condemned to be free", which also means "he is responsible for everything he does".[20] We may even go beyond the external circumstances and choose our own personal mode of life, expressed as "Existence precedes essence." Even if somebody points a revolver at my head as a threat, I have the option to not give in to the pressure. I could say, with Socrates: It is better to suffer from wrongdoing than to do immoral acts. Man the answerer is the polar opposite of man the maker. To be answerable means being accountable to persons and governments, as well as non-governmental organizations and institutions for our conduct. As Allen (2003) writes, "As individuals, couples, families, and communities we live lives enmeshed in webs of accountability" (p. 2). Without accountability it is impossible to conceive of a society and, in particular, a community.[21] Man the maker acts, but man the answerer is held accountable for his actions in the public sphere.

The opposite of freedom is a kind of "must do" behaviour and reveals an agent's, not a maker's, attitude. An agent risks either being swallowed by his "principal's" worldview, or reifying rules or doing whatever is necessary for money, acting as a "hired gun" on behalf of somebody. Submitting to a "must do" behaviour may partly be caused by the anxiety experienced when acknowledging personal responsibility (Kaufmann, 1981). Using terms such as "the agony of thinking", "terrible freedom", and "the torment of choice" existential authors indicate that many people try to avoid an existential situation in which one must choose. One way to avoid the confrontation with difficult choices is through self-deception.[22]

Sartre writes about different types of self-deception. He gives rich and living examples[23] claiming that self-deception "can even be the normal aspect of life for a very great number of people" (Kaufmann, 1981, p. 303). Thus, a person can hide unpleasant truths and live in self-deception, a negative attitude which consists of lying to oneself, even if one is "in a complete position of the truth which he is hiding". The antithesis of self-deception is, according to Sartre, sincerity. But one may also fall into self-deception by being sincere, if the sincerity is enforced and has an artificial character.

What is it to be authentic? It means having the courage to be true to one's self as a moral ideal that implies a standard of what we ought to desire (Taylor, 1991). It means realizing oneself indirectly by seeing oneself in a horizon of important values that go beyond one's individual self. Such *important values could be nature, history, citizenship, or fulfilling the needs of others.* This idea of an extended self transcends narcissism (Lash's term) or hedonism, which are indeed very shallow, narrow, and trivialized modes of being. To be authentic means "finding what is significant in [one's] differences from others" (Taylor, 1991, pp. 35–36). However, it is not an ego trip because it involves a self-fulfillment through others; recognizing ties with others in a more encompassing self (see Gilligan, 1986). To be authentic has a dialogical character, and it is through acquisition of rich human languages of expression that we become fully human beings. Languages are seen in a broad sense, including "the languages" of art, of gesture, of love, and the like.[24] We learn these languages together with others and we define our identity by posing existential questions in dialogue with significant others,[25] "sometimes in struggle against" our parents, friends, or colleagues. The conversations with these significant others continue within us for as long as we live.[26]

We have indicated above that to be authentic is not a quick fix but, rather, something to strive for. The striving is evident when we take authenticity in Kierkegaard and Nietzsche's sense. They hold that authenticity is "nothing but interpretations" that reach no end.[27] This view is congruent with the paradoxical style used by Kierkegaard in his writings, indicating the importance of always being understood in new ways, since new interpretations create new realities. Kierkegaard's style expresses the notion that temporal life can never be understood in a deep sense, because we cannot penetrate it through our own consciousness. Kierkegaard looked upon unlimited reflection as a double-edged sword. Infinite reflection can easily lead to complete ruin because it hinders every decision, and it can become "dialectical twaddle". But reflection can become the condition of authentic *Existenz.*[28] "Infinite reflection [...] is, precisely through its endlessly active dialectic, the condition of freedom."[29] Honesty is an essential attitude and represents the ultimate virtue, which one can and should strive towards. Although seemingly paradoxical, lightness, in addition to sincerity, is a mode that expresses authenticity for Kierkegaard. Dancing is the image used: "I have trained myself [...] always to be able to dance in the service of thought [...]. Everybody is too serious for me".[30] Kierkegaard (and Nietzsche's) dance metaphor is a phenomenon that nowadays is grouped under processes of *free flow.*

Do organizations carry moral responsibility?

Eichmann did not blame himself, and he attempted to place the responsibility for his evildoing elsewhere. This is in line with many managers who, when accused of wrongdoing, blame their organization or the system for the wrongdoing. Therefore, one relevant question emerges: Do organizations carry moral responsibility?

In John Steinbeck's novel *The Grapes of Wrath*, the representative of a bank expels tenant farmers from their homes. As a consequence, the bank representative attempts to explain to the tenants *who is responsible for what they are doing,* and he tries to avoid personal responsibility in this way (Maclagan, 1998).

> We're sorry. It's not us. It's the monster. The bank isn't like a man. Yes, but the bank is only made of men. No, you're wrong there – quite wrong there. The bank is something else than men. It happens that every man in a bank hates what the bank does, and yet the bank does it. The bank is something more than men, I tell you. It's the monster. Men made it, but they can't control it. (Steinbeck, 1951, p. 32)

Who is acting? Is it single individuals, the bank, or the larger capitalistic system? The following was quoted after the Watergate affair, which led to the fall of President Richard Nixon: "There were many bad actions", but no actors. Nobody was willing to be blamed and held accountable. Talking about organizations as if they are able to act, one uses a language that implies anthropomorphism, attributing human abilities to a construct, and seeing an abstraction as a concrete entity is an example of reification.

Organizations cannot be moral agents, only persons can. Corporations are *artificial persons* and "lack the emotional life without which there can be no possession of virtues and vices", as Ewin writes (1991, p. 749). In particular, managers have discretion to enter ethical into deliberation and make choices. Managers are responsible for how they design the "corporate internal decision structure"; that is how they allocate responsibility, define roles and policies, and criteria and procedures for decision making. In this manner managers will influence other persons' lives and actions, as well as the environment.

If Morgan is right in saying that "there is a close relationship between the way we think and the way we act, and that many organizational problems are embedded in our thinking", we should be aware of the metaphors we live by (Morgan, 1986, p. 335). If we take metaphors literally, we are creating myths, and the more conventional myths become the more difficult it is to be free from their grip on us. The danger of using the organization as a moral person is that blame is shifted away from the persons who should be held responsible for the actions done and their consequences on ecological systems and human life (cf. Maclagan, 1998). Organizations are much more than machines, and persons are not puppets on a string. Persons make up the organization, and only persons have the capacity for moral consciousness. Attributing moral responsibility to an organization is a "category mistake".

Professional ethics

An important problem concerns the kind of behaviour that is permitted when acting *qua* a professional role. What does a professional role mean concerning legitimating behaviour? This is a key question in the literature on professional ethics. We will look at problems in professional ethics that fall into the category of role-mediated behaviour. The crucial question is this: Does a professional role create an exemption for personal responsibility?[31] Or is personal responsibility a concern always connected to the person's identity and a virtuous expression of the blossoming of one's character, thus giving a unique and distinct color to every act performed by the agent in his professional roles? This question is frequently treated in ethical juridical literature. I refer to the case of the perverted professionalism of the butler in *The Remains of the Day*.

Kazuo Ishiguro's novel *The Remains of the Day* (1989) gives a strong example of person neutrality in the sense of abdicating one's personal judgment. *The butler, Mr Stevens, holds the belief that the core virtue of a great butler is person neutrality, viz. always be his master's voice.* But Roszak (1981) asks: Is giving up personal responsibility for one's work equivalent to alienation? Roszak maintains that personal responsibility is "the crucial distinction that separates mere employment from true vocation". Vocation, or "calling", refers to the "well crafted and the ethically right". Roszak stresses the personal, responsible concern for others, and for ecological matters. Following Roszak one should ask: *Is the job honest and useful? Is it a real contribution to human need?* (1981, p. 234).

Personal values are one point of departure for personal responsibility, and "personal values are those values that primarily are a result of our upbringing …" (Morsing and Pruzan 1998). Personal values provide the background for the values that we "subscribe and adhere to" in our professional lives. Organizational values are less stable than the "deeper, more 'lived-with' personal values" (ibid., p. 264).

Personal responsibility and advice

What is good advice for someone else's moral dilemma? Jean-Paul Sartre gives an example when he is confronted with a perplexed pupil during the Second World War.[32] The pupil asks Sartre whether to leave his vulnerable mother and join the French liberation forces or stay with his mother and thereby save her life. Sartre's answer is: "*You are free, therefore choose – that is to say, invent.* No rule of general morality can show you what you ought to do." Thus, Sartre is giving complete responsibility to the pupil, inviting him to take on personal responsibility. It is not Sartre's role to take on the moral burden from anyone else. Sartre argues that neither Kantian ethics nor Christian doctrines can help in such choices. When values are "too abstract to determine the particular, concrete case under consideration, nothing remains but to trust in our instincts".

Going beyond Sartre, we may suppose that the pupil is not satisfied with the philosopher's answer and asks his uncle, a military officer, the same question in order to get a second opinion. [33] The uncle answers: "As a military officer I will advise you to do your duty to your country and leave your mother and join the army. However, as your uncle I want you to stay at home and take care of your mother and my sister." This answer is also unacceptable to the pupil. Why is his uncle unwilling to express his own identity by giving him a substantial answer? Why does he *hide behind his professional and social roles?* Why does he not commit *himself as a man*?

Obviously, the uncle is trapped within conflicting roles. A role is a *set of expectations* for how one should behave. A more complex definition is *an implicit social script into whose functions we are socially conscripted.* The behaviour expected by a person is his role. In an organizational context, special obligations are attached to the person who has a role. Following Applbaum (1999, p. 47), the term role may be used to describe "positions with some degree of regularity and durability, and where there are collective expectations, however informal or contested, about the content of the positions, duties, values, and virtues".[34] With this in mind, a manager has a role, but the more arbitrary customer does not.

The officer/uncle's answer is not helpful for the pupil because the elder's personal position is shrouded in different roles. The uncle seems to be locked into various role behaviours, and the pupil may well ask: Is he a real person or is he a mere role construct stuck into different selves?

Generally, a *man's role cannot justify his beliefs* (cf. Coser, 1966–1967). A man cannot cite a role to show that something is acceptable. A certain belief is not attached to a specific role, and to cite one's role is never a good reason for holding a belief. Therefore, the statement: "since I am a leader in a business organization, my main goal is to maximize the organization's profit" is not a valid chain of reasoning. As a leader, one does have a personal responsibility for obtaining the organization's goals, as long as one causes no harm to nature, community, existing people, or future generations in the process. However, being a business leader is not a valid ground for reducing the company's goals to one-sided profit maximization. A person's role is never a reason to believe anything. With reference to Jean-Paul Sartre, we may say that every man is free and is able to transcend his different roles. To hide behind our roles and pretend that we have no freedom is to deceive ourselves as well as others. Finally, the pupil is able to say: "In the end, it is feeling that counts ..." (Kaufmann, 1981, p. 355).

Dirty hands and character

Now and then the leader of a business organization is expected to make hard decisions that influence people's lives in existential ways. Such decisions might be called dirty-hands decisions, of which the downsizing of a plant is an example. The story often goes like this: Company plants that have been an important part of local communities for many years are closed down in so-called high-cost countries, while new plants are erected in countries like India or China. What is the manager's responsibility in those cases? Is it a duty to move the plant in order to maximize the company's profit? Is the exploitation of new business opportunities a primary responsibility or should the manager see her responsibility as making a reasonable profit as long as the company can contribute to the well-being of the local community, the employees, and their families? If the plant is making a reasonable profit, is it the manager's duty to reduce the risk of declining future profits by downsizing and moving the plants to a low cost country? Can the manager justify his decision as an "any-hand" situation: "If I am not doing it, other companies will do it and soon out-compete our company"

Downsizing a plant that makes a reasonable profit involves a kind of technical and instrumental manner of thinking that is problematic for several reasons. First, downsizing is not only about more or less profit and thus exclusively economic in nature. The manager makes a serious fallacy by ignoring or underestimating the systemic, interpersonal, and existential dimensions of downsizing. This fallacy may create grave ethical problems. Using an ethical vocabulary, we may frame the problem in this way: If the manager does not seize the opportunity to maximize the profit by downsizing, does he carry a *negative responsibility* (responsibility for not acting)? Will he be responsible (and in turn held accountable) for not doing the action he "should" have done (the downsizing)?[35]

To downsize a plant in a high-cost country entails the unemployment of many people (often several hundreds) who are often times located in small communities in which the company has been a cornerstone enterprise. In ethical terms, the manager faces a real dilemma between different loyalties – loyalty towards the local community, the employees, local suppliers, and so on, and loyalty to the shareholders. Which group of stakeholders should he favor? Favoring the shareholders' pure economic interests might mean rebuilding the company in China – because of expected lower costs and higher profits. But giving consideration only to the shareholders is a kind of aristocratic thinking. The manager has a choice, and he might well reflect upon how a downsizing decision may influence his own *character.*

Is he ready to look upon himself as a "hired gun" on behalf of the shareholders, or does he wish to be identified with caring for the employees and the community? What kind of person would the manager like to be?

Bernard Williams (1973) discusses the concept of negative responsibility in relation to utilitarianism, using his famous example of Jim and Pedro. Jim is on a botanical expedition and by accident meets the military commander Pedro. Jim suddenly finds himself in the central square of a small South American town and is confronted with a dilemma: Pedro has tied up twenty Indians against the wall and offers Jim a "guest's privilege" of killing one of the Indians himself. By doing this, he will save the nineteen other Indian lives. If Jim refuses, the twenty Indians will be shot by Pedro. What is the right action for Jim in this dirty-hands dilemma?

Should he shoot one Indian and thereby save nineteen lives, or should he care for his character and refrain from killing anybody? Williams uses this example to emphasize that it is not the maximum welfare that is crucial in the world. Jim might have good reason not to shoot one man in order to save nineteen lives. Williams points to the remote effects on Jim's character and states that because his ground project and mental health might be seriously undermined by shooting even one person, his conscience should be clean. Even if Jim happens to be at the wrong place at the wrong time, it is Pedro's, not Jim's, project to kill the Indians. It was not part of Jim's plan to deal with this tragedy – and this is also a reminder that "some situations lie beyond justifications." One serious defect with utilitarianism is that it does not address the character of the agent. With such a perspective we might also, as Williams does, criticize the impartial attitude to the Kantians, who transform man into an abstract monster. Impartiality does not mean impersonality, and Williams argues that the differences of character have a special importance in personal relations because "differences of character give substance to the idea that individuals are not inter-substitutable" (Williams, 1981, p. 15).

Deep attachment to other persons and favoring friends do not mesh with the impartial Kantian view. To treat persons as abstractions from character is a misrepresentation (Williams, 1981, p. 19). Moral thinking that could result in an endless search for justification of one's behaviour has its limitations in tragic situations, like Jim's dilemma. In such situations it is obviously better for a man (and his friends) to have a character that urges him to act in a resolute manner.

But does the distinction between positive and negative responsibility matter? Because we have positive duties to act and negative duties to prevent harm we do not really need the duality (cf. Goodin, 1985). The danger with such a concept is in becoming moral slaves because it requires too much of us.

The concept of negative responsibility is not necessary if we have persons with a virtuous character. This is in line with O'Neill's (1996) writing about virtues and justice. We should not de-emphasize indifference and negligence by declaring such attitudes irrelevant to the moral realm. But when persons have character and see themselves as a part of an extended self, this indifference is not prevalent.

Thus, the notion of virtues may replace the concept of negative responsibility. In accordance with O'Neill (1996), I argue that principled rights, such as justice and virtue, are complementary goods. By justice we may refer to principles that are intended to avoid injury. Such principles require complex institutions and practices that can constrain and guide action and policy. But justice is not enough, because principles of justice do not provide precise instructions; they only set standards we should strive for, and they focus only on the minimalist and formal aspects of institutions. Indifference and neglect are not necessarily unjust. Therefore we need social virtues, not only for action that affects others directly, but also for taking care of the environment upon which human lives depend. In particular, O'Neill (1996) points to executive virtues such as self-respect, self-control, decisiveness, courage, endurance, and autonomy, which may be seen as means to effective actions. These virtues are instrumentally required character traits, and they are needed to implement principles of justice.

People live in a social web with lives infiltrating one another, and these dependencies will typically intensify human vulnerabilities. Thus, we also need the virtues of care, compassion, pride, and shame to selectively take care of others (cf. Solomon, 1993). *Connections are always of a specific kind,* and *this specificity needs a personal approach* (O'Neill, 1996). Social virtues may protect human vulnerabilities from violation. However: Do social virtues demand too little? Do they demand too little guidance?

Unlike a duty, a responsibility carries considerable discretion. The discretion consists in how to best attain some wanted behaviour. Responsibility is judged in terms of the consequences (Jonas, 1984), while duties are judged in terms of the intentions and behaviour. This is in line with Goodin's (1985) work *Protecting the Vulnerable: Acting Responsibly*, which refers to acting with due regard for the consequences of one's actions and choice. In a deep sense, responsibility stems from vulnerability. What may be wounded is vulnerable, and the central issue is to prevent harm. Vulnerability is a relational concept, and the dependency of specific others is crucial. When taking people's welfare or interest as the point of departure, feelings of self-respect and self-image are vital for a person. The more sensitive we are as persons, the more vulnerable we are in how we are treated psychologically.

One main idea behind responsibility is the protection of persons' primary goods (Rawls, 1971), such as food, clothing, shelter, self-respect, and civil liberties. These goods are closely tied to the general interests of people. We should aid those in direct need and intervene on behalf of those who are unable to help themselves (Goodin, 1985, p. 110).

Do we have stronger moral ties to those who are close than those who are far away? Persons who are close to us in space and in time seem more vulnerable to us. We should, therefore, take their interests into consideration. In principle, we have responsibilities to all sentient beings (see Singer for arguments that neither distance nor numbers reduce our moral responsibilities). But with increasing distance, our ability to sustain attention is decreasing.

Systemic responsibility

Is being an alcoholic a core example of incontinence of the addict's will, of not excelling moral authorship, or is the abuse of alcohol a pattern that arises in a bad community? Should we emphasize the interactions between the addict and his/her social relations rather than putting sole responsibility on the addicted individual? Do we have an example of shared, systemic responsibility? It is convenient to take the individual approach because it enables the people close to the addict to look upon themselves as innocent victims. But how are the chain of events and interactions produced, and sustained, between the addict and the people in the community? Have the persons in the local community contributed to the addict's personal life, his maturity, and self-reflexivity in a positive way?

Examples abound that the vicious cycle of alcoholism is often maintained by both the alcoholic's inflated belief in his "incontestable authorship" of his life and the dysfunctional behaviour of his family. If we use the notion of *shared or systemic responsibility* we might take a broader look at the relationship between the addicted person and his social network. We usually see the social system as consisting of weak ties, wherein the addict and the persons in his proximity do not sufficiently engage in each other's lives. In such a situation, we might use Zsolnai's (2003) model of moral responsibility, the 3R model that includes a synthesis of *reverence, rationality, and respect.* In this model, respect means *empathy* toward those with whom we interact. If we have an extended self-concept in which the addict is a part (either as a friend, lover, father, employee, neighbor, etc.), he or she will be part of our self, and we will naturally intervene and help the addicted person.

We may look upon the addict and his surroundings from different perspectives – systemic, existential, interpersonal, or scientific. From the systemic perspective, we regard the acts as parts of a larger pattern. Viewed from an existential (or spiritual) perspective, the alcoholic's problem is a lack of meaning and purpose. Unemployment might contribute to the lack of meaning and purpose. When we look at the addict's situation in a solely technical-scientific light, we tend to distance him from ourselves by suggesting the problem is due to his weakness of will, and we try to cure his sickness with pharmaceuticals (see Schumacher, 1999 for an argument of the positive value of work).

A systemic responsibility requires that those in the alcoholic's social network must assume their share of responsibility for the plight of the addicted. Without such a co-responsibility, we will be in danger of making them all *underresponsible* "in the sense of failing to act on one's own legitimate interests and simultaneously disregarding others" (cf. Harmon, 1995,

p. 137). In a sense, we are all responsible when we have not done what we ought to have done. The interpersonal/social aspects concern the emotions of the persons and families who are affected. It has systemic aspects because the community is affected, and it has existential aspects because alcoholics usually question their own self-worth (Mitroff, 1998, p. 61).

Something can probably be learned from the *Alcoholics Anonymous* twelve-steps program, which gives the freedom and spirit to interpret a higher power as personally conceived. Every step involves a moral reflection with regard to all the different stakeholders. According to Mitroff, Alcoholics Anonymous is a blend of ethics, implementation, and spirituality.

Systemic responsibility implies shared responsibility, and in this sense it contradicts rationalistic thinking about individual agency and moral authorship. Individual actions can be better understood as a matrix of relationships. Therefore, the notion of linear causality has to be replaced by interactive and circular processes. In this case it suggests that many of the addict's stakeholders bear co-responsibility for the tragedy.

Extended self

In this section the reciprocity between character and role is emphasized. Following Luban's (1988, p. 111) argument concerning our moral self-images, we should look upon ourselves as persons first and role occupants second. However, to say that the *person is essential and the role is accidental* goes beyond the limits. Persons and roles are separate, but they are also reciprocal in important senses. In the following *dialectic reciprocity between roles and character* is assumed.[36] According to Applbaum (1999), this separation of roles and identity requires that a person's judgment not be subordinated to the role occupied (p. 75). The danger of role theories is that they reduce a man to the sum of his roles and nothing more. Our conclusion above was that responsibility for an action taken within a role "falls on the person" not on the role, which implies a kind of personal responsibility that lurks behind all actions (positive responsibility), and on some important actions that were not taken but ought to have been taken as a duty (duty replaces the notion of negative responsibility). A prerequisite to professional moral duty that binds all people, regardless of their roles, is the duty not to do harm and not to do acts (for example, lying or manipulating accounting procedures) that may be unfair to other people.

A role typically entails undertaking a special kind of responsibility. Being appointed to a chairmanship implies that you increase your responsibility and take on moral responsibility for the company. And this increased responsibility becomes an element in redefining the chairman's self-concept. The new identity of the person as chairman becomes part of the chairman's self-image. If the chairmanship is a prestigious position, the role will enrich her as a person. "The chairman of Norsk Hydro is me." It means a deeper involvement in Hydro, and this creates a psychological dynamic of the self that enlarges and enriches self-worth and self-image. A prestigious position creates a valuable relationship, an added value for the person as a role taker. In such a perspective we are recreating ourselves all the time by taking on or shirking responsibility. The more responsibility you take on, the more extended will your self become. The important Aristotelian virtue of friendship also implies generating an extended self. This added value is closely connected to the view that there is a community that is an important part of the person's life.

Honor and shame

To be personal is a kind of attitude that secures the best possible action. This attitude involves acknowledging and using one's emotions of caring, sympathy, compassion, honor, and shame. This view is based upon an understanding of the self as being defined by its attachments and affections, and seeing personal relationships as a basic condition for a meaningful life (Solomon, 1993, p. 220; Gilligan, 1986). Aristotle long ago defined a person as *phusei politikon,* by which he emphasized that a man is born to live and blossom in an extended community. Aristotle's concept of friendship suggests that a man has an extended self that includes his friends and his community. In Aristotle's time honor was an important virtue, "something that is always desirable in itself and never for the sake of something else" (Aristotle, NE 1097a33). And honor requires a sense of membership, a belonging, a self that is inseparable from one's identity and one's group or community. Aristotle argues that the motive for courage is a sense of honor, which means living up to the expectations of the group. And honor is not selfish in a narrow sense, because it is never the individual's honor alone under discussion but, rather, the honor of the community. This is the difference between honor and pride, for the latter focuses on an individual self and its accomplishments.

The opposite of honor is shame, and shame is not a virtue in the Aristotelian sense for it is "more like a passion than a state of character". Shame is defined as "a kind of fear of dishonor" (Aristotle NE 1128 bl2). Where there is no honor, there is no shame. To feel shame is to fail in a context of one's extended self. And not to feel shame is evidence of lack of virtue.

Feelings and sentiments as caring, sympathy, devotion, and compassion are important virtues in today's life within a business. These *emotions inspire the search for active ways of helping.* There are contributions on the importance of care in modern management theory (see for example Von Krogh, 1998; Styhre et al., 2002).

Von Krogh and Styhre see care as an activity that "directs oneself towards other human beings, artifacts and other existing things in a heedful manner" (p. 505). Care may be defined as "cautious or watchful attention". Heidegger employed the notion of care (Sorge) in his philosophy, and for him, human beings (Da-sein) essentially exist in the world as beings who "take care".

Compassion is a more specific emotion than caring. Compassion means literally "feeling with" and involves suffering with the other. Blum argues that compassion "involves an active and objective interest in another's welfare, it is characteristically a spur to a deeper understanding than rationality alone could ensure" (cited in Solomon, 1993, p. 230). Overly-strong feelings of compassion make difficult the kind of detached behaviour that may be important in some instances. What we need is compassion and wisdom, not intelligence without compassion. We need the unity of a warm heart and a cold brain to work in a dialectical relationship between proximity and distance.

Tentative implications

Should professionals read more Ibsen or Shakespeare in order to develop a richer vocabulary about everyday life, deepen their feelings, and cultivate their imagination? As long as the professional vocabulary is restricted to machine-age thinking and machine metaphors, the symbolic universe may be so limited that it stifles creativity, compassion, human thinking, and action. The economic language of efficiency, control, and profit is too narrow to help us in sincere and open dialogues about what we as professionals should or should not do (Pruzan, 1998).

Do persons without personal relations lose their motivation and sense of being? By being personal – as professional businesspersons, citizens, or colleagues – one challenge consists of reaching out to fellow beings and caring for them. There are probably many persons who are alienated due to their treatment of self (and in a logical sense, others) as a mere instrument, or who engage in projects that are regarded as harmful and look upon their professional roles as merely accidental. A person with virtuous character – stable across different spheres of life – should be embedded in the practice itself, blossoming when she does the right things in the light of the traditions of that community. Business as a human and existential enterprise is a personal and social activity and in a deep sense consists of webs of interpersonal relationships and widespread cooperation. Emotions are central features of a person's character, perhaps the essence of being a human! It is the emotions and personal values that constitute persons as an extended self and enable us to identify with a larger community.

Eichmann might have been part of an extended, if perverted, self in a minimalist sense when identifying with Hitler. He probably also felt some commitments to his own family. But Eichmann was not a virtuous man in the classical and Aristotelian sense, because Eichmann's attitudes and actions did not match the *unity of virtues* in which the many good things (virtues) require each other in forming a unity. One cannot put aside or deduct those virtues one finds inconvenient. Eichmann lacked several of the essential virtues, not to mention the virtue of justice. He was cruelly unjust with the 6,000,000 Jews he deported to concentration camps. One ingredient to avoiding injustice towards others is the natural feeling of compassion and empathy with others. These feelings are cultivated in a good community. The female whistleblowers originated from small communities. In a small community, every person is important and clearly contributes to the community's well-being.

Did the women's local origins partly explain why they each regarded themselves as part of an extended self? Were they as adults connected to the different stakeholders: customers, shareholders, government, local communities, NGOs, families, and nature in a good relational, virtuous, and dialogical sense? They had obviously not lost their individual characters and spirit. They had not, like Eichmann or Steven the butler, become the voices of their masters. Their personal commitment to the many stakeholders – and not one dimensionally to their leaders – was unconditional. Eichmann gave the impression of acting like a slave, and by his attraction to Hitler he might have swallowed Hitler's (his master's) world completely – his judgment and his style. Eichmann was an important instrument in the genocide; he was excellent in his professional role, but monstrously evil. He had probably stifled his own authentic feelings. In Gestalt psychology, this swallowing of other people's ideas is defined as *introjection,* and this servility blocks genuine contact with others. The opposite vice of losing one's individual identity is being a pure narcissist who, self-absorbed, dwells in his inflated ego, "desperately fending off the anxiety of being alone in the world" (Harmon, 1995, p. 94). We need a complementary relationship between self-reflexivity and personal relations. Within this creative tension, an engaged dialogue between our inner lifeworld and our external world might occur, which would connect us to a community.

Personal responsibility is not reducible to social, professional, or organizational roles, nor to a bundle of duties. In the end its essence is to respond – reply – and answer to different stakeholders through an authentic and active dialogue (Niebuhr, 1999).

Notes

1. From Kirkegaard's *On His 'Mode of Existence'* (from Kaufmann, 1981, p. 88).
2. I am indebted to Laszlo Zsolnai. Without him this chapter would not have been written, and I am grateful for his inspiring comments and constructive suggestions on several drafts. I would also like to thank Zsolt Boda, Odd Langholm, Lars Jacob Tynes Pedersen, and Maria Prestmo for their stimulating comments.
3. Following Davis (2001) there has been a fundamentalist attitude towards responsibility that considers responsibility as pure obedience. "Obedience is the cornerstone of the three monotheistic (Abrahamic) religions. The Koran (4:59) bids the Muslim to 'obey God, his Prophet and those in authority over you.' Islam *means* submission ...and 'Throughout the Hebrew Bible, Yahweh demands obedience from the chosen people.'" (Davis, 2001, p. 14). Davis, in his comparative perspectives of responsibilities, calls this the simple responsibility approach and compares it with a modern, complex, or reflexive meaning of responsibility. While obedience is a heteronomous virtue, modern responsibility presupposes a sense of personal autonomy.
4. There are several Norwegian authors who presuppose that there is such a thing as *common morality*. A recent work is Tranøy (1998), and the Norwegian pioneer in business ethics, Johannes Lunde (1967, 1982), describes in several of his works the *ethical limit of common morality as the "norm horizon"*. Based upon Milton Rokeach's methods, Lunde investigated the ethical values in Norwegian society from 1969 to 1978. His findings demonstrate clearly that common people's beliefs about values concerning honesty, responsibility, reliability, justice, and helpfulness are largely congruent and form what Lunde calls *a common "norm-horizon"* (1988, p.70; cf. also Lunde, 1967, pp. 44–6).
5. Zsolnai (2002) also emphasizes the *process* aspect of moral responsibility in his 3 R model, in which he regards moral responsibility as a combination of three basic components: *rationality, respect, and reverence.* The rationality concept used is process-oriented and includes, among other considerations, care in mapping out alternatives and consequences, as well as attention to details of implementation.
6. Here I follow Martin (2000) whose notion of personal ideals refers to chosen ideals that shape the work without being incumbent on all members of a profession. These values are neither idiosyncratic nor eccentric (p. 5).
7. The viewpoint of the person taken here resembles *personalism* and is inspired by Mounier, in the sense of the reaction to the treatment of a human being "simply as an object of scientific study or reducing her to an economic or socio-political function [...] and [is] closely related to existentialism, reacting against positivism, materialism, and behaviourism [...]" (Whetstone, 2002, p. 385). However, in the present chapter I go beyond an anthropocentric worldview and put greater

emphasis on a deep ecological world view, presupposing an ontology based upon Næss's deep ecological perspective (Næss, 1999; Næss and Rothenberg, 1989). This viewpoint is congruent with deemphasizing technologies and methods, placing "demands on personal competencies and qualities rather than on more traditional managerial skill and technical proficiencies" (Marsing and Pruzan, 2002).

8. The diffusion of responsibility is partly a consequence of a more depersonalized development in economic thinking. For convincing evidence of this development since medieval times see Langholm (1998).

9. Cf. Grit (2004) who discusses how we can strengthen the social responsibility of managers in the perspective of corporate citizenship.

10. "My station and its duties" is a conception of role morality and is a famous phrase of F.H. Bradley (1927) who, in turn, borrowed it from the Anglican catechism. The catechism, in the *Book of Common Prayer,* reads. "*Question:* What is thy duty towards thy Neighbour? *Answer:* My duty towards my Neighbour is to love him as myself, and to do to all men as I would they should do unto me: To love, honour, and succor my father and mother: To honour and obey civil authority: To submit myself to all my governors, teachers, spiritual pastors and masters: To order myself lowly and reverently to all my betters: To hurt nobody by word or deed: To be true and just in all my dealing. And to do my duty in the state of life unto which it shall please God to call me."

11. Eichmann escaped from Germany and arrived in Argentina in 1950 where, under a false identity, he lived with his German family. In May 1960, while living in Buenos Aires, he was captured by Israeli agents and brought to Jerusalem for trial at the age of 55. During the trial he sat in a small, specially constructed cage of glass with 750 spectators in the room. The trial lasted from April 2 until August 14, 1961. Eichmann, together with others, was prosecuted for crimes against the Jewish people (the project termed the Final Solution or, in German terms, "Endlosung") and humanity (the murder of children from Czechoslovakia), and for war crimes. A total of 121 meetings took place during the trial, and the death sentence was pronounced on December 2, 1961. On May 31, 1962, the sentence was carried out; Eichmann was hanged, his body was burned, and the ashes were spread in the Mediterranean Sea.

12. My main source for the story of the three whistleblowers is *Time* magazine December 30, 2002/January 6, 2003, Special Double Issue, "Persons of the Year".

13. Eichmann cook great pride in coining the term "The Final Solution" ("Endlosung"). Lang (1990) gives a number of examples of the different euphemisms the Nazis used to create a technical and impersonal vocabulary for the cruel acts they committed against the Jews and some other minorities. Some of these language games are brilliantly illustrated in Orwell's (1949) *Nineteen Eighty-Four.*

14. "Kadaver-discipline" means the rigid behaviour of one who is trained and drilled to act in blind obedience (see Arendt, 2000, p. 141).

15. Newer research is destroying the picture of Eichmann as a loyal father and husband. There is evidence chat he was unfaithful, emotionally unstable, and crude, and thus quite different from the person Arendt found as normal as anyone (see Miale and Selser, 1975).
16. This view of lack of reflection is in accordance with Svendsen's (2001) *Ondskapens filosofi (The Philosophy of Evil).* Svendsen states that humanity's most important problem is not a surplus of aggression, but rather a deficit of reflection (2001, p. 213). However, the viewpoint of Vetlesen (1994) is that Eichmann's evilness was due to his lack of affection rather than his lack of intellectual capacity.
17. Fascinating illustrations of ethical dilemmas are also given by Gowans (1996), and Mason (1996).
18. Dualism in this context means the notion that body and soul can be disentangled (see Jonas, 1996). Dualistic thinking indicates a kind of rationalistic thinking and has serious limitations. Julie A. Nelson (Chapter 8 in this book) warns us not to attach to either side of such dualistic categories.
19. A person tends to do to others what others have done to him or her (see Vetlesen, 1994, p. 259). Vetlesen cites Horkheimer and Adorno, who see "ego weakness as the psychologically crucial precondition of fascist leanings of hostility toward out-groups or toward otherness ..." (p. 262). This thinking takes the establishment of a solid self, as distinct from a fragmented self, as a point of departure for a healthy human being.
20. Sartre 1946 (see Kaufmann, 1981, p. 353).
21. With the title *Why Privacy Isn't Everything*, Allen (2003) argues that even in the intimate realms of our private life we are held accountable. We are accountable for nominally private conduct that includes "intimacies relating to sex, health, child rearing, finances, and other matters termed private" (p. 2).
22. For an interesting treatise on the role of existentialism in ethical business decision making, see Agarwal and Malloy (2000).
23. A rather comprehensive and detailed example used by Sartre concerns a woman who goes out with a man for the first time. "She knows very well the intentions" of the man but "does not want to realize the urgency; she concerns herself only with what is respectful and discreet in the attitude of her companion." Sartre claims that this is a type of self-deception; "a certain art of forming contradictory concepts which unite in themselves both an idea and the negation of that idea The basic concept which is thus engendered, utilizes the double property of the human being, who is at once a *facticity* and a *transcendence*." (Kaufmann, 1981, pp. 309–10).
24. See Taylor (1991, p. 33).
25. We are introduced into the language that we need in order to define our selves through being with others who are important to us. In Mead's (1934) terms, these are the *significant others*.
26. Taylor, p. 33, in citing Bakthin.

27. I draw here upon a text by Karl Jaspers entitled *Reason and Existenz* (1935), in which he compares Kierkegaard and Nietzsche (see Kaufmann, 1981, pp. 191–2).
28. Kaufmann (1981, p. 195).
29. Ibid., p. 196.
30. Ibid., p. 205.
31. There are interesting legal papers on such issues written; for example, Wendel (1995, 2000).
32. The example is used by Sartre in his famous lecture "*Existentialism is a Humanism*" (see Kaufmann, 1981, pp. 354–6).
33. The pupil's approach corresponds well with Aristotelian chinking, which implies that if you have a moral dilemma, you should ask elder and more mature persons because they are expected to have developed the virtue of practical wisdom (phronesis) from accumulated years.
34. Cf. Coser (1966–1967). To occupy a role means a description of a person that allocates him to a position in the degree that the attribution to him of some rights and/or duties is inseparable from the application of the description.
35. The notion of negative responsibility is emphasized in Lucas (1993).
36. I am indebted to Laszlo Zsolnai for his comments and guidance in this section, in particular on the reciprocity between character and role, and the pride that ensues from being in charge of noble activities.

Bibliography

Agarwal, J. and D. C. Malloy (2000): "The Role of Existentialism in Ethical Business Decision Making," *Business Ethics: A European Review* 9(3): 143–154.

Allen, A. L. (2003): *Why Privacy Isn't Everything. Feminist Reflections on Personal Accountability*. Oxford: Rowman & Littlefield Publishers, Inc.

Applbaum, A. I. (1999): *Ethics for Adversaries. The Morality of Roles in Public and Professional Life*. Princeton, NJ: Princeton University Press.

Arendt, H. (1980): *Eichmann in Jerusalem. A Report about the Banality of Evil* (rev. edn). New York: Penguin Books.

Arendt, H. (2000): *Hannah Arendt. Eichmann in Jerusalem. En rapport om ondskapens banalitet*. ("Hannah Arendt. Eichmann in Jerusalem. A report about the banality of evil with an introductionary essay by Bernt Hagtvet"). Oslo: Bokklubbens dagens bøker, Kulturbibiliotek.

Arendt, H. (2003): *Responsibility and Judgment*. New York: Schocken.

Aristotle (1925, 1980): *The Nicomachean Ethics* (Translated with an introduction by D. Ross). Oxford: Oxford University Press.

Bandura, A., G. V. Caprara and L. Zsolnai (2002): "Corporate Transgressions", in: Zsolnai, L. (ed.): *Ethics in the Economy. Handbook of Business Ethics.* Oxford: Peter Lang.

Bauman, Z. (1989): *Modernity and the Holocaust.* Cambridge: Polity Press.

Bauman, Z. (1993): *Postmodern Ethics.* Oxford: Blackwell.

Burns, T. and G. M. Stalker (1994): *The Management of Innovations.* New York: Oxford University Press (rev. edn, first published in 1961).

Carr, A. Z. (1968): "Is Business Bluffing Ethical?" *Harvard Business Review,* January-February: 143–153.

Cohen, G. A. (1966–1967): "Beliefs and Roles," *Proceedings of the Aristotelian Society* 67: 17–34.

Davis, W. (2001): "Introduction: The Dimensions and Dilemmas of a Modern Virtue", in: Davis, W. (ed.): *Taking Responsibility. Comparative Perspectives.* London: The University Press of Virginia.

Ewin, R. E. (1991): "The Moral Status of the Corporation," *Journal of Business Ethics* 10(10): 749–756.

Ghosal, S. (2005): "Bad Management Theories Are Destroying Good Management Practices," *Academy of Management Learning* & *Education* 4(1): 75–91.

Gilligan, C. (1986): "Remapping the Moral Domain: New Images of the Self in Relationship", in: Heller, T. C., M. Sosna and D. E. Wellbery (eds): *Reconstructing Individualism. Autonomy, Individuality, and the Self in Western Thought.* California: Stanford University Press.

Goodin, R. (1985): *Protecting the Vulnerable.* Chicago: University of Chicago Press.

Goodpaster, K. E. and J. B. Matthews (1982): "Can a Corporation Have a Conscience?" *Harvard Business Review*, 60, January February: 132–141.

Gowans, C. W. (1996): "Moral Theory, Moral Dilemmas, and Moral Responsibilities," in: Mason, H. E. (ed.): *Moral Dilemmas and Moral Theory.* Oxford: Oxford University Press.

Grit, K. (2004): "Corporate Citizenship: How to Strengthen the Social Responsibility of Managers," *Journal of Business Ethics*, 53(1–2): 97–106.

Harmon, M. M. (1995): *Responsibility as Paradox: A Critique of Rational Discourse on Government.* London: Sage Publications.

Ishiguro, K. (1989): *The Remains of the Day.* New York: Vintage.

Jonas, H. (1984): *The Imperative of Responsibility. In Search of an Ethics for the Technological Age.* Chicago & London: The University of Chicago Press.

Jonas, H. (1996): *Mortality and Morality. A Search for the Good after Auschwitz.* Illinois: Northwestern University Press.

Kaufmann, W. (1981): *Existentialism from Dostoyevsky to Sartre.* New York: Meridian.

Lang, B. (1990): *Act and Idea in the Nazi Genocide.* Chicago: The University of Chicago Press.

Langholm, O. (1998): *The Legacy of Scholasticism in Economic Thought. Antecedents of Choice and Power.* Cambridge: Cambridge University Press.

Løgstrup, K. E. (1956): *Den etiske fordring* (2nd edn, 1991), Copenhagen: Gyldendal (in Danish).

Lozano, J. M. (2005): "Towards the Relational Corporation: From Managing Stakeholder Relationships to Building Stakeholder Relationships (Waiting for Copernicus)," *Corporate Governance*, 5(2): 60–77.

Luban, D. (1988): *Lawyers and Justice. An Ethical Study.* Princeton, NJ: Princeton University Press.

Luban, D. (1996): *Stevens's Professionalism and Ours,* 38, Wm. & Mary L. Rev. 297.

Lucas J. R. (1993): *Responsibility.* Oxford: Clarendon Press.

Lunde, J. (1967): *Etiske normer i næringslivet. Forbrukerorientert markedsetikk.* Oslo: Universitetsforlaget (in Norwegian).

Lunde, J. (1982): "Alminnelige menneskers verdisyn i Norge. En intervjuundersøkelse" in *Etikk i økonomikk. Artikler og undersøkelser.* Bergen: Norges Handelshøyskole (in Norwegian).

Maclagan, P. (1998): *Management and Morality. A Development Perspective.* London: Sage Publications.

Martin, M. W. (2000): *Meaningful Work. Rethinking Professional Ethics.* Oxford: Oxford University Press.

Mason, H. E. (1996): "Responsibilities and Principles: Reflections on the Sources of Moral Dilemmas", in Mason, H. E. (ed.): *Moral Dilemmas and Moral Theory.* Oxford: Oxford University Press.

Mead, G. H. (1934): *Mind, Self, and Society.* Chicago: University of Chicago Press.

Miale, F. R. and M. Selzer (1975): *The Nuremberg Mind: The Psychology of the Nazi Leaders.* New York: Quadrangle, The New York Times Book Co.

Mitroff, I. (1998): *Smart Thinking for Crazy Times. The Art of Solving the Right Problems.* San Francisco: Bertt-Koehler Publishers Inc.

Morgan, G. (1986): *Images of Organization.* Beverly Hills, CA: Sage Publications.

Marsing, M. and P. Pruzan (2002): "Values-Based Leadership," in: Zsolnai, L. (ed.): *Ethics in the Economy. Handbook ofBusiness Ethics.* Oxford: Peter Lang.

Næss, A. (1999): *Økologi, samfunn og livsstil. Utkast til en Økosofi.* Oslo: Bokklubbens kulturbibliotek (in Norwegian).

Næss, A. and D. Rothenberg (1989): *Ecology, Community and Life style. Outline of an Ecosophy* (Trans. and ed., David Rothenberg). Cambridge: Cambridge University Press.

Niebuhr, H. R. (1999): *The Responsible Self. An Essay in Christian Moral Philosophy* (originally published in 1963). Kentucky: Westminister John Knox Press.

O'Neill, O. (1996): *Towards justice and Virtue. A Constructive Account of Practical Reasoning.* Cambridge: Cambridge University Press.

Orwell, G. (1949): *Nineteen Eighty-Four.* New York: Penguin Books.

Pruzan, P. (1998): "From Control to Values-based Management and Accountability," *Journal of Business Ethics* 17: 1379–1394.

Rawls, J. (1971): *A Theory of Justice.* Cambridge, MA: The Belknap Press of Harvard University Press.

Roszak, T. (1981): *Person/Planet: The Creative Disintegration of Industrial Society* (Paladin edn). London: Granada (1st edn, 1978).

Scheffler, S. (1992): *Human Morality.* Oxford: Oxford University Press.

Schumacher, E. F. (1999): *Small Is Beautiful: Economics as If People Mattered: 25 Years Later ... with Commentaries.* Point Roberts, WA: Hardey and Marks Publishers.

Solomon, R. C. (1993): *Ethics and Excellence. Cooperation and Integrity in Business.* Oxford: Oxford University Press.

Steinbeck, J. (1951): *The Grapes of Wrath.* Harmondsworth: Penguin Modern Classics.

Styhre, A., J. Roth and A. Ingelgard (2002): "Care of the Other: Knowledge-creation through Care in Professional Teams," *Scandinavian journal of Management* 18: 503–520.

Svendsen, L. F. H. (2001): *Ondskapens filosofi ("The Philosophy of Evil").* Oslo: Universitetsforlaget (in Norwegian).

Taylor, C. (1991): *The Ethics of Authenticity.* Cambridge, MA: Harvard University Press.

Tranøy, K. E. (1999): *Det åpne sinn. Moral og etikk mot et nytt årtusen* ("The Open Mind. Morality and Ethics Towards a New Millennium"). Oslo: Universitetsforlaget (in Norwegian).

Van Krogh, G. (1998): "Care in Knowledge Creation," *California Management Review* 40(3): 133–153.

Vetlesen, A. J. (1994): *Perception, Empathy, and Judgment. An Inquiry into the Preconditions of Moral Performance.* Pennsylvania: Penn State Press.

Vetlesen, A. J. (1997): "Ondskap i Bosnia" ("Evilness in Bosnia"), *Norsk Filosofisk Tidsskrift* 1–2: 71–106 (in Norwegian).

Vetlesen, A. J. (2002): Inn i mørket ("Into the Darkness"), *Samtiden* 1: 48–59 (in Norwegian).

Wendel, W. B. (1995): "Lawyers and Butlers: The 'Remains' of Amoral Ethics," *Georgetown Journal of Legal Ethics* 9: 151.

Wendel, W. B. (2000): "Review Essay: Professional Roles and Moral Agency," *Washington & Lee Public Law and Legal Theory Research Paper Series,* Working paper, No. 00–10, November.

Whetstone, J. T. (2002): "Personalism and Moral Leadership: The Servant Leader with a Transforming Vision," *Business Ethics: A European Review* 11(4): 385–392.
Williams, B. (1973): "Against Utilitarianism," in: Williams, B. and J. C. Smart (eds): *Utilitarianism: For and Against.* Cambridge: Cambridge University Press.
Williams, B. (1981): *Moral Luck. Philosophical Papers 1973–1980.* Cambridge: Cambridge University Press.
Zsolnai, L. (2002): "Responsibility and Profit Making," in: Zsolnai, L. (ed. in co-operation with W. W. Gasparski): *Ethics and the Future of Capitalism.* New Brunswick, London: Transactions Publishers.
Zsolnai, L. (2003): "Decision Making in Multiple Value Perspectives," *International Journal of Value-Based Management* 16: 281–290.

KNUT J. IMS AND LARS JACOB TYNES PEDERSEN

From the Art of Reading to the Art of Leading: An Aristotelian Argument for Why Leaders Should Read Literature[1]

Introduction

This article explores how literature can be a source for developing the ethical competence of leaders. Leadership ethics is an integral part of leadership itself, and it refers to dimensions of ethicality in leadership behaviour, choice, and decisions. A fundamental problem in leadership ethics is *how leaders can develop ethical competence and wisdom*. In this article, we argue that literature can be a valuable source of ethical competence (see for example Ghesquière and Ims, 2010), and we explore several dichotomies of literature that are particularly important from the point of view of ethics in general and the development of ethical competence in particular.

According to Aristotelian ethics, maturity can only be learnt through practice and by studying the practice of wise people. This implies that the road to personal growth and maturity is based on experience and, as such, is consequently thorny and painful. It is painful for many people who are exposed to a dynamic leader's attempts at making his mark on the organization. It is equally painful for the leader who senses that his theoretical knowledge does not convince the employees, but creates friction and opposition. We would like to claim that the leader – inspite of his theoretical knowledge – lacks knowledge of what is special, particular, and unique. He does not realise the limits of theory and fails to see that universal knowledge is too sketchy to give guidelines for good decisions in practical situations.

We argue that literature is a shortcut to insight into the particular; that is, an alternative avenue to personal growth and maturity otherwise arrived at through experience.

Empirical studies of leaders by Henry Mintzberg (1973) concludes that leaders were primarily interested in *real-life stories* that were perceived to be meaningful. Mintzberg made the observation that leaders preferred concrete information, speculations, and rumours to the copious amounts of abstract routine reports in the office. We claim that leadership degrees do not reflect this fact and that literature is the genre that meets the leader's needs in this field to the greatest extent. Literature will make a leader more competent in his execution of leadership.

Leaders are faced with particular ethical challenges as their role grants them the authority and power to lead other people in the organization. Consequently, it is incumbent upon them to manage their role in a responsible manner. Leadership is complex, ambiguous, and conflict prone, and literature may, in all its complexity, inspire and stimulate leaders to execute their roles in a more responsible manner, and to honour their responsibilities when dealing with complex organizational and social problems.

In the following passages we will explore a wide range of arguments stating the reasons why leaders might gain valuable insights from literature. Firstly, we introduce some models for learning through literature, and discuss the distinction between context-dependent particular knowledge and context-independent universal knowledge. Secondly, we explore three relevant dichotomies – (1) proximity and distance, (2) form and content, and (3) explicit and tacit knowledge. Finally, we discuss how literature can simultaneously be viewed as a tool – and as such useful to the leader – and have value beyond utility.

Models for learning from literature

In *Nicomachean Ethics* Aristotle (1985) distinguishes between three intellectual virtues that we develop and use in professional, practical, and social action. *Techne,* that is technical knowledge, is expressed through our ability to choose adequate means by which to attain a given goal. *Techne* encompasses technical and productive skills central to professional work. *Episteme*, that is scientific understanding, refers to theoretical knowledge about the world which enables us to describe and explain incidents and connections. *Episteme* reflects abstract theoretical knowledge of the understanding of phenomena. Such knowledge is highly acknowledged in our culture. The third, intellectual virtue, is deemed to be of the greatest importance by Aristotle, but receives less attention in education and at work. *Fronesis*; that is, practical wisdom, encompasses the aforementioned virtues and is mainly reflected in successful actions. Practical wisdom does not only include consideration of the means needed to attain a particular goal, but also reflects evaluations as to whether the goal itself is desirable. Practical wisdom develops through practical experiences of choosing the middle road ("the golden mean") between extremities in order to realize what is good for its own sake. It is primarily learnt through emulation, that is by observing other wise people's actions and experiences.

Theoretical knowledge is typically acquired and developed through education and is further developed by testing this knowledge against observations and reflections on connections. Literature in a particular field and the teaching of experts are typical sources of theoretical knowledge. Technical skills develop through practice, which might also be a part of an education. Theoretical knowledge is general and describes absolute and more or less universal states and connections, while skills are often specific context-dependent knowledge. Practical wisdom reflects a type of knowledge which, to a great extent, is context-dependent as the individual considers what might be the right thing to do according to the important characteristics of the situation, and according to a holistic and more or less intuitive approach.

Since practical wisdom is developed through practical experience and by observing role models, literature might prove an important source for the development of this kind of action competence. Literature offers rich and in-depth descriptions of literary characters' actions in particular contexts. This could imply precise descriptions of the situation in which the action was taken and the ramifications of the action on the literary character(s), and the wider social and political arena. Literary works offer access to an excess of examples of people, character traits, relations, actions, situations, and worlds, and we can gain insights into experiences that might be inaccessible to us in our own lives. Literature does to a certain extent offer theoretical knowledge and knowledge applicable to skills, yet its main contribution is the vast amount of case studies of social action which might enable the reader to further develop his/her understanding of deep existential issues and interpersonal relationships.

In the reading of literature we distinguish between different types of consciousness. Skarðhamar (2005) postulates three levels of consciousness applied by a reader when reading a piece of literature. Firstly, the reader applies *aesthetic consciousness*; that is, relating to the text as a piece of literature and considering its aesthetic dimensions; for example, linguistic, poetic, or abstract dimensions. The relationship between the reader and the text is, consequently, both intellectual and emotional in the sense that one studies the literary qualities of the text and is influenced by its aesthetic expression. The reader might also apply *social consciousness* by noticing and hearing other voices in the text (cf. Bakhtin, 1981). In this manner one creates, to a certain degree, a virtual relationship between oneself and the literary characters in the text and relates to them intellectually, emotionally, and *empathically*. By exploring the individual characters and their projects one can study their relations and conflicts and thereby gain an understanding of why actions are made and the effect these actions have upon one another. The reader can engage in the life of one or several characters and thereby develop an empathic feeling for their situation. Lastly, the reader can relate to the text at a deeper level; that is what Skarðhamar (ibid.) refers to as existential consciousness, a level at which the reader builds a bridge between the text and his/her own life and relates to the theme of the text, asking the question: *What's in it for me?* The text may have implicit or explicit messages that are experienced as relevant to the reader's own life project and may, in the final analysis, influence the reader's identity. Through existential consciousness the reader may consider the text a source of inspiration. Role models and reported experiences may guide him or her in the development of his or her own competence and in dealing with complex problems of social or environmental nature.

In the following we further investigate how leaders can learn by reading literature, and the manner in which this learning may translate into action competence relevant to leadership.

We start by exploring the first dichotomy – between proximity and distance. Then we will discuss the central role of empathy for a leader and how the reading of literature may stimulate the development of empathy.

Dichotomy 1: Between proximity and distance

A basic dichotomy in the understanding of leadership is that of proximity and distance. A leader needs both perspectives and should alternate between them. As leadership and management degrees tend to focus on the development of analytical skills, and as leaders are recruited according to the possession of such skills, we suspect that the greatest challenge for some leaders is proximity; that is, the ability to be involved in rich and complex interpersonal relationships. Proximity may also apply to animals and nature. We understand proximity as receptivity to another person, that is "receptivity, sensitivity, a multidimensional awareness that is corporal, perceptible, emotional, and cognitive" (Vetlesen and Nortvedt, 1996: 159).

The ability to empathise is especially important. According to Thaulow (1993) empathy is the emotional understanding of another person. It is, however, important not to let your emotions control you. Empathy differs from sympathy (sym-pathos) in that the latter means "to feel or experience the same thing as another person, to become somewhat identical to another person or to share this person's situation" (ibid.: 494). Sympathy makes us subjective and uncritical participants. When we empathise with another person, we are more objective and critical. We try to understand the world as it is perceived by somebody else. "We arrive at empathy by listening to another person, by letting the conversation flow to and from. This is how we correct our preconceived ideas and get a plausible impression of another person's self-concept" (ibid.: 494). An important question to ask in this connection is: *How can we develop empathy*?

Thaulow (ibid.) claims that empathy is either latent or manifest, and that many people experience that their ability to empathise is curbed during their studies as they are not encouraged to respond freely, to be spontaneous, or to rely on their own sound judgment. It is therefore claimed that:

> [E]mpathy is emotional knowledge of another person rather than an intellectual understanding (…) Empathetic understanding does not emerge (…) from countless hours of intellectual pursuit. Empathy manifests itself in consciousness, appearing from nowhere, basing itself on life experiences, ideas and memory (…) deriving from the repository of trivialities and tracks of memory in the brain.(…) we must start from zero and forget our knowledge for a while (…) arguments and empathy may be considered contrasts. (Ibid., 494–495)

Our belief is that this ability may be stimulated by the reading of literature, as good literature is believed to draw us into the story and make us identify with the literary characters. We enter a person's life and build sympathy for or antipathy towards this person's actions and his or her characteristics. Another meaning proximity may have in connection with literature is the emotional mood which makes us open up and become receptive to a message from without. Knud Løgstrup's (1997) comments on the importance of this phenomenon are apt and to the point:

> Philosophy may – at best – make an understanding clear.
> Fiction may make it present.
> Philosophy may clarify, but fiction, a part of literature, may motivate and affect one´s will.

Or, to quote Peer Gynt:

> Yes, think it; wish it, want it as well; –
> But do it! No; This I cannot fathom.

Ibsen (1867) discusses here an important trait associated with human behaviour. It is not enough to wish for something; to execute it is completely different and much harder. The difference between the ability to evaluate and the ability to execute – that is, to evaluate what is valuable and to act in a manner which realises it – is one of the fundamental problems in ethics. It tells us that it is not sufficient to arrive at what is right or what is good. It also questions our culture's deep-rooted belief in knowledge: the belief that pure knowledge – *episteme* – will save the world. Knowledge does not necessarily lead to action and does not always manifest itself as a good ethical action. Literature may lead to more action than specialist texts and philosophy. The latter is characterised by succinct language of a technical nature. An abstract theoretical style appeals to the reader's intellectual abilities and typically points to, denotes, and gives information about reality. Everyday language, on the other hand, unites different levels of reality.

Let us approach the question from the opposite angle: How is it possible to avoid proximity? Vocabulary seems to be of great importance as one could use technical or personal language. Bauman (1993) discusses a wide range of strategies for unsatisfactory meetings, or *mismeetings* (cf. Ims, 1995). In this perspective, distance refers to foreignness and law, whereas proximity encompasses intimacy and morals. We may envisage different strategies for distance. One could, for instance, prevent physical space from becoming social, as this would involve feelings of sympathy and animosity. The main defence against being touched by other people is distance. Pure logic is applicable, but feelings may be "dangerous" in this context. By avoiding the story, we may compartmentalise our lives into separate segments and fragments that do not relate to each other and that do not have a history or a future. If we view our lives as separate segments we may remain at *status quo* and we do not have to worry about or take responsibility for the future. The logic is clear: do not give other people a face – count heads instead. If the other person is given a face, she appears in the foreground, which might entail that we might have to take her life situation into consideration.

Another strategy is to cultivate intellectualism, which means to measure and calculate everything and try to be as accurate as possible. This approach entails letting economic rationality seep into every area of life, including social life. This mode of thinking does not include emotions and impulsive action. Intellectualism may also manifest itself as the abstraction of different phenomena (people, for instance) in order to group or categorise them. Categorisation does, however, entail the reduction of *the other* to the same as *everyone else*. We do not perceive what differentiates this person from other people, but rather reduce *the other* to an element in a category. This is a common distancing strategy which often entails practising justice strictly, that is to treat everyone in the same way (cf. Bandura, 1999). This is a technocratic way of treating people without being sensitive to their individuality. A third strategy connected to the aforementioned strategy is to measure everything in economic terms. By focusing exclusively on money, one can achieve a high degree of precision and one may reduce qualitative values to countable amounts. To immerse oneself in money as an instrument, a goal, and a symbol – to value everything in money and strive towards the ideal of numeric calculation – draws one's attention away from human life.

Distance and intellectualism do, however, establish a necessary detachment to the phenomenon at hand. We need a bird's-eye view to see the whole context of which the phenomenon is a part. The warm heart must connect with the cold brain. If one is deeply involved in an issue or a story one tends to view the situation from one perspective exclusively. Consequently, only certain elements become prominent and decisions are made based on these (Dreyfus and Dreyfus, 1986). In such situations it is of the utmost importance to vary between different perspectives in a dialectic process. Jean Paul Sartre claims that people have the ability to reflect, to take a step back – "*se reculer*" – and in so doing they may see the phenomenon from a new perspective or several new perspectives. This is important in order to be able to envisage different avenues for action.

We could also ask the question: What should come first? Proximity or distance? Is the order of the factors insignificant? Our answer would be probably not. Martin Buber (1958) emphasised the need for I-Thou rather than I-It. The I-Thou relation is characterised by reciprocal immediate concrete experiences, sincerity and sensitivity. An I-It relation is, on the other hand, characterised by the objectification of another person, who is viewed as an external object, as a part of the kingdom of It (cf. Buber, 1958; Bråten, 1983). The two concepts I-Thou and I-It may also qualify the reading experience. The I-Thou experience is such that the reader experiences time standing still, and the presence of and relation to the other person (even though this is merely a literary character) is perceived as central and intense. Simmons (2002) also argues that contact must be established between the author/narrator and the reader/listener before we are able to open up to a story. A small story may on its own establish the interpersonal contact needed for cooperation to be lively and interesting. In this context Skarðhamar's (2005) distinction between aesthetic consciousness, where there is no such contact, and social and existential consciousness, where this is increasingly the case, is to the point.

We believe that reading may help us get in touch with our own feelings, with ourselves, with who we are, and who we want to be. By reading literature, we will primarily learn about ourselves and our own feelings and what it is to be human. Through a text we get access to different types of people, situations, cooperation, and conflicts that may carry us away in endless excitement. We are aided in our attempts to put into words our feelings, moods, and problems. And, after having read a text, we know more, see clearer, and have gained more acute powers of observation. And having put something into words, we are more likely to master it. This is both the Freudian and cognitive-therapeutic way to greater self-fulfilment and liberation (Kelly, 1995).

What makes us read and why are some pieces of literature more important to us than others? How may literature open us up and carry us away from our own situations and give us a strong sense of here and now, where time stands still? What do we find captivating in a good story? Stories give us the opportunity to wonder and stimulate a kind of contemplation that enables us to delve deeper into certain phenomena (see among others, Polyani, 1978). Simmons (2002) claims that stories affect us to such a great extent they temporarily paralyse the logical rational left side of the brain and give the floor to the right side, which is normally in the shadows. While the rational left brain builds walls of logic around certain notions, the right brain may challenge these opinions. Through the reading experience the story may constitute a trance through which our solid self-defence is weakened, which enables us to acknowledge more balanced views. A good story provides proximity to the reading material and leads us to the unconscious and unknown part of ourselves and our lives.

Reading literature normally involves relating to knowledge that is categorised. Much disciplinary and scientific knowledge (*episteme*) is condensed in common terms and is undoubtedly important. This type of knowledge is usually abstract and difficult to acquire in a personal manner and may appear authoritarian, as the researcher and her tradition have the power of definition. Literature is more democratic in that it is up to the reader to build meaningful connections. In literature there are many open spaces where readers may enter and add their own thoughts, perceptions, and feelings. This constitutes an exciting journey in one's self as real life is so rich. It would be a mistake to believe that only certain constructed categories may serve to reflect truth and reality.

Learning from literature is a creative interpretative process according to which every reader enters into a dialogue with a concrete text. It does not, therefore, make sense to postulate a unique model according to which individuals may take part in the learning process. We will, instead, suggest some central characteristics of the relationship between the reader and the text that are deemed essential, such as receptivity, empathy, and involvement. We will first focus on the dichotomy of form and content.

Dichotomy 2: Form and content

Løgstrup (1997: 219) emphasises rigid forms in literature, claiming that certain impressions and experiences cannot be expressed outside these borders as "the sensitivity of impressions and the strength of the experience would be incompatible with such freedom (…) The greater the sensibility and the intensity, the more succinct expression of these is required." In this case, only metaphors, rhymed style, and verse would serve this purpose. Eide (1991: 37) expresses similar views by stating that form carries meaning. Form may counter and neutralise spontaneous resistance and an unwillingness to threatening fantasies and inconvenient realisations by offering a recognisable frame within which a meeting with what is out of reach of our daily consciousness may occur. The form itself may encourage us to let go of what is familiar and enter the story, the picture, and the myth, which are simultaneously parts of ourselves and different from us. Literary effects such as language, words, rhythm, and style may in this way draw us – possibly reluctantly – further towards the unknown. We are, however, free to close the door at any minute (ibid: 37).

In poetics the goal of a story is catharsis. Aristotle's goal is to cleanse us by exposing us to compassion and fear through which we will achieve the catharsis of the soul. Aristotle also claims that tragedy might bring happiness, in the meaning of *hedone*. This may be interpreted in many different ways. Even if it is strenuous and painful to experience a tragedy, the death of a character is not the end.

> After a tragedy one is in balance and yet elevated, thoughts are greater, the soul is more complete, and this is happiness (…) One knows something (…) one has unlimited knowledge (…) This joy is related to realisation (…) artistic realisation. (Ledsaak, in Aristotle, 1989)

Our previous discussion of form has shown the importance of form in relation to content. In a management perspective, especially with regard to ethics, this is connected to the type of knowledge one seeks to acquire. As we have discussed in previous sections of the text, ethical knowledge is to a great extent practical knowledge. This means that ethical competence is mainly tacit (cf. Polyani, 1967), which has major implications for the manner in which this knowledge is acquired. We will, therefore, discuss these avenues in more detail by focusing on the third and final dichotomy of explicit and tacit knowledge.

Dichotomy 3: Explicit knowledge versus tacit knowledge

Much ethical knowledge is tacit, and important parts of knowledge may be viewed as intuition. One of the main arguments is that we cannot teach ethical competence by presenting ethical theories. We need practice; we need knowledge based on experience acquired through practice or literature. One way to argue for this view is to explain the difference between *know-how* and *know that* (cf. Dreyfus and Dreyfus, 1986). In the history of philosophy there are two schools related to knowledge: Plato, representing rationalism, and Aristotle defending *common sense*. Good knowledge should, according to Plato and Socrates, be expressed in explicit principles and definitions. Galileo, Descartes, and Leibniz subscribe to this view. If we, on the other hand, were to ask Aristotle what to do when faced with an ethical dilemma, he would say: *Do not ask the philosophers, ask a wise man*! David Hume is a part of this tradition, as are Pascal and Wittgenstein to a lesser extent. They view fundamental human understanding as an ability similar to that of knowing how to find one's life project. One cannot realise what one deeply appreciates by applying rules on particular instances, or by knowing a great amount of facts and rules that may unite different aspects. One knows how to ride a bicycle because of *know-how* acquired through practice and occasional painful experiences. The fact that one is not able to put what one has acquired into words illustrates that facts and rules do not give access to *know-how*. We also know that innumerable things may not be reduced to *know that*, that is context-free knowledge about how to do something. We are now touching upon the difference between the rule-based "*to know that*" and the experience-based "*to know how*".

Dreyfus and Dreyfus (1986) postulate five levels of competence: (1) novice, (2) advanced beginner, (3) competent problem-solver, (4) knowledgeable practitioner, and (5) expert. The first three steps revolve around the acquisition and application of context-free rules and problem solving from a distance. On the other hand, at the two highest levels intuition, holistic understanding, and *know-how* are essential. These are all employed in handling everyday tasks.

How does one progress from the lowest level to the highest? We suggest working with concrete context-dependent cases rather than trying to decompose and convert holistic experiences into rules. The latter strategy will result in major parts of the rich and diverse content disappearing due to calculating reductionist rationality, which hinders advancement from lower to higher levels of competence. The implication for literature is that reading concrete stories provides insight and understanding not catered for in textbooks, or a particular teaching practice focused on showing and describing definitions in an analytical and precise manner. Considerations based on reflection and distance should, however, not be viewed as the opposite of intuition (Dreyfus and Dreyfus, 1986: 160) as one may consider one's own intuitive experience-based holistic understanding. Thaulow (1993) interprets intuition as an expression of human warmth and recognition of another person and draws the conclusion that after having intuitively decided something, one may adjust the process by means of scientific terms.

How may leaders learn from literature? Firstly, they might learn about wise attitudes. They may also become aware of certain traits characterising literary texts (form versus content) and, lastly, they may reflect upon the competence or knowledge they wish to acquire (explicit versus tacit knowledge). We have now highlighted some of the sides of the object or, to use a less instrumental term, the source and the process. We will now summarise our discussion by returning to the question: *Could reading literature be useful for leaders?*

The utilitarian value of literature and its limits

In a hectic day, and in the light of the utilitarian ideas characterising modern organizations, valuable time should not be squandered on unnecessary tasks or activities. The critical question posed by leaders is therefore: *What is the utility of reading literature?*

Therein lies a paradox as literature may be considered simultaneously of value and completely unnecessary. It is unnecessary, as it is fictional and carries us to a completely different world. It may constitute an aesthetic experience, an example of living art, and a contrast to ordinary daily life. It may, however, also be of value as it provides distance and perspective to intense problems that seem to whirl around in one's head after a long and demanding day at the office. Literature is useful as it may – in the ways mentioned above – help the leader to develop maturity both with regard to leadership competence and with regard to personal development in a more existential sense. Insight gained into social and interpersonal relations and the various implications of actions in different contexts may provide knowledge relevant to leadership, as rich cases and insight into other people's experiences are provided. Literature may also be a mirror to the leader's life and enrich his or her moral universe by giving ideas as to how one *could* live and how one *should* live. Literature may in this way contribute to the development of a leader's identity by stimulating reflection on values, moral horizons, and images of life, through which the creation of a leader's individual character may emerge.

In *The Philosophy of Life* Arne Næss (2005) warns of the dominant technical-economic perspective which permeates our educational institutions. This rationality has lost sight of the goals. We become so obsessed with our projects that we no longer ask what the deeper purpose of these projects is. Næss (ibid.) adequately refers to this mentality as *fluff*, a kind of stolidity that could be rational in the short-term, but is often irrational in the long run. It will be a superficial success at best, and could lead to great mistakes in the future (Ims and Zsolnai, 2006). *Caring about the goals* is more important than rationality. The book may be interpreted as opposition against education that is too analytically oriented, which does not enable us to ask the question: *What is the most important thing in our lives and in our existence?* (Næss, 2005: 60) Such an analytical approach, with its emphasis on theory, may even thwart the joy of reading among literary researchers (Hemmer, 2003: 562–569). Our conclusion is that literature is not of value in a technical reductionist sense. Literature does in many ways transcend purely utilitarian considerations. The main effect of literature is to enhance a person's creativity.

The solution is not to grab a book and assume that this type of learning and development suddenly occurs. Reading requires reading competence, and in order to be able to read a text one must possess a complex structure of abilities and skills. What if the leader does not possess this competence? Is there any purpose in reading literature then? We claim that reading does not require a fundamental reading competence. All understanding is, to a certain extent, constructive work on the part of the reader. Educational institutions have emphasised knowledge of the *know that* type, but the question is whether the definition of an expression is "an exercise in the superficial grammar of the language"? (Danbolt et al., 1979). *Know how*, on the other hand, refers to how words are employed in concrete practical contexts. This cannot be specified, but must be acquired in a particular context. In order to understand an utterance in practice one must place this practice in relation to similar practices. "To have a conceptual grasp of the world is to master a very complicated web of different skills and ways to behave where each concept is identified according to surrounding concepts" (ibid.: 20).

Two concepts that seem essential when discussing the requirements for the ability to learn from literature is receptivity and attentiveness (Vetlesen, 1994). The latter is inadequately translated as it refers to *an inclination to sense something in particular*. Nussbaum (1990: 162) describes this type of sensibility more concretely as an inclination to "be touched by other people's difficulties and problems", and emphasises thus the socio-moral aspect of this type of attentiveness. This requires receptivity, described by Vetlesen (1994: 18) as "an openness to the world and what we experience in it" and as an active willingness to turn one's attention towards and catch and evaluate signals in one's environment. The individual must, consequently, remain open to stimuli in the environment. People vary with regard to openness and attention, and this depends on who we are and our emotional and cognitive inclinations and skills (cf. Pedersen, 2009).

In the context of reading literature the implication is that learning requires receptivity to the possibility of learning from literature, in addition to consciousness and involvement while reading. The reader must actively apply social and existential consciousness to the text, which requires the ability to engage with the text at a deeper level than wanting to know how the story ends. It is, therefore, necessary to reflect on *why* one reads literature; that is, one should ask the question: *What do I hope to achieve*? In a study on knowledge and culture in the upper middle class in Norway, Skarpenes (2007) discovered that economists dominated the group reporting using literature as entertainment only, which suggests that leaders, to a lesser extent, approach literature with social and existential consciousness. In our model (Skarðhamar, 2005) this is a restricted point of view which may not release and provoke the full potential of reading. We know that economists do not read much literature and, combined with relatively one-dimensional reading, economists as a reading group have great potential. As we see it, consciousness and reflection on the purpose of reading is also very important because it may guide the reader's attention towards certain traits, and may encourage involvement on the part of the reader. In this way the reader may actively relate to his or her reading competence and its purpose, in addition to which texts are best suited to contribute to the development of the reader's competence and learning.

Conclusion

In this article we have explored how leaders may learn from literature and the manner in which literature constitutes a valuable source of learning for leaders. We have, in particular, focused on emotional and empathetic development through proximity and distance. We have also explored the relationship between form and content, the different types of knowledge that may be acquired, and the utility of literature and its limits.

There are many approaches and stages in the reading process. One may focus on the reading experience and be thrilled by the superficial traits of the story by applying aesthetic consciousness. Social and existential consciousness, on the other hand, leads to the reader actively listening and letting himself/herself be immersed in the different voices and perspectives in the text, feeling sympathy for or antipathy towards the literary characters.

Lastly, a text may give insight and build bridges between itself and the reader's life. In this manner, the text may provide experience, involvement, and insight for leaders.

This type of learning may be more or less directly relevant to leadership as the leader develops his or her identity and character by being inspired by other people's experiences, by role models, and by emotionally developing antipathy towards evil and self-destructive behaviour, which also occurs in literature. Literature typically offers a broader range of expressions of thoughts and feelings than does factual prose. This is especially important for leaders who normally come across technical vocabulary and logical reasoning related to economics and profit.

We believe that the virtue of practical knowledge may develop if one registers and is inspired by other people's experiences, by the life projects of literary characters, and the rich and many-faceted descriptions of their actions. A book will shed light on the long-term effects of actions and the implications these actions may have on the character's personal life. Working life is characterised mainly by leaders who have obtained degrees from theoretical and analytically oriented educational institutions. Literature may, therefore, constitute a complementary source of learning, inspiration, and creative development. The danger of traditional education is that imagination is stifled and people's creativity is consequently reduced. We believe that leaders should be encouraged to handle the polyphony of the environment, and may advance from a reductionist to a holistic understanding of complex issues, fine-tuning their feelings and sensitivity. This will enable them to spot the moral, social, and existential dimensions of problems previously only conceptualised as technical or economic. In this way literature may contribute to the enrichment of a leader's life and to new perspectives on leadership. It may even contribute to *improved* leadership.

Note

1. The authors are listed alphabetically. The article is written in dynamic collaboration and the authors are jointly responsible for its content. The article is a refined synthesis of two articles previously published in Norwegian (Ims and Pedersen, 2009a; 2009b). We are greatly indebted to Assistant Professor Agnete Løvik, for her help in translating our manuscript from Norwegian.

Bibliography

Aristotle (1968), *Poetics (with introduction, commentary and appendixes by D. W. Lucas)*. Clarendon, Oxford.

Aristotle (1985), *Nicomachean Ethics (translated by T. Irwin)*. Hackett, Indianapolis.

Aristotle (1989), *Om diktekunsten (Oversatt fra gresk med innledning og anmerkninger avSam Ledsaak med et essay om å lese poetikken av Jostein Børtnes)*. [*Poetics (translated from the Greek with an introduction and comments by Sam Ledsaak with an essay about how to read poetics by Jostein Børtnes*]. Dreyers Forlag, Oslo.

Bakhtin, M. M. (1981), *The Dialogical Imagination: Four Essays*. University of Texas Press, Austin, TX.

Bandura, A. (1999), Moral Disengagement in the Perpetration of Inhumanities. *Personality and Social Psychology Review*, 3(3), 193–209.

Bartley, W. W. (1973), *Wittgenstein*. Open Court, LaSalle, IL.

Bauman, Z. (1993), *Postmodern Ethics*. Blackwell, Oxford.

Bloom, H. (1994), *The Western Canon: The Books and School of the Ages!* Riverhead Books, New York, NY.

Bråten, S. (1983), Modeller av dialogiske systemer: Tid og selv-refleksjon [Models of Dialogical Systems: Time and Self-Reflection]. In: B. W. Hennestad and F. E. Wenstøp (eds): *Bedriftsøkonomi og vitenskapsteori* [*Business economics and the Theory of Science*]. Universitetsforlaget, Oslo.

Buber, M. (1958), *I and Thou*. Scribner, New York, NY.

Danbolt, G., K. S. Johannessen and T. Nordenstam (1979), *Den estetiske praksis* [*The Practice of Aesthetics*]. Universitetsforlaget, Bergen.

Dreyfus H. L. and S. E. Dreyfus (1986), *Mind over Machine: The Power of Human Intuition and Expertise in the Era of the Computer*. The Free Press, New York.

Eide, T. (1991), *Outsiderens posisjoner: Axel Jensens tidlige forfatterskap* [*The Outsider's Positions: The Early Works of Axel Jensen*]. Universitetsforlaget, Oslo.

Ghesquière, R. and K. J. Ims (2010), *Heroes and Anti-heroes: European Literature and the Ethics of Leadership*. Garant, Antwerp-Apeldoorn.

Hemmer, B. (2003), *Ibsen: Kunstnerens vei* [*Ibsen: The Path of the Artist*], Vigmostad Bjørke, Ibsen-museene i Norge, Bergen.

Ibsen, H. (1867), *Peer Gynt: et dramatisk digt* [*Peer Gynt: A Dramatic Poem*]. Gyldendal, København.

Ims, K. J. (1995), Mismeetings: An Essay about Money, Urban Life and the Ethics of Proximity. In T. B. Holmesland, K. J. Ims and A. Pedersen (eds): *Essays in Marketing and Management: A Festschrift in Honour of Kjell Grønhaug*. Fagbokforlaget, Bergen.

Ims, K. J. and L. J. T. Pedersen (2009a), Å omsette lesekunst til lederkunst: "Peer, du lyver!" og lederkompetanse [Translating the Art of Reading to the Art of Leading: "Peer, You Are Lying!" and Leadership Competence], *Magma: Tidsskrift for bedriftsøkonomi* [*Magma Business Journal*], 5.

Ims, K. J. and L. J. T. Pedersen (2009b), Fortellingens fortrylling: å lese Peer Gynt – nyttig? [The Wonder of the Story: Reading Peer Gynt – Useful?], *Magma: Tidsskrift for bedriftsøkonomi* [*Magma Business Journal*], 6.

Ims, K. J. and L. Zsolnai (2006), Shallow Success and Deep Failure. In: L. Zsolnai and K. Ims (eds): *Business Within Limits: Deep Ecology and Buddhist Economics*. Peter Lang, Oxford.

Ims, K. J. and L. Zsolnai (2010), Self-realization in Business: Ibsen's *Peer Gynt*. In: R. Ghesquière and K. J. Ims (eds), *Heroes and Anti-heroes. European Literature and the Ethics of Leadership*. Garant, Antwerp-Apeldoorn.

Kelly, G. (1955), *The Psychology of Personal Constructs*, Vol. I and II. Norton Company, Inc, New York, NY.

Løgstrup, K. E. (1997), *The Ethical Demand*. The University of Notre Dame Press, NotreDame, IL.

Mintzberg, H. (1973), *The Nature of Managerial Work*. Harper & Row, New York, NY.

Næss, A. (2005), *Livsfilosofi: et personlig bidrag om følelser og fornuft* [*The Philosophy of Life: A Personal Contribution about Emotions and Reason*]. Universitetsforlaget, Oslo.

Nussbaum, M. C. (1990), *Love's Knowledge: Essays on Philosophy and Literature*. Oxford University Press, New York.

Pedersen, L. J. T. (2009), *Making Sense of Sensitivity: Moral Sensitivity and ProblemFormulation in Business*. Ph.D. Dissertation. Norwegian School of Economics and Business Administration, Bergen.

Polanyi, M. (1967), *The Tacit Dimension*. Routledge & Kegan Paul, London.

Polanyi, M. (1978), *Personal Knowledge: Toward a Post Critical Philosophy*, Routledge and Kegan Paul, London.

Schumacher, E. F. (1977), *A Guide for the Perplexed*. Cape, London.

Simmons, A. (2002), *Historiefortelling* [*Storytelling*]. N. W. Damm & Søn, Oslo.

Skarðhamar, A. K. (2005), *Kunsten å lese skjønnlitteratur: om lesestimulering og lesekompetanse* [*The Art of Reading Literature: About the Stimulation and Competene of Reading*]. Universitetsforlaget, Oslo.

Skarpenes, O. (2007), Den «legitime kulturens» moralske forankring [The moral anchoring of the «legitimate culture»], *Tidsskrift for Samfunnsforskning* [*The Journal of Societal Research*] 04, 531–558.

Thaulow, J. F. (1993), Om samtalen [About Conversation]. *Tidsskrift for Norsk Lægeforening* [*The Journal of the Norwegian Association of Medical Doctors*] 4(113), 494–495.

Tønneson, J. L. (2008), *Hva er sakprosa?* [*What is Prose?*] Universitetsforlaget, Oslo.

Vetlesen, A. J. (1994), *Perception, Empathy and Judgment: An Inquiry into the Preconditions of Moral Performance*. The Pennsylvania University Press.

Vetlesen, A. J. and P. Nortvedt (1996), *Følelser og moral* [*Emotions and Morality*]. AdNotam Gyldendal, Oslo.

KNUT J. IMS AND LARS JACOB TYNES PEDERSEN

Personal Responsibility and Ethical Action

Introduction and central issue

In this chapter we discuss the nature of personal responsibility and ethical action – two concepts that are central to business ethics, both in theory and practice. A main concern of business ethics is to distinguish between ethical and unethical decisions or actions in business; that is, identifying ethically relevant differences between acts that are justifiable from an ethical point of view and those that are not. One of the objectives of such analyses is to assess whether a particular act constitutes *responsible behaviour.*

Assessments of responsibility are often retrospective, in the sense that to a large extent they focus on the *accountability* of the decision maker for decisions and activities that have been carried out. The focus is typically the degree to which they: (1) comply with or breach important moral principles (deontological ethics); (2) produce a positive or negative surplus of consequences (utilitarian ethics); or (3) are consistent or inconsistent with conceptions of virtuous character and practice.

However, in line with Hans Jonas's (1984) emphasis on the continuous increase of power that follows from technological progress and the corresponding vulnerability of nature, business ethics also needs to focus on the *prospective responsibility* of individuals and organizations. Prospective responsibility, argues Zsolnai (2008), implies that the moral actor has a duty or obligation to ensure that something occurs. Hence, responsibility is *future-directed* – it deals with the question of *what to do* in order to achieve certain desired outcomes, or to avoid certain undesirable outcomes.

Already at this point in distinguishing between ethically justifiable and unjustifiable acts, we can see the close connection between questions of ethicality and questions of responsibility – notions that are inextricably intertwined. As this chapter progresses, we will further explore the question of how responsibility and ethical action relate to each other.

We will also emphasize how an action perspective should be distinguished from a decision perspective in the sense of processes and outcomes. There may be decisions without actions and actions without any prior decisions. While the decision paradigm indicates rational oriented actors, an action paradigm is more mundane in the sense of assuming non-heroic individuals who may be seriously torn between conflicting claims and duties, and ultimately have to choose one of them. Such processes of choice are not necessarily driven by one-sided rational thoughts, but may be partly based on deep emotions like empathy and compassion.

We aim to give an insight into the role of personal responsibility in promoting ethical action in business organizations. Moreover, we will discuss several different strategies for assuming personal responsibility in order to shed light on the different meanings and ways of acting responsibly in business. We will discuss how this involves acting in a manner that is both (1) other-directed, in the sense that it takes into account important norms and the interests of stakeholders, and (2) self-directed, in the sense that it is compatible with the personal values and life-project of the individual moral actor. This leads us to the central issue: how individuals in organizations can act responsibly with regard to the organizations of which they are a part, the stakeholders who are influenced by their decisions and acts, and towards themselves. Often, their objectives will conflict, creating dilemmas and loyalty problems for the individual. This is a core problem when attempting to act responsibly. For what and towards whom are you responsible – and how do you make priorities in practice?

This chapter is structured as follows. In section "State of the art of current theories" we outline contemporary theories of responsibility and what they tell us about what constitutes ethical action. In section "New approaches and solutions" we discuss novel perspectives on promoting personal responsibility in organizational life. In section "Real-world examples" we present real-world examples in order to illustrate the theoretical points discussed throughout the paper. For purposes of illustration, we choose the individual's decision on whether or not to blow the whistle following the realization that one's organization is doing something blameworthy. We will discuss three different strategies – exit, voice, and loyalty (cf. Hirschman, 1970) – as decision alternatives facing the individual and discuss them in the light of personal responsibility and the ethicality of the action. Finally, we briefly outline our conclusions (section "Conclusion").

State of the art of current theories

In this section, we outline central contemporary perspectives on responsibility and ethical action. First, we discuss the nature of ethical action. Second, we outline various conceptions of responsibility. Third, we discuss a framework for responsible decision making. Thereby, we provide the reader with a broad perspective on responsibility in organizational life.

Ethical and unethical action

Ethical actions are typically exercised in situations where there are multiple values at stake and where important values can be realized or protected by the choice and behaviour of the individual. This means that the individual as a moral actor needs to understand that he or she is faced with an ethical problem and has to make a responsible decision and carry out that decision in actual behaviour (cf. Vetlesen 1994). First, this requires moral sensitivity, the ability to recognize that important values are at stake in the situation and that the action alternatives between which he or she must choose have different value implications for the various involved parts or stakeholders (Pedersen, 2009). In such a situation, the person considers (1) what is at stake and (2) for whom. Thus, he or she identifies stakeholders – among them *himself* or *herself* – to which he or she must attend, as well as their diverse interests or needs. Moreover, in order to act ethically, the individual needs the ability to reflect on the ethically relevant dimensions of the problem and arrive at a justifiable decision about what should be done. Finally, the individual needs the moral character to carry out the decision, even if it entails personal cost. As discussed by Zsolnai (2002), moral character is necessary for ethical action in most situations since there is often a high cost in acting ethically. This is a characteristic of many, but not all, ethical actions. The fact that costs are often associated with such an act implies that the most comfortable course of action is often not the ethical one.

Another important characteristic of both ethical and unethical action is that the individual may *unconsciously* carry out an act. This implies that he or she may act ethically or unethically without considering the act from an ethical point of view at all, or equally the (un)ethical act may be the outcome of a conscious, deliberative process. Opdal (2011) describes the different types of unconscious motivations for individuals and propagates the importance of knowing more about such motives in order to tackle important problems, such as the climate crisis. He draws the conclusion that we do not like to be responsible when the consequences of our decisions are likely to be unacceptable to us (Opdal, 2011, p. 36). Generally, we hold individuals responsible for the choices they make and the acts they carry out, whether or not those acts are the outcome of a conscious ethical judgment. In the following paragraphs we outline relevant perspectives on responsibility in organizational life.

Various conceptions of responsibility

Responsibility is a concept at the heart of ethics. However, it is a concept with multiple meanings, which are interchangeable and overlapping. According to Bovens (1998, pp. 24–26), five forms of responsibility can be distinguished and give an indication of the broadness of the concept. First, we may talk of *responsibility as a cause*; that is, people, things, or circumstances may cause the emergence of certain outcomes. We refer to this as causal responsibility. Second, we may talk of *responsibility as accountability*, that is moral, political, and/or legal liability for the outcomes of a behaviour carried out by the agent. This means that moral responsibility is closely linked to the concept of accountability, which implies giving an account of why a given behaviour was carried out (Messner, 2009). Using the terms of Jonas (1984), this might be called formal responsibility. Third, we may talk of *responsibility as capacity*, that is the responsibility (or lack thereof) that follows from being *able to* exercise a certain amount of responsibility (either due to power, knowledge, or mental ability, or the lack thereof) (Sen, 2002). This is what is reflected in the notion "noblesse oblige" or "with great power comes great responsibility". It implies that an individual's responsibility is a function of the degree to which he or she has the means to take responsibility. Fourth, we may talk of *responsibility as task*; that is, the obligations or functions that follow from having a particular role or position. This is closely related to role-mediated or professional responsibility. Finally, we may talk of *responsibility as virtue*; that is the value-laden individual character trait that inclines the individual to act responsibly.

As the five forms of responsibility indicate, we variously use the concept of responsibility to denote features of: (1) the individual and his or her capacities and character, (2) his or her role or professional affiliation, or (3) his or her agency in a particular situation. However, Jonas (1984) uses several distinctions that should be mentioned. For one thing, we have contractual responsibility, where we assume that two partners are more or less on an equal footing. This implies a kind of horizontal responsibility. In addition, we also have a kind of vertical, non-reciprocal responsibility, of which the parent-child relationship is the paradigmatic case. The same reasoning should be used for man's relationship with nature and future generations. Jonas's (1984) central concept is *substantive responsibility*, which is for what matter you are responsible. In this way, Jonas extends the horizon of responsibility and imposes a positive duty of power on human beings to take wider responsibility. For the remainder of the chapter, we will focus on the individual's *moral responsibility* and will build on several of the above different forms in our discussion. Moreover, as noted in the introduction, it is useful to distinguish between retrospective and prospective responsibility (Jonas, 1984). In our discussion of personal responsibility we will focus on the latter, because personal responsibility in organizational life is largely oriented toward acting in a manner that ensures that acts and outcomes are aligned with the individual's personal beliefs, values, and life-projects.

Responsible decision making

In light of the multitude of meanings of responsibility, there is a need for a concrete and more operational framework for making responsible decisions. Zsolnai (2008) offered a comprehensive framework for responsible decision making that comprises three dimensions of responsibility for managerial decision making. According to Zsolnai responsible decisions integrate *rationality* in goal attainment, *respect* for stakeholders, and *reverence* for ethical norms.

This framework implies that decision makers should balance three different objectives in order to promote responsible decisions. First, the decision maker must formulate the goal and find the most efficient means to achieve it; that is maximize the goal attainment value of the decision problem. Second, the relevant ethical norms related to the decision must be identified and the manner in which the norms constrain the decision must be taken into account. Third, the relevant stakeholders and their interests must be identified and the implication of respecting their interests for the decision must be understood.

Decision makers make trade-offs along these three dimensions, which are seldom in harmony. According to Zsolnai's model (2008, p. 116), responsible decision making involves a multidimensional evaluation of any decision alternative based on three types of value. First, the decision should be evaluated with regard to whether it is rational for adequately achieving the desired goal. Second, it should be evaluated based on the degree to which it complies with relevant ethical norms. Finally, it should be evaluated based on the degree to which the decision respects the interests of affected stakeholders. The decision rule that is suggested in order to ensure the most responsible decision is the *maximin* rule (cf. Rawls, 1971); that is, the decision maker should choose the least worst alternative when all three dimensions are taken into account. This means that the worst aspect of the chosen alternative is better than the worst aspect of any other decision alternative.

Zsolnai's (2008) framework for responsible decision making provides an insight into the multidimensional nature of responsible decisions as ethical action, and it makes explicit how trade-offs between different dimensions of one's responsibility can be carried out in practical decision making.

New approaches and solutions

In this section we outline novel perspectives on responsibility – and in particular *personal responsibility*. First, we discuss personal responsibility as the individual's active responsibility – taking in organizational life. Second, we develop this perspective on personal responsibility by discussing the concept of *shared responsibility* and what it implies for personal responsibility.

Personal responsibility

In the business ethics literature, the main emphasis is on corporate social responsibility, which deals with questions relating to the responsibility of business organizations. Central questions in this regard are: (1) For what are they responsible, that is their domain of responsibility? (2) Towards whom are they responsible, that is the scope of relevant stakeholders? and (3) How far does this responsibility extend, that is their horizon of responsibility?

These are important issues in contemporary business. However, there is a danger that such strong emphasis on this organizational responsibility displaces important questions about the *personal responsibility* of individual organizational members. Most importantly, it is a mistake to reduce the personal responsibility of the individual actor to the sum of role-defined responsibilities that he or she assumes as an agent for the organization. While considering the concept personal responsibility, we will move beyond this narrow understanding of the individual's responsibility.

Ims (2006) proposed a threefold model of responsibility, which distinguishes between role-mediated responsibility, common morality, and personal responsibility (see Figure 1, p. 725). In a business context, the individual's *role-mediated responsibility* is often the focus and refers to the responsibilities assumed by the individual when he or she takes on a given role in the organization (cf. Bovens's (1998) concept of responsibility as task). These role-mediated responsibilities are professional responsibilities, and they give allowances and limitations to the individual. For example, a surgeon is allowed to cut open other people's bodies because of the surgeon's role, but this role also limits the domain of acceptable actions, such as not being allowed to enter into a romantic relationship with a patient. Therefore, the role determines acceptable and unacceptable types of behaviour and places a set of responsibilities on the individual who takes up the role.

Secondly, Ims (2006) points out that any individual is bound by the *common morality* of his or her community or society. This means that there are moral values and corresponding moral obligations that are shared by a community and society, and the individual is partially bound by these values and obligations. In the debate about professional responsibility there has been strong support for a type of ethics within business that might not be restricted by common morality. A little bluffing in business, perhaps, should be allowed according to some experts, because everyone in business does it. However, as a general rule common morality trumps professional ethics when professional and societal values are at stake. Common morality is constantly evolving through dialogue between members of the community and as a result of societal changes.

Finally and most importantly, however, the individual has a profound *personal responsibility* that is independent of the individual's roles and of his or her being part of a given community. Moreover, this personal responsibility sets the limits for the individual's role-mediated responsibilities, in the sense that the individual evaluates what types of behaviour he or she is willing to engage in as the performer of various roles. Similarly, the personal responsibility of the individual is in constant tension, both with common morality and role-mediated moralities. The individual's moral life evolves in the tension between adhering to the moral tradition of the community and making radical value choices that sometimes may conflict with that tradition, which is how common morality evolves over time (cf. Kekes, 1991). Similarly, the individual's personal values are in continuous tension with the values inherent in the tasks carried out as part of his or her professional role. Therefore, when the individual experiences a mismatch between the values upon which he or she exercises professional responsibility and his or her own values – or similarly between the values the community expects him or her to adhere to and his or her own values – it is necessary to critically reflect on the conflicts and prioritize before choosing and acting. Personal responsibility exists in context, and reflections on one's own norm-horizon should be seen in relation to shared norm-horizons.

Ims's (2006) triangle illustrates a central challenge with regard to acting responsibly in organizations, namely that decision makers typically experience *conflicting loyalties*. In such situations the individual's personal responsibility may clash with his or her professional responsibility, or with a sense of responsibility towards the broader society. For some these two responsibilities may overlap such as in the case of auditors whose professional responsibility largely coincides with their social responsibility, which is to ensure that the financial information communicated to societal stake holders is accurate and trustworthy. However, there are many examples of professionals whose organizational responsibilities clash with their own personal beliefs or values. A contemporary example is a group of Norwegian medical doctors who refused to act in line with official governmental policy and recommend pregnant patients to take the swine flu vaccination during the recent epidemic. Even though public health organizations instructed them to give this advice, they refused on the grounds

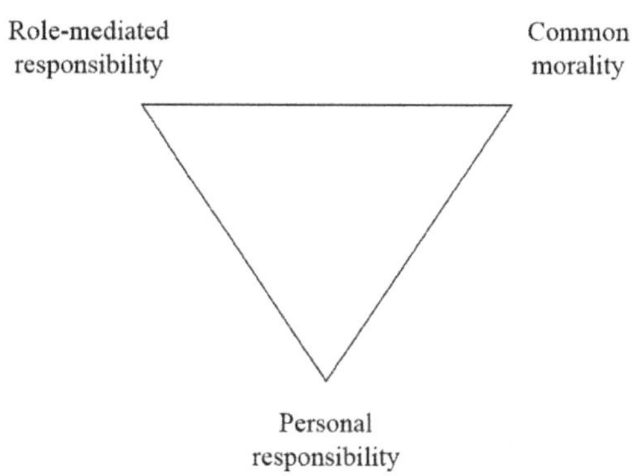

Figure 1. The triangle of responsibility. *Source*: Ims (2006).

that they felt personally responsible towards the patients and that they were not convinced that existing research gave adequate reassurance that the unborn children would not be harmed.

This cross-pressure of conflicting loyalties can be illustrated in a minimalistic stakeholder model that we may call *the triangle of professional ethics* (see Figure 2). In the model, the individual agent or professional is at the center of the triangle, and arrows indicate his or her responsibilities towards other key stakeholders. As the model suggests, at any one time a professional is typically responsible in different ways and to different degrees to at least one client (for example, customer–patient), to one's employer, and to society at large. In the case of the medical doctors, their responsibility to the client seemed more important than their employee responsibility to their employer, or their social responsibility for acting in line with publicly regulated guidelines.

In light of Zsolnai's (2008) framework for responsible decision making, we can interpret the doctors' decision as an application of the maximin principle. In this case, the doctors promoted the goal attainment value of the decision (protecting the health of the pregnant women and their unborn children). Moreover, they prioritized a "do-no-harm" norm over the norm of following public health guidelines – both of which may be seen as important norms. Finally, they made an assessment of the stakeholders involved and their interests, and gave priority to the interests of

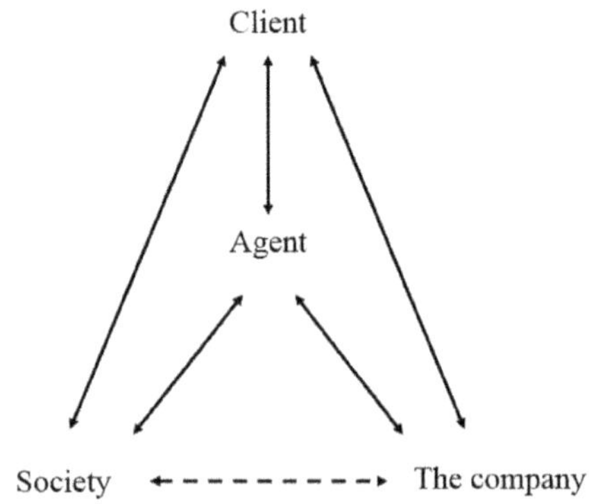

Figure 2. The triangle of professional ethics.

the pregnant women who were at risk of becoming infected by swine flu (rather than the stakeholders). In this way, we can see how doctors' attempts to resolve conflicting loyalties are compatible with the principles of the responsible decision making framework.

Shared responsibility

Somewhat paradoxically, the perspective on personal responsibility can be broadened by taking into account the concept of *shared responsibility*. At first glance, this concept may appear contrary to the notion of personal responsibility. However, we will argue that there are important implications for personal responsibility by assuming a perspective of shared responsibility.

The basic idea of shared responsibility is that individuals within a community (for example, within an organization) should see themselves as sharing the blame (that is being accountable) for harms perpetrated by, or occurring within, that community with or without their active participation (May, 1992). This extends the individual's accountability to include acts that he or she has not carried out, but on which the individual may have had some ability to exert influence. Seeing responsibility as shared in a community or an organization implies awareness that one's personal responsibility transcends one's own choices and behaviour. This may cause us to consider our roles, omissions, and attitudes in light of our ethical standards in a similar way to how we consider our actual behaviour (May, 1992). Hence, such an approach to responsibility implies extending our horizon of responsibility. The perspective is in line with Hannah Arendt's (2003) proposition that to obey and obediently take part in organizational activities should be seen as an active support of that activity. This perspective is also in opposition to Jean-Paul Sartre's existential perspective, wherein all responsibility is laid on the shoulders of the single individual actor.

Thinking in terms of shared responsibility means ascribing responsibility in part based on the *attitudes* of individuals; for example, the individual organizational member's attitude of *lenience* toward organizational activities that he or she is not a part of, but does not speak out against.

This is a strong form of responsibility, in the sense that it assumes the perspective that individuals who have some degree of ability to influence the activities carried out in an organization thereby carry a responsibility for doing so if it can lead to avoidance of harm or transgressions. As such, this notion gives attention to the fact that we can be personally responsible for acts we contribute to, or facilitate, only in an indirect sense (May, 1992). Causally speaking, the individual is part of the causal chain that creates or facilitates the outcome and, therefore, has a share in the responsibility for that outcome.

Of course, this does not imply that an individual who is part of an organization or community automatically shares the blame for any wrongdoing perpetrated by that group. As discussed by Williams (1981) and Ims (2006), an individual should not take responsibility for the projects of others, which may conflict fundamentally with his or her own projects. One crucial issue here relates to the individual's degree of power and freedom, that is the degree to which the individual could have prevented or influenced the act in question in order to hinder negative outcomes. In such a prospective sense the extent of the individual's share in the responsibility varies according to the role he or she could have played in avoiding the negative outcome (May, 1992). Another crucial issue is: What's at stake? In general, we cannot require supererogatory behaviour in the name of ethics – for example, require a person to sacrifice his or her own life in order to do what might be right from a utilitarian viewpoint. We should warrant that the agent takes his or her ground projects into account before considering an act that might fundamentally conflict with his or her own deep convictions.

Real-world examples

In this section, we outline and analyze some real-world examples that illustrate our perspective on personal responsibility and ethical action. For the purposes of illustration, we focus on the individual's decision of whether or not to blow the whistle following the realization that his or her organization is doing something blameworthy. We will discuss three different strategies – exit, voice, and loyalty (cf. Hirschman, 1970) – as decision alternatives facing the individual, and discuss them in light of personal responsibility and the ethicality of the action.

Exit as responsible action

At the end of 2008, Inge Lisenka Wallage, who at the time was vice president of international communications in the Norwegian oil company StatoilHydro (hereafter referred to as Statoil, which is the company's current name) made a radical and surprising decision. During her two years in communications at Statoil, Wallage had spent a large amount of her time justifying and defending Statoil's controversial oil sands venture in Canada, but she had grown increasingly uncomfortable with the task. In spite of her powerful and prestigious position within a major oil company, a role one might imagine many business students and practitioners would envy, Wallage came to an inevitable decision. She resigned from her position in Statoil and started working for Greenpeace – Statoil's main antagonist in the oil sands venture.

Reactions to Wallage's decision were plentiful. Statoil demanded that she left her post immediately. Former Statoil executives questioned her direct transfer to "the enemy", while business professors hailed Greenpeace's new recruit as a strategic coup (Andreassen, 2009). Wallage herself explained her decision in an interview, stating that she wanted to be able to look back at her life and say that she had contributed to making the world a better place. Moreover, she asserted: "We only have one planet, and we have to treat the globe with respect" (Kongsnes, 2009, p. 6). She emphasized that she had entered her role in Statoil in order to contribute to the sustainable development of the company's operations, but that she had failed to do so. Finally, she explained her choice with reference to her responsibility towards her children and said: "I hope that my children will grow up able to live on a planet that is still intact and that I have contributed to this through my work life" (Kongsnes, 2009, p. 7).

Wallage's decision was – for her – the only ethical action in her situation. She had tried and rejected other action alternatives that were compatible with her fundamental objective of contributing to a sustainable future. Her ethical action was grounded in a deep sense of personal responsibility: (1) *towards* various stakeholders – most notably her children, nature, and future generations, and/or (2) specific outcomes – most notably a sustainable future. The action alternative that, for her, was the ethical one was to take her competence to what Statoil considered "the enemy" – Greenpeace – where she could engage in work that was consistent with the values toward which she felt a profound obligation.

We can interpret Inge Lisenka Wallage's reaction to working with Statoil's oil sands venture as resulting from her ability to see the non-technical and ethical dimensions of the case. While Statoil approached the oil sands venture in a techno-centric manner – highlighting their technological advantage over their competitors and de-emphasizing the arguably grave influence of the project on vulnerable ecosystems – Wallage sensed the non-negligible moral implications of the activities of which she was a part. For Wallage, an important ethical issue needed to be addressed, which meant that ethical action was required.

In light of Hirschman's (1970) distinction between exit, voice, and loyalty, Wallage considered the exit strategy to be the most responsible action. To use voice – for instance by criticizing Statoil's practices publicly – would be in conflict with her organizational role as vice president of communications, and would as such be unsustainable and disloyal to her employer. Likewise, the loyalty strategy, which would imply staying in the organization and perhaps aiming to change the organization's practices from the inside, probably appeared unrealistic. In light of this situation, leaving the organization meant that she no longer actively supported the of oil sands practices by being part of the organization (cf. Arendt, 1963). At the same time, when leaving the company it allowed her to send a message publicly, which can be powerful, and it also allowed her to use her competence at Greenpeace, where there was alignment between the organization's and Wallage's goals. In this way, Wallage resolved her ethical problem by stepping away from the scene of the problem.

Whistleblowing as responsible action

In 2000, Per-Yngve Monsen was employed as a project economist in Siemens, which is a company with a global presence with approximately 70 billion euros in revenue, 500,000 employees, and representation in more than 190 countries. Monsen was employed in the defense division of the subsidiary Siemens Business Services (SBS). (This section is based on Monsen's book, *A Mole in Siemens* (2008).

In his first months, Monsen and other economists in the organization saw only red numbers on the bottom line. The economists even reported that the result for Siemens was "catastrophically bad". The Norwegian Defence (Norwegian Armed Forces) invited a tender for a new IT structure for the Norwegian Defence as a whole. The contract would be one of the largest IT contracts in Norway with a value of more than 100 million euros. During the process, there seemed to be active dialogue between the director in SBS and the Norwegian Defence officials. Monsen thought it was strange that one Defence official had a previous career in Siemens. Obviously, the parties had a good relationship and had even shared an apartment.

The German headquarters subsequently fired the Norwegian CEO due to bad financial results. Monsen was surprised by the use of invoices in December (the invoices meant that the client was debited for undelivered goods). However, due to this "Christmas shopping" the Norwegian Defence was able to use its whole budget for the year. Also, in the following years, Monsen witnessed similar "Christmas shopping". However, Monsen knew that the Norwegian government auditor (Riksrevisjonen) does not permit public institutions to pay in advance.

In late 2000, SBS won the defence contract, which meant that SBS would be a secure work place for many years. A premise in the contract was that SBS would calculate a maximum margin of 8 percent on products and that SBS would be paid for services related to planning, installations, the mapping of IT material and logistics, etc. SBS employed a new director who changed the organization, and Monsen was promoted to division economist for the new product division. However, Monsen noticed that the sellers in SBS did not adhere to the 8 percent margin. Many of the products were sold for a larger profit and, at the same time, Monsen watched the new financial director attempting to hide the illegal profit in the accounting systems by "cooking the accounting books". Inside the organization spirits were high, not least because of the bonuses, which Monsen also received. However, he felt no joy in receiving money from the dirty arrangement between SBS and Norwegian Defence.

The arrangement continued through 2001 and 2002, and Monsen was able to see where the illegal profit was accounted for. The amount was now significant, since it was more than 2 million euros. He decided to ask the financial director about it, although he knew that criticism of management was a delicate matter at Siemens. The financial director's response was that this was a matter for management and that he would speak with the client at the appropriate time. The tone of the director's response was not at all friendly. Monsen watched the continuation of overbilling and saw how SBS soon disposed of 3 million euros – an amount that in reality belonged to Norwegian Defence. Monsen again asked the director, who told him it was a matter for senior management.

Subsequently, the corporate headquarters developed business conduct (ethical) guidelines that were to be signed by all employees. It was stated that the consequences of breaking the guidelines could include loss of job and being sued. In addition, all employees received a book entitled *Trust,* which contained ethical and moral admonishments, encouragements, and sayings. Monsen understood that the top executives in Siemens took their ethical codes seriously, and if the overbilling of "Defence" continued, it would become a serious problem for him. According to the ethical guidelines, he had a duty to report such matters. If he did not sign the guidelines, it would be the same as handing in his resignation. Monsen experienced a "Catch-22" situation, but signed and submitted the guidelines.

According to Monsen, subsequent annual reports were fraudulent. He was aware of bribery payments from SBS to Defence public officials. In 2004, Monsen decided to write an anonymous letter about the fraud to Siemens headquarters in Germany. Then, Siemens started "hunting the witch". Monsen was interrogated in several meetings and felt a huge pressure, culminating in not being able to sleep and it affecting him physically and mentally. A delegation came from Germany to Norway to investigate the fraud and Monsen wrote a new letter to Germany in which he condemned the way the delegation carried out the investigation. Monsen believed it would destroy the company, and threatened to report what he knew to Siemens AG – the corporate management group.

Soon afterwards, Monsen felt he was in the midst of a tornado. Knowing that he was approaching his last days at SBS, he decided to prepare for probable court proceedings and started printing documentation for his case. Close colleagues pointed him out as the whistleblower, and condemned him for being disloyal towards management and for destroying the company. In 2004 Monsen was dismissed due to "internal organizational change". Monsen maintained that the dismissal was impartial and sued SBS. In the subsequent year, the story broke in the Norwegian media and readers were told that Norwegian Defence had been overcharged by 50 million kroner. The case was dealt with in the Norwegian parliament, and both the bribes and the "Christmas shopping" were out in the open.

In April 2005, Monsen won in court and received a compensation of 1.5 million NOK from Siemens due to his dismissal. Norwegian Defence dismissed officials who were bribed. Monsen sent all his correspondence to the German Stock Exchange authorities. In Norway, a governmental committee investigated the story and concluded that the allegations were true. A consequence was that Norwegian Defence stated that Siemens was excluded from future public tenders. In 2007, whistleblower Monsen received an anonymous letter that was seen as a serious threat to him personally, and so was investigated by the police. In the same year he received two prizes for his whistleblowing, and also published the book *A Mole in Siemens,* where he told the story of his experiences in Siemens SBS.

Why did Monsen blow the whistle? Clearly, he showed moral courage. But why did he do it? Was he driven by his own deep moral emotions, such as empathy, compassion, and justice? Monsen admits that he could not enjoy a bonus that he perceived as unfair, and it is obvious that he was driven by deep and powerful motives. It appears that for Monsen unconscious motivation was an important part of the picture. In an interesting passage in Monsen's book, he reflects upon the question as to whether he would have done the same after all the pain he had suffered as a whistleblower. Monsen's answer is:

> It is difficult to answer. It is about what one experiences as right and wrong, and which aspects one emphasizes in oneself. I was concerned about my own self-esteem, and for my part this was connected with the beliefs I have of right and wrong. Maybe there might be quite different and less noble feelings which had the same importance, for example pride, stubbornness, and anger. Nevertheless, it was unwise to do it considering my role as breadwinner in the family – and not least considering my own health. (Monsen, 2008, pp. 159–160; authors' translation)

For Monsen, blowing the whistle was the responsible action. He attempted to solve the problem from inside the organization, but did not succeed. Thereby his loyalty to the organization was also undermined. He ended up blowing the whistle first internally, to higher layers of the organization, and subsequently externally to Norwegian Defence and to the media. In light of Hirschman's (1970) distinction between exit, voice, and loyalty, voice was the only option for Monsen, and it proved to be a powerful strategy. However, it should be noted that there was a correspondingly heavy personal cost for Monsen following the use of this strategy. This is the pattern that is often seen in whistleblowing cases – the strategy is effective, but may destroy the individual's professional and personal life. Monsen's strategy was to tackle the ethical problem by confronting it head on and becoming a crusader for the sake of justice. His story, like the story of many other whistleblowers, shows us the price that is often paid by those who challenge unethical action in business.

Moral whispering as responsible action

In a sense, the concept of moral whispering refers to a weaker form of whistleblowing. Moral whispering implies that the individual, rather than using the "voice" strategy in order to reveal a blameworthy practice, chooses to discreetly communicate his or her concerns within the organization in order to attempt to influence practice in a positive direction (cf. Bird, 2002). As such it is a subtle influence strategy by means of communication and social interaction. Such behaviour is arguably plentiful in organizations, and it is a manifestation of how moral awareness among organizational members may contribute to the organization's avoidance of ethical and reputational crises.

A defining characteristic of moral whispering is its discreetness, because the very strategy it entails is to attempt to promote moral practice without creating noise and uproar outside the organization itself. For this very reason, there are, of course, fewer known cases of such behaviour. Successful moral whispering makes it unnecessary to talk about the case beyond the organization's boundaries. For this reason, we discuss this strategy by means of the example of shareholder activism. Shareholder activism refers to the practice of using shareholder rights and privileges to try to make companies improve their social, ethical, and/or environmental performance (Sandberg, 2011, p. 52). Typically, this strategy is used in order to influence the company so that its practices become more responsible. It is generally executed in one of two ways: either by raising concerns at the company's general assembly or by contacting management directly with one's concerns or claims. The latter form of shareholder activism can be seen as a sort of moral whispering.

In asset management companies of a certain size, shareholder activism is becoming increasingly widespread. Among the companies that have employed this strategy successfully is the Norwegian financial institution, Storebrand. The shareholder activism in Storebrand includes traditional approaches such as voting and asking questions at general assemblies. Also, however, Storebrand has a more direct and comprehensive approach that involves contacting the management group of companies directly to demand that they alter their irresponsible practices. Storebrand has had considerable success with this strategy, and currently contacts several hundred companies directly each year as part of their shareholder activism strategy (Meisingset and Norum, 2011).

Storebrand's approach to shareholder activism can be interpreted as moral whispering if we consider Storebrand to be an internal stakeholder by virtue of their role as a stakeholder. In this sense they can act as a sort of organizational actor much like an employee. Therefore, in light of Hirschman's (1970) distinction between exit, voice, and loyalty, Storebrand could similarly choose any of the three. Shareholder activism is often done in a very public way, for example at general assemblies or by means of communication in the media. Such a strategy would be considered as "voice", much like whistleblowing for the organizational actor. In this context the "exit" strategy would be to sell their shares in the companies they consider irresponsible. Storebrand's discreet influence strategy towards the companies in which they invest is, rather, a form of loyalty, whereby they remain investors in the company but act internally in order to promote ethical business practices (Meisingset and Norum, 2011).

Moral whispering can be a powerful strategy in order to develop an ethical culture inside a company. In their shareholder strategy, Storebrand raise concerns, ask difficult questions, and ultimately giving companies an incentive to move their business activities in a responsible direction. Otherwise, a company runs the risk of Storebrand selling their share of the company and justifying it with reference to the unethical practices of the company, which from a reputational point of view can be disastrous in Zsolnai's (2002) vocabulary, Storebrand are increasing the relative cost of *unethical action,* thereby making it more rational for the company to act responsibly.

The case of shareholder activism in Storebrand is, of course, special in the sense that Storebrand has a great deal of power as a big shareholder in the company. As such, making moral whispering effective must be done in different ways by individual organizational members. Regardless, the Storebrand case illustrates how the loyalty strategy can also constitute responsible action in organizations, at least when it is combined with an active attempt at changing existing practices through more or less discreet forms of interaction and communication inside the organization.

Conclusion

We have discussed the nature of personal responsibility and ethical action and the relationship between the two. We emphasized the desirability of personal responsibility in organizations and proposed a broader notion of personal responsibility that includes the notion of shared responsibility; that is, that individuals in organizations are in part responsible for acts they do not actively carry out, but on which they have some ability to exert an influence. Moreover, we have explored the organizational challenge of promoting personal responsibility and ethical action in organizations and the different forms that personal responsibility can take in organizational life. By discussing responsible action in the light of Hirschman's (1970) distinction between exit, voice, and loyalty we have also analyzed how different strategies of personal responsibility can be enacted in organizations. We have argued that individuals can assume responsibility in organizations in ways that differ along several important dimensions, but can all be effective in promoting responsible business practices.

The ethical action perspective is important because it illuminates the value of deep moral emotions like empathy and compassion as a foundation for courageous acts like whistleblowing. Jonas's (1984) substantive responsibility is important for Wallage, who left Statoil to join Greenpeace, and the Norwegian whistleblower, Per Monsen, whose acts cannot be explained unless we see him as partly driven by his deep emotions. To act ethically in a situation where the organization actively and brutally tries to stifle one's voice is extremely demanding. For Monsen there was no money reward waiting for a successful voice – only personal sacrifice. For Monsen, however, his integrity was at stake. Integrity is for some people more important than external rewards or punishment. In some sense Monsen did what was good for its own sake. He took personal responsibility in pursuing the idea of justice in society. We should not forget the subjective side of ethics. Reason has to complement emotion. Emotions are crucial in moral life, and their role is primary in acts of personal responsibility.

Bibliography

Andreassen, K. (2009) Oppsiktsvekkende kritikk, Eyeopening criticism. *Stavanger Aftenblad*, 9 June.

Arendt, H. (1963) *Eichmann in Jerusalem: A Report on the Banality of Evil.* New York: Viking Press.

Arendt, H. (2003) *Responsibility and Judgment.* New York: Schocken.

Bandura, A. (1999) Moral Disengagement in the Perpetration of Inhumanities. *Personality and Social Psychology Review,* vol. 3, no. 3, pp. 193–209.

Bird, F. B. (2002) *The Muted Conscience: Moral Silence and the Practice of Ethics in Business.* Westport, CT: Quorum.

Bovens, M. (1998) *The Quest for Responsibility: Accountability and Citizenship in Complex Organizations.* Cambridge: Cambridge University Press.

Hirschman, A. O. (1970) *Exit, Voice, and Loyalty: Responses to Decline in Firms, Organizations, and States.* Cambridge, MA: Harvard University Press.

Ims, K. J. (2006) Take It Personally. In Zsolnai, L. and Ims, K. J. (eds) *Business within Limits: Deep Ecology and Buddhist Economics.* Bern: Peter Lang.

Ims, K. J. and Zsolnai, L. (2006) Shallow Success and Deep Failure. In Zsolnai, L. and Ims, K. J. (eds) *Business within Limits: Deep Ecology and Buddhist Economics.* Bern: Peter Lang.

Jonas, H. (1984) *The Imperative of Responsibility: In Search of an Ethics for the Technological Age.* Chicago: University of Chicago Press.

Kekes, J. (1991) *Moral Tradition and Individuality.* Princeton, NJ: Princeton University Press.

Kongsnes, E. (2009) StatoilHydro-sjef ville ikke forsvare oljesand-prosjekt [Statoil Hydro executive would not defend oil sand project]. *Stavanger Ajtenblad,* 8 June, pp. 6–7.

May, L. (1992) *Sharing Responsibility.* Chicago: University of Chicago Press.

Meisingsec, C. T. and Norum, D. (2011) Bæekrafftige investeringer [Sustainable Investments]. *Praktisk Økonomi og finans,* vol. 27, no. 3, pp. 19–27.

Messner, M. (2009) The Limits of Accountability. *Accounting, Organizations and Society,* vol. 34, pp. 918–938.

Monsen, P.-Y. (2008) *Muldvarp i Siemens: En dokumentar* [Mole in Siemens: A Documentary]. Oslo: Spartacus.

Opdal, L. C. (2011) Responsibility and the Unconscious. In Jakobsen, O. and Pedersen, L. J. T. (eds), *Responsibility, Deep Ecology and the Self.* Oslo: Forlag1.

Pedersen, L. J. T. (2009) *Making Sense of Sensitivity: Moral Sensitivity and Problem Formulation in Business.* PhD dissertation. Bergen: Norwegian School of Economics and Business Administration.

Rawls, J. (1971) *A Theory of Justice.* Cambridge, MA: Harvard University Press.

Sandberg, J. (2011) Changing the World through Shareholder *Activism?, Etikk i praksis – Nordic journal of Applied Ethics,* vol. 5, no. 1, pp. 51–78.

Sen, A. (2002) *Rationality and Freedom.* Cambridge, MA: Harvard University Press.

Vetlesen, A. J. (1994) *Perception, Empathy, and Judgment: An Inquiry into the Preconditions of Moral Performance.* University Park: Pennsylvania State University Press.

Williams, B. (1981) *Moral Luck: Philosophical Papers* 1973–1980. Cambridge: Cambridge University Press.
Zsolnai, L. (2002) The Moral Economic Man. In Zsolnai, L. (ed.), *Ethics in the Economy: Handbook of Business Ethics.* Bern: Peter Lang.
Zsolnai, L. (2008) *Responsible Decision Making.* New Brunswick, NJ: Transaction Publishers.

KNUT J. IMS AND LARS JACOB TYNES PEDERSEN

Personal Responsibility for the Greater Good

Introduction

This chapter deals with the phenomenon of personal responsibility in organizational life. Specifically, we aim to shed light on the process by which individuals in organizations arrive at radical acts of personal responsibility despite inflicting considerable costs on themselves (and, potentially, on others). Thus, we investigate a particular variety of personally responsible behaviour – acts that are characterized by autonomous, value-driven choice. This chapter relates personal responsibility to moral concerns for the broader implications of organizational action. As such, we address the relationship between personal responsibility and conceptions of the greater good. In doing so, we explore the nature of personal responsibility and, in particular, how the individual translates one's moral values into action.

The chapter builds on previous work by Ims and Pedersen (2013; Ims, 2006), who discussed strategies of personally responsible behaviour in organizations. Depending on the circumstances, Ims and Pedersen (2013) argue different strategies can be considered as instances of personal responsibility. Voicing concerns to managers or colleagues when it may be personally costly to do so, exiting the organization, or staying loyal to the organization while attempting to change its practices from the inside can all constitute responsible acts depending on the situation (cf. Hirschman, 1970). In this chapter, we investigate the case of *exiting the organization as responsible action* by providing an in-depth analysis of a top manager's choice to resign from an organization for ethical reasons.

Our empirical setting is the case of Inge Wallage, who used the *exit strategy* in order to take personal responsibility. Wallage left her former position as vice president of communications in the Norwegian oil company Statoil, and instead chose to take on the role as communications director of Greenpeace. She later also left Greenpeace in order to join the non-profit IWA (International Water Association). We discuss Wallage's decision to take personal responsibility by leaving Statoil, as well as how her decision relates to her conceptions of a greater good. The empirical analysis builds both on several personal interviews with the case subject – both in person and via email – as well as with various sources of secondary data. In addition to the Wallage case, we also draw on some published accounts of people who spoke out about wrongdoings in their workplace. Thereby, we also shed light on some similarities and differences between exit and voice as strategies for personally responsible action (cf. Hirschman, 1970).

We structure the remainder of the chapter as follows. First, we explore the concept of personal responsibility and its relation to role-mediated responsibility and common morality. Therein, we discuss the nature and scope of personal responsibility. Second, we investigate the case of leaving the organization as an example of taking personal responsibly. Thereby, we discuss how acting personally responsibly may relate to promoting or protecting the common good.

Personal responsibility

What does personal responsibility entail? In an important sense, personal responsibility is about the individual's realization of his or her own beliefs and values in action (cf. Ims, 2006). This can refer both to how the individual responds and acts to situations as they occur and to how the individual over a longer time acts in order to live in a manner he or she can justify. Moreover, personal responsibility is, crucially, about reflecting on the limits of role-mediated behaviour. This means that in an organizational context personal responsibility deals with (1) the individual's attempt to exercise professional roles in a manner that is consistent with values that are important for him or her, and (2) the individual's concern for the broader implications of his organizational acts; that is, how these acts promote or undermine the greater good. In order for responsible action in organizations to mean something more than simple obedience, it must be based on a certain *sense of responsibility*. That is, the individual actively reflects on and takes seriously the degree to which a given action transgresses a relevant norm or implies unacceptable consequences (Bovens, 1998; Zsolnai, 2008).

As proposed by Ims (2006) it is useful to distinguish between role-mediated responsibility, common morality, and personal responsibility (see Figure 1). The individual's *role-mediated responsibility* refers to the responsibilities assumed by the individual when taking on a given role in the organization. These responsibilities are professional responsibilities and they provide allowances and limitations to the individual. For instance, a doctor *is allowed* to conduct medical examinations of patients and prescribe medications due to his or her role as a doctor, but the role also *limits* what are acceptable actions for a doctor, such as entering a romantic relationship with a patient. Hence, the role determines which types of behaviour are acceptable and unacceptable, and it places a set of responsibilities on the individual who takes on the role. Taking on the role thus implies accepting these responsibilities.

Second, the individual is bound by the *common morality* of her community or society. Communities have moral values and corresponding moral obligations that are relatively widely shared, and the individual is partially bound by these values and obligations. There is an interesting tension between common morality and professional responsibilities, as the ethical standards of professional groups may sometimes differ from those that are reflected in common morality. Common morality is constantly evolving through dialogue between members of the community because of societal changes and, as professional roles change, they may also influence common morality.

Finally, the individual has a *personal responsibility* that is independent of the individual's role and of her being part of a given community. This personal responsibility relates to the individual's role-mediated responsibilities in the sense that the individual's willingness to take on and to perform according to the expectations of a professional role depends on the degree to which the role is compatible with his or her personal morality. The question is which types of behaviour the individual is willing to engage in as the performer of various roles.

The personal responsibility of the individual is also in tension with both common morality and role-mediated moralities. The individual's morality in part adheres to the community's moral tradition and sometimes implies making radical value choices that can conflict with that tradition (cf. Kekes, 1991). In a similar way, the individual's personal values are in tension with the values that are reflected in his or her professional role. Any mismatch between these two sets of values necessitates a critical reflection on these value conflicts and a prioritization between them in choice and action. Thus, personal responsibility relates to one's own norm-horizon as well as to shared norm-horizons.

From a psychological point of view, we consider a type of personally responsible action that can be considered as *self-determined behaviour.* It is intrinsically driven because it relates to issues relevant to the individual's conscience or, more broadly, has implications for his or her life project (see Williams, 1981). Williams writes that a ground project is one of several projects that are closely related to one's existence and which to a significant degree gives a meaning to one's life (Williams, 1981, p. 12). The idea

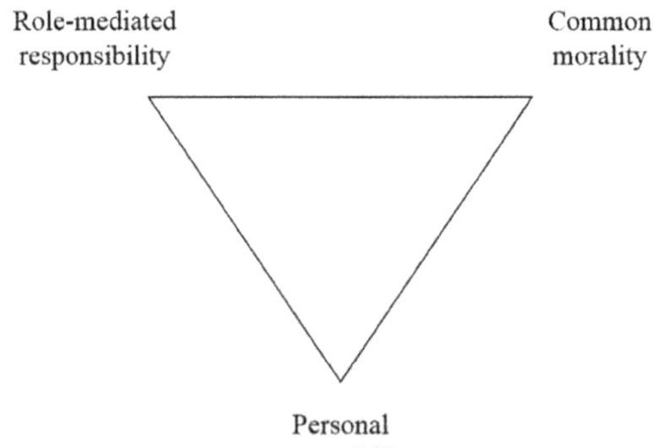

Figure 1. The triangle of responsibility. *Source:* Ims (2006).

of a person's ground project provides "the motive force that propels him into the future and gives him a reason for living" (p. 13).

We may assume that there may be a struggle between a person's internal world and the objective external world. Emphasizing the internal subjective world, we may localize responsibility in "Man the Maker" (actor/agent). In the internal world, we perceive, interpret, and judge what is going on in the objective world around us. Distinguishing between the internal world (lifeworld) and public world, we may see it as a bipolar concept, each pole named "Man the Maker" and "Man the Answerer", respectively (Niebuhr, 1999; Harmon, 1995; Ims, 2006; see Table 1). Thus, we imagine a *dialectic between moral authorship and accountability.*

As human beings, we have an internal world within which ethical reflection can take place. The human freedom is the capacity to mentaly withdraw, and having freedom implies responsibility. One's ethic is social in meaning and in context, but personal in judgment and action. According to Kierkegaard, there are no genuine tests in morality, and our moral standards can only be chosen. The authority is that a person has chosen to utter his morality ("the subjectivity is the truth"). Rational arguments can do no more than to present us with alternatives. In any chain of reason we finally reach a point where we have *to choose* to take a stand. Personal responsibility as construed in this chapter involves moving beyond narrow accountability and viewing one's actions in a horizon of a greater good.

Table 1. A dialectic between moral authorship and accountability

Man the Maker	Man the Answerer
Internal world	External world
Subjective	Objective
Feelings	Rationalistic thinking
Can do	Must do
Self-reflection/authenticity	Self-deception/inauthenticity

The case of Inge Wallage: Personally responsible action in organizational life[1]

In 2006, the Norwegian oil company StatoilHydro (hereafter referred to as Statoil, which is the company's current name) hired Inge Wallage. By the end of 2008 she was the company's Vice President of International Communications, at which point she made a radical and surprising decision. In her two years working with communications in Statoil, Wallage had spent a large part of her time justifying and defending Statoil's controversial oil sands venture in Canada and had grown increasingly uncomfortable with this task. Even though she had a powerful and prestigious position within a major oil company, Wallage decided to resign from her position in Statoil and instead start working for Greenpeace – a main antagonist to Statoil in the oil sands venture.

The reactions to Wallage's decision were varied. Statoil demanded that she should leave her post immediately. Former Statoil executives questioned her direct transfer to "the enemy", while business professors hailed Greenpeace's new recruit as a strategic coup (Andreassen, 2009). Wallage explained her decision in an interview, wherein she stated that she wanted to be able to look back at her life and say that she had contributed to making the world a better place. Moreover, she asserted:

> "We only have one planet, and we have to treat the globe with respect" (Kongsnes, 2009, p. 6). She emphasized that she had entered her role in Statoil in order to contribute to a sustainable development in the company's operations, but that she had failed to do so. Finally, she explained her choice with reference to her responsibility towards her children, and said: "I hope that my children will grow up able to live on a planet that is still intact and that I have contributed to this through my work life." (Kongsnes, 2009, p. 7)

According to Wallage, there were a couple of triggering events behind her remarkable change in employment. First, Wallage was contacted by a recruiter who was wondering if she was interested in being hired by Greenpeace International. It would mean that her salary would go down 60 percent compared with her salary in Statoil. This served as an opening for a new career, but with a considerable decrease in salary. Second, Wallage recalls having the "feeling of being 'the odd one out' inside of Statoil" and this feeling "grew continuously". She felt that in Statoil they tried "to make [her] more and more Norwegian", even when she was hired for her international expertise and experience. Third, Wallage recalls, "[i]t was hard for me that my colleagues were treating Mother Earth as something that is not alive and other parts of the world as "undeveloped" areas that were waiting for the "Norwegian model", as if their societies did not have their own values that we should respect … the lack of openness for different non-Norwegian views was hard for me to swallow". Moreover, Wallage had grown tired of commuting. She was working in Norway during the week, with her family based in Amsterdam. "My children were turning into young teenagers and I needed to be home more. My son was quite outspoken about this. "When will you come home and work in Amsterdam, mummy?"

As this overview shows, the picture painted by Wallage is one of complex motivations, which included moral, cultural, practical, and identity issues. Reflecting back on those events, Wallage emphasizes that the recruiter from Greenpeace arrived at the right time in her life. She had tried to get a job at Greenpeace in 2003, but was unsuccessful due to her lacking experience with NGOs and civil society organizations.

Was Wallage in doubt of whether her choice was the right one? Did she consider alternative options?

> "As soon as I was really with Greenpeace, I knew it was the right choice. I felt home very soon. I felt home in a way I had never experienced in my corporate years, not with any company I had been with. I could be myself, use my professional experience, but allowing for my gut to contribute too". She admits to have been quite wary once I had taken the decision to sign on the dotted line of the Greenpeace employment contract, as I knew I was changing worlds I realized that it is pretty hard to move from the corporate world to the NGO world. I knew this was my chance and I better grab it. I could always go back to the corporate world.
>
> Wallage further admits that even if Greenpeace would not have worked out for her, she would have wanted to leave Statoil. And resigning was not something she considered as radical. She had resigned quite a few times before, moving from one company to the next.

An important question with regard to Wallage's strategy is why she did not deliberate on other action alternatives within Statoil as a responsible choice. Wallage's perception is that she "might have considered staying if I could really have been part of contributing to the new business area of green energy/renewable. The problem with all traditional energy companies, not just for Statoil, is that exploration and production is their main business."

Wallage had tried and rejected other action alternatives that were compatible with her fundamental objective of contributing to a sustainable future. Her ethical action was grounded in a deep sense of personal responsibility towards stakeholders like her children, nature and future generations, and more generally for contributing to a sustainable future. The change of jobs would enable her to engage in work that was consistent with the values to which she felt a profound obligation.

Wallage further elaborated on what personal responsibility means for her. According to Wallage:

> [E]veryone in this world has a responsibility to live and act in a way that does not harm others. That is the bare minimum. To me it means that I need to have done as much as I can to make this world a better place. I don't understand how people can look into the mirror and live with themselves knowing that what they do is not really good.

Moreover, she added: "[t]he impact – environmental, but also social – to some of the (oil) exploitation that is happening through/around the business of exploration and production of oil and gas is incredible. We (the world) are out of balance completely. The world cannot handle this anymore, the rising temperatures should tell it all. Because there is not enough resources, we need to share and come up with new solutions."

Wallage wonders why people in the world of business apparently do not feel the urgency like it is felt in civil society – that the world needs to be "changed in order to be environmentally and socially fair". Wallage concludes that she feels "an ever-increasing sense of responsibility to work hard to find different ways to get people on the journey for a sustainable world. I feel a responsibility to the future generations, to reference the

Native American proverb, "we don't inherit the Earth from our ancestors, we borrow it from our children".

When asked about her decision to leave Greenpeace after a few years and start working for IWA, Wallage offered the following explanation:

> [T]here were a variety of factors that played a role in my decision, one that I did not take lightly as I had so enjoyed my time and contribution to Greenpeace and its vision/mission. However, it became clear that after 4.5 years my time as a change agent had come to an end … In addition, Greenpeace was going through changes, a reorganization and decentralization with more power to the national and regional offices and with its centre/ headquarters (which is where I worked) becoming much smaller. Even though I agreed with the overall approach, I did not agree with the diminishing role of the global communications function at the central HQ or at the most senior level. This was not in line with what I express overall (at communications conferences and such). And though I would have been happy to stay at Greenpeace longer, I wanted the choice to stay to be a positive one.

Wallage also clearly expressed that there was an attraction to the work IWA offered:

> The fact that I joined the IWA was for a variety of reasons: first of all, I wanted to continue my quest to contribute to a better world. Secondly, I found working for water extremely meaningful … And lastly, the Executive Director had clearly indicated that I would not only join as the Communications and Engagement Director, but as a management team member who would work on the overall IWA strategy and implementation.

Thus, were there relevant similarities or differences between the choice of leaving Statoil and the choice of leaving Greenpeace? Wallage elaborates:

> The similarity of leaving Greenpeace was that I was into something new. However, it is much easier to think of the difference of leaving Statoil and joining Greenpeace. Leaving Statoil and joining Greenpeace was a personal "Great Transition" for me in joining the quest to actively contribute to a sustainable world. However much as Statoil and its employees like to think they do so too, it is not in the same way one does when working in civil society. Furthermore, Statoil is engaged in activities that cannot be part of a sustainable world. I could no longer be part of that; I had to change/leave. Hence, joining Greenpeace meant actively working to change the world for the better and try to end wrong practices. The choice of leaving Greenpeace and joining the IWA was, therefore, a continuation of the quest to contribute to a better world.

Discussion: Personal responsibility and the greater good

In this section, we discuss Wallage's account of her choices in light of the notion of a personal sense of responsibility. One way to understand Wallage's choices is through the lens of Taylor's (1989) concept of strong evaluations. Taylor explored the background of our moral and spiritual intuitions. He coined the term strong evaluations, which refers to evaluations that are closely connected to a person's identity. One might say that they are related to the question "what kind of person do I want to be?" Strong evaluations are the second order assessments of our first order desires and preferences. That is, our strong evaluations reflect our moral horizon and the manner in which we assess our desires and preferences from a moral point of view; thus changing our behaviour in ways that are more in line with our desired identity. The contrast to strong evaluations are so-called weak evaluations, which refer to following our immediate desires and preferences. By doing so, Taylor argues, we may easily end up not fulfilling our life. Strong evaluations are thus central to the realization of moral values that we hold to be important.

What is Wallage's moral framework or moral horizon? She points to the values of respecting Mother Earth and future generations and thereby ensuring a sustainable future. Taylor (1989) describes three axes relating to our strong evaluations: respect, fulfilment, and dignity. Taylor argues that there are differences even within given cultures with regard to how respect, fulfilment and dignity are understood and acted upon. We believe that all the three axes are relevant in the case of Wallage. With regard to the first axis, respect for ecology and nature is at stake in how Wallage sees her practice and her choices. However, the second axis, fulfilment, is perhaps the most relevant in her case. A number of questions may be asked with regard to what constitutes a fulfilled (as opposed to a meaningless) life. When Wallage decided to leave her prestigious position in Statoil and get a position in Greenpeace, she knew that this would mean a considerable cost for her personally. Some doors would be closed for her in the future – not only in Statoil but, perhaps, big corporations in general. How can a director in a big oil corporation who chooses to move directly to the big enemy be trusted?

In Taylor's (1989) sense, persons have an orientation towards the good and this is central to our identities. Persons define themselves by reference to where they stand with respect to their orientation to this good. The identity of the individual serves as a frame of reference that provides a horizon of meaning. This moral space defines what is good or bad, what is worth doing, and what is not (Taylor, 1989, pp. 25–32). Asking questions within this framework about what is good helps us understand who we are. The articulation of this framework is never complete and can, rather, be seen as continuous work-in-progress. According to Taylor (1989, p. 42), one of the most basic aspirations of human beings is the need to be connected to what we see as good and what is of fundamental value. What kind of life is worth living and what would be a meaningful life?

In a Taylorian sense we may ask: What are Wallage's strong evaluations? Did she search for a life of unity or purpose? What kind of life does she consider worth living? When Wallage points to the fundamental value of ecological sustainability it is clear that she is not reducing her moral motivation to strive for only one good: she sought pleasure and happiness in a deep sense (Taylor, 1989, p. 332). While shallow happiness makes all human desires equally worthy of consideration and thus eliminates "strong evaluations". Wallage's actions are not reducible to pleasure-seeking. On the contrary, she makes choices that may hinder future career opportunities, which may lead to direct and significant personal and social costs as she criticized her former employer. But in doing so, she was able to act in accordance with her standards for what constitutes a justifiable work life.

Reflexivity is central to our moral understanding and the danger is to think of ourselves as the center of the world (Taylor, 1989, pp. 138–139). We can only see ourselves properly if we see ourselves in relation to some standard. However, in the modern world it is problematic to determine what constitutes this standard and this horizon. Nietzsche coined the term "God is dead" to wipe away the whole horizon. In this context Augustine is the moral philosopher *par excellence*, who opens up the inward space and engages in radical reflexivity. Augustine accentuates the conflict between one's knowledge about the good, the immediate desires, and the will. The will is not simply dependent on knowledge. This is the space of "moral choice", and a main problem is the phenomenon of weakness of the will – "akrasia".

The problem is how to create a meaningful life in harmony with one's desires? In Wallage's case, it may be more meaningful to apply Aristotle's concept of happiness as "eudaimonia". The main content in the Aristotelian concept of eudaimonia is that it is based on activity and the judgment of a whole life. It is a teleological concept viewing human life as consisting in the pursuit of ends. While individual actions and projects may be judged as being virtuous or not, a single good action is not enough to evaluate a man's goodness. A complete life requires virtuous activities and, in such a view, pleasure cannot be the sole purpose of man. Pleasure that accompanies virtuous activities may be seen as "added perfection – like the bloom of youth to hose in their prime" (Aristotle, 1935/1951, p. 259).

Wallage was able to remain independent of the opinions of her colleagues and openly protested against the exploitation of oil sands in Canada. What was at stake? Wallage was concerned about being an instrument for activities that she saw as destructive. She was concerned that the requirements of her role were incompatible with her personal sense of responsibility as well as the moral standards of significant others – primary among them her children. According to Vetlesen (2004) we can fear being transformed into mere means when we make efforts to adapt to the ever-increasing demand for growth and change. Sennett (1998) writes about the "corrosion of character" and also points to how the modern workplace and society may easily make us into "flexible instruments". This may lead to the dissolution of the vitality of the self and make us into commodities. "The attitude to the self in such a situation is one of ruthless exploitation" (Vetlesen, 2004, pp. 138–139).

Wallage seems to have a clear ground project in Bernard Williams' (1981) sense. She repeatedly emphasizes the importance to take responsibility for Mother Earth. This appears to be closely related to her existence and gives her a reason for living. The choice of Greenpeace as an employer for Wallage was not accidental. She had actively attempted to join Greenpeace earlier and her wish to join had been strong for years. However, Wallage admits that even without the possibility of a position in Greenpeace she would have left Statoil. Her life respect could not be reconciled with Statoil's conventional exploration and production of fossil energy, and her choice allowed her to reconcile her own standards with those of a new employer. The decisive moment in the long process of leaving Statoil for Greenpeace was, perhaps, Wallage's walk in a park in Amsterdam. She tells: "It was as a sudden revelation. I cannot stay with Statoil any longer. I have to do something else with my life." This was a moment of awakening for Wallage.

An interesting comparison can be made to Emma O'Reilly's (2014) story, which is about organizational voice – an alternative approach to responsible action (Hirschman, 1970; see also Ims and Pedersen, 2013). O'Reilly revealed the inside story of being the person closest to Lance Armstrong in his early cycling career. In the meantime, Emma learned about several young cyclists who died due to doping, "senseless, tragic deaths of talented riders". She had experienced the doping culture from the inside and in 2001 she chose to speak out to a journalist.

O'Reilly claimed that her aim was to clean up the sport. However, to confront the world's most successful cyclist, Lance Armstrong, for systematically doping, Emma had to pay her price. She was attacked for her allegations by Armstrong's lawyers, and she revealed that blowing the whistle had made her feel disloyal to her former colleagues and, in particular, to Armstrong himself. She thus felt that she deserved every thing she got, including harassment and being sued for her disloyalty. According to O'Reilly (2014, p. 194), "[t]his time, along with the numbness came a new, overpowering emotion that ate away deep inside. Guilt." As in the case of Wallage, both prior to and after the choice, there were strong emotions accompanying her reflections of the situation and what was at stake.

What was O'Reilly's moral horizon? She was "brought up to tell the truth, that lying was a sin", and said that "[f]or years I had felt guilty … I even doubted the values I'd been brought up to believe in: a sense of honesty at all costs" (O'Reilly, 2014, pp. 276–277). However, our interpretation of Emma's whistleblowing is that its not the truth as such, but rather the people involved that ultimately were at stake. Primarily, Emma seems to consider the fate of the young riders who die because of doping. In fact, Emma stands in a web of human relationships that she has to manage and protect.

Wallage also told about feelings of guilt. In her case, it is not so clear from where this feeling of guilt came. In this context, we can perhaps interpret Wallage's guilt as a deep inner voice that spoke to her – stemming from deep moral sources that place a demand on us when we do not live up to our own standards. In Wallage's case there was a considerable discrepancy between her ideals and the feeling of being complicit in destructive oil sands extraction in Canada. According to Taylor (1989) we all have a demand laid upon us from our deep inner sources. We may think of this as a kind of moral intuition. We feel the demand like a burden when we violate moral standards. A similar phenomenon is coined "the ethical demand" by the Danish philosopher Knud Løgstrup (1956/1991). It emerges from the spontaneous life utterances of human beings – like trust, openness, and compassion.

Earlier in the chapter, we introduced a dialectic concept of responsibility, which included the tension between self-reflection and authenticity on the one hand versus self-deception and inauthenticity on the other (Harmon, 1995; Niebuhr, 1999). There are different degrees of self-deception. Myopic vision or tunnel vision is one example, while an emphasis on means at the expense of ends may be another. Moreover, role-mediated responsibilities may lead individuals into self-deception, as most acts and even transgressions may be justified with reference to the role-holder's accountability to superiors in the organization. Bandura et al. (2002) discussed how corporate transgressions often take place when otherwise prosocial managers adopt socially injurious corporate practices. This is possible by construing situations and decisions in a manner that downplays their agency in the situation, and thereby their responsibility for eventual adverse outcomes.

We may view Wallage's exit from Statoil as her unwillingness to accept the notion that using oil sands to generate energy usage morally justifies the detrimental environmental effects. She expresses how no longer holding her former position feels like breaking free from a practice she could not support. Again, an interesting parallel from the Lance Armstrong story can shed light on such a feeling of emancipation. Tyler Hamilton, who was Lance Armstrong's "trusted lieutenant" with the US Postal Service team, and a member of his inner circle, has also admitted the truth about substance abuse (Hamilton and Coyle, 2012). Hamilton won an Olympic gold medal but was later found guilty of doping and exiled from the sport. He wrote:

> The world works in strange ways. I know that old saying that when God closes a door He opens a window. I think that saying is really talking about the resilience of truth. I've come to learn that truth is a living thing. It has a force inside it, an inner springiness. The truth can't be denied or locked away, because when that happens, the pressure builds. When a door gets closed, the truth seeks a window, and blows the glass clean out. (Hamilton and Coyle, 2012, p. 271)

An interesting question is whether the aftermath of such personally responsible action that Wallage, O'Reilly, and Hamilton carried out leads them to act personally responsibly in other instances. In the quote above, Hamilton refers to the cathartic and energizing feeling of breaking out of his prior behavioural patterns. Perhaps are there even long-lasting behavioural changes when individuals are able to live lives that are more authentic and that are in alignment with moral values that are important to them?[2]

Concluding remarks

In this chapter, we have explored personal responsibility and the greater good as a dual perspective, primarily by shedding light on Inge Wallage's journey from a public role-mediated defender of oil sands in Statoil to becoming an adversary to the same practice as an executive in Greenpeace.

Wallage's attitudes and acts appear to be defining characteristics of her as a person. These responses refer to much more than one single act. They may, rather, be seen as long and more or less continuous processes across time. Emotional and cognitive impulses led her to transcend her own narrow self-interest in response to different stakeholders through authentic and active dialogues; dialogues outside of her respective working environment – with her partners and close friends.

The actions of Wallage cannot be reduced to a performance of social, professional, or organizational roles. On the contrary, she went beyond her roles as a good team player or communication director. She transcended those roles in order to realize what she saw as essential values tied to a greater good. For Inge Wallage, this related to her life project and to ecological sustainability and future generations. Similarly, in the case of Emma

O'Reilly, the value of truth and the concern for young cyclists at risk led her to sacrifice her loyalty to her employer and her colleagues. The values at stake were so pressing for Wallage and O'Reilly that they had to be acted upon – even at the foreseen high personal and social cost, which became a sort of self-inflicted pain.

Both Wallage and O'Reilly had developed reflexivity about the ethical norms of their societies and a capability to display empathy or compassion with others with whom they shared a common environment (cf. Zsolnai, 2008, p. 166). These type of actions could be interpreted as reflecting a type of personal development that might be called secure autonomous persons with a strong sense of self. Emotions are a central feature of a person's character and together with personal values they constitute persons as extended selves who identify with a larger community. The challenge is to go "beyond self" when the self is defined in a narrow way. Arne Næss (1999) writes about the extended self, clearly influenced by Gandhi's view of the oneness of sentient beings. According to this view, we are all part of a web of life and, consequently, what we do to others we do to ourselves. Wallage clearly expressed dismay about her complicity in the destruction of the environment and how she could not defend standing idly by while the future planet of her children was placed at risk.

The famous organization scholar, James March, has said that undermining the self-interest doctrine may be the most important project of the twenty-first century (Zsolnai, 2014). In business, economics, politics, and everyday life the doctrine of self-interest is pervasive. It even demonstrated that in business people who act altruistically do not have the courage to talk about their true motivation, but camouflage it by using the vocabulary of self-interest (Ferraro et al., 2005). Singer (2010) similarly argues for the need to challenge the norm of self-interest in order to obtain a better life. More altruistic behaviour will lead to more flourishing and a higher quality of life for both the altruist and the beneficiary, argues Singer. The major problem is that the norm of self-interest is an ideological belief "resistant to refutation by the behaviour we encounter in everyday life" (Singer, 2010, p. 77). We tell stories about our acts of compassion and put a self-interested face on them. The norm of self-interest is self-enforcing and socially pernicious. According to Singer, the necessary change is to develop a culture of giving, which requires more openness, putting a face on the needy, and a better understanding of the greatly underplayed pleasure of altruistic behaviour (Singer, 2010).

To use Henrik Ibsen's famous concept, Wallage had to face the "compact majority" in the leadership group in Statoil when she resigned and told them that she would take on a new position in Greenpeace. For Wallage, her new position in Greenpeace – and later IWA – again enabled her to flourish in her work. Her competency could be used for a greater good and her sense of personal responsibility and the work that was required of her in her new roles were aligned. Personal responsibility assumes that each individual may make a difference. The story of Wallage – much like the story of O'Reilly – illustrates that personal responsibility ultimately is connected to freedom and dignity, and the ability to live one's life autonomously and in pursuit of the realization and protection of important values.

Notes

1. This section is based on qualitative interviews with Inge Wallage, both in person and via email. Also, parts of the case description are based on Ims and Pedersen (2013).
2. In order to strengthen the validity of our interpretations we used communicative validation (for example, Ims, 1987, pp. 183–7). This implies examining whether or not the analyzed individual can recognize the manner in which he or she is depicted and interpreted, or if it is far-fetched or "off the mark." This kind of communicative validation also has an important ethical aspect, since it necessitates that the interviewee is given the opportunity to comment, correct, or deepen the answers given during the interview process. For this reason, we sent a complete draft of this chapter to Inge Wallage, who has given feedback on our use and interpretation of the data analyzed in this chapter. She had no objections to our interpretations.

Bibliography

Andreassen, K. (2009), "Oppsiktsvekkende kritikk [Eyeopening criticism]", *Stavanger Aftenblad,* June 9.

Aristotle (1935/1951), translated by P. Wheelwright (the translation contains selections from seven of the most important books of Aristotle, among other from *The Nicomachean Ethics* which is the central source in this chapter), New York, NY: The Odyssey Press.

Bandura, A., G.-V. Caprara and L. Zsolnai (2002), "Corporate Transgressions", in L. Zsolnai (ed.), *Ethics in the Economy – Handbook of Business Ethics,* Oxford, UK: Peter Lang, pp. 151–164.

Bovens, M. (1998), *The Quest for Responsibility: Accountability and Citizenship in Complex Organizations,* Cambridge, UK: Cambridge University Press.

Ferraro, F., J. Pfeffer and R. I. Sutton (2005), "Economics Language and Assumptions: How Theories can Become Self-Fulfilling", *Academy of Management Review,* 30 (1), pp. 8–24.

Hamilton, T. and D. Coyle (2012), *The Secret Race. Inside the Hidden World of the Tour De France: Doping, Cover-ups, and Winning at all Costs,* London, UK: Bantam Press.

Harmon, M. M. (1995), *Responsibility as Paradox: A Critique of Rational Discourse on Government,* London, UK: Sage Publications.

Hirschman, A. O. (1970), *Exit, Voice, and Loyalty: Responses to Decline in Firms, Organizations, and States,* Cambridge, MA: Harvard University Press.

Ims, K. J. (1987), *Leder i dialog. En studie av informasjonssøk med metoder for personlig utvikling* (Manager in Dialogue: A study of Information Search with Methods for Personal Development). Bergen: Universitetsforlaget.

Ims, K. J. (2006), "Take It Personally", in L. Zsolnai and K. J. Ims (eds), *Business within Limits: Deep Ecology and Buddhist Economics,* Bern: Peter Lang, pp. 219–268.

Ims, K. J. and L. J. T. Pedersen (2013), "Ethical Action and Personal Responsibility", in L. Zsolnai (ed.), *Handbook of Business Ethics: Ethics in the New Economy*, Oxford, UK: Peter Lang, pp. 127–150.

Kekes, J. (1991), *Moral Tradition and Individuality*, Princeton, NJ: Princeton University Press.

Kongsnes, E. (2009), "StatoilHydro-sjef ville ikke forsvare oljesand-prosjekt [StatoilHydro executive would not defend oil sand project]", *Stavanger Aftenblad,* June 8, pp. 6–7.

Løgstrup, K. E. (1956/1991), *Den etiske fordring* (The Ethical Demand), Copenhagen: Gyldendal.

Næss, A. (1999), *Økologi, samfunn og livsstil, utkast til en økosofi*, Oslo: Bokklubbens kulturbibliotek.

Niebuhr, H. R. (1999), *The Responsible Self. An Essay in Christian Moral Philosophy* (originally published in 1963), Louisville, KY: Westminster John Knox Press.

O'Reilly, E. (2014), *"The Race to Truth". Blowing the Whistle on Lance Armstrong and Cycling's Doping Culture*, London, UK: Bantam Press.
Sennett, R. R. (1998), *The Corrosion of Character: Work in the "New" Capitalism*, New York, NY: W. W. Norton & Co.
Singer, P. (2010), *The Life You Can Save. How to Do Your Part to End World Poverty*, New York, NY: Random House Trade Paperbacks.
Taylor, C. (1989), *Sources of the Self. The Making of the Modern Identity*, Cambridge, UK: Cambridge University Press.
Vetlesen, A. J. (2004), *A Philosophy of Pain*, London, UK: Reaction Books.
Williams, B. (1981), *Moral Luck: Philosophical Papers 1973–1980*, Cambridge, UK: Cambridge University Press.
Zsolnai, L. (2008), *Responsible Decision Making*, New Brunswick, NJ: Transaction Publishers.
Zsolnai, L. (2014), "Economics, Ethics and Spirituality", in *Beyond Self: Ethical and Spiritual Dimensions of Economics,* Oxford, UK: Peter Lang.

KNUT J. IMS, TRULS LILAND, AND MAGNE SUPPHELLEN

The Importance of Calling in Realization of Life Projects: The Case of Maverick and Serial-Entrepreneur Hans Nielsen Hauge with Implications for Business Education

Introduction

One of the most influential business leaders and entrepreneurs in Norwegian history, Hans Nielsen Hauge (1771–1824), was also a lay preacher. Based on his values, moral convictions, and his Christian faith, he devoted himself to a life of service of the common people in Norway through his entrepreneurial business endeavors, his prolific authorship, his preaching, and establishment of communities of faith all over the country. Hauge was able to inspire and motivate his followers as he empowered them to lead and develop new business ventures. The movement he initiated made a substantial and lasting impact on Norwegian society for several generations, and its influence is still seen in Norway today.

In this chapter we briefly describe Hauge's experience of calling and how his life project emerged from, and was motivated by, this experience. The letters he wrote to managers of all the businesses he established are a rich source of information on Hauge's reflection on this experience and his business philosophy in general. In the midst of his active life, Hauge was arrested and put in prison. Also, the extended and usually comprehensive trial was a special source of information. The process involved almost 300 witnesses from all over Norway, and Hauge himself were interrogated. Without a verdict, Hauge was kept in prison for almost ten years. There are protocols from a much delayed trial, see Breistein (1955). We use this historical example to inform a principal discussion in nature and relevance of the concepts of calling and life projects in a modern business context, and how Hauge as a role model of the will to serve can inspire to a redirection of current business education.

Methodology

This chapter will use the case method as a research and learning tool. In business and leadership studies it is of paramount importance to select and study cases that may be regarded as paradigmatic cases in the meaning that "they shine". According to Kuhn (1987), a scientific discipline without a large number of thoroughly executed case studies is a discipline without a systematic production of exemplars. The literature needs exemplars that can serve as role models, that can inspire, and be imitated. Therefore, the strategy of this article will be to find a lightening case that can show how the best leader(s) lead. In earlier case studies, we selected SEKEM and its founder Ibrahim Abouleish (2012) as a paradigmatic case (Ims and Zsolnai, 2015), and also the company Patagonia and its entrepreneur and leader, Yvon Chouinard (Ims, 2018). It may be fruitful to extend our study of contemporary entrepreneurs to explore different cases – even historical cases to gain deeper insight and accumulate knowledge about successful entrepreneurs. Flyvberg (2006) writes that it is not possible to follow fixed criteria to select a paradigmatic case, but experts can use their intuition and their experiences based upon thousands of other cases in an informed selection of cases. Case studies often contain narrative elements. A good narrative is complex with surprises, paradoxes, and complexities. It makes the narrative difficult to summarize into general propositions and theories, but a thick narrative should not be understood as a problem, but a "rich problematic" (Flyvberg, 2006, p. 237).

We expect that by selecting Hans Nilsen Hauge as an example of successful entrepreneur leadership, and enrich theoretical concepts about callings and life projects in the context of entrepreneurship we may draw some implications for modern business education. The story about Hauge illustrates that leaders with a profound will to serve may both succeed in the marketplace and contribute to a favorable social transformation of society.

The context of Hauge

Hans Nielsen Hauge was born on a medium-sized farm in the eastern part of Norway in 1771 and grew up with nine sisters and brothers. Later Hauge expressed he was proud of his background as a peasant, even when obtaining the official letter of being a citizen of Bergen in 1801, with all the privileges that such citizenship involved. The letter required an oath to the king and gave Hauge the opportunity to travel around in Norway between different regions without getting allowances from the local police. As a peasant, he had not enjoyed this privilege, since a farmer's role was to stay at his property and use all his efforts to cultivate his land. Thus, farmers had very restricted opportunities to move around and to do business.

In order to illuminate the courage and innovativeness of Hauge, we have to emphasize the very rigid political, economic, and social context that existed. Norway had been under the rule of a Danish King since 1523, and in 1537 Martin Luther's Reformation was imposed in Norway. Then Denmark-Norway became an absolutist and hereditary monarchy in 1661 with an administrative system organized into departments and Norway divided into counties. Each county was led by a district governor, and further subdivided into bailiwicks. A number of laws and decrees regulated the economic and social relationship between the different social classes. Protestantism remained a very powerful national institution, and the Danish king became only responsible vis-à-vis God, not the pope. The king gave a number of significant privileges to the elite: the nobility, money collectors, and the sheriffs, and not the least the priests who were all educated in Denmark. The state church was influenced by rationalism, with many clergy rejecting miracles and the physical resurrection of Christ, the Danish king tried to hinder the layman's movement by enforcing the The Coventicle Act of 1741 (Konventikkelplakaten) that prohibited unauthorized religious gatherings without clerical approval or oversight.

Denmark-Norway was a feudal society with a mercantilist economic system characterized by a number of regulations and privileges. Trade was restricted via monopolies and tariffs. During the seventeenth and eighteenth centuries, up to two-thirds of Norway's national income was transferred to Copenhagen. Due to its location, Bergen was twice as large as Oslo (Christiania) in terms of population, with a long-standing shipping tradition from being a Hanseatic city.[1] Fish was an important export article and a subsidiary income for farmers along the coast. The need for imported salt was strong.

Denmark-Norway became allied with France during the Napoleonic Wars. The British blockaded the Norwegian Sea and closed the Norwegian access for import of grain and other important goods. Many merchants went bankrupt during the English blockade. In 1814, Denmark-Norway was on the losing side of Napoleonic Wars and Norway was ceded to the king of Sweden, which gave Norway more national independence. In the last decades of the eighteenth century, the poor classes were suppressed, and Hans Nielsen Hauge started the Haugian movement. Norway was ready for change.

Hauge's calling

The starting point of Hauge's extraordinary life project was a spiritual breakthrough. He was out working in the field as a peasant and singing a hymn when he suddenly had an overwhelming experience of God's presence and grace. This experience was followed by a strong urge to tell his fellow citizens about this grace and assist them into a closer relationship with God. Later, Hauge described this experience in a letter to his followers:

> It resounded in my inner being, You will confess my Name to (your) people, exhort them to repent and seek me while I am to be found, call on me when I am near and touch their hearts, so they can turn from darkness to light. (Hauge, 1910)[2]

Later, Hauge gave a shorter account of this calling:

> [T]o love God and my neighbor. (Source: The newspaper: *Bergenske Adressecontoirs Efterretninger*, 10. juli 1802)

When we compare Hauge's experience to modern research on callings in an organizational setting (for example Conway et al., 2015; Haney-Loehlein et al., 2015), we find that Hauge's experience resonates well with current definitions of the concept of calling:

> (a) a transcendent summons, experienced as originating beyond the self;
> (b) to approach a particular life role in a manner oriented toward demonstrating or deriving a sense of purpose or meaningfulness that;
> (c) holds other-oriented values and goals as primary sources of motivation. (Dik and Duffy, 2009, p. 427)

This literature also shows that the religious content of Hauge's calling is not to be considered a special historical case with little relevance for current organizational practices. Modern leaders have similar experiences. For instance, Haney-Loehlein et al. (2015) studied the callings of fifty-nine American leaders from various types of organizations, including commercial firms. Haney-Loehlein et al. observed that, for several participants in this study, religious messages and a sense of awareness of God's guidance were core aspects of the calling experience.

The concept of calling as defined by Dik and Duffy (2009) is related to, but conceptually distinct from, the notion of *defining moments*. A defining moment is a special event or experience which puts things in a new perspective and sheds new light on life in a fundamental and inspirational way. For instance, the *Merriam- Webster Dictionary* defines it as "the time that shows very clearly what something is really about".[3] *The Oxford Dictionary* focuses on the capacity of defining moments to frame peoples thinking: "An event which typifies or determines all subsequent related occurrences" (Oxford Dictionary).[4] A calling experience is normally based on a defining moment, but defining moments *do not necessarily* lead to a calling experience, which always contains an element of action calling the called to a certain role or class of action. For instance, an individual may listen to a sermon, read a book, or watch a play and experience a defining moment in terms of suddenly realizing in a new way that all humans have equal value. Still, this experience may have no impact on the individual's plans or behaviors. If a calling follows from the experience of a defining moment there is a higher probability that the new insight leads to action.

However, callings are often abstract and typically do not include specific directions regarding the *type of action* needed to answer the calling. For instance, callings such as "helping the poor" or "serving the fellow man" are mainly *behavioral goals* that could be obtained in a number of ways. Without further detailing of specific actions, the called may end up disoriented and frustrated (see Berkelaar and Buzzanell, 2015).

Thus, callings do not necessarily lead to practical action if the abstract behavioral element of it does not translate into concrete principles of action. This is why we need a second concept to understand the potential impact of callings; the concept of *life project.* This is a concrete plan with explicit long-term goals and dear implications for behavior. The same concept is termed ground project by Williams (1973), who connects it closely to integrity defined as "entireness, wholeness: the unimpaired state of anything". According to Williams a ground project is one or several projects that are closely related to one's existence and which, to a significant degree, give meaning to one's life, something that a "person is identified with", and entails that his actions flow from projects or attitudes which he takes seriously at the deepest level "as what his life is about" (1973, pp. 116–117). For Williams the assessment of a person's ends in action is important. A slaveholder and a Nazi cannot act from integrity given the depravity of their ends, but cleverness is compatible with bad ends (Thomas, 2015, pp. 3–4).

Hauge translated his calling "to love God and my neighbor" into a life project with three core elements: preaching the gospel to the nation, establishing religious communities all over the country, and starting up a large number of businesses in order to finance the movement and putting people to work. Notably, Hauge's life project is different from the ones of other historical figures with similar calling experiences. For instance, John Wesley has a similar experience as Hauge, but Wesley's life project was partly different and did not include practical entrepreneurship (Dalgaard and Supphellen, 2011). As a theologian Wesley focused on preaching and developing a new theology. However, Hauge and Wesley had much in common. Both were preachers and concerned about the poor. We suggest that, in the process of translating callings into life projects, the called use available information to come up with concrete implications.

Hauge was an uneducated peasant but had experience in small-scale entrepreneurship before the calling experience. Scaling up this entrepreneurial effort was a natural response to the calling of doing well for the fellow man. To Wesley, an educated theologian, the natural response was to address poverty in sermons, stimulate prosocial behavior, and develop a new theology. The common involvement of the two leaders in improving conditions for the poor resulted from similar observations of inequality and suppression in Norway and Britain at the time. Hauge observed these problems mainly in rural areas of Norway, whereas Wesley's attention was directed primarily at the extreme poverty in the urban areas of Britain. Thus, for Hauge and Wesley, both internal self-perceptions and aspects of the historical context probably played crucial roles as they translated their calling into a life project.

Importantly, the life project of Hauge was not a clear-cut and fully specified constant appearing right after the calling experience. On the contrary, it was flexible and dynamic. The basic elements of his project were formed early (preaching and putting people to work through entrepreneurial activity), but new elements were added over time, such as the comprehensive training of the poor in reading and writing, his own authorship of books, and his extensive publishing activity (Daalgaard and Supphellen, 2011). In the letters to his followers we find that when explaining the importance of new types of activities he repeatedly went back to the concrete practical implications of the calling over time, based on new experiences and the discovery of new opportunities.

Realization of Hauge's life project

During the period 1796–1804, Hauge covered more than 15,000 kilometers by foot, establishing small communities of believers all over the country. Hauge's spiritual message focused on the individual's personal relationship with God. He argued that nobody could rely on the faith of the clergy. Everyone was personally responsible before God and needed a personal relationship with Him. This message also contained a strong element of empowerment. The personal relationship to God gave believers a sense of self-worth. It also gave them access to divine support and comfort in their daily lives, and hence a new belief in their ability to make a difference and change local conditions for the better. This belief or mentality was a major force in the Hauge movement. Hauge's groups formed a large national network and become a highly influential movement in Norway. Moreover he was, via like-minded friends in business (see Grytten, 2013), involved in more than 150 business projects where he among other roles served as an advisor, co-investor, and co-founder. Hauge himself founded at least thirty companies in various industries and was involved in businesses in industries as diverse as grain mills, fish processing, paper mills, printing houses, brick factories, shipping, property development, merchant houses, trade stations, salt distillers, and fish processing (Rødal and Kiplesund, 2009; Grytten, 2013).

Most of these companies were founded during an intense period, 1801–1804. Even today we find several companies in Norway with roots in the Hauge movement, especially on the west coast within industries like furniture production and shipping (Rødal and Kiplesund, 2009). Clearly, Hauge was a social entrepreneur who established businesses with a higher purpose, focusing both on social and financial goals. Thus, he was actively engaged in fighting poverty, improving literacy, building caring and stimulating communities, and providing equal job opportunities to all, even those with disabilities and others who were not considered fit for the workplace. Dørum and Kringlebotn (2016) has demonstrated that the followers of Hauge had a strong impact and influence on different local communities in Norway. Thus, in these areas the predominant view regarding business and commerce was that they, in addition to being profitable, should serve society, provide opportunities to do good to others and, in sum, behave like good corporate citizens. The Hauge-inspired businesses certainly had a purpose that went beyond profit-maximization as they sought to improve society in several ways (Supphellen and Liland, 2017).

In all the places where Hauge established communities and businesses, he also initiated extensive training in reading and writing (Haukland, 2014). The purpose was to empower ordinary people to read the Bible and Hauge's own writings. Hauge himself was a prolific writer, writing more than thirty books, and was by far the most read author of his time in Norway. Many of his friends were later recruited into leading positions in the local communities, and this movement played a decisive role in supporting liberation from Denmark.

Hauge is the most prominent serial-entrepreneur in Norwegian history (Dalgaard and Supphellen, 2011). According to leading historians (for example, Grytten, 2013), Hauge's contribution to the spiritual awakening and empowerment of ordinary people was instrumental to the democratization of Norway in the nineteenth century.

Hauge clearly saw his business endeavors as a way to express his faith and to make the world a better place. Through these business activities, he would finance the printing of the literature he wrote (spiritual literature), create new jobs, alleviate poverty, and work toward achieve other social goals. The businesses were in many ways seen as arenas to develop people's abilities and talents while helping prevent idleness and laziness, especially among the poor. Furthermore, the business activities were venues for interaction and community building, where the workers became acquainted and achieved results together. Hauge aimed to create profitable enterprises, and the Haugian businesses were sometimes even more profitable than other companies in the same industries at the time. However, profitability was seen as a means for a higher purpose and thus not an end in itself. This approach to business is compatible with modern research on management and corporate development where employee motivation is enhanced when business is driven by a higher purpose than just profits (Supphellen and Liland, 2017).

As we have seen, the Haugian companies shared a deep commitment to establish and develop socially responsible businesses. The profits they generated were reinvested in the business or other like-minded businesses, and thus it was spent to create better working conditions or to help the poor and disadvantaged. A strong social commitment and poverty alleviation were therefore two key characteristics of Haugian businesses. His commitment to fight for equality and fairness was deeply influenced by his faith, which led him to a radical inclusion policy – including hiring those with disabilities of various kinds as equal workers in their companies, and by appointing female business leaders when they had the necessary skills. It was certainly remarkable for this to happen 200 years ago, and this way of thinking has influenced the way several local communities thought about businesses, corporate social responsibility, and social entrepreneurship in Norway (Dørum, 2016).

Concrete business cases

After being prevented by the authorities from establishing a paper mill in Aadalen in eastern Norway, Hans Nielsen Hauge and his brother, Mikkel Hauge, founded a new paper mill and factory in Eiker, close to Christiania and Drammen in 1802. The employees considered themselves one big close-knit family, sharing meals and living together in a community setting. The company had up to fifty employees during its most productive period. The paper they produced was considered high-quality and was useful for Hauge because he was writing different books and pamphlets (Grytten, 2013). A similar production unit was established in Fennefoss in the southern part of Norway. This was a very profitable business and it also became an important community-oriented and missions-oriented base for the Hauge movement in the area, which was not uncommon for these businesses.

Because of his prolific literary production, Hauge was in constant need of paper and printing house capabilities. In 1803, he purchased a printing house in Kristiansand in the southern part of Norway in order to publish some of the books he had written, which were gaining popularity as the Hauge movement grew. Also, they published different magazines and a local newspaper called *Christiansand Adressecontoirs Efterretninger.* They struggled a while to make the business profitable. Hauge saw the potential in a young man in his movement, Christoffer Grøndahl, who together with his family were able to make this publishing house profitable and, according to Grytten (2013), made it "one of the most prominent in Norway". Later, it moved most of its operations to the capital, Christiania, later named Oslo (Breistein, 1955).

Before Hauge was arrested in 1804, traveling throughout the country he conducted some of his religious gatherings at the former brick factory at Eeg, close to Kristiansand. The former owners of the factory had not been able to run the business at a profit, but Hauge soon appointed a young man who was part of the movement, Ole Eyelsen. He and his family were able to turn the business around and make it profitable. Hard work, perseverance, and competent leadership were important factors contributing to this positive transformation of the brick factory at Eeg.

Hauge established his merchant base in Bergen, from where he and his friends purchased several trade stations along the coast of southern Norway (Breistein, 1955). Among those, the merchant house at the fishermen's village Sør-Gjæslingan in Trøndelag is perhaps the best known. Here Hauge, supported by one of his best friends, the lay preacher, merchant, and industrialist, Arent Solem, convinced him to buy the place. Thus, it served as a key base in a network of trade stations along the coast ensuring fairly good control of the distribution of products traded by Haugians.

In Bergen, Hauge was supported by and received help from his brother-in-law in Bergen who already was a merchant, Johan Nikolai Loose. With help from Loose, Hauge also built a career as an innovative shipowner, however, not always with success. Hauge served as both a ship broker and a shipowner. The idea was to raise capital by ship broking in order to invest in ships for ownership. These could also be used to transport products produced or traded within his business network.

Hauge purchased and sold a number of ships. Even though some of this trading was successful, and definitely innovative, other purchases became very problematic and Hauge ended up losing money. During the spring of 1804, Hauge really wanted a ship for long voyages to buy hemp and canvas and purchase goods like salted herring and fish. Unfortunately, this trade was not successful. The captain was not able to sell the herring and, on short notice, he had to get 1,500 riksdaler (rd) *in* cash to pay for the hemp. Also, this time his brother-in-law helped Hauge.

In April 1804 Hauge bought the frigate "Christiane Margareth" from a shipping company in Bergen. Hauge was deceived, he paid 6,200 rd but the ship had been rebuilt in Bergen twice, and when arriving from Spain in 1805 the ship was in very bad condition. It became clear that Hauge was deceived by the traditional shipping families in Bergen. Hauge was alert about goods and prices and reported to his friends about actual prices and the market situation in Bergen. Among other things, he reported that, in 1804, a significant number of goods were stored in the city, prices were falling, and additional price decreases were expected. Hauge's brother in Kristiansand could tell that prices of liquor "and similar commodities" were better in Bergen, and he asked Hauge to send liquor. But Hauge was not a simple liquor speculator. He worked seriously to get people, *in* general, to be aware of the bad effects of intoxicating liquor. The spiritual revival he had stimulated led to less drunkenness. However, the general conception was that liquor was a commodity of necessity. One of Hauge's captains that sailed north of Bergen traded fish and paid with liquor and grain.

According to Grytten (2013), Hauge and *his* closest business partners kept their network of trade through these trade stations, and they also appointed managers and provided the necessary credits. Moreover, they exchanged business letters containing both spiritual and economic reaching. The buildings at the trade stations were often used as halls for evangelical meetings.

During the years 1803 and 1804 Hauge invested in no less than nine sailing vessels, two of which were frigates (Grytten, 2013). The ships were called Forsøg (Attempt), Anna Helena, Haabet (The Hope), Nicolai, Aurora I, Aurora II, Aurora III, Stadt Bergen (City Bergen), and Christiane Margaretha. Ship broking and owning was a very risky business during the

Napoleonic wars. Hauge experienced both heavy losses and impressive profits. The broking activity was additionally an important tool in order to raise quick money for investments (Breistein, 1955).

How were Hauge's entrepreneurial activities initiated? What were the dynamics behind his business and how did he finance these endeavors? Hauge had no inherited money, but had saved about 300 rd. To print his books he used his savings, but had to borrow money and pay it back in the form of printed books. Hauge also received money from sponsors, and the money he collected for his books was reinvested. Hauge could bind the books himself, even if the professional book binders in Bergen sued him because Hauge did not have the required privilege. Many of Hauge's books were given away in the service of his mission. With the income from book sales Hauge wished to stimulate "diligence" and useful enterprises. He bought herring from the west coast of Norway and sold it in the eastern part of Norway, bringing grain back. There were no problems selling the grain. He bought the ship "Haapet" for 480 rd. Less than a year later he was able to buy a property in the city of Bergen for 5,500 rd while many of the old merchant families suffered from economic problems. Thus the commercial elite in Bergen posed the enigmatic question: How is it possible for a maverick to pay for such valuable real estate properties? Buying this property, Hauge also took over an ongoing business with a rich assortment of goods (from the best type of "old cheese" to "ovens and chalks in a number of barrels").

During 1802 Hauge had received significant amounts of money as loans. Hauge received 500 rd from his housekeeper and 200 rd from one of his captains. Moreover, others lent him more than 1,500 rd. Around New Year 1803, Hauge had borrowed more than 4,000 rd from friends. With the money, Hauge bought real estate and some small vessels. All free capital was invested in commodity turnover. The property bought in Bergen became the shared home for Hauge and his friends while staying in Bergen. The Society of (Haugianer) Friends became for many a spiritual enrichment and an impulse to continued effort and spiritual the revival. It was a unique combination of active entrepreneurship and a deep social-ethical reasoning behind trade, and by the power of the example it stimulated individual diligence to follow his worldly calling (Breistein, 1955).

All the time Hauge practiced an extremely frugal lifestyle. In his home the inventories were cheap, so spartan that it created wonder among the rich merchants in Bergen. In terms of silverware, Hauge and his friends were more than frugal, borrowing spoons from a friend when Hauge's captains were in the city and visited his home. Typically, his Gesell (steward) was a "friend" which meant a faithful follower.

Hauge's success and entrepreneurial self-efficacy

Hauge was very successful in realizing his life project. The impact of his movement probably far exceeded his own expectations. Why was he so successful? There are many factors. Based on the history of his movement and modern research on management and psychology we have identified some phenomena and concepts. We also believe that Hauge's entrepreneurship may partly be explained by the concepts of self-determination, meaning, and persistence.

Self-determination theory makes a distinction between intrinsic and external motivation. Intrinsic motivation is a stronger force and leads to higher efforts, persistence, and performance (Deci and Ryan, 2000; Ryan and Deci, 2000). Modern research of small-scale entrepreneurs in Africa has found that the sense of being divinely called to do business has a positive effect on intrinsic motivation (Supphellen, 2014). Hauge's calling probably had a similarly strong effect on his intrinsic motivation. Part of the increased intrinsic motivation of the called probably comes from a greater sense of meaning. Research shows that, when individuals define work goals and manage to structure/prioritize their goals, they find work more meaningful and rewarding (Barrick et al., 2013). A divine calling probably has a significant effect on the goals and goal structure of the called.

Do we find a decisive factor behind the founder of so many businesses?

Can it be explained by character or context? The hypothesis set forward by Weber (1930) concerning the cultural individualist orientation is not easily supported because we could not expect an individualistic culture in Norway with a strong entrepreneurial spirit among the farmers in Denmark-Norway at that time. During Hauge's life Norway was a class regulated society. It is difficult to find beliefs in the self-made man like we have in the American dream, emphasizing that anybody has a fair opportunity to become a millionaire. Individuals that succeed are characterized by strong independency and a strong drive to achievement (see Cox and Jennings, 1995). Clearly, Hauge fits well into one of the three classifications, namely the "elite independent entrepreneurs".

Reading about Hauge, we find that Bandura's concept of locus of control and self-efficacy makes sense (Bandura, 1977; Zhao et al., 2005). Hauge was uneducated, but based on his achievements and the sources we have describing his life it seems highly probable that he had more than a fair share of entrepreneurial talent and social intelligence. His businesses often had a humble and needs-based beginning, and they emerged gradually when devoted and talented managers from Hauge's movement took the reins and worked hard to grow them. Hauge needed better tools for his enterprises, and he knew he was a "handyman". Thus, he often provided original and practical solutions to the challenges they faced. According to the theory, when an agent's locus of control is internal and he or she has the knowledge and capabilities to carry out the tasks, the more acute and sustained will be their alertness to opportunities (Harper, 2003). A major source of self-efficacy is the entrepreneur's perception and evaluation of his or her resource base. A strong resource base leads to greater entrepreneurial self-efficacy. When God calls, He does not withdraw from the scene and leave it to the individual to do the job alone. Rather, the notion of calling in the Haugian perspective meant that God actively supported the called with power, wisdom, and comfort. Thus, Hauge and his followers probably felt empowered by a sense of God's support in their business endeavors (Dalgaard and Supphellen, 2011). Also, the extensive networks of communities and followers represented financial and human resources which probably stimulated the entrepreneurial self-efficacy of Hauge and his people.

A major component of Hauge's calling was to love his fellow man. His calling experience made him more aware of the needs and problems of his people. In turn, his calling experience probably made the virtues of care and compassion more salient in Hauge's identity. The results of this would be more identity-consistent behavior and more care and compassion in personal interactions with the people he met and recruited to his movement (see Dutton et al., 2010). The letters Hauge wrote to his followers support this reasoning. Hauge's care and compassion for people, and for his followers in particular, certainly earned him a lot of trust and respect. Based on his biblical understanding that all men are created equal, he promoted the self-worth of people and assigned them all, rich and poor, to work-tasks that matched their talents (Dalgaard and Supphellen, 2011). This is an early version of what is termed servant leadership in modern leadership theory (Greenleaf, 2002). Here the leader is primarily motivated by, and directs his efforts toward, the employees' well-being and development, as well as serving the purpose of the company. Based on the available sources we know that Hauge was a strong leader, having a clear sense of how he wanted to develop the various businesses he was involved in (Supphellen and Liland, 2017). However, he was not driven by power, ego, or personal ambition, but by an important component of his calling and, indeed, his conduct as a Christian was to love others. He put the employees' needs above his own. Characteristically, Hauge sought to invest in his employees' development and education, which obviously also benefitted the businesses as the leaders and employees gained more competence and capability in key business areas. One employee and future leader Hauge recruited and invested in was the young Christopher Grøndahl. As mentioned above, he became a publisher of many of Hauge's books and started Grøndahl & Son Forlag AS, a famous Norwegian publishing house. Another example was Peter Møller, a chemist and pharmacist and the inventor of the product "Møller's cod liver oil". Møller's business evolved into what today is known as the Orkla-owned company, Lilleborg AS. Hauge paid for these men's education to prepare them for their future business development.

Implications for management, management research, and current business education

We need a type of business education and business training that assists students in defining their life goals and life projects. This will likely be regarded as positive for the organizations in which they work and might lead to happier and more fulfilled businesspeople, happy in the sense of eudaimonia. This means that business education needs a profound change. An important part of Hauge's life was to build communities and be inspired together with other like-minded people. Pruzan (2015) addresses the need to move beyond purely rational perspectives and sheds light on how spirituality may enrich business ethics as a field and as a practice. According to Pruzan the major challenge should be to "help transform the mindset of leaders to a higher level of conscious awareness" (2015, p. 265). His concern is that we should reverse the direction, from focusing on unethical behavior to the overall goal of contributing to the common good, to act for the well-being of the organization's stakeholders. For Pruzan this reorientation will "require that decision makers supplement their traditional rational analysis with spiritual-based guidance from their 'inner world' ". The condition will be a focus on emancipating and empowering the inherent source of their ethicality. One of the consequences will be that the "ego-based aspirations for financial wealth and power", greed and lack of sensitivity has to stop because it "undermines our trust in our business enterprises, their leaders and the institutional frameworks" they are a part of (Pruzan, 2015, p. 266).

Pruzan's emphasis on spiritual-based guidance is of great relevance to the discussion of the need for a mentor in education. Pruzan advocates a relationship with a master or a "mentor" who can guide leaders or a leader *in spe* in her or his development (2015, p. 138). For Pruzan's own spiritual path, central factors were attempts to operationalize concepts of virtue, ethics, values, and responsibility, and his living in an ashram where he experienced the guidance of a spiritual teacher/master. Pruzan experienced the value of living with like-minded people and the development of the state of mind where he could begin to emancipate and empower "our embodied ethicality". Aphorisms such as "Love all, Serve all", "Help ever, Hurt never", and "Hands that serve are holier than lips that pray" provided guidance for him at the mental level. Positive activities, such as "reading elevating literature, prayer and seeking the company of wise and compassionate people", led to avoidance of negative states of mind, "including greed, anger, hate, pride, jealousy, fear, violence, and an attachment to my deeds and their fruits" (Pruzan, 2015, p. 139).

According to Lozano (2015) business education has been much more business than education, emphasizing that we have confused offering business ethics courses with being trained in business ethics. His viewpoint is that business ethics should focus more on the processes of learning and personal development. Training is more of the same, but education at its best is a transformation of mindsets and attitudes, which is much more than referring to techniques, tools, and procedures. In this sense, business education means that we should build into the student the personal integration of fragments from different disciplines. It would entail moving in the direction of transdisciplinary and attempting to articulate wholeness· Lozano (2015) also points to the importance of creating spaces of freedom in business schools. The education of the future management world requires new voices to be heard in business schools, "other than those that come from the world of management". We should not facilitate "management education as a closed loop, in which management only speaks *of* management *for* management" (2015, p. 273). Voices from culture, politics, and spirituality should also be heard in management. This would open up for spaces of freedom, "not of joining in and replicating what goes on in the silos of repetition" (Lozano, 2015, p. 273).

As educators we cannot tell students what their calling is, but we can promote the idea of callings and assist them in the search for one. But how? We may learn something of vital significance from Buber's life and pedagogy, in particular the differentiation between I-It and I-Thou.[5] If we take a look on Buber's "conversion", as he himself called it, we may say that he met a young student in a polite way, but without being there in spirit. Buber says: "I certainly did not fail to let the meeting be friendly; … I conversed attentively and openly with him – only I omitted to guess the questions which he did not put. Later, not long after, I learned from one of his friends – he himself was no longer alive[6] – the essential content of these questions; I learned that he had come to me not casually, but borne by destiny, not for a chat but for a decision … What do we expect when we are in despair and yet go to a man? Surely a presence by means of which we are told that nevertheless there is meaning" (Buber, 1965, pp. 13–14). Buber treated the young student politely, but as an it-an object, a thing. After Buber's conversion he changed his pedagogy and the base for all his relationships became the living dialogue not the "dead" monologue. In fact, I-It and I-Thou represent two modes of life. In an I-Thou relationship I am related as a subject to subject, and what happens is not in me, or in her, it is in the "between". It is a mutual relationship and, as a teacher, I enter the relationship with the whole of my being and relate to the whole of the being of the other. Another distinction that Buber draws is between the role of the educator and the propagandist. When one tries to impose himself, his opinion, and his attitude on the other he takes the role of the propagandist, which is a kind of manipulation. The educator's task is to affect the other by helping the other to unfold his potentiality.

Buber stresses the importance of affirming the other even though the teacher may think the student is wrong. For Buber it became important to heighten the awareness of the other "in his uniqueness and concreteness", and not reduce the other to an abstraction. In an I-Thou dialogue the heart of the dialogue is to respond to the other fully and authentically. Gradually through the dialogue, trust will be built and I and Thou can begin to reflect on hidden and deeper questions. Some critics have said that one of the problems with Buber's approach is that it is empty in terms of content. "When we are discussing mentorship, we believe that the mentor should have broad life-experiences and some professional competence. Good mentorship requires wisdom".

In the literature we also find discussion of the *sometimes-dark side of life projects.* Williams discusses the situation that arises when a life project conflicts with the social responsibility for other human beings by using the painter Gauguin as an illustrating example. Gaugin left his wife and small children to go to Tahiti for several years to paint the nature and people there. Was Gauguin's ground project worthy? According to Williams, it depends upon its success. When we pose this question on Hauge, we find that his personal sacrifice was disastrous because he was jailed for many years and went through a most brutal detention, with the result that he destroyed his health the rest of his life. In this sense we might compare Hauge with Nelson Mandela or Mahatma Gandhi. The latter was imprisoned a number of times because of his struggle to liberate India and for his innovative and powerful nonviolent actions. Gandhi also fasted many times using his own suffering as one of his weapons in the political struggle. Hauge also struggled for the right to do business and for evangelical preaching. In Denmark-Norway there was a law that prohibited laymen from preaching, and when Hauge violated this law a number of times he was arrested. Hauge's impact on people throughout Norway was *so* considerable in a positive sense, that the powerful elite feared a revolution from beneath. Our conclusion is that Hauge's life project, due to its far-reaching impact in a multitude of ways, can be assessed as a formidable and long-lasting success.

Lozano makes a distinction between training and education, and Buber stresses that education should be education of character. It means to impress the pupil or student as a whole, both in actuality and in his possibilities; that is, what she or he can become (Buber, 1965, p. 104). The Greek word character means impression. It means that actions and attitudes will be impressed on the student and influence his character. This deep influence needs an atmosphere of confidence, which is won by direct and ingenuous participation in the life of the persons one is dealing with.

The case of Hauge demonstrates the importance of trust-building skills. Hauge's character, charisma, and preaching lead to a huge number of followers and friends who contributed in many profound ways in the Norwegian community. There are a lot of examples where people, after a meeting with Hauge, became so fascinated and inspired that they felt a great need to learn more from him and become a "Haugianer". Hauge benefitted from his own charisma by being able to do business and lend money without written documents. Hauge's word was regarded as trustworthy. Can we develop similar trust-building skills in our students and, if yes, how can we do it? Notably, modern management research tells us that trustworthiness has three main sources: competence/skills, integrity, and care/benevolence (Mayer and Gavin, 2005).

How can business schools develop integrity and the ability to care, not just analytical skills in their students? Integrity in Williams's sense is a person's ability to originate actions in himself/herself and promote own initiatives, purposes, or concerns. When you have a ground project you will not be an easy subject of manipulation by others who have ground projects. For Williams it is important that an agent should not align himself only with ground projects of "the impartial point of view", because then he will not be an agent.[7] Clearly, Hauge's ground project went beyond a utilitarian view, even if he contributed to the creation of businesses over almost all of Norway. Hauge's ground project was to preach the gospel to new people so that they would no longer be sinners, but true believers in Christian pietistic sense.

As we teach business school students about ethics our curriculums should not exclusively contain content like ethical theories, how leaders or philosophers have behaved in the past, and how to face ethical dilemmas. We should also include training that is actually helping students to develop their characters and moral capacity. Developing moral capacity does not happen overnight, or even during one semester. Students need to reflect on their own ethical and moral choices in different contexts over an extended period of time and learn from them as they interact with other students under faculty supervision, not just learn theories during lectures. Based on the definitions of Hannah et al. (2011), moral capacity consists of moral conation, that is "the capacity to generate responsibility and motivation to take moral action in the face of adversity and persevere through challenges", and moral maturation, "the capacity to elaborate and effectively attend to, store, retrieve, process, and make meaning of morally relevant information". The Hauge School of Management[8] has adopted this approach for five years now by offering a course in ethics throughout the bachelor's program. Here the students interact, reflect, and discuss ethical dilemmas as they connect them to their moral values and convictions. Observations so far indicate that the students are growing and expanding their moral capacity as they go through this extended course.

Further research

In this chapter, we have emphasized the two concepts of calling and life projects, related them to a business context, and drawn some implications for current business education. We believe that we need a principal discussion of one more concept, namely "defining moment". The nature and relevance of the concepts of calling and life projects should be complemented with the concept of defining moment. We suggest that these concepts are related to major motivational factors and essential outcome variables in modern leadership theories. A next step would be to develop a model consisting of these three concepts. Then we might obtain more precise implications for leadership theory, mentoring of managers, and business education.

Notes

1. In 1801 the number of citizens in Norway was 882,300, with ca. 18,000 inhabitants in Bergen. Of the people, ten percent lived in the towns, and the rest were farmers and land laborers living in rural Norway.
2. The original text in Norwegian, "Det gjenlød i mitt Indre: Du skal bekjende mitt Navn for Menneskene, formane dem at omvende sig og søke mig, meden jeg findes, kalde pa mig, naar jeg er og røre deres Hjerter, sa kunde de omvende seg fra Mørker til Lyse" (Hauge, 1910).
3. <https://www.merriam-webster.com/dictionary/defining%20moment>.
4. <https://en.oxforddictionaries.com/definition/defining_moment>.
5. For a short but nice summary of some of Buber's work I refer to P. LeFevre MCMLXVI, *Understandings of Man,* Philadelphia, The Westminster Press.
6. The young man's name was Mehe, and he died two months later in the trenches of World War 1 (committed suicide).
7. The historical origin in the Latin of integrity is *in-teger,* meaning what is not touched, taken away from, or interfered with. See also Chappell, Sophie Grace, "Bernard Williams", *The Stanford Encyclopedia of Philosophy* (Spring 2015 Edition), Edward N. Zalta (ed.), <https://plato.stanford.edu/archives/spr2015/entries/williams-bernard/>.
8. NLA University College in Oslo, Norway, where two of the authors are engaged, Truls as managing dean, and Magne as one of the founding fathers.

Bibliography

Abouleish, I. (2012) SEKEM. *En veiviser for det nye Egypt* (A Guide for the New Egypt) Sverre Dahl, trans. (Original title: Die Sekem Vision – Eine Begegnung von Orient und Okzident veranderc Egypten), Oslo, Antropos.

Bandura, A. (1977). *Social Learning Theory.* Englewood Cliffs, NJ: Prentice Hall.

Barrick, M. R., Mount, M. K., & Li, N. (2013). The Theory of Purposeful Work Behavior: The Role of Personality, Higher-Order Goals, and Job Characteristics. *Academy of Management Review*, 38, 132–153.

Berkelaar, B. L., & Buzzanell, P. M. (2015). Bait and Switch or Double-edged Sword? The (Sometimes) Failed Promises of Calling. *Human Relations*, 68(1), 157–178.

Breistein, D. (1955). *Hans Nielsen Hauge. "Kjøbmand i Bergen" ("Grocer in Bergen").* Bergen: John Griegs Forlag.

Buber, M. (1965). *Between Man and Man.* New York, NY: Macmillan Publishing Company.

Conway, N., Clinton, M., Sturges, J., & Budjanovcanin, A. (2015). Using Self Determination Theory to Understand the Relationship between Calling Enactment and Daily Well-being. *Journal of Organizational Behavior*, 36(8), 1114–1131.

Cox, C., & Jennings, R. (1995). The Foundations of Success: The Development and Characteristics of British Entrepreneurs and Intrapreneurs. *Leadership & Organization Development Journal*, 16(7), 4–9.

Dalgaard, B., & Supphellen, M. (2011). Entrepreneurship in Norway's Economic and Religious Nineteenth-Century Transformation. *Scandinavian Economic History Review*, 59, 48–66.

Deci, E. L., & Ryan, R. M. (2000). The What and Why of Goal Pursuits: Human Needs and the Self-Determination of Behavior. *Psychological Inquiry*, 11, 227–268.

Dik, B. J., & Duffy, R. D. (2009). Calling and Vocation at Work: Definitions and Prospects for Research and Practice. *The Counseling Psychologist*, 37(3), 424–450.

Dørum, Knut & Kringlebotn, H. S. (2016). *Mellom gammelt og nytt:. Kristendom i Norge pa 1800- og 1900-tallet. ("Between old and new: Christianity in Norway during the 1800s and 1900s").* Bergen: Fagbokforlaget.

Dutton, J. E. Roberts, L. M., & Bednar, J. (2010). Pathways for Positive Identity Construction at Work: Four Types of Positive Identity and the Building of Social Resources. *Academy of Management Review*, 35(2), 265–293.

Flyvberg, B. (2006) Five Misunderstandings about Case-Study Research. *Qualitative Inquiry*, 12(2), April, 219–245.

Greenleaf, R. K. (2002). *Servant Leadership: A Journey into the Nature of Legitimate Power & Greatness* (3rd edn). Mahway, NJ: Paulist Press.

Grytten, O. (2013). The Protestant Ethics and the Spirit of Capitalism: Entrepreneurship of the Norwegian Puritan Leader Hans Nielsen Hauge. *Review of European Studies*, 5(1), 31–44.

Haney-Loehlein, D. M., McKenna, R. B., Robie, C., Austin, A., & Ecker, D. (2015). The Power of Perceived Experience: Events that Shape Work as a Calling. *The Career Development Quarterly*, 63, 16–30.

Hannah, S. T., Avolio, B. J., & May, D. R. (2011). Moral Maturation and Moral Conation: A Capacity Approach to Explaining Moral Thought and Action. *The Academy of Management Review*, 36(4), 663–685.

Harper, D. A. (2003) *Foundations of Entrepreneurship and Economic Development* London: Routledge.

Hauge, H. N. (1910). *Udvalgte Skrifter ("Selected Writings")* Bergen, Norway.

Haukland, L. (2014). Hans Nielsen Hauge. A Catalyst of Literacy in Norway. *Scandinavian Journal of History*, 39(5), 539–559.

Ims, K. J. (2018) *Caring Entrepreneurship and Ecological Conscience Illustrated by Yvon Chouinard and Patagonia* London: Palgrave.

Ims, K., & Zsolnai, L. (2015). Social Innovation and Social Development in Latin America, Egypt, and India. In G. Enderle & P. E. Murphy (eds), *Ethical Innovation in Business and Economy.* Cheltenham, UK; Northampton, MA: Edward Elgar.

Kuhn, T. H. (1987). What Are Scientific Revolutions? In L. Kruger, L. J. Daston, & M. Heidelberger (eds), *The Probabilistic Revolution,* vol. 1: *Ideas in History* (pp. 7–22). Cambridge, MA: MIT Press.

Lozano, J. (2015). The Future of Business Ethics: A Structured Dialogue between the Participants. In K. Ims & L. J. Pedersen (eds), *Business and the Greater Good. Rethinking Business Ethics in an Age of Crisis.* Cheltenham, UK: Edward Elgar.

Mayer, R. C., & Gavin, M. B. (2005). Trust in Management and Performance: Who Minds the Shop While the Employees Watch the Boss? *Academy of Management Journal*, 48, 874–888.

Pruzan, P. (2015) The future of Business Ethics: A Structured Dialogue between the Participants. In K. Ims & L. J. Pedersen (eds), *Business and the Greater Good. Rethinking Business Ethics in an Age of Crisis.* Cheltenham, UK: Edward Elgar.

Rødal, Vegard Tafjord & Kiplesund, Andreas (2009). *Hans Nielsen Hauge: Entrepreneur, Banker and Industrialist.* Bergen: Masteroppgave, NHH.

Ryan, R. M., & Deci, E. L. (2000). Self-Determination Theory and the Facilitation of Intrinsic Motivation, Social Development, and Well-being. *American Psychologist*, 55, 68–78.

Supphellen, Magne (2014). Effects of Religious Attitudes on the Self-Determination of Small-Scale Entrepreneurs in the Slum Areas of Nairobi. In Tomas Drønen (ed.), *Religion and Society in Africa.* Bern, Switzerland: Peter Lang.

Supphellen, M., & Liland, T. (2017). Kallsorientert seriegrunder. Hans Nielsen Hauge i møte med dagens næringsliv. *Trådene I samfunnsveven. Hva har reformasjonen betydd for Norge?* Nils Ivar Agøy, Knut Edvard Larsen og Chr. Anton Smedshaug (editors). Verbum. Oslo, Norway. 125–140.

Thomas, A. (2015). Williams on Integrity, Ground Projects and Reasons to be Moral. In B. Himmelmann & R. B. Louden (eds), *Why be Moral?* De Gruyter, pp. 249–271.

Weber, M. (1930). *The Protestant Ethic and the Spirit of Capitalism.* Translated by Talcott Parsons. New York: Routledge/Taylor & Francis e-Library, 2005.

Williams, B. (1973) Against Utilitarianism. In B. Williams & J. J. C. Smart (eds), *Utilitarianism: For and Against.* Cambridge: Cambridge University Press.

Zhao, H., Seibert, S. E., & Hills, G. E. (2005). The Mediating Role of Self Efficacy in the Development of Entrepreneurial Intentions. *Journal of Applied Psychology*, 90(6), 1265–1272.

<https://www.merriam-webster.com/dictionary/defining%20moment.> The Oxford Dictionary focuses on the capacity of defining moments to frame peoples thinking: */.In event which typifies or determines all subsequent related occurrences”* (Oxford Dictionary, <https://en.oxforddictionaries.com/definition/defining_moment>. A calling 11).

Notes on Contributors

JOHANNES BRINKMANN is Professor Emeritus in business ethics at the BI Norwegian Business School in Oslo, Norway. His PhD is in sociology, obtained in 1975 from the University of Münster, West Germany. His main professional networks are EBEN, with regular attendance of most Annual and Research conferences, on the one hand, and related to sabbaticals and co-authorship, mostly abroad, on the other. Special interest areas: marketing and consumer ethics, industry and (business) professional ethics, non-naïve business ethics teaching, inconvenient/skeptical business ethics, risk and insurance business ethics, dialogue approaches, literature as a source of ethical reflection. He has published ten books (textbooks and monographs) in Norwegian, more than forty articles in refereed journals (fourteen of them in the JBE), and many book chapters; see his BI website for a complete publication list and further information.

KJELL GRØNHAUG is Professor Emeritus at the Norwegian School of Economics (NHH) and was graduated from the school as "siviløkonom." From the same institution he also obtained his graduate degree after several years in business. He also holds a master's degree in sociology from the University of Bergen, and did his postgraduate studies emphasizing quantitative methods at the University of Washington, Seattle. He has been preoccupied with a broad range of problems related to marketing and strategy as well as methodological issues. He has served as a visiting professor at several foreign institutions both in the US and Europe and always emphasized the importance of cooperation. Over the years he has published widely including eighteen books and more than two hundred articles in leading European and American journals. He has also served on the editorial board of several outlets. Grønhaug is the recipient of NHH's prize for excellence in research awarded every fifth year. Grønhaug is an honorary doctor at Turku school of economics and business administration, Gothenburg University and Stockholm school of economics.

MORTEN HUSE is Professor Emeritus at BI Norwegian Business School in Oslo. He has obtained his PhD (Dr. Oecon.) from NHH in Bergen. Morten has been working at several institutions in Norway as well as in other countries, including The Church of Norway, Nordland Research Institute, Centre for Church Research (Oslo), and Witten/Herdecke University. He is committed to follow a holistic Humboldtian educational philosophy, and he has signed the DORA declaration and endorsed RRBM. His most recent books are about "Resolving the Crisis in Research," value-creating boards and the human side of corporate governance, and about Gender issues – all within stakeholder and sustainable society perspectives. He is a former President.

KNUT J. IMS is Professor Emeritus at the Norwegian School of Economics (NHH) in Bergen, which is his Alma mater since 1972 (bachelor's degree/1976 and HAE degree/1979). He obtained his PhD from the School of Economics and Legal Science, Gothenburg University, and has been an active member of the business ethics faculty group of (CEMS) – Global Alliance for Management Education for two decades. He is Fellow of the European SPES Institute and has been visiting scholar at number of universities (USA, Argentina, Europe). He has been a member of the committee for research misconduct at the University of Bergen and is current member of the board of Center for Ethics and Economics at NHH (since 2001), member of the fair trade board of the municipality of Bergen, and a board member of a local green NGO. From 1985 he has annually presented his research at international scientific conferences from Hawaii to Corvinus, Oxford, Cambridge, Hong Kong, Pharo in Bhutan, Tel Aviv, Bologna, and more. He has published several books and book chapters as well as articles in a variety of international journals.
Knut Ims's books include the following:

Business within Limits. Deep Ecology and Buddhist Ethics. 2006, Oxford and Bern, (coedited with Laszlo Zsolnai)

Heroes and Anti-heroes. European Literature and the Ethics of Leadership. 2010, Garant, (coedited with Rita Ghesquiere)

Art, Spirituality and Economics. Liber Amicorum for Laszlo Zsolnai. 2018, Springer, (coedited with Luk Bouckaert and Peter Rona).

OVE D. JAKOBSEN is Professor in Ecological Economics at Business School Nord University in Bodø, Norway. He holds master's degrees in Marketing (SLHK, 1982), Business administration (HAE at NHH 1987), and Philosophy (University of Bergen 1991). He obtained his Dr. Oecon. from the Norwegian School of Economics (NHH) in 1989). Jakobsen has been a member of a number of national committees, amongst others The National Committee for Research Ethics in the Social Sciences and the Humanities (NESH). He is one of the founders and leader of Centre for Ecological Economics and ethics at Nord University. Jakobsen has been a visiting scholar at Stockholm University (Sweden), Business School in Århus (Denmark), and Corvinus University, Budapest, (Hungary). He has published a number of book chapters and articles in a variety of national and international journals. Jakobsen is often invited as a speaker at seminars and conferences.
Ove D. Jakobsen has published seventeen books including the following;

Jakobsen, Ove (2018); *Anarchism and Ecological Economics – A Political Platform for Ecological Economics*, Routledge

Jakobsen, Ove and Zsolnai, Laszlo (2017); *Integral Ecology and Sustainable Business*, Emerald Publishing

Jakobsen, Ove (2017); *Transformative Ecological Economics – Process Philosophy, Ideology and Utopia*, Routledge

Ingebrigtsen, Stig and Jakobsen, Ove (2007); *Circulation Economics – Theory and Practice*, Peter Lang

TRULS LILAND is a research scholar at the Norwegian School of Economics and served until August 2021 as Dean at the Hauge School of Management at the NLA University College in Oslo. He has several publications on Haugean stewardship and management. He teaches courses in innovation, entrepreneurship, and leadership, and has developed the school's unique internship program, focusing on innovation projects. Liland has experience as a manager in the telecom and banking industry, and has held several board positions, His educational background is in marketing, management, and theology.

LARS JACOB TYNES PEDERSEN is Associate Professor and Head of the Centre for Sustainable Business at NHH Norwegian School of Economics. He holds a PhD degree from NHH (2009). He does research on the design and innovation of sustainable business models, and natural field experiments on pro-social and pro-environmental behaviors. His research has been published in journals such as *Management Science* and *Journal of Business Ethics* and by publishers such as Palgrave, Edward Elgar, and Wiley. Pedersen is an active public speaker and strategic advisor for business organizations and other institutions. He has also co-founded the company Nordic Circles, which upcycles maritime steel into environmentally friendly building materials. The company was included on the Fast Company *World Changing Ideas* list in 2021.

MARIA PRESTMO is a Bergen-based photographer and gallery director. Her work has been exhibited in galleries and festivals such as Galleri Ramfjord (Oslo), Hardingar (Utne), Hydrogenfabrikken Kunsthall (Fredrikstad), Hardanger Musikkfest (Lofthus), Banksmidja Galleri (Voss), and Galleri Laszka (Oslo).

She has a master's degree (2006) from NHH, and her master thesis was written within the field of business ethics. She worked as a student assistant for Knut J. Ims during this period and has collaborated with Lars Jacob Tynes Pedersen and Knut J. Ims on several academic projects later.

MAGNE SUPPHELLEN is Professor of marketing and Chair of BRANDINNOVA, Center for brand research, at the Norwegian School of Economics (NHH) in Bergen, Norway. He has published widely in consumer psychology, marketing, and brand management journals, including the *Journal of Marketing*, *Journal of Consumer Psychology*, *Journal of Business Research*, and *Public Opinion Quarterly*. He is also involved in research on empirical ethics. His work in this area has appeared in journals such as *BMC Medical Ethics, Clinical Ethics*, and *Journal of Empirical Theology*. Professor Supphellen is a consultant for Norwegian and international companies and a board member of several commercial and non-profit organizations.

JUDITH A. WHITE is a retired professor of business ethics and organizational behavior. She received a PhD from Case Western Reserve University, an MA from Stanford, and an MS from San Francisco State University. She held academic positions at Loyola Marymount University, St. Mary's College of California, the University of California Santa Cruz, and Santa Clara University. She is the author or coauthor of articles published in *The Journal of Management Inquiry, The Journal of Management Education, Thunderbird International Business Review, Journal of Corporate Citizenship, Journal of Business Ethics Education, Business Ethics Quarterly, Journal of Civil Society, International Journal of Social Science Research, Organizational Development Journal, Journal of Management Development, The Learning Organization, International Journal of Value-Based Management, Social Behavior and Personality, Management Learning, Psychological Reports, and Journal of General Internal Medicine.* From 1997 to 2013 much of Dr. White's research focused on the moral courage of leaders of the human rights and democracy movement in Burma. She lives in Berkeley, California.

LASZLO ZSOLNAI is Professor and Director of the Business Ethics Center at Corvinus University of Budapest, Hungary, and Associate Member of Las Casas Institute for Social Justice, Blackfriars Hall, University of Oxford, UK. He also serves as President of the European SPES Institute in Leuven, Belgium. He has been a guest professor/visiting scholar at the University of Cambridge, the University of Oxford, the University of California at Berkeley, Georgetown University, the University of Richmond, Concordia University, Montreal, University of St. Gallen, Bocconi University Milan, and the Netherlands Institute for Advanced Study.

Laszlo Zsolnai's latest books include "Ethics, Meaning and Market Society" (2018, Routledge), "Caring Management in the New Economy"

(2019, Palgrave), "The Routledge International Handbook of Spirituality in Society and the Professions" (2019, Routledge), and "Responsible Research for Better Business" (2020, Palgrave). His website: <http://laszlo-zsolnai.com>

Index

Frontiers of Business Ethics

Series Editor
LASZLO ZSOLNAI
Business Ethics Center
Corvinus University of Budapest

This series is dedicated to alternative approaches that go beyond the literature of conventional business ethics and corporate social responsibility. It aims to promote a new ethical model for transforming business into humanistic, sustainable and peaceful forms. The series publishes monographs and edited volumes with fresh ideas and breakthrough conceptions relevant for scholars and practitioners alike.

VOLUME 1

Laszlo Zsolnai and Knut Johannessen Ims (eds):
Business within Limits: Deep Ecology and Buddhist Economics. 2005.
324 pages. ISBN 3-03910-703-8

VOLUME 2

Luigino Bruni and Stefano Zamagni:
Civil Economy: Efficiency, Equity, Public Happiness. 2007.
282 pages. ISBN 978-3-03910-896-1

VOLUME 3

Stig Ingebrigtsen and Ove Jakobsen:
Circulation Economics: Theory and Practice. 2007.
349 pages. ISBN 978-3-03911-089-6

VOLUME 4

Luk Bouckaert, Hendrik Opdebeeck and Laszlo Zsolnai (eds):
Frugality: Rebalancing Material and Spiritual Values in Economic Life.
2008. 322 pages. ISBN 978-3-03911-131-2

Volume 5

Peter Verhezen:
Gifts, Corruption, Philanthropy: The Ambiguity of Gift Practices in Business. 2009. 343 pages. ISBN 978-3-03911-842-7

Volume 6

Henri-Claude de Bettignies and François Lépineux (eds):
Business, Globalization and the Common Good. 2009. 447 pages. ISBN 978-3-03911-876-2

Volume 7

Peter Pruzan:
Rational, Ethical and Spiritual Perspectives on Leadership: Selected Writings. 2009. 340 pages. ISBN 978-3-03911-907-3

Volume 8

Josep M. Lozano:
The Relational Company: Responsibility, Sustainability, Citizenship. 2010. 198 pages. ISBN 978-3-03911-940-0

Volume 9

Antonio Tencati and Laszlo Zsolnai (eds):
The Collaborative Enterprise: Creating Values for a Sustainable World. 2010. 429 pages. ISBN 978-3-0343-0159-6

Volume 10

Simon Robinson and Jonathan Smith:
Co-Charismatic Leadership: Critical Perspectives on Spirituality, Ethics and Leadership. 2014. 334 pages. ISBN 978-3-0343-0216-6

Volume 11

Hendrik Opdebeeck (ed.):
Responsible Economics: E.F. Schumacher and His Legacy for the 21st Century. 2013. 363 pages. ISBN 978-3-0343-1707-8

Volume 12

Laszlo Zsolnai:

Beyond Self: Ethical and Spiritual Dimensions of Economics. 2014.
203 pages. ISBN 978-3-0343-1772-6

Volume 13

Knut J. Ims:

Deep Ecology, Business Ethics and Personal Responsibility. Selected Papers (1988–2020). 2022.
840 pages. ISBN 978-1-80079-229-6

www.ingramcontent.com/pod-product-compliance
Ingram Content Group UK Ltd.
Pitfield, Milton Keynes, MK11 3LW, UK
UKHW020837290926
15470UKWH00004B/1400

9 781800 79229